Terrestrial Environment and Ecosystems of Kuwait

Majda Khalil Suleiman • Shabbir Ahmad Shahid
Editors

Terrestrial Environment and Ecosystems of Kuwait

Assessment and Restoration

Editors
Majda Khalil Suleiman
Desert Agriculture and Ecosystems Program, Environment and Life Sciences Research Center
Kuwait Institute for Scientific Research
Safat, Kuwait

Shabbir Ahmad Shahid
Desert Agriculture and Ecosystems Program, Environment and Life Sciences Research Center
Kuwait Institute for Scientific Research
Safat, Kuwait

ISBN 978-3-031-46261-0 ISBN 978-3-031-46262-7 (eBook)
https://doi.org/10.1007/978-3-031-46262-7

Cover illustration: *Vachellia pachyceras (Al Talha)* is the only native tree of Kuwait. The credit for this photo goes to Dr. Majda Khalil Suleiman.

This Springer imprint is published by the registered company Springer Nature Switzerland AG
The registered company address is: Gewerbestrasse 11, 6330 Cham, Switzerland

Paper in this product is recyclable.

Dr. Mahdi Abdal

This book is dedicated to memorialize Dr. Mahdi Abdal's scientific efforts accomplished at the Arid Land Agriculture Greenery Department (AAGD), and Environment and Life Sciences Research Center of Kuwait Institute for Scientific Research. He served as the Department Manager of AAGD and as a Senior Scientist for many years. His research accomplishments are highly appreciated by the national and international scientific communities. He was a dedicated scientist, strategic thinker with a practical and solutions oriented outlook, and program developer and possessed exceptional visionary leadership in agrifoods-focused research and advocacy. During his extensive research carrier, he valued integrity and loyalty to the utmost degree as it is the most fundamental basis for success. Dr. Mahdi left behind enormous scientific knowledge to be followed by future generations for the welfare and development of Kuwait, and he will always be remembered as a kind person.

Rhanterium epapposum (arfaj), a national flower of Kuwait. The credit for this photo goes to Dr. Majda Khalil Suleiman.

Foreword

Kuwait is a hyper-arid country vulnerable to climate change impacts leading to the degradation of its ecosystems and the reduction of its functions. Nowhere on the globe are the impacts of human activities more apparent than in the terrestrial ecosystems, and the same applies to Kuwait, where during the Iraqi invasion and occupation, the worst man-made environmental disasters occurred. The critical role that the environment plays in maintaining and improving the ecosystems, functioning, and providing ecosystem services from the well-managed soils and nutrients that underpin food production to the critical improvement of biodiversity is well established. Therefore, it is imperative to understand how human actions affected the natural processes of the terrestrial ecosystem, and an even greater need to evaluate the consequences of these changes. To understand the insights into the terrestrial environment, it is notable to examine the current status, identify the causes and threats (natural and anthropogenic), and formulate regenerative and climate-smart management practices to sustain the ecosystems for long time management without compromising the environmental quality. Significant research has been accomplished in Kuwait in the past many years and active rehabilitation of deserts has progressed to restore the degraded desert ecosystems. This book highlights the exciting work that was carried out to date and the new challenges to sustain the terrestrial ecosystems being altered as a result of natural impacts and human activities. These include loss of vegetation, destruction of habitat, change in land cover, increase in plants and wildlife species extinction rate, increase in invasive species, degradation of natural resources, and many more.

The overarching objective of this book is to present the main components of the Kuwait terrestrial environment, understand the cross-sectoral structure, properties, and functions of terrestrial ecosystems, their interactions and mutual benefits that ultimately lead to environmental protection, maintain healthy ecosystems through protecting biological diversity, and improve deserts. The contents of the book are very comprehensive presenting the multi-discipline approach to complete the entire cycle of the terrestrial environment and ecosystems of Kuwait. Sixteen cross-sectoral chapters are distributed into four themes covering the complete aspects of the book. The four themes are: (i) Soils, Geographical and Climatic Features, Eco-resources, and Desertification; (ii) Water Resources, Salinization Aspects, and Modeling; (iii) Terrestrial Ecosystems and Their Management; and (iv) Agriculture, Food Security, and Water Footprint of Crops.

This book is regarded as a landmark and likely to set the precedence to understand and manage desert ecosystems for healthy and sustainable management. A key message of this book is for the users to have increased access to reliable environmental information from Kuwait by identifying issues as they emerge and developing effective actions and policies to respond to them. Hence, this book is prepared to facilitate terrestrial environmental data flow for rapid knowledge sharing and wider accessibility. The potential users of this book are encouraged to explore the information embedded to support their research activities in the advancement of environmental protection for sustainable development.

Environment and Life Sciences Research Center
Kuwait Institute for Scientific Research
Safat, Kuwait

Sameer Al-Zenki

Preface

Ecosystem is a functional unit consisting of living organisms, their nonliving environment and the interactions within and between them. A healthy and well-functioning ecosystem provides four services, such as supporting, provisioning, regulating, and cultural. Globally, ecosystems have been degraded due to desertification to a great extent, and no continent is invulnerable to environmental and anthropic impacts. The drylands are more prone to desertification through wind erosion, contrariwise, humid areas are highly susceptible to water erosion; however, in both the cases, degradation is irreversible and decline soil fertility. Drylands have a ratio of average annual precipitation to potential evapotranspiration of less than 0.65. In drylands, the principal degradation processes are vegetation degradation, wind erosion, loss of biodiversity, and salinization. In future, it is likely that desertification will be prompted due to rising temperatures as a consequence of greenhouse gasses emission and prolonged droughts. In Kuwait, degradation of desert lands is momentous due to the impact of Gulf war, when deserts were degraded due to human impacts. This has led to the reduction of ecosystem functions and services. Considering the above dynamics, it is timely to assess the levels of degradation of Kuwait terrestrial environment and ecosystems and identify the constraints and develop rehabilitation and restoration based on informed decisions. The objectives of restoration should be to assure that the degraded lands are reverted back to the original state and to safeguard the terrestrial environment and minimize the natural and anthropic impacts.

This book is an outcome of many years work in the desert environment of Kuwait and presentation into a book format for easy access to the latest information for potential stakeholders who respect deserts integrity for their conservation and long-time ecosystem services without conceding their quality. The book includes 16 multidiscipline chapters, distributed into four themes, covering the complete presentation of desert environments and ecosystems of Kuwait. Theme 1 covers soils of Kuwait and threats to their uses and sustainable management, geographical features, climate and climate change aspects, ecological footprint and biocapacity of Kuwait and comparison to GCC countries and the world, as well as status of desertification and highlighting efforts made in the past for deserts rehabilitation. Theme 2 covers presentation of water resources and challenges for their sustainable uses, salinization as key threat to indigenous ecosystems, numerical model development and use to predict the behavior of salt-brackish water transition zones during scavenger well pumping. Theme 3 addresses multiple aspects of the terrestrial ecosystems and their management, including terrestrial habitats and ecosystems, native vegetation and flora, wildlife, threats to the ecosystems and potential conservation strategies, and forecasting the climate change impacts on desert ecosystems. Theme 4 comprehensively addresses the agricultural farming both in the open field and under temperature controlled protected agriculture and describes energy-food-water-climate nexus for a better understanding of interdependency of each sector. Finally, a dedicated chapter is on agriculture water footprints (green, blue, gray) of major crops in Kuwait and their comparison to the average footprints in the world.

Although the editors have followed *double-blind peer review* process for all chapters from internationally renowned scientists and edited the book, the contents of the chapters and the ideas expressed are the exclusive responsibility of the authors.

Safat, Kuwait

Majda Khalil Suleiman
Shabbir Ahmad Shahid

Acknowledgements

The book is a contribution of eminent cross-sectoral scientists engaged in different aspects of terrestrial environment and ecosystems of Kuwait. We are thankful to the management of Kuwait Institute for Scientific Research (KISR) including Dr. Sameer Al-Zenki, Acting Executive Director, Environment and Life Sciences Research Center (ELSRC), for his support and approval to proceed for book preparation and publication. We wish to thank Dr. Afaf Al-Nasser (DD/OD-ELSRC) for her interest and encouragement in publishing the book.

We also thank the international reviewers who participated in the peer review process of the book chapters. In addition, the Springer management, especially Margaret Deignan and the editorial team, deserves thanks for accepting the proposal for book publication. Ms. Saritha Sarath deserves special thanks for her help in formatting the book chapters. Finally, we would like to thank all the scientists for their valuable contribution that made it possible to publish this valued book.

Safat, Kuwait

Majda Khalil Suleiman
Shabbir Ahmad Shahid

Contents

Part I Soils, Geographical and Climatic Features, Eco-resources, and Desertification

1 **Potential Threats to Soil Functions and Mitigation Options for Sustainable Uses** 3
Shabbir Ahmad Shahid

2 **Perspectives of Geography, Environment, and Physiography of Kuwait** 21
Hebah Jaber Kamal, Megha Thomas, and Ahmed Abdulhadi

3 **Climate and Climate Change Aspects of Kuwait** 57
Amal J. Alkandari

4 **Ecological Footprint and Biocapacity of Kuwait and Proposed Eco-resources Management Strategies: A Review** 93
Shabbir Ahmad Shahid and Majda Khalil Suleiman

5 **Desertification: A Central Problem to Restoring Ecosystems** 119
Shabbir Ahmad Shahid

Part II Water Resources, Salinization Aspects and Modeling

6 **Current Status, Challenges, and Future Management Strategies for Water Resources of Kuwait** 141
Khalid Hadi

7 **Groundwater Salinization in Kuwait: A Major Threat to Indigenous Ecosystems** 171
Dalal Sadeqi, Amjad Sami Aliewi, Habib Al-Qallaf, and Tareq Rashed

8 Predicting the Behaviour of the Salt/Fresh-Brackish Water Transition Zone During Scavenger Well Pumping: 1. Numerical Model Development and Testing 197
Amjad Sami Aliewi

9 Predicting the Behaviour of the Salt/Fresh-Brackish Water Transition Zone During Scavenger Well Pumping: 2. Model Application in Kuwait and Pakistan 223
Amjad Sami Aliewi

Part III Terrestrial Ecosystems and Their Management

10 Terrestrial Habitats and Ecosystems of Kuwait 247
Tareq A. Madouh and Ali M. Quoreshi

11 Native Vegetation and Flora of Kuwait 265
Arvind Bhatt

12 Wildlife of the Terrestrial Ecosystems of Kuwait 311
Matrah Abdulrazag Al-Mutairi

13 Major Threats to the Terrestrial Ecosystems of Kuwait and Proposed Conservation Practices 329
M. Anisul Islam and Sheena Jacob

14 Kuwait Deserts and Ecosystems in the Context of Changing Climate 341
Ali M. Quoreshi and Tareq A. Madouh

Part IV Agriculture, Food Security and Water Footprint of Crops

15 Prospective of Agricultural Farming in Kuwait and Energy-Food-Water-Climate Nexus 363
Majda Khalil Suleiman and Shabbir Ahmad Shahid

16 Agricultural Water Footprint of Major Crops Grown in Kuwait Compared to the World Average: A Review 393
Majda Khalil Suleiman and Shabbir Ahmad Shahid

Index 415

Editors, Contributors, and International Reviewers

About the Editors

Majda Khalil Suleiman is a Research Scientist and Program Manager of Desert Agriculture and Ecosystems Program at the Environment and Life Sciences Research Center of Kuwait Institute for Scientific Research. She has over 34 years research and development experience in agriculture and urban landscape development and management. She earned her PhD degree in the year 2016 "*Restoration Ecology*" from the University of Western Australia. She earned a BSc degree in Botany/Plant Biology from Kuwait University in 1989, and started her career as Botanical Researcher at the Public Authority for Agriculture Affairs and Fish Resources of Kuwait. She is a Desert Landscaper certified by Desert Botanical Garden Phoenix, Arizona, USA. She is the recipient of nine achievement awards from different organizations, including Kuwait University Poster Day (2018–2019), KNPC HSE performance award (2018–2019), five Scientific Achievement Awards graced by KISR (2009–2015), and two Awards of Excellence from *Euro-Mediterranean Institute for Sustainable Development* in 2014 and 2015. She is member of Society for Ecological Restoration (SER), and the Convention on International Trade in Endangered Species (CITES) of Wild Fauna and Flora. She organized several in-house and *on-the-job* trainings in her field of specialization at KISR and attended over 50 training courses locally and internationally. She has led several research projects and was part of the team of many national projects covering wide range of topics, such as restoration of coastline through mangroves plantation, greenery master plan, and urban demonstration garden. She participated as Principal Investigator in 7 and Task leader in 16 projects. She is prolific author/co-author of 120 scientific papers published in international refereed journals (58), conference proceedings, abstracts, and book chapters (62). In addition, she published eight books and 137 technical reports. Currently she is leading a Government Initiative Food Security project and coordinating three projects related to desert rehabilitation through Kuwait Environment Remediation Program (KERP) funded by Kuwait National Focal Point.

Shabbir Ahmad Shahid was embraced with prestigious Sir William Roberts award to pursue PhD degree in Soil Science specialization in Soil Micromorphology of Salt-affected Soils at the University of Bangor, Wales, UK, completed in 1989. He earned BSc (Hons) and MSc (Hons) degrees in Soil Science from the University of Agriculture Faisalabad, Pakistan, in the years 1977 and 1980, respectively. He has over 42 years' experience as a Soil Scientist in Pakistan, the UK, Australia, the United Arab Emirates, and Kuwait. Currently, Dr. Shahid is a Research Scientist, Desert Agriculture and Ecosystems Program, Environment and Life Sciences Research Center of Kuwait Institute for Scientific Research. He led several projects in natural resources assessment and management. He was a technical coordinator in multi-million-dollar national soil surveys of the State of Kuwait and Abu Dhabi Emirate and developed the soil survey action plan for the Northern Emirates of UAE and the Republic of Mauritania. Dr. Shahid, with his co-associates, discovered anhydrite soil which is formally added in the 12th edition of US Keys to Soil Taxonomy as a diagnostic horizon, mineralogy class and subgroups in the Salids suborder of the order Aridisols. He is also the principal author of *United Arab Emirates Keys to Soil Taxonomy* and *Kuwait Soil Taxonomy* books published by Springer in the years 2014 and 2022, respectively. In addition, Dr. Shahid is a creator and co-founder of the Emirates Soil Museum launched in 2016 at the International Center for Biosaline Agriculture, Dubai, United Arab Emirates. He is a prolific author of over 200 scientific papers published in peer-reviewed scientific journals, book chapters, conference proceedings, and newsletters. As editor/co-editor/principal author, he published eight books through professional publisher Springer. Dr. Shahid's Research Interest Score (RIS) is 2,739, h-index 23, i10-index 48 with 2,731 citations, and 176,087 reads globally. His RIS is higher than 97% of ResearchGate members. He is recipient of David A Jenkins Award (1995) graced by the Soil Mineralogy and Micromorphology, Society of Pakistan.

Contributors

Ahmed Abdulhadi Systems and Software Development Department (SSDD), Science and Technology Sector (STS), Kuwait Institute for Scientific Research, Safat, Kuwait

Amjad Sami Aliewi Water Research Center, Kuwait Institute for Scientific Research, Safat, Kuwait

Amal J. Alkandari Desert Agriculture and Ecosystems Program, Environment and Life Sciences Research Center, Kuwait Institute for Scientific Research, Safat, Kuwait

Matrah Abdulrazag Al-Mutairi Desert Agriculture and Ecosystems Program, Environment and Life Sciences Research Center, Kuwait Institute for Scientific Research, Safat, Kuwait

Habib Al-Qallaf Water Research Center, Kuwait Institute for Scientific Research, Safat, Kuwait

Arvind Bhatt Desert Agriculture and Ecosystems Program, Environment and Life Sciences Research Center, Kuwait Institute for Scientific Research, Safat, Kuwait

Khalid Hadi Water Research Center, Kuwait Institute for Scientific Research, Safat, Kuwait

M. Anisul Islam Desert Agriculture and Ecosystems Program, Environment and Life Sciences Research Center, Kuwait Institute for Scientific Research, Safat, Kuwait

Sheena Jacob Desert Agriculture and Ecosystems Program, Environment and Life Sciences Research Center, Kuwait Institute for Scientific Research, Safat, Kuwait

Hebah Jaber Kamal Systems and Software Development Department (SSDD), Science and Technology Sector (STS), Kuwait Institute for Scientific Research, Safat, Kuwait

Tareq A. Madouh Desert Agriculture and Ecosystems Program, Environment and Life Sciences Research Center, Kuwait Institute for Scientific Research, Safat, Kuwait

Ali M. Quoreshi Desert Agriculture and Ecosystems Program, Environment and Life Sciences Research Center, Kuwait Institute for Scientific Research, Safat, Kuwait

Tareq Rashed Water Research Center, Kuwait Institute for Scientific Research, Safat, Kuwait

Dalal Sadeqi Water Research Center, Kuwait Institute for Scientific Research, Safat, Kuwait

Shabbir Ahmad Shahid Desert Agriculture and Ecosystems Program, Environment and Life Sciences Research Center, Kuwait Institute for Scientific Research, Safat, Kuwait

Majda Khalil Suleiman Desert Agriculture and Ecosystems Program, Environment and Life Sciences Research Center, Kuwait Institute for Scientific Research, Safat, Kuwait

Megha Thomas Systems and Software Development Department (SSDD), Science and Technology Sector (STS), Kuwait Institute for Scientific Research, Safat, Kuwait

Reviewers

Mahmoud A. Abdelfattah Soils and Water Department, Faculty of Agriculture, Fayoum University Fayoum, Fayoum, Egypt

Ali Al-Keblawi Department of Applied Biology, College of Sciences, University of Sharjah, Sharjah, United Arab Emirates

Narayana R. Bhat Desert Agriculture and Ecosystems Program, Environment and Life Sciences Research Center (Ex-Employee), Kuwait Institute for Scientific Research, Safat, Kuwait

Redouane Choukr-Allah Agricultural Innovation and Technologies Transfer Center, University Mohammed VI Polytechnic, Ben Guerir, Morocco

Jagdish Dagar Indian Council of Agriculture Research, New Delhi, India

Zineb Elmouridi National Institute of Agriculture Research, Rabat, Morocco

Salim Javed Environment Agency Abu Dhabi, Abu Dhabi, United Arab Emirates

Damase Khasa Laval University, Quebec, Canada

Henda Mahmoudi International Center for Biosaline Agriculture, Dubai, United Arab Emirates

Raafat Misak Desert Research Center, Cairo, Egypt

Ijaz Rasool Noorka College of Agriculture, University of Sargodha, Sargodha, Pakistan

Asad Sarwar Qureshi International Center for Biosaline Agriculture, Dubai, United Arab Emirates

Iftikhar Hussain Sabir Alberta, Canada

Pandi Zdruli International Center for Advanced Mediterranean Agronomic Studies (CIHEAM) Mediterranean Agronomic Institute of Bari, Bari, Italy

Acronyms and Abbreviations

AGEDI	Abu Dhabi Global Environmental Data Initiative
AQI	Air Quality Index
ASL	Above Sea Level
AW	Available Water
BC	Biocapacity
bcm	Billion Cubic Meter
BNF	Biological Nitrogen Fixation
BWF	Blue Water Footprint
CA	Conservation Agriculture
CBD	Convention of Biological Diversity
CE	Circular Economy
COP	Conference of Parties
CSA	Climate Smart Agriculture
DAEP	Desert Agriculture and Ecosystems Program
DEM	Digital Elevation Model
ECe	Electrical Conductivity of Extract from Saturated Soil Paste
EF	Ecological Footprint
ELSRC	Environment and Life Sciences Research Center
EPA	Environmental Public Authority
ESP	Exchangeable Sodium Percentage
EWF	External Water Footprint
FAO	Food and Agriculture Organization
FC	Field Capacity
FO	Forward Osmosis
IWF	Internal Water Footprint
GCC	Gulf Cooperation Council
GFN	Global Footprint Network
gha	Global Hectare
GHG	Greenhouse Gasses
GI	Government Initiative
GWF	Green Water Footprint

GrWF	Gray Water Footprint
IEA	International Energy Agency
IoT	Internet of Things
IPCC	Intergovernmental Panel on Climate Change
ITPS	Intergovernmental Technical Panel on Soils
ITS	Increased Tariff Slab
IUCN	International Union of Conservation Nature
KEPA	Kuwait Environmental Protection Authority
KERP	Kuwait Environmental Remediation Program
KI	Keyword Inter-linkage
KISR	Kuwait Institute for Scientific Research
KNAP	Kuwait National Adaptation Plan
LD	Land Degradation
LDN	Land Degradation Neutrality
mbgl	Meters Below Ground Level
MCM	Million Cubic Meter
MD	Membrane Distillation
MEA	Millennium Ecosystem Assessment
WDM	Water Demand Models
MED	Multi-Effect Distillation
MLD	Minimum Liquid Discharge
MPW	Ministry of Public Works
MSF	Multi-Stage Flash Distillation
MWTP	Municipal Wastewater Treatment Plant
NDVI	Natural Deviation Vegetation Index
NFA	National Footprint Accounts
NPP	Net Primary Productivity
NRCS	Natural Resources Conservation Service
OECD	Organization for Economic Co-operation and Development
PV	Photo Voltaic
PWP	Permanent Wilting Point
RO	Reverse Osmosis
ROWA	Regional Office of Waste Asia
SAANR	Sabah Al Ahmad Nature Reserve
SDGs	Sustainable Development Goals
SLR	Sea Level Rise
SSM	Sustainable Soil Management
TDS	Total Dissolved Solids
TSE	Treated Sewage Effluent
TWW	Treated Wastewater
UNCC	United Nations Compensation Commission
UNCCD	United Nations Convention to Combat Desertification
UNEP	United Nations Environment Program
UNFCCC	United Nations Framework Convention on Climate Change
USDA	United States Department of Agriculture

OPEC	Organization of the Petroleum Exporting Countries
UXO	Unexploded Ordnance
WASH	Water Sanitation and Hygiene
WDT	Water Desalination Technology
WF	Water Footprint
WMO	World Meteorological Organization
WWTP	Wastewater Treatment Plant
ZLD	Zero Liquid Discharge

Part I
Soils, Geographical and Climatic Features, Eco-resources, and Desertification

Chapter 1
Potential Threats to Soil Functions and Mitigation Options for Sustainable Uses

Shabbir Ahmad Shahid

Abstract The soils of Kuwait are mostly coarse textured presenting four out of the twelve soil texture classes recognized globally. The dominant soil texture class is sand. The sandy soils are simple in composition, but there are challenges to manage for sustainable uses. These soils are low in moisture and nutrient retention capacities and present high infiltration rates, thus require costly investments for productive and sustainable uses. The inherent and dynamic soil qualities with respect to Kuwait desert soils are discussed. Of the 12 soil orders recognized globally, in Kuwait only two (Aridisols and Entisols) are identified. Soil formation factors and processes operating in the desert landscapes of Kuwait are highlighted and described. The calcification, gypsification, and salinization are the main pedogenic processes diagnosed in the desert environment. Without going into taxonomic details, the soils are grouped and described into general categories (sandy, gypsic, saline, and calcareous), the main threats are highlighted and mitigation options for sustainable uses are discussed. Erosion is the main cause of unstable and infertile soils in the deserts of Kuwait. Threats to soil functions are highlighted, and mitigation actions are anticipated for sustainable uses under arid environmental conditions.

Keywords Soil formation · Calcification · Homogenization · Salinization · Erosion · Classification · Management · Soil functions · Soil threats

S. A. Shahid (✉)
Desert Agriculture and Ecosystems Program, Environment and Life Sciences Research Center, Kuwait Institute for Scientific Research, Safat, Kuwait
e-mail: sshahid@kisr.edu.kw

M. K. Suleiman, S. A. Shahid (eds.), *Terrestrial Environment and Ecosystems of Kuwait*, https://doi.org/10.1007/978-3-031-46262-7_1

1.1 Introduction

Soil delivers multiple functions; however, different soils vary inherently in their capacity to function; therefore, quality is specific to each kind of soil. In this regard, soil possesses two types of qualities, *inherent quality* and *dynamic quality*. The soils having good inherent quality are basically productive. For example, soil texture and mineralogy are determined by the nature of the parent material the soil is established. The soil texture and mineralogy are considered inherent soil property and are determined by factors of soil formation i.e., parent material, climate, topography, vegetation, and time. These, almost permanent soil properties determine different uses of soils and differentiate other soils for specific use. In the case of Kuwait, the soil texture is mainly sandy and has low water retention capacity; in contrast a loamy soil can retain high water content; therefore, sandy soil has low *inherent quality*. This is the reason why global soil science community developed consensus that sandy soils have poor quality to deliver functions. The inherent soil quality can be rephrased to soil capability. The soil attributes which determine inherent soil quality are mostly based to name soil taxa for soil classification purposes (Soil Survey Staff 2022; Shahid and Omar 2022).

The second type of quality is named as *dynamic quality* which changes with management practices and how these soils are handled for different uses. The use of organic manures or organic fertilizers can increase organic matter which results into increased water and nutrient holding capacity and thus have positive impact on soil quality. Both the inherent and dynamic soil qualities are important to be considered when the soils are used for specific purpose; however, it should be kept in mind that inherent soil property such as soil texture is fixed and not easy to change, unless significant quantity of soil from an external source is exported to the site. The dynamic properties change with the level of management for a specific use; therefore, the user should forecast the consequences of management practice.

Soil tillage when wet can cause significant compaction that seeds may not germinate; in contrast tilling soil when dry can create soil pores and perform better for seed germination and crop production. The question is why we assess soil quality, and the answers are straight: (i) to assess the capability of soil for a certain use, (ii) how it behaves to certain management practices leading to better performance for specific use, and (iii) monitoring trend of change in soil properties with management practices to diagnose any problems that may cause hindrance in achieving the objective. For a better assessment of dynamic soil quality, it is appropriate to consider soils physical, chemical, mineralogical, and biological properties and their interaction in the soil system during the operation of management practices. For example, the use of brackish water with salts may create soil salinity which will hinder plant growth and a sodic water degrade soil structure; however, the use of same brackish water in two different texture soils will have different impact to affect soil behaviors, e.g., low effect on coarse textured (sandy, loamy, sand) soils than clayey soil in terms of structure degradation and infiltration rate. The integrated approach of soil properties assessment is a better approach than measuring soil property separately. During the use of soil for a specific purpose, soil quality management is an effective approach to resource use and conservation.

1.2 Soils of Kuwait

The soils of Kuwait have been surveyed using the standard soil classification systems over the past 50 years. These surveys covered the entire Kuwait and selected areas at scales ranging from 1:500,000 to 1:5000. At the broadest reconnaissance scale, the FAO used the USDA methods (1951) to survey and map the soils of Kuwait (Ergun 1969). Soil information also reported by Omar et al. (1986, 1988), where Omar et al. (1986) studied the relationship of edaphic factors to plant and animal communities. In addition, Omar et al. (1988) mapped the soils in the Al-Wafra, Al-Sulabiya, and Al-Huwaymiliyah areas, where the soils were mapped at subgroup level using the US soil taxonomy (Soil Survey Staff 1975). The latest soil survey was completed in two phases during 1995–1999 (KISR 1999a, b; Omar and Shahid 2013).

A further survey of three demonstration farm sites (50 ha each) was completed (Shahid and Omar 1999; Shahid et al. 2004) at 1:10,000 scale (first-order level of USDA-NRCS system). Recently, Kuwait soil taxonomy (Shahid and Omar 2022) is published, where the results from both surveys (KISR 1999a, b) were correlated to the latest standards (Soil Survey Staff 2022; Soil Science Division Staff 2017). The recently published Kuwait Soil Taxonomy (Shahid and Omar 2022) is considered a milestone and an *easy-to-understand* user friendly soil classification guide for potential stakeholders as they strive to better understand and educate about this vital natural resource “the soil.”

The above survey revealed the soils of Kuwait are mostly coarse textured including mainly 4 out of 12 soil texture classes (Omar and Shahid 2013; Soil Science Division Staff 2017). Sandy (Fig. 1.1a), loamy sand, and sandy loam are the main soil texture classes. In general, the soils of Kuwait are calcareous, gypsiferous, salty, and gravelly. In depressions, water after rain accumulates finer soils (clayey) are encountered, but to a limited extent in area (Fig. 1.1b). The surface soil features of Kuwait deserts are determined by the action of wind erosion, and the sand is continually moving, accumulating around plants developing into “nebkhas” (Fig. 1.1a); low height sand dunes accumulate on road sides and even block the highways and interrupt the traffic.

In this chapter, soils of Kuwait are described in general without going into details as given in soil survey reports, threats to the soils are identified, and management options are discussed for the soil information seekers to use the soils for various purposes on sounds ground basis.

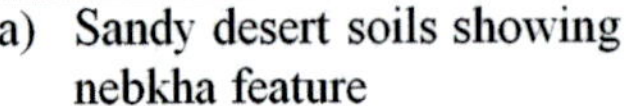

a) Sandy desert soils showing nebkha feature

b) Clayey soil in depression

Fig. 1.1 Different textured soils in Kuwait

1.3 Functions of Soils

The soil is a product of five soil formation factors and four processes. The soils are not static but dynamic and changing due to natural and anthropogenic impacts. Based on the soil parent material the soils are developed on, they have inherent capability related to parent material. Soils have diversified functions; however, from the farmer's perspectives, it is medium to support plants.

Soil performs multiple functions based on the parent materials they are developed, the environmental conditions of the area, and the adoption of specific management practices for specific uses. The FAO (fao.org/soils-2015) described soil functions at higher levels without going into further details. These functions vary from place to place mainly due to variations in environmental factors; therefore, these should be considered as guideline to make plans for different uses of diverse soils. The general soil functions include provision of food, fiber and fuel, carbon sequestration, water purification and soil contamination reduction, climate regulation, nutrient cycling, habitat for organisms, flood regulation, source of pharmaceuticals, foundation for human infrastructure, provision of construction materials, and cultural heritage.

1.4 Average Composition of Soil

Soil contains four main components; their average composition is shown in Fig. 1.2. This is the ideal situation for better soil services. However, the soils in the arid region present different average composition, that is, organic matter about 1%, porosity 45%, water (1%), and mineral matter 53% (Shahid and Omar 1999).

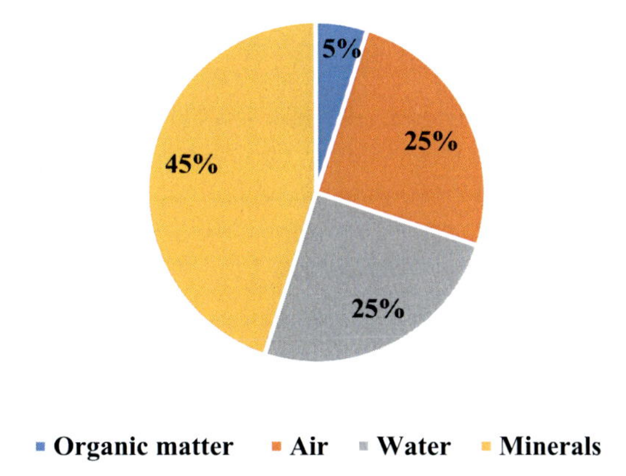

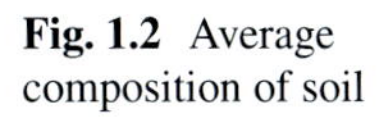
Fig. 1.2 Average composition of soil

1.5 Soil Formation in Kuwait in the Context of Soil Formation Factors and Processes

Soils are the important components of the terrestrial ecosystems. Soil properties change differently under different climatic conditions, however, soil change by humans is often more rapid and far-reaching (Sandor et al. 2022). In Kuwait, due to the arid climate, it is nearly impossible for the soil to form in the deserts, and strong wind erosion also limits soil formation. For soil formation the soil material has to go through physical, chemical, mineralogical, and biological changes over a period of time; therefore, the phenomenon of soil erosion has to be stopped, which is not possible under drifting sand environment (Misak et al. 2009, 2013); however, drifting sand can be reduced after significant interventions including but not necessarily limited to creation of shelter belts *wind breakers* and soil stabilization (Misak et al. 2013).

In the dry desert areas, the soils are least developed without the development of subsurface diagnostic horizons, and these soils are classified in the soil order "Entisols" (Soil Survey Staff 2022; Shahid and Omar 2022). However, in places diagnostic horizons (argillic, calcic, gypsic, petrogypsic, salic) have been recognized to classify these soils in the Aridisols order. Five soil formation factors (climate, organisms, relief, parent material, and time) cumulatively operate to form soil. The intensity of their operation varies in different places. The soil formation is faster in humid region compared to the hot arid region with low rainfall. Cumulatively these soil formation factors can be expressed as (cl, o, r, p, t). It stands for *climate*, *organisms*, *relief*, *parent material*, and *time*. These five soil formation factors affect all of the soils on the planet and describe the physical characteristics that have contributed to the development of soils over time. The soil formation concepts in use today and many years in the past are derived from the earlier work of Jenny (1941, 1980) and Hilgard (1921) whose work is considered classic in the field of soil formation.

The role of soil formation factors in developing soils of Kuwait is described in the following sections.

- Parent material (geological rocks, precursors to the soils)
- Climate (temperature, rainfall, and evaporation)
- Organisms (vegetation, microbial populations, soil fauna, flora)
- Relief (slope, slope aspect and gradient, landscape position, level, inclined)
- Time (period to undergo soil formation)

While contributing a chapter on *soils* in Physical Geography Sector chapter of Abu Dhabi Global Environment Data Initiative (AGEDI), Shahid (2006) has described the formation of soils of Abu Dhabi emirate in the context of soil formation factors. The soils of Abu Dhabi emirate are similar to the soils of Kuwait (Shahid and Abdelfattah 2008; Shahid et al. 2013; Omar and Shahid 2013; Shahid and Omar 2022), although the intensity of Entisol (varying heights of sand dunes and sand sheet) occurrence is higher in Abu Dhabi emirate than Kuwait.

The soil scientists agreed that soil formation is a complex process, and the formed soils are a product of the interaction of soil formation factors (*clorpt*). Among all factors of soil formation, *climate* plays a dominant role in soil formation. Whereas, high temperature and low rainfall restricts soil formation in the desert areas like Kuwait. Soils tend to develop rapidly under warm, humid, and forested conditions, which do not prevail in Kuwait, where soils develop slowly under hyper-arid conditions. In humid regions, low temperature and high rainfall accelerate soil formation due to dissolution and translocation of soil material to subsurface to develop pedogenically developed B horizon.

The sandy deserts soils are also highly vulnerable to wind erosion, and thus surface sediments are continually moving and shaping the landscape (aeolian soils) through erosion and accumulation. In contrast to terrestrial sandy desert soils, the soils in the coastline are developed mainly on the coastal calcareous shells (*parent material*) and are also rich in soluble salts due to sea water intrusion and subsequent evaporation developing into strongly saline soils locally called "Sabkha" Omar et al. (2002). *Topography* plays great role in soil formation, as the soils are well developed at the foothills due to deposition of material and low level of development on the slopes due to water erosion. *Time* is the fifth factor affecting significantly the soil formation and determines the age of the soil. Over time, living organisms, vegetation, and climate act on parent material and topography. The degree and intensity of soil formation over time depend on other four soil formation factors. In Kuwait, in deserts the soil formation is very slow owing to unstable soil surface consisting of moving sand under desert environment. Following are the soil formation factors which are considered to slow down the soil formation in Kuwait:

- Low organic matter and biological activity
- High-calcium carbonate equivalents and quartz (SiO_2) contents not readily soluble in water to enhance transformation as a part of soil-forming processes
- High temperature, prolonged drought, and scanty rainfall
- High erosion rate

1.5.1 Soil Formation Processes

It is evident that the surface sediment in the deserts of Kuwait is not static but dynamic and changing the soil surface due to desert conditions and the action of wind. There are four main soil formation processes:

- Additions
- Losses
- Translocation
- Transformations

Following soil formation processes have been observed in the soils of Kuwait:

1.5.1.1 Calcification and Decalcification

Calcification is a soil-forming process, whereby dissolved calcium and carbonates are precipitated into calcium carbonates. In Kuwait both processes have been identified through observing the accumulation of calcium carbonates masses in the B horizons of the profile (calcification), while in the A horizon, calcium carbonate masses were either absent or significantly less than the B horizon (decalcification), from where (A horizon) the calcium carbonates were dissolved, and reprecipitated in the B horizon and developed as calcic horizon [$(CaCO_3 + H_2O + CO_2 \leftrightarrow Ca(HCO_3)_2$]. The *calcification* occurs when H_2O or CO_2 is removed from the system and the reaction moves to the left; decalcification occurs when $H_2O + CO_2$ are present and the reaction moves to the right. In addition, the direct source of calcium carbonate in soil could be due to the calcareous parent material (limestone) on which soils are developed; this is inherited property and not a pedogenic process. The decalcification and calcification can be considered *transformation* process.

1.5.1.2 Gypsification

Similar to calcification and decalcification, gypsum ($CaSO_4.2H_2O$) accumulation has been observed in the lower soil depth of the profile developing into "gypsic horizon" through gypsification process. Gypsum was identified in four soil taxa of Kuwait (*Leptic Haplogypsids*, *Typic Haplogypsids*, *Typic Calcigypsids*, and *Typic Petrogypsids*). The gypsification can be considered a *transformation* process.

1.5.1.3 Salinization

Salt accumulation has been observed in the terrestrial land especially farming areas where the farmers are using saline/brackish water to irrigate their crops. The mismanagement and improper use of brackish water transform normal soils to saline soils. Based on the root zone soil salinity level, the choice of crops become limited, and if salinity is not managed, significant yield loss of crops yield can occur and even death of plants at maximum salinity level of specific crop (Zaman et al. 2018a). A large saline area in Kuwait is mapped in the coastline of Kuwait due to sea water intrusion and subsequent evaporation (Omar and Shahid 2013; Shahid and Omar 2022). The salinization can be considered a *translocation* process.

1.5.1.4 Sodification

Accumulation of sodium on soil exchange complex has also been observed both in the farming and coastal areas affected by soil salinity. The soil sodicity affects the soils in two ways, (i) by degrading soil structure and (ii) through Na-ion toxicity to plants.

1.5.1.5 Erosion

Erosion is a dominant soil-forming process in the desert soils of Kuwait. Wind erosion is common in the deserts, whereas, water erosion is limited to highlands e.g., *Jal-az-Azor* in Kuwait and in the Wadis due to flash floods (Fig. 1.3a). The addition of fallen dust and plant leaves on soil surface can be considered an *addition* process of soil formation. The removal of soil from an area is considered a *losses* soil-forming process.

1.5.1.6 Homogenization and Oxidation

Processes are at low level due to poor biological activity (rodents, ants) (Fig. 1.3b). Mottling (red soil spots in soil) has been observed in coastal soils, red soil color shows *oxidation* process when element loses electron ($Fe^{++} \rightarrow Fe^{+++} + e^{-}$; where

a. Soil loss through water erosion

b. Homogenization process (rodents activity)

Fig. 1.3 Water erosion in the sloping landscape *Jal-e-Azor* (**a**) and soil homogenization by ants (**b**)

e^- = electron transfer), and green, blue-green, and grayish color indicate *reduction* process (low oxygen supply, or waterlogged soils).

1.6 General Description of Soils

The soils of Kuwait regardless of strict soil taxa names can be grouped into four general categories, defined below.

1.6.1 Sandy Soils

Soils are classified for soil texture (% distribution of sand, silt, and clay) after placing the % values on the soil textural diagram (Soil Science Division Staff 2017). The soils of the world are classified into 12 textural classes (Schoeneberger et al. 2012). These 12 classes can be broadly grouped into *coarse*, *medium*, and *fine textures*. In Kuwait the soils are mainly classified under *coarse* category including three soil texture classes (sand, loamy sand, and sandy loam) as shown on the lower left end of soil textural triangle. These soils are mostly loose and/or poorly cohesive, with high infiltration rate and low moisture and nutrient retention capacities. These soils have low production capacity, and this constraint is compensated by adding mineral and organic fertilizers for crop production (Farrar et al. 2018). From soil classification point of view, sandy soils are grouped under the soil order *Entisols* provided no subsurface diagnostic horizons are identified in the soil profile.

1.6.2 Gypsic Soils

These soils contain gypsum ($CaSO_4.2H_2O$) to various extents and at different depths. Although these soils have a commercial value, as gypsum is the main constituent of cement which is commonly used in the construction industry. Gypsum slabs are commonly used to divide rooms into small units. The soils rich in gypsum are not suitable for crop production. The presence of gypsum at subsurface can cause buildings subsidence when water become in contact with gypsum. In Kuwait gypsum is identified as noncoherent masses in the soil profiles (*Typic Haplogypsids*, *Leptic Haplogypsids*, *Typic Calcigypsids)* and subsurface hard pan *(Typic Petrogypsids)*. The gypsic soils are not suitable for agriculture uses, unless gypsum accumulation is below 1 m from soil surface, where short root crops can be grown.

1.6.3 Saline Soils

Soil salinity in the soils of Kuwait can be divided into two types, primary (dryland) and secondary (human-induced). The primary salinity is due to the salts released from the soil parent material after dissolution and/or the salts added into the soil due to sea water intrusion and subsequent evaporation "no human influence." However, the secondary salinity is caused due to human activities, such as the use of brackish /saline water to irrigate crops for food security or on urban landscapes with impeded drainage. Most of the desert soils (native soils) are nonsaline (ECe <2 dS/m) due to the presence of quartz and calcite (very low solubility) in the main matrix of desert soils. Whereas, the soils in the coastal areas are strongly saline and supports specific type of plants which are highly salt tolerant *halophytes*. When the soil salinity reaches to maximum to form sabkha (salt-scald), then hardly any vegetation is observed, but thick salt-crusts are observed.

1.6.4 Calcareous Soils

The soils of Kuwait are developed from sandy and calcareous parent materials, where coastal shells are the common soil constituents. The calcium carbonates equivalents material is ubiquitous in the soil environment of Kuwait. It occurs as soft segregations (*Typic Haplocalcids*) and as hard pan (*Typic Petrocalcids*) in the soil profiles. The *Typic Petrocalcids* soils are unsuitable for crop production due to calcareous hardpan at different depths creating insufficient soil depth for plants growth and causing hindrance to root penetration and water movement. In addition, calcareous soils have high buffering capacity (resist to change in soil pH) and high pH; therefore, there is general problem in nutrients availability to plants.

1.7 General Threats to Desert Environment and Proposed Measures

Multiple threats are identified to desert soils functions limiting their uses to their full capability. The threats are briefly described below with proposed actions.

1.7.1 Erosion

Soils of Kuwait are sandy, loose and fragile, and vulnerable to both wind and water erosion. Millions of tons of sand are shifted from the desert landscape along with organic matter and nutrients and leaving behind infertile soils to provide healthy ecosystem services. The use of organic soil amendment can bind soil particles into soil structure to increase particles cohesiveness and to improve nutrients and moisture retention capacities. Wind breakers and shelter belts can decrease the wind effects and reduce sand drifting. In addition, soil soilization using plants based soil binder "carboxy methyl cellulose-CMC" which is an eco-mechanical solution to desertification (Yi and Zhao 2016) where sand is continuously transform between the rheological state and the solid state. This eco-friendly approach has the potential to reduce soil erosion from the desert environment of Kuwait.

1.7.2 Contamination

Soils are contaminated with residues of pesticides and insecticides used to control the insect-pest infestations on crops. In addition, during the Gulf war a large area of Kuwait was contaminated with oil due to oil wells fire. This has affected the quality of soils to provide healthy ecosystem services. Such soils require treatments through bioremediation/ phytoremediation, to bring the contaminates to a safe level for specific uses, following by revegetation (Al-Baroud 2021) of these areas using well-adopted plants to the desert environment of Kuwait.

1.7.3 Salinization

Seawater intrusion and evaporation have salinized the coastal lands to a great extent. These areas are either devoid of vegetation or support only salt-loving plants called halophytes. The soils where salt appears to a great extent should be either left on nature, or salt harvesting can be accomplished using appropriate structures. Agricultural farming areas in Al-Abdally and Al-Wafar are also salinized due to the use of brackish water to irrigate crops. In these salinized farms, an integrated soil reclamation program is needed to be implemented including salts leaching, growing salt-tolerant crops, and using crop specific modern irrigation systems (drip, sprinkler) as well as draining the water from water logged soils. In this regard, currently a national level project has been started to assess farm land and groundwater salinity to develop a salinity management action plan for Kuwait (Shahid et al. 2023).

1.7.4 Waterlogging

Coastal soils are naturally waterlogged due to seawater intrusion. The inlands are waterlogged due to the occurrence of gatch (near surface hardpan) causing impedance to water. Waterlogging is evident after heavy rain and overuse of water in farming areas. Such features provide insufficient oxygen for plants roots to be able to adequately respire. Draining water using appropriate location-specific drainage system from such waterlogged areas is the best option. However, there are some plants which consume huge volumes of water for their growth and hence are considered bio-drainage plants. In many countries *Eucalyptus* is being used for bio-drainage; however, in many countries, it is banned as it lowers the water table and threat agricultural farms survival.

1.7.5 Biodiversity

Soil contamination and poor soil health due to low moisture retention and organic matter reduced the soil biodiversity of micro- and macroorganisms present in the soil. The loss of soil biodiversity has reduced many biological functions performed by the microbial community. The use of organic and bio-fertilizers and arbuscular mycorrhizae (AM) is the viable option to improve organic matter in soil to improve microbial activities and gain biodiversity.

1.7.6 Texture

Soils of Kuwait are dominant in sand (2.0 to 0.05 mm) and minor in silt and clay (<0.05 mm) fractions. Three soil texture classes (sand, loamy sand, and sandy loam) are common and present high porosity, low moisture, and nutrients retentions, leading to significant losses of water and nutrients and high costing to farmers. The investment in the soils of Kuwait to improve cohesiveness of soil particles and to increase clay contents is required. This can be accomplished by introducing nanoclay technology into the soils to improve soil texture, at least the surface soil depth-active root zone to improve soil texture and plant growth. The use of bentonite clay at the rate of 1.25 tonnes per hectare resulted in an average rice yield increase of 73% (IWMI 2010). Currently a proposal preparation is underway to invest in soils of Kuwait through introducing clay into the root zone of soil to improve soil tilth and plant growth (Alkandari and Shahid 2023).

1.7.7 Organic Matter

The removal of surface productive layer is common due mainly to wind erosion process. Wind erosion occurs when dry and loose soils are subjected to strong wind leading to lose productive soil and organic matter. The loss of organic matter is linked to poor soil properties and low soil biodiversity (microbial community). Organic matter can be increased by using organic fertilizer-compost and manures.

1.7.8 Sealing

Urbanization is taking land for roads, buildings, and other infrastructures. These soils are permanently sealed and not available for further uses. Thus, national soil resources are shrinking and per capita share is decreasing. The sealing of soils is likely to continue simultaneous to increase in urbanization. It is suggested to use soil information (KISR 1999a; Omar and Shahid 2013), and where prime arable lands are located, these areas to be left for future use and conserved to expand agriculture and where marginal lands are located and these soils are permanently unsuitable for agriculture may be exploited for urbanization. With this strategy we can save soils for future generations.

1.7.9 Compaction

The naturally compacted subsurface soil layer "gatch" at different depths impairs the functions of both the top- and subsoil, to support plants growth, and impedes root penetration, water infiltration, and gaseous exchange. In addition, off-road vehicles maneuvering and camping also created surface compaction. The subsurface hardsetting-gatch to be shredded using subsoiler and the surface sealing to be broken using tined based tractor to improve surface soil porosity. The use of organic fertilizers will lead to develop soil structure for better performance.

1.7.10 Dumping

Dumping and/or disposing waste, e.g., construction material and trash in deserts occupy land and reduce area for other uses. The organic based materials to be recycled for use in soils to improve organic matter. Dumping/landfilling to be done on marginal lands and arable land is conserved.

1.8 General Soil Types, Threats to Functions, and Mitigation Actions

There are multiple threats to the soils of Kuwait for their sustainable uses to have ecosystems services including crop production to their maximum capacity. These potential threats are highlighted here along with proposed management options for their sustainable uses (Table 1.1).

Table 1.1 General soil types, threats, and options for sustainable uses

General soil types	Potential threats to soil functions	Mitigation actions for sustainable uses
Sandy soils	Loose and highly vulnerable to wind and water erosion	Erect shelter belts and wind breakers of Kuwait environment adopted trees and shrubs (Misak et al. 2013) Revegetate the deserts with native plants Improve surface soil aggregation and soil structure development using eco-friendly plant-based and inorganic soil conditioners (Tahir and Marschner 2016; Yi and Zhao 2016) Create barriers (fences, dikes) to halt soil erosion and water harvesting for use or aquifer recharge
	Very low moisture and nutrients retention capacities, high infiltration	Use mineral soil amendments (clay-based) to improve soil structure leading to increased moisture/nutrients retention capacities and decrease infiltration rate (Farrar et al. 2018) Increase the use of inorganic soil conditioner and organic fertilizers to improve soil carbon sequestration and soil fertility (Alshankiti et al. 2015) Use conservation tillage practices (low to no till farming) to conserve soil moisture and control evaporation losses (Benites 2008; Goddard et al. 2008) Use clay and clay-based soil amendments (nano-clay technology) to improve soil fertility (IWMI 2010; Peng and Sun 2012) Use modern irrigation systems (surface and subsurface drip, sprinkler, sprayer, and central pivot) to improve water-use efficiencies
	Low organic matter and clay (poor inherent soil fertility)	Return crop residues into soil to improve organic matter Use plants based biochar to improve soil tilth and increase soil moisture retention (Abel et al. 2013) Use organic fertilizers (compost, manures) to increase soil organic matter
	High soil pH	Use organic fertilizers and humic acids to lower soil pH Use mineral fertilizers (NH_4-based) that release H^+ in soil through nitrification process and lower soil pH
	Low fertility	Use NH_4-based fertilizers to reduce NO_3 leaching in coarse texture soil Use 4R nutrient management strategy (Shahid 2018) including: Right type of chemical fertilizers (ammonium versus nitrate-based fertilizers) Right rate of fertilizers based on the soil testing and target yield Right time of fertilizer application at the right growth stage Right location of fertilizer application (root zone area for best use of nutrients)

Saline soil	High salts	Saline soils in the coastal area to be avoided for crop production Coastal lands to be revegetated with halophytes (salt-loving shrubs) to protect the coast land from erosion Cultivate salt-tolerant crops in the agriculture farms based on the irrigation water salinity (Zaman et al. 2018a, b) Leach the residual salts from the root zone using reclamation leaching concept prior to seeding Use extra water above the crop demand to leach the salts from the root zone and to maintain the root zone salinity at or below the crops' threshold salinity level Blend brackish/saline water with fresh water to reduce water salinity level suitable for the crop grown Adopt conjunctive water-use concept (alternate irrigation with brackish-salty and fresh water) Use integrated soil reclamation procedures, including physical, chemical, hydrological, and biological methods (Shahid and Rehman 2011; Shahid et al. 2018; Zaman et al. 2018a, b) Laser land leveling is recommended to avoid water ponding and development of high soil salinity zone Use fresh water for irrigation purposes to mitigate salinity effects
Calcareous soils	Calcareous hard pan and high calcium carbonate	Not recommended for open field agriculture but can be used to develop greenhouse as a part of protected agriculture (soilless, hydroponics, etc.) Excellent source of calcium carbonates for commercial uses Shallow hardpan soils to be avoided for agriculture purposes. Hardpan can impede plants roots movement, restrict water movement, develop waterlogging, and affect gaseous exchange for roots respiration Develop drainage system above the hardpan to avoid waterlogging If hard pan soils are to be used, it should be broken using chisler or a subsoiler If hardpan is occurring deeper (1–2 m), these soils can be used for shallow rooted crops High-calcium carbonate soils without hardpan can be used for agriculture purposes. See interpretation in high pH soils Calcareous soils can fix phosphorous and micronutrients (Fe, Cu, Zn, Mn). The micronutrients can be used as foliar application to avoid interaction with calcareous soils, as these are needed by plants in traces Phosphorous to be applied in band placement by placing suitable quantity near to the seed
Gypsic soils	Gypsic hard pan and high gypsum contents	Not recommended for open field agriculture but can be used to develop greenhouse as a part of protected agriculture (soilless, hydroponics, etc.) Excellent source of gypsum for commercial uses Shallow gypsic hardpan soils to be avoided for agriculture purposes High gypsic soils without hardpan can be used for agriculture purposes, especially shallow rooted crops are the viable option, provided gypsic horizon occurs below 1meter depth.
Contaminated soils	Oil and heavy metals	Bioremediation and phytoremediation are recommended to reduce hydrocarbon and heavy metals levels

1.9 Conclusions and Recommendations

The soils of Kuwait are dominantly coarser in soil texture (sand, loamy sand, and sandy loam). These soils are simple in terms of composition but very difficult to manage for their sustainable uses. Reasons being the loose sandy soils are always in mobile mode and present high infiltration rates. Both characters require significant investments to improve sandy soils qualities. Two soil orders (*Aridisols and Entisols*) have been mapped in Kuwait. These soils present significantly different soil properties and suitability for their uses. For a specific use, soil quality must be determined and any constraint for a particular use to be diagnosed and rectified before putting into a specific use. The soil taxonomy system provides soil-landscape relationships to facilitate soil management in a sustainable manner. The soils of Kuwait are moderately suitable for agriculture uses mainly due to sandy texture and low clay and organic matter contents. Therefore, efforts to be made to improve water-use efficiencies in these constrained soils. It is recommended to consult soil survey of Kuwait reports for rationale uses and as a science-based tool for land-use planning and policy development for Kuwait. It has been realized that the soils of Kuwait are under natural and anthropogenic threats and limiting the sustainable use and soil threat-specific mitigation actions to be adopted for sustainable ecosystem services including food security.

References

Abel S, Peters A, Trinks S, Schonsky H, Facklam M, Wessolek G (2013) Impact of biochar and hydrochar addition on water retention and water repellency of sandy soil. Geoderma 202–203:183–191

Al-Baroud AS (2021) Revegetation in remediated soil at south-east Kuwait and revision of remediation standards – a case study from desert environment. Int J Curr Res 13(4):16981–16992

Alkandari AJ, Shahid SA (2023) Investing in the soils of Kuwait to save agriculture water and boost crop production for food security. Proposal. Desert Agriculture and Ecosystems Program, Environment and Life Sciences Research Center, Kuwait Institute for Scientific Research Kuwait. p 37

Alshankiti A, Shahid SA, Gill S, Jarrar M (2015) Evaluating green compost and zeolite to enhance soil quality for Dubai sustainable city. ICBA-TADWEER Joint Publication, pp. xv+39. Dubai, United Arab Emirates

Benites JR (2008) Effect of no-tillage on conservation of the soil and soil fertility. In: Special Publication No 3, published by The World Association of Soil and Water Conservation (WASWAC) pp 59–72

Ergun HN (1969) Reconnaissance soil survey of Kuwait. FAO Project FAO/KU/TF-17. Food and Agriculture Organization, Rome

Farrar M, Gill S, Shahid SA, Babiker K (2018) Water saving innovation in urban landscaping and irrigated agriculture by using AustraBlend multimineral root zone conditioner. Asian J Sci Technol 9(12):9092–9100

Goddard T, Zoevisch M, Gan Y, Ellis W, Watson A, Sombatpanit S (eds) (2008) No-till farming systems. Special Publication No. 3, Published by The World Association of Soil and Water Conservation (WASWAC), p 538

Hilgard EW (1921) Soils: their formation, properties, composition, and plant growth in the humid and arid regions. Macmillan, London

IWMI (2010) Improving soils and boosting yields in Thailand. Success stories Issue – 2. International Water Management Institute (IWMI)

Jenny H (1941) Factors of soil formation. A system of quantitative pedology. McGraw-Hill; Dover, Mineola

Jenny H (1980) The soil resource-origins and behavior, Ecological Studies, vol 37. Springer, New York

KISR (1999a) Soil survey for the State of Kuwait: reconnaissance survey, vol II & III. Kuwait Institute for Scientific Research, Kuwait

KISR (1999b) Soil survey for the State of Kuwait: semi-detailed survey, vol IV & V. Kuwait Institute for Scientific Research, Kuwait

Misak RF, Al Sudairawi M, Al-Dousari A, Al Gamilly H (2009) Long term national program for managing the hazard of shifting sands in terrestrial environment of Kuwait. Proposal EUD, Kuwait Institute for Scientific Research Kuwait

Misak RF, Khalaf KI, Omar SAS (2013) Managing the hazards of drought and shifting sands in dry lands: the case study of Kuwait. In: Shahid SA, Taha FK, Abdelfattah MA (eds) Development in soil classification, land use planning and policy implications: innovative thinking of soil inventory for land use planning and management of land resources. Springer, Dordrecht, pp 707–729

Omar SAS, Shahid SA (2013) Reconnaissance soil survey for the State of Kuwait. In: Shahid SA, Taha FK, Abdelfattah MA (eds) Developments in soil classification, land use planning and policy implications: innovative thinking of soil inventory for land use planning and management of land resources. Springer, Dordrecht, pp 85–110

Omar SA, Al-Sdirawi F, Agarwal V, Hamdan A, Al-Bakr D, Al-Shuaibi F (1986) Criteria for establishment of Kuwaiti's first national park/nature reserve, Inventory and zoning, vol 1. Kuwait Institute for Scientific Research, AG-51, Final Report, Kuwait

Omar SA, El-Bagouri I, Anwar M, Khalf F, Hashash M, Nassef A (1988) Measures to control mobile sand in Kuwait. Kuwait Institute for Scientific Research, Report EES78, vol 1. KISR 8790, Kuwait

Omar SAS, Misak RF, Shahid SA (2002) Sabkhat and halophytes in Kuwait. In: Barth HJ, Boer B (eds) Sabkha ecosystems, pp 71–81

Peng Y, Sun Y (2012) Resources characteristics and market situation of bentonites at home and abroad. Metal Mine 4:95–99

Sandor JA, Burras CL, Thompson M, Wills SA (2022) Factors of soil formation—human impacts. In: Reference module in earth systems and environmental sciences. Elsevier. https://doi.org/10.1016/B978-0-12-822974-3.00029-X

Schoeneberger PJ, Wysocki DA, Benham EC, Soil Survey Staff (2012) Field book for describing and sampling soils, version 3.0. USDA–NRCS. National Soil Survey Center, Lincoln

Shahid SA (2006) Soils. In: Sector paper on physical geography. State of Environment Report, Abu Dhabi Global Environment Data Initiatives (AGEDI), Environment Agency, Abu Dhabi, pp 6–41

Shahid SA (2018) Salt-affected soils—4 R nutrient stewardship. Farming Outlook 17(2):19–22

Shahid SA, Abdelfattah MA (2008) Soils of Abu Dhabi Emirate. In: Perry R (ed) Terrestrial environment of Abu Dhabi Emirate. Environment Agency, Abu Dhabi, pp 71–91

Shahid SA, Omar SAS (1999) Order 1 soil survey of the demonstration farm sites with proposed management, Kuwait Institute for Scientific Research, Kuwait, p viii + 144. KISR no 5463. ISBN 0 957700369

Shahid SA, Omar SAS (2022) Kuwait soil taxonomy. Springer, p 149

Shahid SA, Rehman K (2011) Soil salinity development, classification, assessment and management in irrigated agriculture. In: Passarakli M (ed) Handbook of plant and crop stress, 3rd edn. CRC, pp 23–39

Shahid SA, Omar SAS, Jamal ME, Shihab A, Abo-Rezq H (2004) Soil survey for farm planning in northern Kuwait. Kuwait J Sci Eng 31(1):43–57
Shahid SA, Abdelfattah MA, Othman Y, Kumar A, Taha FK, Kelley JA, Wilson MW (2013) Innovative thinking for sustainable use of terrestrial resources in Abu Dhabi emirate through scientific soil inventory and policy development. In: Shahid SA, Taha FK, Abdelfattah MA (eds) Developments in soil classification, land use planning and policy implications: innovative thinking of soil inventory for land use planning and management of land resources. Springer, Dordrecht, pp 3–49
Shahid SA, Zaman M, Heng L (2018) Salinity and sodicity adaptation and mitigation options. In: Zaman M, Shahid SA, Heng L (eds) Guidelines for salinity assessment, mitigation and adaptation using nuclear and related techniques. Springer, pp 55–88
Shahid SA, Burezq HA, Baron HJ (2023) Farm land salinity risk assessment to develop a management strategy for food security. Proposal. Kuwait Institute for Scientific Research, Kuwait. KISR # 17167. Kuwait Foundation for the Advancement of Sciences-KFAS code:PN22-41SE-1668. p 51
Soil Science Division Staff (2017) Soil survey manual United States Department of Agriculture Handbook No 18
Soil Survey Staff (1975) Soil taxonomy: a basic system of soil classification for making and interpreting soil surveys. Soil Conservation Service US Department of Agriculture Handbook 436
Soil Survey Staff (2022) Keys to soil taxonomy, 13th edn. US Department of Agriculture, Natural Resources Conservation Service, US Government Printing Office, Washington, DC
Tahir S, Marschner P (2016) Clay addition to sandy soil – effect of clay concentration and ped size on microbial biomass and nutrient dynamics after addition of low C/N ratio residue. J Soil Sci Plant Nutr 16(4):864–875
Yi Z, Zhao C (2016) Desert "Soilization": an eco-mechanical solution to desertification. Engineering 2:270–273
Zaman M, Shahid SA, Heng L (2018a) Irrigation systems and zones of salinity development. In: Zaman M, Shahid SA, Heng L (eds) Guidelines for salinity assessment, mitigation and adaptation using nuclear and related techniques. Springer, pp 91–111
Zaman M, Shahid SA, Heng L (2018b) Irrigation water quality. In: Zaman M, Shahid SA, Heng L (eds) Guidelines for salinity assessment, mitigation and adaptation using nuclear and related techniques. Springer, pp 113–131

Chapter 2
Perspectives of Geography, Environment, and Physiography of Kuwait

Hebah Jaber Kamal, Megha Thomas, and Ahmed Abdulhadi

Abstract Terrestrial environment ecosystems vary according to their geographical location on the earth. The State of Kuwait has a dry and hot desert ecosystem and lies in the hyper-arid climate zone in the Arabian Peninsula. In Kuwait's ecosystem, the biotic and abiotic components interact to adapt to the desert's main characteristics, including lack of fresh surface water resources, high temperatures, and evaporation rates. Despite the harsh conditions, some plants and animals can adapt to the desert environment in Kuwait. Lands in Kuwait are devoid of rugged or prominent terrain such as mountains or high rocky cliffs, or prominent terrain, nevertheless interspersed with some desert geomorphological landforms, such as wadis, ridges, desert depressions, sand dunes, and sabkhas. This publication discussed Kuwait's terrestrial environment from a physical geography perspective, including location, climate, geology, desert and coastal landforms, terrestrial biodiversity, natural resources, and economy, through combined authenticated and the latest published references. Finally, the chapter covers the most critical environmental challenges and threats resulting from the expansion of anthropogenic activities, urbanization, and climate variations and how these activities led to the ecological imbalances in the terrestrial environment of Kuwait.

Keywords Climate · Landforms · Biodiversity · Natural resources · Desert ecosystem · Geological strata · Ecological imbalance

---

H. J. Kamal (✉) · M. Thomas · A. Abdulhadi
Systems and Software Development Department (SSDD), Science and Technology Sector (STS), Kuwait Institute for Scientific Research, Safat, Kuwait
e-mail: hbaron@kisr.edu.kw; mthomas@kisr.edu.kw; aabdulhadi@kisr.edu.kw

M. K. Suleiman, S. A. Shahid (eds.), *Terrestrial Environment and Ecosystems of Kuwait*, https://doi.org/10.1007/978-3-031-46262-7_2

2.1 Introduction

Organisms on earth live in different ecosystems according to their geographical location on the earth's surface, and the biotic and abiotic components interact for living in various ways in this ecosystem. Each ecosystem has distinct components of abiotic features, such as temperature and humidity, and others are biotic such as plants, animals, and soil. Thus, organisms in each ecosystem adapt to the surrounding conditions in specific geographical regions. Different ecosystems exist on earth, such as forests, tundra, grasslands, deserts, and so on. This publication highlights the terrestrial environment and desert ecosystem in the State of Kuwait from the physical geography perspective. Physical geography is concerned with describing and studying earth's surface features, in addition to earth's geological structure and the factors or forces affecting surface formation or geomorphological landforms evolution. Physical geography also covers natural aspects, including climate, weather, biodiversity, and natural resources. Geography is no longer limited to the descriptive approach; modern geography studies and analyzes the patterns of relationships and interactions between humans, society, and nature. It contributes to understanding human impact on the terrestrial environment, climate, and living organisms and vice versa, thus scientifically interpreting the transformations emerging on earth and assisting in adaptation or control of it effectively. Geographical information system (GIS) tools and remote sensing (RS) techniques are utilized to map, extract, interpret, and analyze spatial data, supporting the research process.

Kuwait is located in the middle of the ancient world and the Arabian Peninsula of the Asian continent. The country is situated at the head of the Arabian Gulf. The State of Kuwait lies between latitudes 28° 30′ and 30° 06′ N and longitudes 46° 30′ and 48° 30′ E, occupying an area of 17,818 km^2. The Republic of Iraq is located on the north and northwest of Kuwait, and on the south and southwestern borders is the Kingdom of Saudi Arabia. Kuwait is a Middle Eastern country and a member of the Gulf Cooperation Council (GCC) established in 1981 that includes six countries, Saudi Arabia, Oman, United Arab Emirates, Kuwait, Qatar, and Bahrain. The system of government in Kuwait is a hereditary constitutional system that derives its legitimacy from the Kuwaiti constitution drawn up on November 11, 1962, and ratified by Sheikh Abdullah Al-Salem Al-Sabah (Al-Sewaji 1999). Kuwait is an oil-rich country that depends primarily on producing and exporting oil and gas, and it is a member of the Organization of the Petroleum Exporting Countries (OPEC). It is considered the fifth in the world regarding oil reserves and one of the largest crude oil and oil products exporters, as reported in the State of Kuwait Second National Communication submitted by Environmental Public Authority (EPA 2019a).

Kuwait's geographical location in the Arabian Peninsula shaped its ecosystem (Fig. 2.1), as the hyper-arid climate played a significant role in drawing the main characteristics of the desert ecosystem, where the long, hot, and dry summer and the low precipitation rates in the cold winter are prevailed. During summer, the average temperatures range between 42 and 48 °C; in winter, temperatures reach below

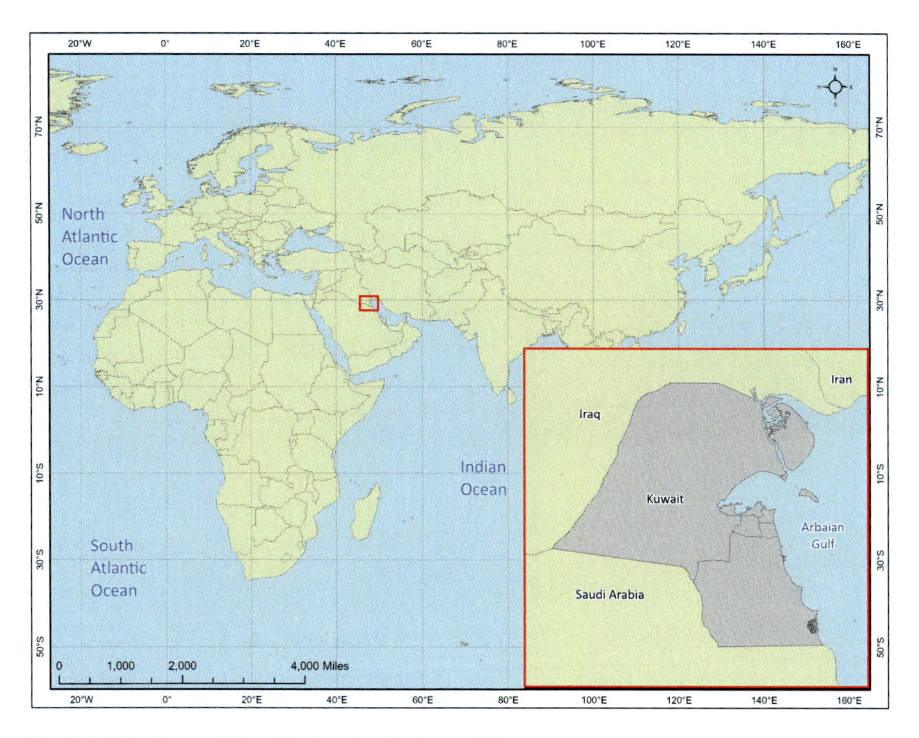

Fig. 2.1 The State of Kuwait's location in the world

zero °C at times, with low precipitation rates. The prevailing winds are northwest, continental, and dry; mild in the summer and cold in the winter (Al-Dosari 2021).

According to the prevailing climatic conditions and the scarcity of water resources, in addition to the soil properties lacking organic matter and clay, the vegetation cover is considered scarce and usually associated with the spring season. Nevertheless, few plants can adapt to Kuwait's harsh desert ecosystem and tolerate the drought conditions. According to the geomorphological surface, most lands are plain interspersed with ridges and depressions (Fig. 2.2). The high-elevation lands could be seen in the southwest part of Wadi Al-Batin (>250 m) and then lands gradually decline toward the coast to the east. Nevertheless, a group of geomorphological features is formed in Kuwait's desert, such as the Jal Az Zour escarpment, Al Ahmadi ridge, dry wadis, sand dunes, coastal and inland sabkhas, as well as desert depressions such as the Umm al-Rimam.

Kuwait's geographical and environmental aspects have been published over the past years. These studies dealt with various fields in physical and human geography. Some studies analyzed and discussed a specific geographical feature in the form of scientific research papers. In contrast, others generally discussed Kuwait's physical or human geography through books. For example, but not necessarily limited to, Kleo et al. (2003) studied the geomorphological landscapes in Kuwait, such as of Jal Az Zour escarpment, Wadi Al-Batin, Umm Al-Rimam depression, nebkha,

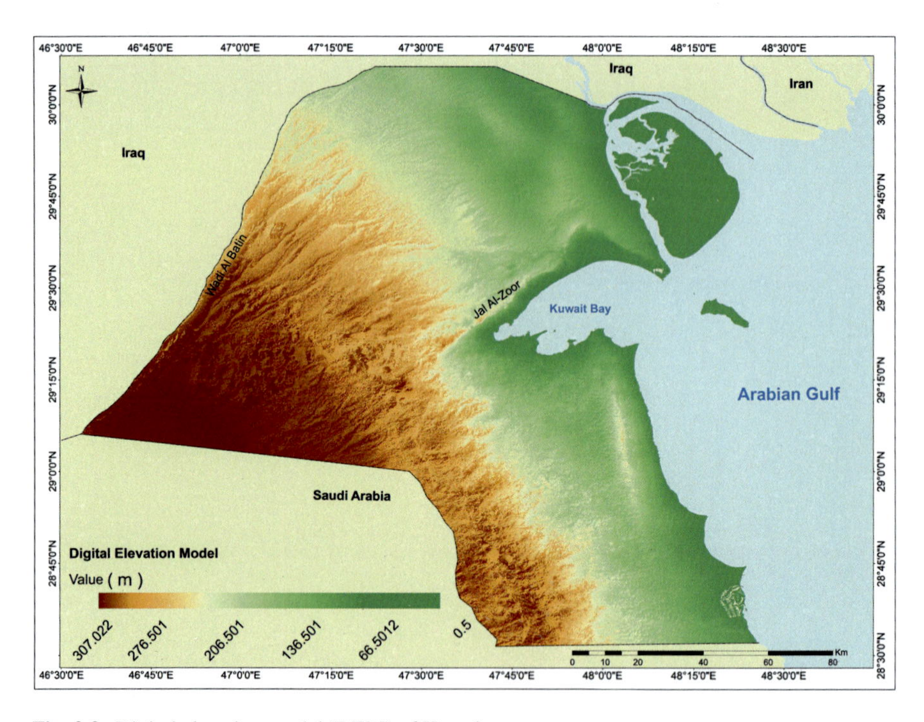

Fig. 2.2 Digital elevation model (DEM) of Kuwait

playa, and other geomorphological features. Each book chapter included detailed explanations about the feature's formation, geology, and surface sediments. Additionally, Al-Hurban and Gharib (2004) described the geomorphological characteristics of Kuwait's southern sabkhas; collected samples to analyze the mineral, structural, and sedimentary characteristics; and provided detailed descriptions of the coastal and inland southern sabkhas. Moreover, Al-Sarawi (1995) presented scientific research dividing Kuwait's geomorphological surface into four main sections: coastal region, wadis and ridges, depressions (lowlands), and dune fields, using morphological and stratigraphic observations, air photo analysis, and field examination. The paper also discussed different surface geomorphological features of Kuwait, origins, geological formations, and surface deposits. Al-Dosari (2021) dealt with various topics related to Kuwait's physical geography, such as terrain, climate, and economic geography, including natural resources, road networks, and traffic. Finally, the book discussed the most critical challenges of sustainable development in Kuwait. Last but not least, one of the most recent published works was the Geology of Kuwait book, which discussed Kuwait's geology through ten scientific chapters, including marine and terrestrial geology, petroleum geology, earthquakes, and groundwater in Kuwait (Abd el-aal et al. 2023).

This publication aims to collect and present the geographical aspects of Kuwait's terrestrial environment and its desert ecosystem based on a set of authenticated and latest published work. This includes terrestrial and coastal geomorphology, surface sediments, geological formations, biodiversity, and the distribution of Kuwait's natural reserves. Moreover, this publication briefly reviews the water resources in Kuwait, agricultural areas, and economic resources. Finally, it discusses the most prominent challenges and difficulties threatening Kuwait's ecosystem due to natural and anthropogenic activities leading to the depletion and destruction of environmental and natural resources.

2.2 Description of Kuwait Geographical Location and Prominent Terrestrial Features

The State of Kuwait location is in the heart of the old world in the Asia continent, specifically on the northern edge of the Arabian Peninsula (Fig. 2.1). The eastern borders of Kuwait are entirely marine, extending in a straight line for about 170 km along the Arabian Gulf coast. In contrast, land borders extend to about 422 km north and northwest with Iraq and south and southwest with the Kingdom of Saudi Arabia.

Kuwait is a small country compared to its neighboring countries, and its area is about 17,818 km^2. The land is primarily flat plain interspersed by prominent terrains such as ridges, rocky edges, and dry wadis. Kuwait's coast is predominantly straight and free of complex meanders, except Kuwait Bay, which appears as a small gulf in the middle of the land. Additionally, Kuwait owns some islands in the Arabian Gulf, according to size from the largest to the smallest, Boubyan (about 972 km^2), followed by, Warba, Failaka, Umm Al-Namel, Miskan, Umm Al-Maradeim, Kubbar, Auha, and finally Qaruh (0.033 km^2), the smallest (Mufadhal 2021).

The harsh conditions of the desert ecosystem in Kuwait such as scarcity of vegetation and drought directed people to the eastern coast toward the Arabian Gulf, as a means of living by building sailing ships for fishing, pearl diving, and trading. In the 1930s of the last century, it was discovered that Kuwait lies in a land floated on underground oil lakes. This discovery changed the entire future of the people and the state.

Today, Kuwait is divided into six administrative governorates (Fig. 2.3), each involving several residential areas. Administrative governorates are primarily Kuwait City, the capital governorate, including the old Kuwait City, Kuwait's old walls, the seat of government, the Kuwait National Assembly, and the national landmark of the country, Kuwait towers in front of Dasman Palace. Secondly, Al Ahmadi governorate in the south comprising the largest oil field, Burgan, and Al-Wafra, the most prominent agricultural areas on the southern border with Saudi Arabia. Thirdly, the largest governorate, Al-Jahra, includes big oil fields such as Al-Sabriya, and Al-Ratqa. The last three governorates are Mubarak Al-Kabeer, Hawalli, and Al-Farwaniya, which are primarily residential. Hawalli and Al-Farwaniya are among the most densely populated governorates.

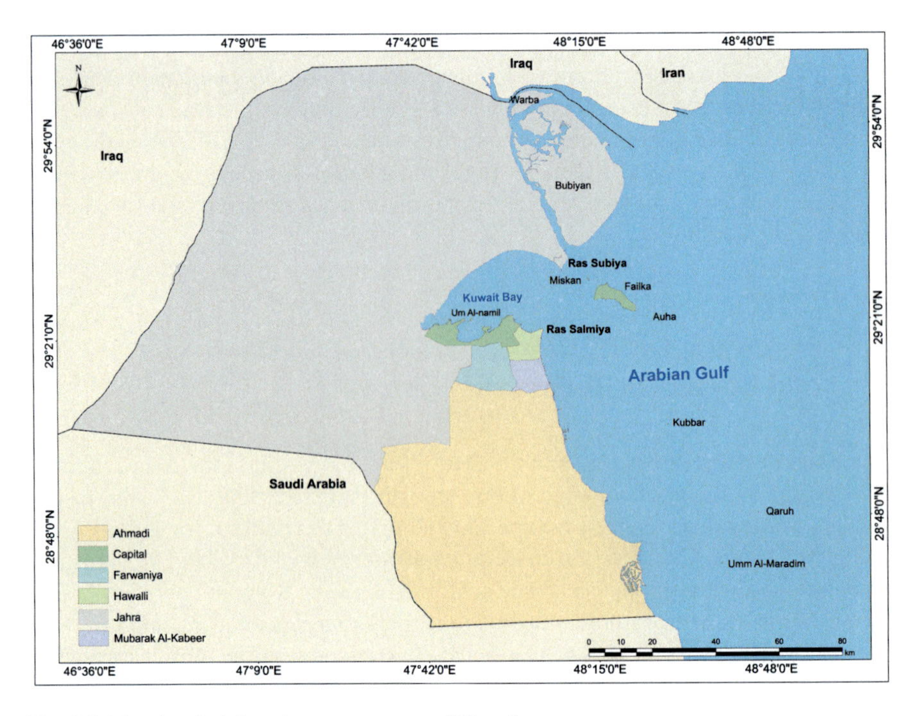

Fig. 2.3 The six administrative governorates of Kuwait

2.2.1 *Climate and Weather*

The location of Kuwait in the desert region made it one of the hottest countries in the world, where the climate is hyper-arid, where the long summer and short winter are the two main seasons. The temperature in summer is incredibly high, reaching the maximum in July and August, to record more than 48 °C (EPA 2019b) due to the intensity of the solar radiation and long daylight hours. By contrast, winter is cold and usually dry, extending from December to February. January is the coldest, with an average temperature of 13 °C (Al-Dosari 2021). Precipitation in Kuwait is scarce, as drought is a prevalent climatic feature in the desert ecosystem. However, Kuwait receives 75–150 mm of rain annually. Furthermore, the neighboring areas are a direct factor affecting the local weather. Kuwait's prevailing winds are northwesterly continental, up to 120 km/h (Al-Dosari 2021). This wind is dry but ordinarily causes cooling weather in summer and increasing coldness in winter, while dust storms in summer are a regular weather feature in Kuwait.

2.2.2 *Kuwait Geomorphology, Topography, and Geology*

The topography of a desert's ecosystem is usually characterized by several terrestrial terrains or particularly natural landforms. Desert forms are created when climatic conditions such as winds, temperatures, and sometimes rain interact with the land's geological and structural conditions such as surface sediments, rocks, geological formations, and soil; therefore, the Kuwait desert had its scenario. The surface in Kuwait is flat plain devoid of rugged terrains such as mountains, high rocky cliffs, or prominent terrain. The surface is a gradient from west to east toward the coast (Fig. 2.2). According to Al-Dosari (2021), based on the surface sediments and geomorphological landforms, Kuwait geomorphology was divided into four main sections: (1) the western region, including Wadi Al-Batin, the highest region above sea level; (2) the northern region, the largest region (5200 km^2), and its morphology were affected by river torrents and sediments; (3) the coastal plain region, formed by Arabian Gulf waves and currents; and finally (4) the southern region, including the southern coastal strip, Ahmadi plateau, and the western part of Wadi Al-Batin.

Although Kuwait's lands are dominated mainly by plain character, it interspersed with some prominent terrain resulting from some geomorphological processes that occurred above the land surface, through, (1) weathering and erosion forming desert depressions, as Umm Al-Rimam depression in the north of Kuwait Bay; (2) sedimentation, forming sand dunes, desert pavements, and sand sheets; (3) chemical weathering which happened when seawater interacts with coastal strip's rocks; and (4) due to underground tectonic movements, the instability of Kuwait's eastern and northeastern region extends to the Zagros Mountains region in Iran, as the area is exposed to severe folding operations (Al-Sarawi 1995), such as the Jal Az Zour escarpment.

Moreover, some surface topographies were formed by completely different climate than the present. The large amount of rains prevalent in the Pleistocene contributed to excavating a great wadis network, appearing clearly until now in the desert, especially in Wadi Al-Batin.

Kuwait stratigraphy formed during different geological epochs, starting from Eocene to Pleistocene, until the current Holocene. The three main physiographic formations in Kuwait were named "Kuwait Group" by Owen and Naser in 1958 (cf. AlRefaei et al. 2023) when the first stratigraphic study of Kuwait was conducted. Kuwait Group is overlaid on Dammam Formation, and from the oldest to recent are Ghar, lower Fars, and Dibdibba. Owen and Naser described Ghar formation as sands correlated with subordinate gravels and occasional clays, lower Fars composed of anhydrite, gypsum, marls, and limestones. Dibdibba formation consist of sands, gravels, and subordinate marls (cf. AlRefaei et al. 2023).

The surface deposits in Kuwait, aged from Middle Eocene to Holocene, are mostly siliciclastic sediments, sedimentary rock units, and carbonate deposits in the south (AlRefaei et al. 2023). Furthermore, AlRefaei et al. (2023) stated that Dammam formations sediments in Ahmadi Quarry and southern areas of Kuwait are considered the oldest surface sediments. Dammam formation is a part of the Hasa group, formed in the Eocene, and is described by Al-Sulaimi and Al-Ruwaih (2004) as nummulitic shale topped by silicified limestone. Further, Hassan et al. (2021) divided the surface sediments of Kuwait into nine deposits: aeolian and marine sand, coastal, desert floor, sabkha, Dammam, Fars, Ghar, Lower Dibdibba, and Upper Dibdibba formations. Figure 2.4 presents the surface geomorphology units as mapped in Kuwait Integrated Environmental Information Network (KIEIN) project database 2007. Figure 2.5 presents geological map of Kuwait.

2.2.3 *Most Prominent Landforms in the State of Kuwait*

In this section the most prominent landforms including Wadi Al-Batin, Jal Az Zour Escarpment and basic landform units are described.

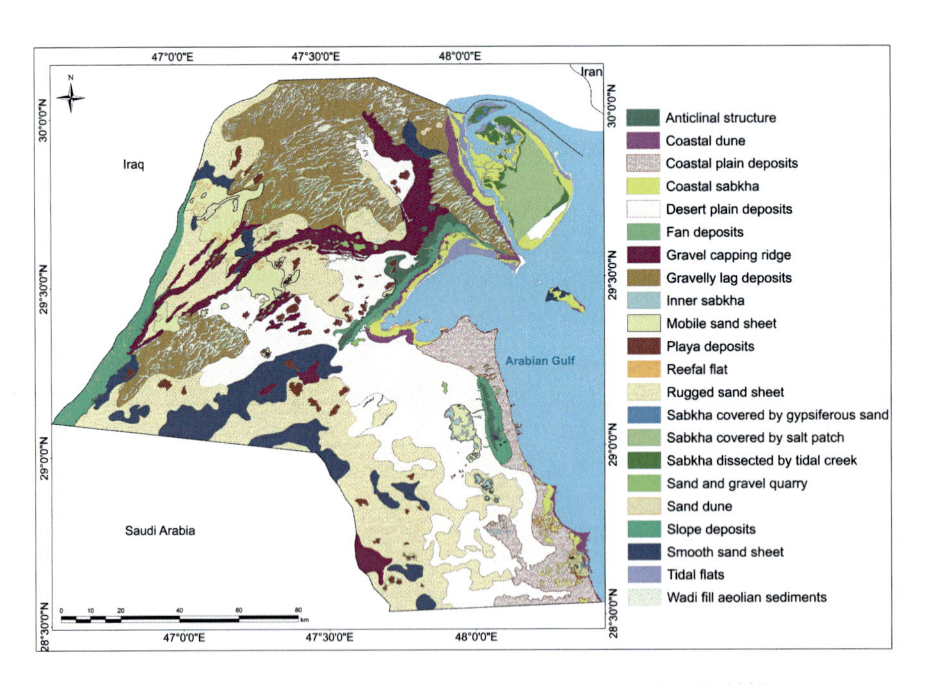

Fig. 2.4 Kuwait geomorphology units. (*Source*: KIEIN Phase III (El-Gamily 2007))

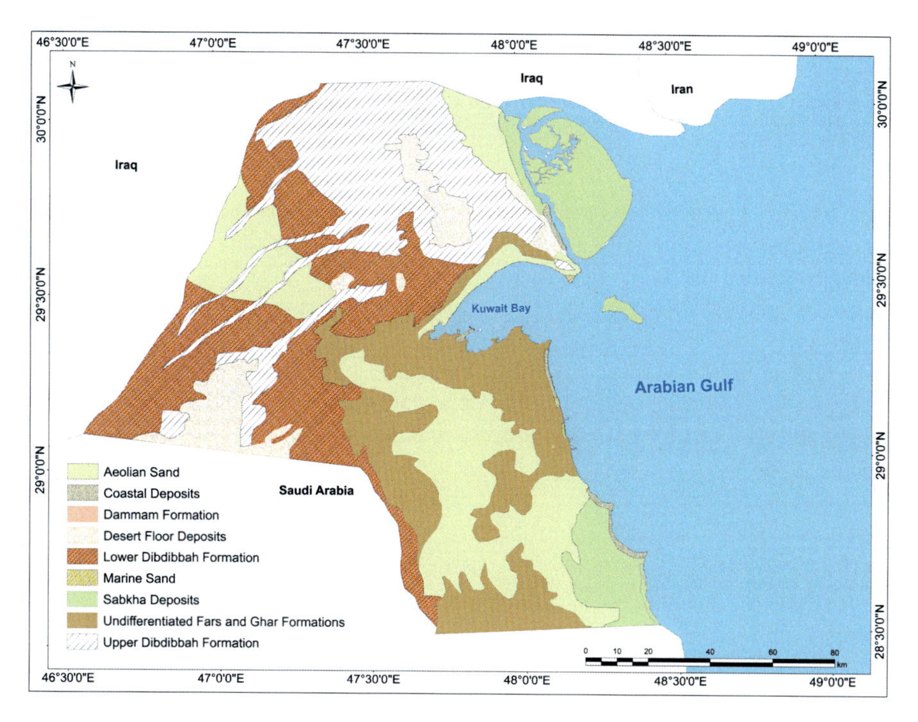

Fig. 2.5 Geological map of Kuwait. (*Source*: Hassan et al. 2021)

2.2.3.1 Wadi Al-Batin

Wadi Al-Batin is one of the most prominent and the largest morphological forms formed during the Pleistocene. The wadi's geomorphology is full of evidences indicating humidity, heavy rains, and dense vegetation, showing a completely different climate than the current dry climate. Wadi Al-Batin is the lower section of the large Wadi Al-Rima which stems from the Al-Hijaz mountains. According to Kleo et al. (2003), Wadi Al-Batin extends to about 130 km within Kuwait from Al-Salmi in the west to Al Abdali in the east, along Kuwait's northwestern and southwestern borders (Fig. 2.2) and between 5 and 7 km width as reported by Al-Sarawi (1995). The basin of Wadi Al-Batin embraced heavy rainwater and fluvial deposits during the Pleistocene, and the wadi's semi-delta is located in Al-Abariq area. Sand, pebbles, alluvial sediments, channels, and dry wadis cover the wadis surface. Terraces appear on both wadi sides, resulting from the water's continuous sculpture process. Moreover, the caliche deposits (accumulations of pedogenetic calcium carbonate) are a prevalent and distinctive feature in Wadi Al-Batin, particularly along the gullies formations (Al-Sarawi 1995).

In Kuwait, the highest point of the wadi is the domed hills 290 m above sea level (ASL), while the lowest point is in Al Abdali area, about 25 m ASL. The wadi's geology is a part of the unstable shelf, which was slightly elevated during the Miocene and inclined the rock layers northeast (Kleo et al. 2003). According to Al-Dosari (2021), Wadi Al-Batin formed due to a large 25–35 m fault in depth during the Miocene to Eocene. Kleo et al. (2003) mentioned the geological strata and the surface sediments of Wadi Al-Batin as:

The Lower Third Geological Time Formation (Lower Tertiary) It includes three formations, namely, (1) Umm Ramdhah, a Marly limestone and dolomite with 200–450 m thickness; (2) Um Al-Ros, limestone, and anhydrite with 80–130 m thickness; and (3) Dammam rocks limestone with a thickness of 250 m.

The Upper Third Geological Time Formation (Upper Tertiary) Known as Neogene formations, a cohesive sedimentary formation that includes Ghar (quartz sandstone) and Fars formations (siliceous and the white sandstone immersed in the middle of kaolin clay), which lies between Dammam calcareous dolomitic formations and the sedimentary beds of the fourth geological time.

Surface Sediments Kleo et al. (2003) also described the recent surface sediments of Wadi Al-Batin during rainy periods. Water laden with flood sediments flowed from the Wadi Al-Batin semi-delta forming the fluvial sediments beds, which are considered the oldest sediments in the wadi, overlaid with recent sediments such as fine aeolian sand deposits, active sand sheets, sand dunes, and sands with coarse and meandrous surfaces, in addition to other deposits of sabkhas and playas.

2.2.3.2 The Current Scenario of Wadis

Kuwait desert is covered by a well-defined wadis network dug by Pleistocene's heavy rains on the Arabian Peninsula. Due to the current dry conditions and the high temperatures, these wadis are dried up and left their marks clearly on the land. A GIS software was used to draw the current scenario of the dry wadis network and the drainage basins in Kuwait (Fig. 2.6), through applying the hydrology toolset of the spatial analyst tools box.

2.2.3.3 Jal Az Zour Escarpment

Jal Az Zour escarpment is one of Kuwait's most prominent geomorphological landforms. It extends along the northern shores of Kuwait Bay for 60 km (Al-Sarawi 1995), starting from the north in Jahra and ending at Ras Al-Sabiyah (Fig. 2.2). Jal Az Zour is a group of sedimentary rocky edges standing in the center of the surrounding depression lands, slopes cover a large part of its surface, and the height of the edge decreases toward Al-Subiya. The rocky edges of Jal Az Zour appear as a series of connected hills sloping steeply toward the coastal plain. Kleo et al. (2003) reported the highest peak around 145 m ASL. The edge is interspersed with a large

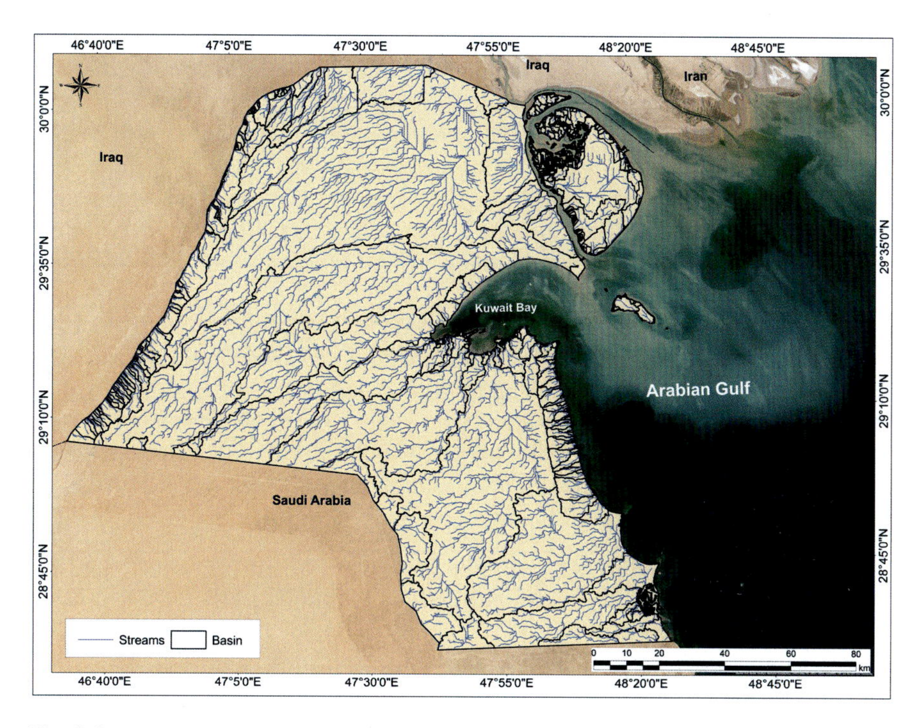

Fig. 2.6 Dry wadis and drainage basins in Kuwait map produced using hydrology toolset in ArcMap software

number of dry wadis of varying length (1.5 to 4 km), sloping from the top to flow into sabkhas and playas down to the slope's feet (Kleo et al. 2003).

Kleo et al. (2003) mentioned that scientific theories conflicted about interpreting Jal Az Zour's genesis if it arose due to above-ground processes such as weathering and erosion or due to tectonic movements. Kleo et al. (2003) refer to Jal Az Zour's origin as surface processes rather than the subsurface tectonic movement, and justified the interpretation with the presence of sea terraces that indicate sea level changes in the Pleistocene. However, many researchers attributed escarpment's origin to the subsurface tectonic movement or faulting operations (Al-Sarawi 1995; Amer and Al-Hajeri 2020; Al-Dosari 2021; AlRefaei et al. 2023). Amer and Al-Hajeri (2020) tried to determine faults by utilizing shallow seismic data to study the subsurface geology formation of Jal Az Zour. A complex duplex system of three folds and a thrust fault in the upper Dammam formation below Jal Az Zour were delineated. The results indicated the presence of folds in Dammam formations increasing toward Jal Az Zour, in addition to early stages of faults in the Kuwait group, interpreting that it resulted from lifting the Kuwait Arc toward the south of the escarpment. Amer and Al-Hajeri (2020) proposed that the friction in the Dammam formations promoted the proliferation of the fold toward the south, leading to the Jal Az Zour formation.

However, the geological formations of Jal Az Zour are considered modern; the base rocks date back to the Oligocene, and the oldest rock layers are classified as Kuwait Group according to the geological column of the Jal Az Zour region (Kleo et al. 2003). In Jal Az Zour, Dibdibba formations are described as conglomerates, gravel, sand, igneous, and metamorphic rock residues, marls, and gypsum. In contrast, lower Fars formations consist of soft and coarse calcareous sandstone, marl, red and green clay, and sandy limestone. Finally, Ghar formations are coarse pebble sandstone with sandy limestone and marl back to Oligocene and lower Miocene (Kleo et al. 2003).

Jal Az Zour's surface is covered by Pleistocene's Dibdibba deposits, river deposits from Al-Hijaz Mountains, pebbles of different sizes such as quartz, and some metamorphic and igneous rocks, in addition to fine sediments of gypsum, clay, and sand. Additionally, modern Holocene sabkha sediments and muddy coastal swamps are transported by wind and floods. According to Kleo et al. (2003), Jal Az Zour is divided into five parts or basic morphological units, and each one has its unique characteristics, which are listed as follows:

The Back Slope It extends along the edge from the southwest to the northeast and southeast, slopes slightly by 5° toward the northwest and starts widening in the southwest until reaching its maximum 9 km west of the Al-Mutlaa hills. Gravel covers the slope's surface, and some shallow wadis descend from the top, where rainwater flows to settle in some depressions down the slope to evaporate or precipitate, forming playas or sabkhas.

The Crest The high peaks appear in the southwestern region of the edge, about 164 m ASL, and peaks are convex with a slight slope of 0–5°. The height of the peaks gradually decreases to the northeast to Al-Subbiya to reach 32 m.

The Free Face This part is located directly after the crest, slopes around 50–90°, and comprises Fars and Ghar hard rocks. The weathering process and the sliding rocks contributed to face shaping, while the ridge front was slashed with several short and steep wadis. The winds formed descending sand dunes covering many areas on the front edge.

Debris Slope The area between the edge free face and the coastal plain, crossed by dry wadis, are descending from the edge front. Rock fragments continuously falling from the edge are accumulated on the slope surface. Debris slope has two types of sediments, different rock sizes of gravels resulting from weathering processes or rockslides and sand transported by wind or flaking of rocks.

The Coastal Plain The coastal plain, located 10 m ASL, extends from the end of the debris slope to the seacoast, described as uneven breadth and sandy deposits gathering around plants and forming the nebkha desert feature (Fig. 2.7),

Fig. 2.7 Nebkha feature in the desert: form where sands deposits accumulate on plant

in addition to the rock fragments brought by sloping wadis or by winds. Moreover, coastal sabkhas are spread in some areas in the coastal plain (Khuwaisat, Kadhma, and Bahra), and the tidal flats appear in the lowest lands of the plain, about 2–4 m ASL.

2.2.4 Sabkha

Sabkha is a desert geomorphological semblance formed in arid and semiarid regions. In Kuwait and Arabian Gulf countries, many sabkhas are spread, especially along the coastal strip. Sabkha can be defined as flat muddy areas formed when the collected rain, seawater, or rising groundwater evaporates during the hot, dry season or after drains into the ground, leaving a salt crust on the surface. Figure 2.8 shows the geographical distribution of sabkhas in Kuwait. Sabkhas are distributed in the north and south areas and are divided into two main types, inland and coastal sabkhas (Al-Hurban and Gharib 2004).

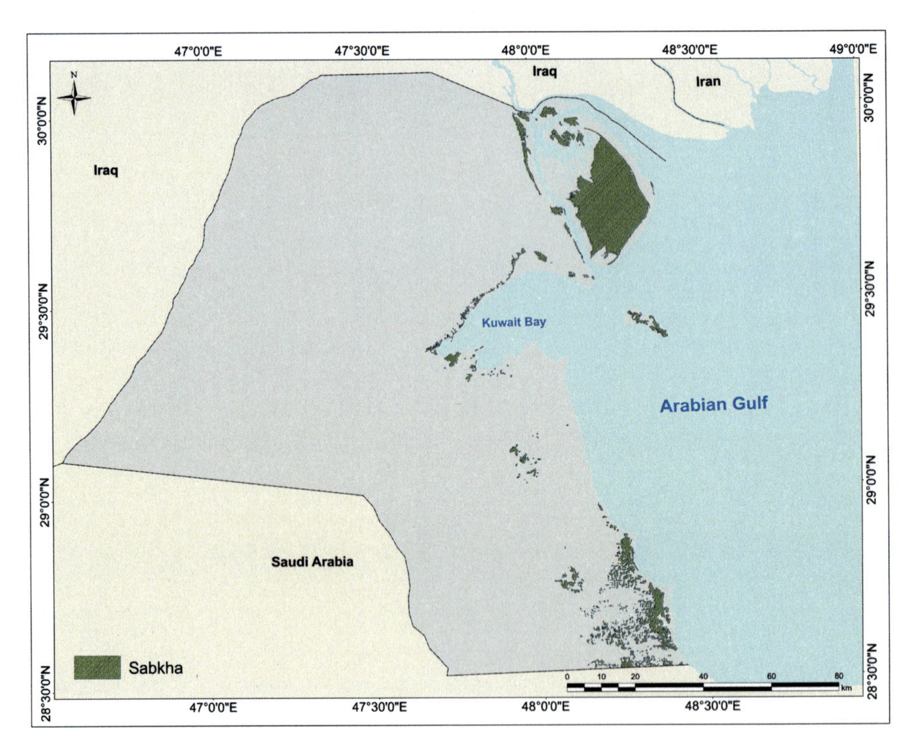

Fig. 2.8 Coastal and inland sabkhas distribution in Kuwait. (*Source*: KIEIN Phase III (El-Gamily 2007))

2.2.4.1 Northern Coastal Sabkhas

Northern coastal sabkhas were described in detail by Kleo (2006). These sabkhas are very close to the high tide line, below 10 m. Therefore, they are wet and often flooded with seawater. Their water salinity ranges between 45,000 and 221,000 mg/l. The northern coastal sabkhas have a longitudinal, regular, and continuous morphological shape. Most of its sediments sources are Shatt Al-Arab, Arabian Gulf, and Jal Az Zour wadis, like small Dibdibba's gravel.

The formations of the northern coastal sabkhas are linked to two morphological patterns, first is the coastal interferences, such as the round bays, called Doha in Arabic, and the inlets. Doha is a shallow land with lower currents, and wave movement forms small lakes. Then, marine and desert sediments accumulate in these shallow lakes and turn into wet sabkhas. Most of these sabkhas sediments are gypsum and salt, such as Umm al-Kishish, Al-Khuwaisat, and Al-Mumalaha sabkhas. Moreover, Al-Bhaith inlet carried the sediments of both Shatt Al-Arab and the sandy northwestern winds, creating the largest northern sabkha in Kuwait "Al-Bhaith." The second pattern of the northern coastal sabkha formation is linked to the tidal flats, which are the vast majority of the northern sabkhas. These sabkha's sediments are fine sand and alluvial sediments carried by sea currents from Shatt Al-Arab. Some plants are growing on its surfaces, such as *Phragmites australis*, *Lycium shawii*, and *Seidlitzia rosmarinus*, forming nebkhas.

The northern sabkhas were divided in terms of sediment type into three zones. The northern and southern zones have more fine sand deposits than gravel and clay, which could be related to the muddy tidal flats and Shatt Al-Arab fine sediments. The middle zone deposits are mostly sandy between very coarse and coarse sand deposits due to their location next to the adjacent continental sand sources, such as Jal Az Zour descending dunes and sand sheets.

Regarding mineral properties, the main mineral in the northern sabkhas is quartz from sandstone and calcareous sandstone of the Fars, Ghar, and Dibdibba formations. Other minerals are gypsum and anhydrite, due to the presence of limestone in the region, in addition to the impact of high groundwater contents of sulfates and chlorides deposited, and finally, calcite and dolomite minerals as a result of limestone dissolving by water (Kleo 2006).

2.2.4.2 Southern Inland and Coastal Sabkhas

In the view of Al-Hurban and Gharib (2004) and Kleo et al. (2003), two types of sabkhas are spread in the southern area, coastal and inland. Coastal sabkhas are close to the shoreline in the lowlands (1–5 m ASL) with variable sizes of 1–5 km^2. This type is scattered in Al-Khiran and Al-Jailaiaha areas. Some plants that can tolerate high-salinity environments grow on their surfaces, such as *Zygophyllum coccineum* and *Halocnemum strobilaceum*. The second type is the inland sabkhas, located more than 10 km from the coastline on multiple heights (80 m ASL) and spread in Al-Maqwa, Urafjan, and Al-Gurain areas.

Both inland and southern coastal sabkhas are scattered and irregular in shape, consisting of medium to fine grains of sand and mud, rich carbonates deposited from the seawater (coastal sabkha) and from the high groundwater level (inland sabkha). Regarding mineral properties, the inland sabkhas are high in quartz and feldspar and contain rock fragments. In contrast, the coastal sabkhas include gypsum and calcareous grains due to their proximity to fossil limestone ridges. However, the inland and coastal sabkhas contain approximately similar heavy minerals 0.4–0.5% (particle density > 2.95 g/cm^3), such as hornblende, apatite, zircon, spinel, epidote, and magnetite (Al-Hurban and Gharib 2004; Shahid and Jenkins 1990).

2.2.5 *Sand Dunes and Sand Sheets*

The land surface in Kuwait is either primarily vegetation-free most of the year or presents low vegetation cover. Dust storms are natural weather phenomena in the desert ecosystem. Al-Sulaimi and Mukhopadhyay (2000) mentioned that 50% of Kuwait's surface deposits are aeolian sand. The northwest winds carried the sand from Iraq and Saudi Arabia desert plains to Kuwait (Al-Sarawi 1995). Sand dunes are among the most prominent terrestrial desert terrain that arose from wind deposits. In Kuwait, 2304 sand dunes (Al-Dousari et al. 2023) are widespread in the northern and southern deserts and are 5–8 m in height (Al-Sarawi 1995). Sand dunes are formed when dusty wind speed slows down or collides with a barrier or obstacle. Desert and coastal sand dunes in Kuwait are in many forms; according to Al-Dousari et al. (2023), six types of sand dunes have been identified in Kuwait, Barchan or crescentic dunes, which are the most widespread, especially in Al-Huwaimiliyah and Jal Az Zour, Domal dunes, falling dunes, climbing dunes Barchanoid dunes, and nebkha.

The northwestern winds blowing on Kuwait from Iraq and Saudi Arabia desert plains greatly contributed to the emergence, growth, and development of the northern sand dunes chain, where most sand dunes in Kuwait are in the Al-Huwaimiliyah, Al-Atraf, and Um Urta regions (Fig. 2.9). Al-Huwaimiliyah area includes the longest chain of Barchanoid dunes, 2400 m. Al-Dousari et al. (2023) in their investigation of the dunes fields in Al-Atraf, Buhaith, Huwaimiliyah, Kabd, and Um Urta stated that the northwest wind pushes the dunes to the south-east at a rate of 14 m/year.

Dunes are also observed in the coastal areas of Al-Subiya and Jal Az Zour, where falling dunes are formed on the edges (Al-Dousari et al. 2023). The coastal dune's height is 2–3 m and 2–4 m long (Al-Sarawi 1995). Moreover, crescentic white dunes are extended along the southern coast of Dhubaiyah due to white limestone

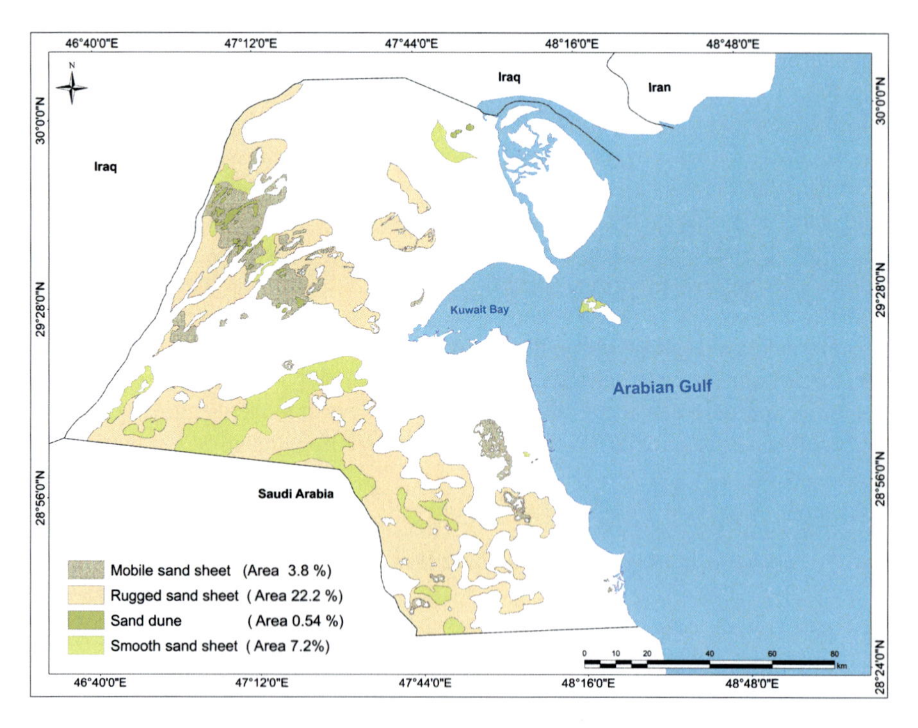

Fig. 2.9 Sand dunes and sand sheets locations in Kuwait. (*Source*: KIEIN Phase III (El-Gamily 2007))

ragstone eroded by wind (Misak et al. 2003). According to Misak et al. (2003), the anchored dunes or nebkhas are found on the northern and southern coasts of the Kuwait-Saudi border in the south. These dunes cannot move due to plants acting as stabilizers to the base of the dune preventing movement, and the dune head appears as a triangle extending toward wind direction. Nebkhas differ in height; big ones of 3 m are usually formed around the *Nitraria retusa* plant, whereas the small nebkhas are formed around plants such as *Zygophyllum qatarense* and *Panicum turgidum*, and are short of less than 1 meter height.

In addition to dunes, Al-Sulaimi and Mukhopadhyay (2000) explained the formation of the sand sheets from wind deposition (aeolian) processes in Kuwait. Sand sheets arise from fine sand wind deposition on a flat surface free of meanders or coarse pebble within desert depressions or as a consequence of the coalescence of the small dunes. Sand sheets are formed in the southern, northern, and western parts of the desert (Fig. 2.9), where the smooth and flat surfaces enable the wind to unload sand smoothly, especially in the flat surfaces of the southern area (Khalaf 1989). The active movement of sand belts in Al-Huwaimiliyah and Al-Qashaniyah in the north, including many dunes, is considered a significant sand funder for the south and southeast toward Al-Sulaibiya area and the seacoast. According to Khalaf (1989), the northern desert surface contains limited amounts of sand in the northwest and southeast, form Barchan dunes in Al- Qashaniyah and Al-Huwaimiliyah's sand belts. The surface in the north is dominated by residual gravel sheets protecting the dune sediments from wind activity and limiting sand areas spread.

2.3 General Geographical Features of Kuwait Coast

Kuwait is benefited from its geographical location in Arabian Gulf in the past in trade and as an economic resource for fishing and pearl diving. Kuwait relies mainly on seawater desalination to obtain water for drinking, various consumptions, and irrigation. Furthermore, Arabian Gulf includes large offshore oil fields within Kuwaiti maritime territorial borders.

Arabian Gulf is in the east of Kuwait, and the marine border of Kuwait extends in a straight line for about 170 km. Arabian Gulf is one of the warmest water bodies in the world, its geographical location in the desert region raises the surface water temperature to more than 35 °C in summer (cf. Karam 2023).

The Arabian Gulf borders Kuwait along the eastern border for 474 km. Kuwait's coast has minor tortuous, including some bays and land capes; the most famous are Ras Al-Sabiyah and Ras Al-Salmiya, which embrace Kuwait Bay, a large bay in the middle of the coast and the largest terrain in Kuwait's coastal strip. Moreover, Kuwait's territorial water includes nine islands of different sizes. The largest is the mud island of Boubyan (972 km^2), whereas the smallest is coral Qaruh island (0.033 km^2) (Mufadhal 2021).

2.4 Kuwait Islands Features

According to Mufadhal (2021), Kuwait islands are varied in size; few are significant, and the majority are small, less than 1 kilometer square, such as Qaruh, Umm Al-Maradim (0.158 km^2), and Kubbar (0.138 km^2). The islands are not very high than sea level, as most of their lands immersed by water during the high tide and are devoid of prominent terrain, except for Qaruh, where a small part of its land rises to approximately 35 meters ASL. These islands are uninhabited due to their structural and geomorphological conditions or small area.

Warba and Boubyan islands are alluvial sediments formed by the Tigris and Euphrates rivers sediments that flow into the Arabian Gulf through Basra Canal from Shatt Al-Arab. Warba and Boubyan are very low-level islands; the maximum height in Warba is 10 m ASL, while Boubyan lands are much lower (less than 4 m). The impact of high tidal water is tremendous, submerging its lands to form swamps, lagoons, and sabkhas, where 70% of Boubyan area is sabkha deposits (Fig. 2.8).

This muddy environment has prepared an integrated ecosystem for many settled and migratory birds. In Boubyan, more than 60 species of birds, such as *Platalea*, *Dromas ardeola*, and *Ciconia ciconia*, have been observed, in addition to *Oxudercinae* fish in Warba. Plants are few and sporadic, due to the high salinity, except for some plants that tolerate salinity, such as the *Zygophyllum qatarense* in Warba and the *Bienertia cycloptera* and *Seidlitzia rosmarinus* in Boubyan.

Failaka (46.6 km^2) Island is one of the most important islands in Kuwait and the only one inhabited before the Iraqi invasion in 1990, and it has a unique historical importance dating back to B.C. Moreover, Failaka has sandy, clay soil suitable for cultivation, and its coasts are rich in fish and freshwater wells. Some geomorphological phenomena are formed in Failaka, such as nebkhas, sand dunes, and sabkhas covering 46% of its area. Failaka is considered an extension of Jal Az Zour escarpment, but it is separated due to the tectonic movement that led to the fault that formed Kuwait Bay. The abundance of plants, such as a *Malva parviflora*, *Lotus tetragonolobus*, *Cyperus*, and *Terfeziaceae*, are growing in Failaka and some migratory and endemic birds considering the island as a primary destination, such as the *Stercorarius parasiticus*, the *Chroicocephalus ridibundus*, and the *Thalasseus sandvicensis*.

In addition to a group of small islands, Miskan (0.36 km^2), also another extension of the Jal Az Zour, has some plants such as the *Rhanterium epapposum* and *Zygophyllum qatarense*; as a result, some birds flock to it. Umm Al-Namil (0.62 km^2) is the only island inside Kuwait Bay; its rocks are calcareous and have coarse sandy beaches. Auha island is very small (0.115 km^2), has coarse sandy soil, and its southern coasts are rocky, low-high cliffs.

As for Kubbar, Umm Al-Maradim, and Qaruh islands, located in the southern part of the Kuwait territorial waters, beaches are flat and sandy, surrounded by coral reefs of picturesque nature. Umm Al-Maradim blooms with plants in winter and birds such as *Phoenicopterus roseus* and *Larus* flock to it. Kuwait island's environments provide habitats for breeding some kinds of rare sea turtles such as *Eretmochelys imbricata* and *Chelonia mydas*, especially in the coral islands in addition to Auha, Miskan, and Failaka, where sandy beaches and seaweeds are available near the coast.

2.5 Coastal Geological and Morphological Features

According to Al-Qasabi (2014), Kuwait's coastal strip consists of three main formations, which are indicated in Table 2.1. The coastal strip permeates various surface features that differ according to their geographical location and the geomorphological conditions of formation. The northern part includes tidal flats, sandy mud beaches, and sedimentary mud islands and a series of coastal sabkhas (Fig. 2.10). The mud flats, plains, cliffs, and coastal dunes are manifest on the central coast, specifically in Kuwait Bay. Finally, studies by Papathanasopoulou and Zogaris (2015) and Al-Qasabi (2014) show that the semi-smooth southern coast has sandy beaches, coastal dunes, sabkhas, reefs, and coral islands (Fig. 2.10).

Types of sediments on the north coast are different from the south. The northern beaches are located within the shallow depression where fine river materials from Shatt Al-Arab gather and fill the coast with sandy, muddy, and silt sediments. On the southern beaches, where the coastal plain is located, sediments are of the dry desert wind forming coastal sand dunes, gravel, and coral fragments.

The hydrodynamic factors shaped the morphology of the Kuwait coast through three main processes. First: the waves. In Kuwait's sea, the northwestern waves of

Table 2.1 The three main formations of the coastal strip

Geological group	Epoch	Formations	Location	Rate
Surface sediments	Holocene	The deposits contain high proportions of calcium carbonate and silicates Varies between coarse and soft sand Are mostly aeolian sand deposits	Majority of coastal line	49.11%
Dibdibba formations	Pleistocene	Coarse sand and gravel of varying sizes Metamorphic igneous rocks Some fine sediments such as gypsum, clay, and sand	Ras Al-Sabiyah and the northern coast	22.85%
Fars and Ghar formations	Oligocene and lower Miocene	Clay sands intertwined with pebble sandstone, and limestone Some fossils are appeared	Ras Al-Sabiyah to umm Al-Hayman in the south	28.04%

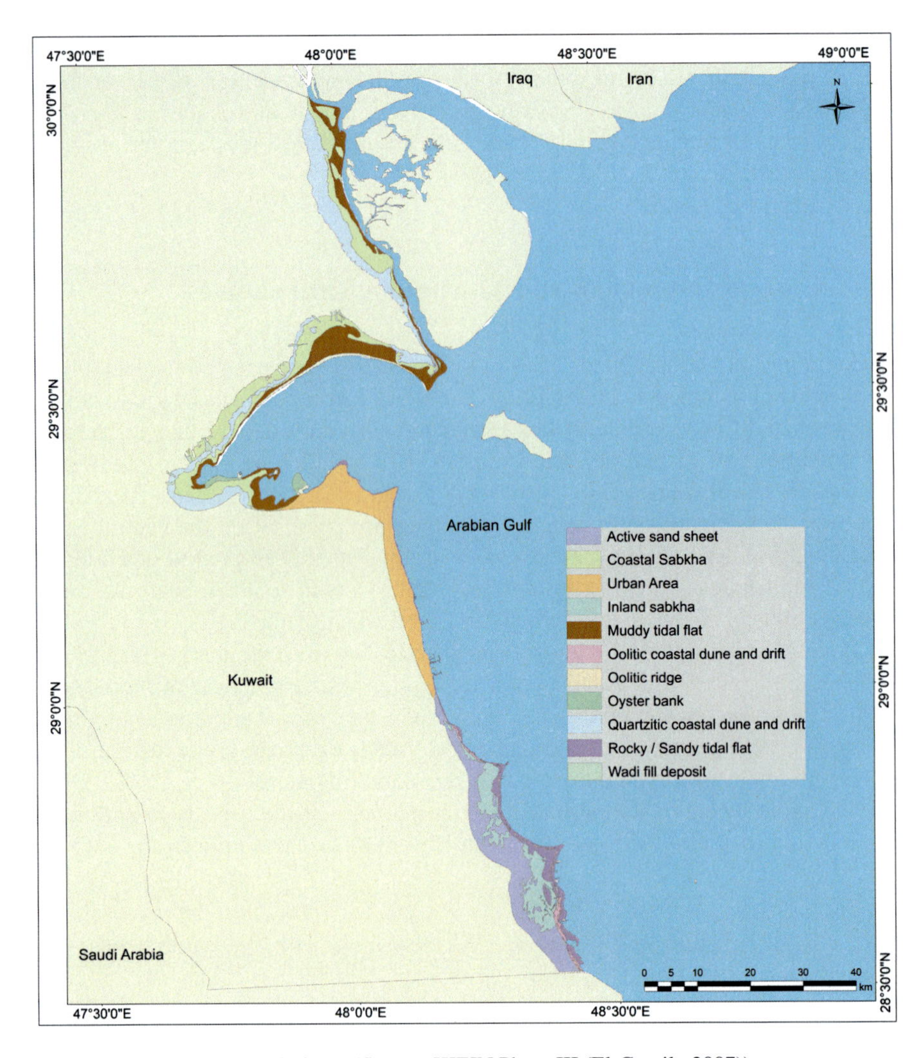

Fig. 2.10 Coastal geomorphology. (*Source*: KIEIN Phase III (El-Gamily 2007))

average height about 3 m are prevailing. Waves impact on the coast vary based on its location, stronger in the southern coast, where the average waves' height reaches 4.5 m (Al-Qasabi 2014). Second: the tidal, which occurs mainly twice a day in Kuwait (every 6 hours), directly contributing in currents generating, forming various coastal geomorphological features such as, tidal flats, sabkhas, and temporary lakes. The third process is the currents contributing to coast formation, seawater transfer, and sedimentation, especially on muddy northern beaches. In Kuwait, currents are counterclockwise and vary in speed from one area to another; for instance, it is 120 cm/sec in the Al-Subiya creek (khor) and 26 cm/sec in Kuwait Bay, where the depth is shallow (Mufadhal 2021).

2.6 Biodiversity

The terrestrial environment of Kuwait is embraced with different types of organisms. Groups of insects, desert animals, and plants between perennials and annuals find Kuwait a suitable environment for life. Moreover, Kuwait is a major destination in the itinerary of migratory birds; the following lines present an overview of Kuwait's terrestrial biodiversity.

2.6.1 Plant Biodiversity

Despite the arid climatic conditions and soil deficiency of organic matter for plant growth, some plants are still growing and spreading in some desert areas, adapted to the ecosystem. These plants are considered original country plants (indigenous and native) and a part of Kuwait's natural heritage, such as *Haloxylon salicornicum*, *Rhanterium epapposum*, *Panicum turgidum*, and *Zygophyllum qatarense*. These plants can adapt to high temperatures in most months, the large daily and seasonal diurnal temperature variation, and drought. Still, undoubtedly, growth rates vary according to rain amount each year. Moreover, some plants growing in saline environments, such as sabkhas and salt depressions, mainly include *Nitraria retusa*, *Suaeda aegyptiaca*, and *Salsola imbricata*. The National Strategy for Biodiversity of the State of Kuwait 2011–2020 (EPA 2011), indicated 385 species of uncultivated wild and endemic vascular plants between annuals and perennials in Kuwait.

2.6.2 Animal Biodiversity

According to Jamaan and Mickenz (1998), animals in Kuwait's environment belong to two main phyla, namely, the Arthropoda, which includes all crustaceans, arachnids, and insects. The black scorpion and spiders, such as the black widow, are the most widespread. As for insects, 648 winged and non-winged insects were counted, including *Locusta*, beetles, butterflies, and bees, as reported by EPA (2011). The second phylum is the chordates comprising all terrestrial and marine vertebrates. The widespread vertebrates are about 30 species of reptiles, including geckos of different kinds, *Chamaeleo chamaeleon*, *Uromastyx*, mice snake, numerous types of lizards such as *Acanthodactylus scutellatus* hardyi and ciliated-toed, and amphibians like farm and playa frogs, ten types of rodents of which the most famous are mice and rats. Furthermore, 28 types of mammals, such as camels, sheep, and goats, in addition to some mammals, unfortunately, are extinction to threatened, such as *Vulpes vulpes*, *Mellivora capensis*, *Felis silvestris gordoni*, and *Herpestes edwardsi* (EPA 2011).

2.6.3 Natural Reserves

The idea of natural reserves has emerged to preserve the environmental heritage and Kuwait's terrestrial landforms, biodiversity, and natural ecosystem environment components. Natural reserves are defined as lands surrounded by precise dimensions, including the basic natural terrain and different types of animals and wild plants, to live in peace and preserve the unique environmental heritage. Thus, KISR conducted a feasibility study to achieve this mission. During the 1980s of the last century, areas of environmental and natural importance were identified and studied. In 1986, lands were allocated to be converted into natural reserves for various purposes involving scientific research, protecting geological formations, maintaining biodiversity, and rehabilitating land and marine environments (EPA 2011).

Accordingly, the first two wild reserves were established in Kuwait in 1987: Sabah Al-Ahmad Nature Reserve (325.03 km^2), including a group of terrains such as hills and sabkhas and groups of animals and plants. The second reserve was Al-Jahra (presently Al-Khuwaisat), occupying an area of 17.91 km^2, followed by the Doha wild reserve in 1988 for protecting endangered animals and migratory birds.

Respectively, the other terrestrial reserves, such as Umm Al-Niqqa, Umm Qdair, the biggest wild reserve of Wadi Al-Batin with an area of 520.62 km^2, and Al-Huwaimliyah, were established in 2011. Marine natural reserves such as Mubarak Al-Kabeer was established in Boubyan island in 2010 and is consider the second biggest natural reserve in Kuwait with an area of 510.22 km^2 (Fig. 2.11). The established natural reserves are owned and supervised by different entities, mainly the EPA and the Public Authority for Agriculture Affairs and Fish Resources (PAAFR).

Presently, the total number of natural reserves is 12, occupying an area of 2072.94 km^2 of Kuwait's lands. Moreover, a proposal of allocating the atolls, of Qaruh, Umm Al-Maradim, and Kubbar, as natural reserves, was included in the National Strategy for Biodiversity of the State of Kuwait 2011–2020 (EPA 2011).

2.7 Water Resources and Agricultural Aspects

The GCC countries and the desert ecosystem generally suffer from drought and freshwater resource scarcity. Kuwait relies on Arabian Gulf water desalination as a primary source of fresh water through seven main stations: Northern Shuaiba, Southern Shuaiba, Eastern Doha, Western Doha, Southern Zour, Shwaikh, and Subbiyah. Desalinated water is used for drinking and various forms of domestic consumption, such as irrigating gardens, etc. The annual consumption of potable water in 2019 reached 149.760 million gallons, as stated in the Annual Statistics Abstract (2021) by the Central Administration of Statistics, Kuwait (CSB 2021).

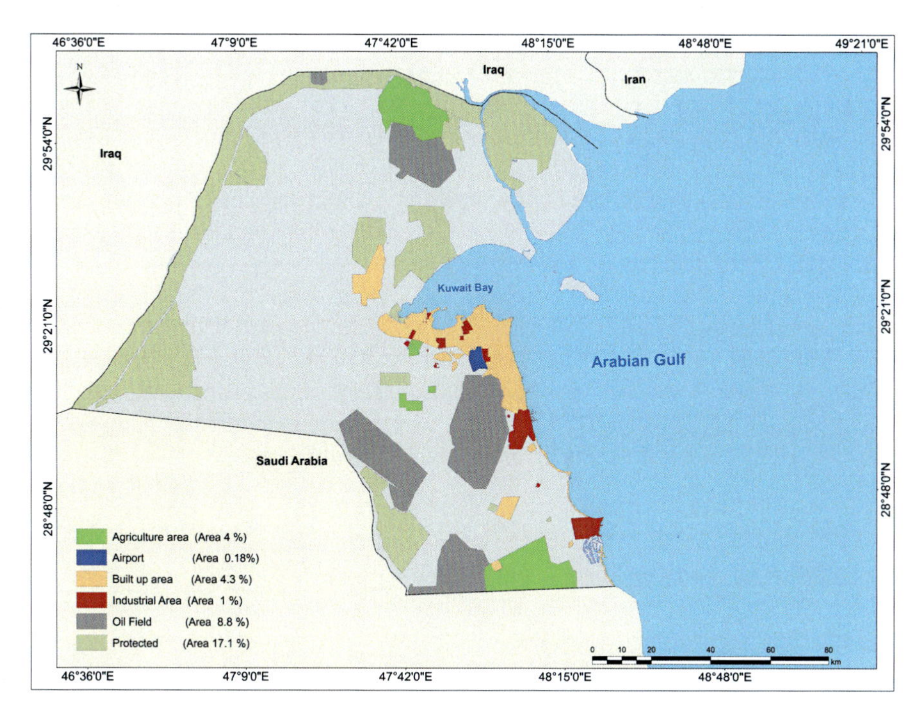

Fig. 2.11 Land-use/land cover (LULC) map

Furthermore, the fresh and renewable groundwater (TDS 600 to 1000 mg/liter) preserved in the calcareous rocks of the deep aquifers in the Al-Rawdatain and Umm Al-Aish regions in the north is a secondary source of fresh water for drinking. Still, this source is at most 5% of the annual consumption. The brackish groundwater reservoirs (TDS ranges from 3000 to 10,000 mg/liter), more abundant in Kuwait aquifers, are often used in irrigation in Al-Wafra and Abdali farms and in oil depressurization by Kuwait Oil Company (KOC). Additionally, Kuwait uses treated waste water (TWW) as a third means of fresh water supply. Wastewater treatment in Kuwait began in the last century; nowadays 90% of the wastewater is treated through five treatment plants: Al-Riqqa, Umm Al-Haiman, Kabd, Al-Wafra, and Al-Khiran using the rapid filtration of sand and chlorination treatment, besides using reverse osmosis and ultrafiltration technology in Sulaibiya the sixth plant (EPA 2019a). The TWW is used in agricultural and production fields (restricted use for crops) and in aesthetic landscaping.

Agricultural land occupies only 4% of the total land and is located in Al-Wafra, Abdali, and Sulaibiya areas (Fig. 2.11). In general, the soils of Kuwait are sandy, poor in organic matter and clay contents, and present high infiltration rate (Omar and Shahid 2013). Furthermore, the presence of gatch or calcic and gypsic pan, precipitated and cemented carbonate, or gypsum at different depths in Kuwait lands prevents plant roots from penetrating deeply into the soil. In Kuwait's western side, gatch deposits are in a shallow depth (within 50 cm) and a moderate depth (within 50 to 100 cm) and are deeper (below 100 cm) in the east (Fig. 2.12) (KISR 1999). Based on the soil survey of Kuwait (KISR 1999; Omar and Shahid 2013; Shahid and Omar 2022), the soils of Kuwait are classified into two soil orders, Aridisols and Entisols, occupying 70% and 30% surveyed area of Kuwait, respectively. The soils of the Abdali farming area contain gypsum and calcium carbonates rich subsurface diagnostic horizons, and the soils in the Al-Wafra farming area are sandy (Shahid and Omar 1999; Shahid et al. 2004).

Despite the arable land and water scarcity, Kuwait farms contribute significantly to financing the local market with various crops such as tomatoes, cucumbers, potatoes, leafy vegetables, and dates. Additionally, poultry and dairy farms are supported by PAAFR through providing subsidies to some agricultural activities. According to the statistics issued by the Central Administration of Statistics of Kuwait, agricultural production in 2019 reached 866,862 tons.

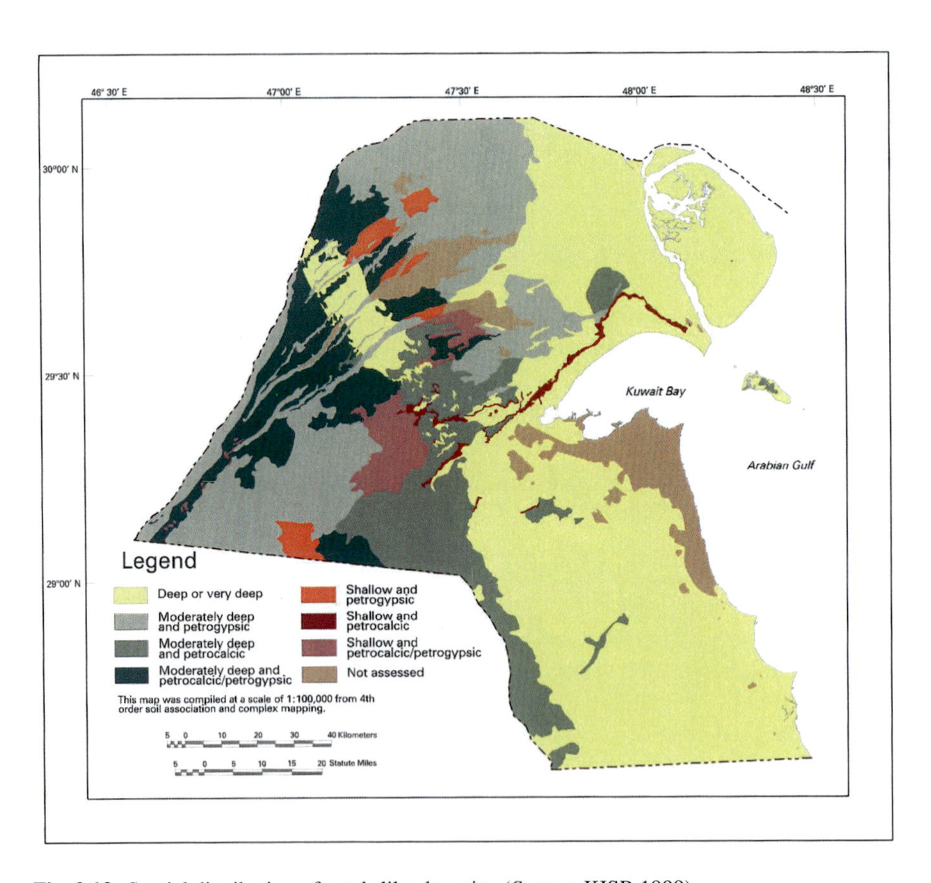

Fig. 2.12 Spatial distribution of gatch-like deposits. (*Source*: KISR 1999)

2.8 Kuwait Oil Resources and Economy

Due to the drought conditions and high temperature in the desert, in addition to the soil's unsuitability for agriculture, people in the past were directed to the marine side. The economy was based on tradecraft. Besides fishing and pearl diving, peoples used large ships to travel to many destinations to import fabrics, spices, and other commercial goods. In return, they sold the natural pearls extracted from the Arabian Gulf. After the oil discovery in the late 1930s, the state worked to profit from the export of oil and the revenues of its products to improve people's living standards in education, health, and economic aspects.

Kuwait is the third largest oil-producing country in OPEC. The country owns 8% of the global oil reserves (Al-Hemoud et al. 2019), which are estimated at 101 million barrels (OPEC Annual Statistical Bulletin 2017) reflected on the per capita gross domestic product (GDP) of Kuwait, to become about $29,904.8 in 2018 (CSB 2021). Kuwait's oil industry recorded a production of 2439 barrels/day in 2020, while the marketed natural gas amounted to 13,952 million m^3/year, according to the latest statistics (CSB 2021). The global oil demand and the fluctuations in international oil markets directly influence Kuwait's economy.

However, some services contributed large shares to the GDP in 2015; first, social services, with a rate of 17%, involve essential government services such as health care; second, financial services such as banks, insurance, and real estate with a rate of 14%; third: transportation and communications 6%; and, finally, the petrochemical industries sector, building materials, mining, and steel production, which contribute 5% to the GDP (EPA 2019a).

2.9 Ecological Imbalances

The harmful effects of natural or human activities on the environment lead to various ecological imbalances. Kuwait is a desert country known for its harsh, hyperarid, and hot environment leading to various environmental threats. The country's rapid population growth demanded facilities and increased urbanization, leading to an imbalance in Kuwait's terrestrial and coastal environment. Climate change, including temperature rise and changes in precipitation patterns and intensity, is another factor for ecological imbalances. Alahmad et al. (2022) predict a 1.8 °C to 2.57 °C rise in Kuwait's temperature by mid-century in association with the 2000–2009 baseline, with no mitigation occurring in the current climatic trends. While rich in terrestrial and marine biodiversity, these systems are fragile and highly susceptible to climate variations. Kuwait is also top among the water-stressed countries, having the lowest per capita renewable internal freshwater availability, demanding extensive seawater desalination to meet water needs. The significant factors affecting the environmental stability of Kuwait can be explained in detail under the following topics:

2.9.1 Urbanization and Population Growth

Urbanization reflects the relationship between humans and their environment by transforming land cover into urban areas for human use. Kuwait occupies a 17,818 km^2 area, while most of the urban development is on the coastline. The urbanization resultant growth and development of the southern coastal zones left extreme stress on Kuwait's fragile near-shore environment. The south coastal region is exposed to coastal water pollution due to oil spills, industrial wastes, thermal contamination from various industries, fecal coliform, and solid garbage. The toxins impact the health and functioning of the coastal and marine ecosystems (Alajmi et al. 2021). Air pollution is another factor related to oil consumption in vehicles and other uses. It releases more greenhouse gases (GHGs) and heat waves, thus creating a temperature rise resulting in the urban heat island (UHI) formation in the region, which would intensify the air temperature of the built-up zone by 2–8 °C as stated by Oke in 1982 (cf. Kamal et al. 2022). Also, vast areas of the Kuwaiti desert were converted as landfill sites for dumping municipal solid waste, construction materials, and tires, generating large amounts of toxic gases and are at the risk of spontaneous fire (Soleimani et al. 2021), adding to the terrestrial environment's economic catastrophe (Al-Salem et al. 2020).

2.9.2 Anthropogenic Activities' Impacts on the Coastal Areas

The rapid growth and development of the manufacturing industry in Kuwait are primarily focused along its shoreline borders. Human activities and development have influenced Kuwait's coastal areas over the last 60 years. In coastal regions with high population density, the anthropogenic impact generally focuses on converting land cover to land use for urban, industrial, and other developments.

Some coastal areas in Kuwait, mainly the southern part, are experiencing coastal erosion owing to the natural and anthropogenic effects of developing infrastructures near the high water level without considering the impact of sea level rise (SLR). Certain coastal areas, specifically the northern part of the Kuwaiti coast, are experiencing notable sedimentation due to highly suspended sediments in the seawater. Al-Dalamah and Al-Hurban (2019) stated that anthropogenic activities in urbanization had affected the southern Kuwait coastal sabkhas Al-Jailiaha, Al Zour, and Al-Khiran. Moreover, urbanization caused adverse alterations in the sabkhas' characteristics, area, and geomorphology. Rapid advancement in the real estate business to endorse the tourism sector and several waterfront developments built in Kuwait adversely affected the natural ecosystem of coastal area.

2.10 Climate Change and Global Warming

Climate change has turned out to be a critical threat to social development. Due to global warming and climate variation, hazards including temperature rise, drought, flash floods, and SLR have become the major threats to terrestrial ecosystems, human settlements, and surroundings in recent years. Climate variation in Kuwait has ended in temperature rise, frequent dust storms, obvious precipitation changes, and SLR. More details on the effect of climate variations on Kuwait's environment are discussed below.

2.10.1 High Temperature

Kuwait's hyper-arid desert climate is exceedingly hot throughout the long summer. Summer average maximum temperatures are above 45 °C, and during heat waves, the daytime temperature frequently exceeds 50 °C, and the night time temperature remains above 30 °C. On July 21, 2016, the World Meteorological Organization (WMO) stated Kuwait as the hottest location globally, with an extreme temperature reading of 54 °C at Mitribah, Kuwait (Merlone et al. 2019). The alteration and reformation of the natural land cover vary the urban landscape's thermal characteristics. The impacts of the LULC pattern on the thermal characteristics of Kuwait were studied by Al-Dousari et al. (2022) and found that the conversion of 27.24% of the open area and 5.43% of the cultivated area into the urban area increased the mean temperature by 5 °C contributing to an increase in the UHI values by 0.861 from 1991 to 2021. The heat islands were mostly focused on industrial areas, oil fields, and airports. According to Kamal et al. (2022), the rural area's land surface temperature (LST) reached 41.47 °C in 2020, 4 °C higher than in 2013. A strong positive correlation between the land surface and the air temperature confirms the strong positive influence of the increased surface temperature on the growing air temperature in the region.

The increasing temperatures impact Kuwait's terrestrial and coastal ecosystems. High temperature increases evaporation rate and water requirement, adding pressure on the energy and water demand. Increased temperatures may indirectly affect water desalination, including rising seawater salinity and water irrigation demands. Extremely high temperatures and prolonged heat waves can damage the country's agricultural production. Hence, high temperature and humidity might be among the leading parameters responsible for the country's potential, terrestrial, and biodiversity loss.

2.10.2 Dust Storm

Sand and dust storms (SDS) arise from wind erosion due to natural or human activities. Kuwait's location in the northeast of the Arabian Peninsula makes it close to dust sources. According to Cao et al. (2015), most sandstorms touching Kuwait initiate from Iraq and transfer dust and sand particles to neighboring countries, including Saudi Arabia and Iran (Fig. 2.13). The storm trajectories also transfer particles from Iran and Egypt. Northern Kuwait marked the highest wind speed and lowest visibility rate. The storm obstructs urban areas or regions with a considerably prevailing level of activities than the other sites with fewer infrastructure or populations. In the view of Al-Hemoud et al. (2018), the small geographic area and similar climatic conditions across the country are the reason for the limited variation in dust storm events in Kuwait. The dust particles exist for 2 days, even after the dust storm. Sand and dust

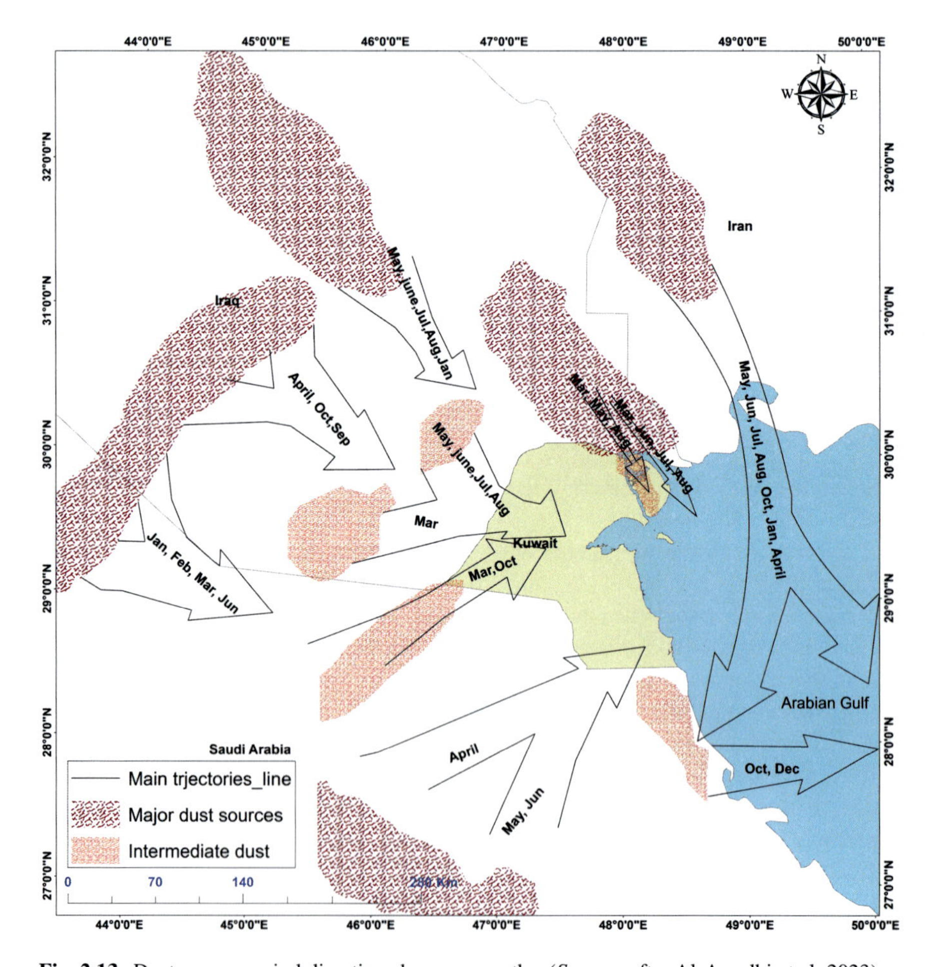

Fig. 2.13 Dust sources, wind directions by year months. (*Source*: after Al-Awadhi et al. 2023)

storms affect the economy directly or indirectly due to the drift of sand dunes on roads, power plants, and oil refineries. Topsoil erosion can also lead to agricultural land degradation even though the cultivated regions are less in the country.

2.10.3 Change in Rainfall Pattern

Kuwait is recording a difference in the rainfall pattern in recent years due to climate change impacts. Sudden spells of rainfall accompanied by heavy storms and long dry intervals have recently existed in Kuwait's rainfall pattern causing severe damage to infrastructure and human life. The annual rainfall average from 1962–2017 is 112 mm. Kuwait experienced heavy rainfall with an average rate of 151 mm and a great flash flood from November 4–14, 2018 (Al Jassar et al. 2023). According to Al-Qallaf et al. (2020), the total yearly rainfall value in 2018 reached 344 mm, nearly three times the average annual rainfall rate, thereby increasing the yearly mean from 1962–2018 by 4 mm to reach 116 mm. Flooding often occurs in the state due to the country's flat landscape, poor water drain-off plan, and sewage discharge. The rainfall is greater than the sewage and drainage systems' capacity. The flood caused severe water erosion, resulting in soil removal by runoff water, agricultural damage, disruption and wearing away of desert surface, and damage to wildlife habitats. Al-Hemoud et al. (2023) identified Kuwait City and suburban areas, coastal areas, Abdaly in the North, and Sabah Al-Ahmad and Wafra in the South as flood-prone areas.

2.10.4 Sea Level Rise (SLR)

A long shoreline features Kuwait's mainland and 61% of its borders share with the Arabian Gulf. Al-Mutairi et al. (2021) state that Kuwait's coastal region is around 1315 km^2, ranging from 10 to 20 km from the coastline, and the coastal zone houses most of the urban area and about 90% of the total population. Kuwait's coastal area is tremendously susceptible to SLR, resulting in ecosystem and geomorphological component variations. The country's coastal regions will be sensitive to coastal flooding, enhanced beach erosion rates, deterioration of groundwater quality and the agricultural regions, and some key installations and infrastructure destruction. Al-Mutairi et al. (2021) studied the impacts of SLR in Kuwait's coastal region in different circumstances of 0.5 m, 1 m, 1.5 m, and 2 m rise in sea level by using the Bruun Rule and identified Hawalli and Capital governorates to be highly involved regarding the number of victims and the housing areas' financial losses. Several oil facilities, including refineries, pipelines, petrochemical plants, harbors, and factories, are situated on the country's coastline and, hence, exposed to the danger due to SLR. The vulnerability assessment performed by Hassan and Hassaan (2020) identified four regions of the Kuwait zone that would be highly under pressure by SLR: Boubyan and Failaka islands, the northern shore of Kuwait Bay, and Al-Khiran area in the southern coastal region. Raised sea level is anticipated to end in the damage of tidal flat habitat and an alteration in the marine ecosystem's natural equilibrium. Kuwait Bay houses four desalination plants that are Kuwait's major freshwater sources, and any disturbance to those amenities owing to SLR would significantly affect human health and security. The SLR would considerably damage the majority of the sandy shores in the country.

2.10.5 Limited Water Resources and Soil Degradation

Kuwait has an arid desert ecosystem with no permanent surface water source. Kuwait heavily depends on desalinated water, groundwater, and TWW to meet its local water requirements. Climate change and population growth cause severe water stress due to the significant drop in per capita water resources and hike in per capita consumption (PCC) with lower freshwater availability worldwide. The usable groundwater varies from brackish to saline, whereas the fresh groundwater availability is limited (EPA 2019a). The extended groundwater abstraction reduces aquifer storage capacity and causes a reduction in the groundwater levels and an increase in the water's salinity.

The excess use of irrigation water is the country's leading cause of soil salinity, extremely damaging the terrestrial environment. Due to the lack of drainage system in the farms, the water got collected in the low-lying areas making the soil over there saline due to high temperature and evaporation rate. The sandy and salty soil with less organic content and moisture due to the country's harsh climatic nature and the limited water resources are the main constraints faced by the agricultural sector of Kuwait. Misak et al. (2014) identified that the salinity had adversely affected the long-term productivity of a few farms in the Al Wafra agricultural region. The topsoil removal due to heavy wind and dust storms, increased evaporation rate due to high temperature, and precipitation changes are the other factors responsible for limited agriculture and green cover, thereby enhancing the food security threat in the region.

2.10.6 *Land Degradation* **(Drought, Desertification, Camping)**

UN Convention to Combat Desertification (UNCCD 1994) defines land degradation as the reduction or loss in arid, semiarid, and dry subhumid regions of biological or economic productivity and complexity of rainfed cropland, irrigated cropland, or range, pasture, forest, and woodlands resulting from a combination of natural and human factors. Al-Awadhi et al. (2003) report that Kuwait's annual average of desertified lands is estimated at 285 km^2, where lands and vegetation are exposed to certain factors leading to deterioration. These factors can be natural or unrelated to human attitudes, such as the harsh climatic conditions, the geographical location, scarcity and irregular rain, prolonged drought periods, and the strength of prevailing northwestern winds. Human activities that exacerbated desertification and worsened the situation included military activities that compressed the soil mechanically through heavy military vehicles during the Gulf War. It results in soil disturbance and excessive exploitation of sand and gravel sources, exposing fine sediments to wind erosion. Excessive grazing leads to biomass depletion, forage loss, soil trampling, and sediment destabilization. Furthermore, intruding on animals, plant habitats, and the desert through unaware human practices, especially during the spring camping season adversely affect the desert ecosystem (Al-Awadhi et al. 2003). Besides, excess artesian well digging affects the underground water chemistry and leads to high groundwater levels, causing salinization, hindering plant growth, and thus land degradation.

Al-Awadhi et al. (2005) monitored seven land degradation parameters in Kuwait by studying Al -Mutlaa, Ras Al-Sabiyah, Sulaibiya, and Al–Dhahr areas. Indicators were defined as: soil loss by wind and/or water, a drop of green cover, soil crusting and sealing, soil compaction, soil pollution by oil, and soil salinization. The study concluded that the main reason for desertification in Kuwait was overgrazing, followed by military operations, quarrying, and camping. The effects of desertification in the desert lands are the lack of vegetation cover and thus stirring dust, which causes serious environmental, economic, and health problems. Al-Awadhi et al. (2014) conducted a study on the rate of dust falling in Kuwait for the period 2009–2011 and recorded an average monthly dust falling of 18 g/m^2 in open areas. Land degradation indicators in Kuwait have also been identified by Shahid et al. (1999). Salinization has been declared an early warning of land degradation (Shahid et al. 1998), and the causes of land degradation in the arid environment of Kuwait were highlighted in Shahid and Omar (2001).

2.11 Conclusions and Recommendations

During different geological epochs, Kuwait's geomorphological terrestrial and coastal landforms were formed by surface and subsurface processes. Surface sediments are aged from Middle Eocene to Holocene and are mostly siliciclastic and sedimentary rock units. Kuwait stratigraphy formed during different geological epochs, starting from Eocene to Pleistocene, until the current Holocene. The three main physiographic formations in Kuwait are Kuwait Group and lies above Dammam Formation. Kuwait's organisms are adapted to water scarcity and high temperatures, where the mud flats are a fertile and attractive environment for many migratory birds, and the salt flats are an excellent environment for some kinds of plants (halophytes). Kuwait relies on the Arabian Gulf desalination as a significant resource for drinking and human uses. The agricultural sector in Kuwait is very limited, but it contributes well to supplying markets with cultivated products. Kuwait's economy is based mainly on the exports of oil and the revenues of its products, as it is directly affected by the global demand for oil and the fluctuations in the global oil markets.

Furthermore, the manuscript discussed the natural and anthropogenic factors directly or indirectly responsible for degrading the terrestrial and coastal environments of Kuwait. Hence, the authorities must consider mitigation and adaptation measures by improving renewable energy utilization to reduce greenhouse gas effects and protect natural resources. Also, establishing artificial lakes and water channels within urban area can help in lowering the air temperature. The treated wastewater usage should be supported more for landscaping and gardening purposes in urban areas, conserving desalinated water and saving energy. Authorities should invest in innovative technologies to save water and improve soil quality leading to intensify farms productivity for food security. Moreover, salt-tolerant native plants can be utilized to develop the country's green cover. Also, measures that would include protection against rising sea levels and making buildings less energy-intensive should be taken in the coming decades to save the terrestrial environment of Kuwait from ecological imbalances. Finally, the management and protection of the natural reserves must be carefully and consciously activated to preserve biodiversity, terrestrial landforms, and the natural ecosystems of Kuwait.

References

Abd el-aal AK, Al-Awadhi JM, Al-Dousari A (eds) (2023) The geology of Kuwait. Reg Geol Rev. https://doi.org/10.1007/978-3-031-16727-0_1

Alahmad B, Vicedo-Cabrera AM, Chen K, Garshick E, Bernstein AS, Schwartz J, Koutrakis P (2022) Climate change and health in Kuwait: temperature and mortality projections under different climatic scenarios. Environ Res Lett 17(7):074001. https://doi.org/10.1088/1748-9326/ac7601

Alajmi AJ, Al-Rashed AR, Al-Hamad YM (2021) Cusum usage for environmental assessment of Kuwait south coastal waters. Egypt J Geol 65(1):155–167. https://doi.org/10.21608/egjg.2022.111994.1014

Al-Awadhi JM, Misak RF, Omar SAS (2003) Causes and consequences of desertification in Kuwait: a case study of land degradation. Bull Eng Geol Environ 62:107–115. https://doi.org/10.1007/s10064-002-0175-0

Al-Awadhi JM, Omar SAS, Misak RF (2005) Land degradation indicators in Kuwait. Land Degrad Dev 16(2):163–176. https://doi.org/10.1002/ldr.666

Al-Awadhi JM, Al-Dousari AM, Khalaf FI (2014) Influence of land degradation on the local rate of dust fallout in Kuwait. Atmos Clim Sci 4:437–446. https://doi.org/10.4236/acs.2014.43042

Al-Awadhi JM, AeK A e-a, Misak R, Abdulhadi A (2023) Geo-and environmental hazard studies in Kuwait. In: The geology of Kuwait. Springer, Cham, pp 171–198. https://doi.org/10.1007/978-3-031-16727-0_8

Al-Dalamah AK, Al-Hurban A (2019) The impact of urbanization expansion on the geomorphology of the southern coastal sabkhas from Ras Al-Jailiaha to Al-Khiran, South Kuwait. J Geogr Inf Syst 11(05):609–632. https://doi.org/10.4236/jgis.2019.115038

Al-Dosari N (2021) Geography of Kuwait and the Arabian Gulf, 4th edn, Kuwait. ISPN 978-99966-1-066-0 (Arabic Translation)

Al-Dousari AE, Kafy AA, Saha M, Fattah MA, Almulhim AI, Al Rakib A, Jahir DM, Rahaman ZA, Bakshi A, Shahrier M, Rahman MM (2022) Modelling the impacts of land use/land cover changing pattern on urban thermal characteristics in Kuwait. Sustain Cities Soc 86:104107. https://doi.org/10.1016/j.scs.2022.104107

Al-Dousari AM, Al-Sahli M, Al-Awadhi J, Al-Enezi AK, Al-Dousari N, Ahmed M (2023) Sand dunes in Kuwait, morphometric and chemical characteristics. In: The geology of Kuwait. Springer, Cham, pp 51–81. https://doi.org/10.1007/978-3-031-16727-0_3

Al-Hemoud A, Al-Dousari A, Al-Shatti A, Al-Khayat A, Behbehani W, Malak M (2018) Health impact assessment associated with exposure to PM_{10} and dust storms in Kuwait. Atmosphere 9(1):6. https://doi.org/10.3390/atmos9010006

Al-Hemoud A, Al-Dousari A, Misak R, Al-Sudairawi M, Naseeb A, Al-Dashti H, Al-Dousari N (2019) Economic impact and risk assessment of sand and dust storms (SDS) on the oil and gas industry in Kuwait. Sustainability 11(1):200. https://doi.org/10.3390/su11010200

Al-Hemoud A, Al-Enezi A, Al-Dashti H, Petrov P, Misak R, AlSaraf M, Malek M (2023) Hazard assessment and hazard mapping for Kuwait. Int J Disast Risk Sci 14:143–161. https://doi.org/10.1007/s13753-023-00473-2

Al-Hurban A, Gharib I (2004) Geomorphological and sedimentological characteristics of coastal and inland sabkhas, Southern Kuwait. J Arid Environ 58(1):59–85. https://doi.org/10.1016/S0140-1963(03)00128-9

Al Jassar H, Petrov P, Al Hemoud A, Al-Enezi A, Alsaleh A (2023) Applications of remote sensing in Kuwait. In: The geology of Kuwait. Springer, Cham, pp 215–242. https://doi.org/10.1007/978-3-031-16727-0_10

Al-Mutairi N, Alsahli M, El-Gammal M, Ibrahim M, Abou Samra R (2021) Environmental and economic impacts of rising sea levels: a case study in Kuwait's coastal zone. Ocean Coast Manag 205. https://doi.org/10.1016/j.ocecoaman.2021.105572

Al-Qallaf H, Aliewi A, Abdulhadi A (2020) Assessment of the effect of extreme rainfall events on temporal rainfall variability in Kuwait. Arab J Geosci 13:1129. https://doi.org/10.1007/s12517-020-06086-z

Al-Qasabi A (2014) Human influences on the geomorphology of the coasts of the State of Kuwait, 1st edn, Kuwaiti Research and Studies Center, Kuwait. ISBN 978-99906-94-56-7 (Arabic translation)

AlRefaei Y, Najem A, Amer A, Al-Qattan F (2023) Surface geology of Kuwait. Geol Kuwait:1–26. https://doi.org/10.1007/978-3-031-16727-0_1

Al-Salem SM, Zeitoun R, Dutta A, Al-Nasser A, Al-Wadi MH, Al-Dhafeeri AT, Karam HJ, Asiri F, Biswas A (2020) Baseline soil characterisation of active landfill sites for future restoration and development in the State of Kuwait. Int J Environ Sci Technol 17:4407–4418. https://doi.org/10.1007/s13762-020-02774-1

Al-Sarawi MA (1995) Surface geomorphology of Kuwait. GeoJournal 35:493–503

Al-Sewaji HN (1999) Democracy in Kuwait. Army War Coll Strategic Studies Inst Carlisle Barracks PA US Army War College pp 27–30

Al-Sulaimi JS, Al-Ruwaih FM (2004) Geological, structural and geochemical aspects of the main aquifer systems in Kuwait. Kuwait J Sci Eng 31(1):149–174

Al-Sulaimi J, Mukhopadhyay A (2000) An overview of the surface and near-surface geology, geomorphology and natural resources of Kuwait. Earth Sci Rev 50(3–4):227–267. https://doi.org/10.1016/S0012-8252(00)00005-2

Amer A, Al-Hajeri M (2020) The Jal Az-Zor escarpment as a product of complex duplex folding and strike-slip tectonics; a new study in Kuwait, northeastern Arabian Peninsula. J Struct Geol 135. https://doi.org/10.1016/j.jsg.2020.104024

Cao H, Amiraslani F, Liu J, Zhou N (2015) Identification of dust storm source areas in West Asia using multiple environmental datasets. Sci Total Environ 502:224–235. https://doi.org/10.1016/j.scitotenv.2014.09.025

CSB (2021) Central Administration of Statistics, Kuwait. Annual Statistical Abstract 2019–2020, edition 54, https://www.csb.gov.kw/Pages/Statistics?ID=18&ParentCatID=2. Accessed 16 June 2023

El-Gamily HI (2007) Kuwait Integrated Environmental Information Network (KIEIN) phase III; report no KISR 8735. Kuwait Institute for Scientific Research, Kuwait

EPA (2011) National Strategy for Biodiversity of the State of Kuwait 2011–2020. https://www.fao.org/faolex/results/details/en/c/LEX-FAOC195042/. Accessed 6 May 2023

EPA (2019a) State of Kuwait Second National Communication. The United Nations Framework Convention on Climate Change (UNFCCC). https://epa.gov.kw/Portals/0/PDF/kuwaitSNC.pdf

EPA (2019b) Kuwait National Adaptation Plan 2019–2030. https://unfccc.int/sites/default/files/resource/Kuwait-NAP-2019-2030.pdf

Hassan A, Hassaan MA (2020) Potential impact of sea level rise on the geomorphology of Kuwait state coastline. Arab J Geosci 13:1139. https://doi.org/10.1007/s12517-020-06084-1

Hassan A, Alfaraj M, Fayad M, Allen CD (2021) Optimizing site selection of new cities in the desert using environmental geomorphology and GIS: a case study of Kuwait. Appl Geomat 13:953–968. https://doi.org/10.1007/s12518-021-00403-1

Jamaan S, Mickens R (1998) Biological diversity in the Kuwaiti environment: a field study, 1st edn, Kuwaiti Research and Studies Center. ISPN 60999-32-17-00 (Arabic translation)

Kamal H, Aljeri M, Abdelhadi A, Thomas M, Dashti A (2022) Environmental assessment of land surface temperature using remote sensing technology. Environ Res Eng Manag 78(3):22–38. https://doi.org/10.5755/j01.erem.78.3.31568

Karam QE (2023) Marine geology of Kuwait. In: The geology of Kuwait. Springer, Cham, pp 83–97. https://doi.org/10.1007/978-3-031-16727-0_4

Khalaf FI (1989) Desertification and aeolian processes in the Kuwait desert. J Arid Environ 16(2):125–145. https://doi.org/10.1016/S0140-1963(18)31020-6

KISR (1999) Soil survey for the State of Kuwait: reconnaissance survey, vol II & III. Kuwait Institute for Scientific Research, Kuwait

Kleo A (2006) The marshes of the northern coast in the State of Kuwait (distribution-origination-characteristics), Geographical Letters, Issue 318, Department of Geography/Kuwait University and the Kuwaiti Geographical Society (Arabic translation)

Kleo A, Al-Anein H, Al-Husseini A, Al-Asfour T, Al-Sheikh M (2003) Selected studies in the geomorphology of Kuwaiti lands, 1st edn, Kuwaiti Research and Studies Center, Kuwait. ISBN 99906-32-94-4 (Arabic translation)

Merlone A, Al-Dashti H, Faisal N, Cerveny RS, AlSarmi S, Bessemoulin P, Brunet M, Driouech F, Khalatyan Y, Peterson TC, Rahimzadeh F, Trewin B, Abdel Wahab MM, Yagan S, Coppa G, Smorgon D, Musacchio C, Krahenbuhl D (2019) Temperature extreme records: World Meteorological Organization metrological and meteorological evaluation of the 54.0 C observations in Mitribah, Kuwait and Turbat, Pakistan in 2016/2017. Int J Climatol 39(13):5154–5169. https://doi.org/10.1002/joc.6132

Misak R, Mahfouz S, Al-Asfour T (2003) The desert environment in the State of Kuwait (general features – deterioration causes - rehabilitate ways), 3rd edn, Kuwait research and studies center, Kuwait. ISBN 99906-32-21-9 (Arabic translation)

Misak R, El Gamily H, Hussain W (2014) Threats to agriculture lands at Al-Wafra, southern part of Kuwait. J Arid Land Stud 24(1):9–12

Mufadhal W (2021) Kuwait islands and its coasts: environmental and geomorphological features, and sources of wealth and distinction, 1st edn, Kuwait Research and Studies Center. ISBN 978-9921-750-15-7 (Arabic translation)

Omar SAS, Shahid SA (2013) Reconnaissance soil survey for the State of Kuwait. In: Shahid SA, Taha FK, Abdelfattah MA (eds) Developments in soil classification, land use planning and policy implications: innovative thinking of soil inventory for land use planning and management of land resources. Springer, pp 85–107. https://doi.org/10.1007/978-94-007-5332-7_3

OPEC Annual Statistical Bulletin (2017) Organization of the Petroleum Exporting Countries: Helferstorferstrasse 17, A-1010 Vienna, Austria. ISSN 0475-0608. p 8. https://www.opec.org/opec_web/static_files_project/media/downloads/publications/ASB2017_13062017.pdf

Papathanasopoulou N, Zogaris S (2015) Coral reefs of Kuwait. Cyprus: Kuwait Foreign Petroleum Exploration Company (KUFPEC), Biodiversity East, Cyprus. ISBN 978-9963-2811-2-1

Shahid SA, Jenkins DA (1990) Heavy minerals separation and their study by the optical and scanning electron microscopy. Pak J Agric Sci 27(4):422–428

Shahid SA, Omar SAS (1999) Order 1 soil survey of the demonstration farm sites with proposed management. Kuwait Institute for Scientific Research, Kuwait. p 144, KISR 5463. ISBN 0 957700369

Shahid SA, Omar SAS (2001) Causes and impacts of land degradation in the arid environment of Kuwait. In: Bridges EM et al (eds) Response to land degradation, pp 76–77

Shahid SA, Omar SAS (2022) Kuwait soil taxonomy. Springer, p 149. ISBN 978-3-030-95296-

Shahid SA, Omar SAS, Grealish G, King P, El-Gawad MA, Al-Mesabahi K (1998) Salinization as an early warning of land degradation in Kuwait. Prob Desert Dev 5:8–12

Shahid SA, Omar SAS, Al-Ghawas S (1999) Indicators of desertification in Kuwait and their possible management. Desertif Control Bull 34:61–66

Shahid SA, Omar SAS, Jamal ME, Shihab A, Abo-Rezq H (2004) Soil survey for farm planning in northern Kuwait. Kuwait J Sci Eng 31(1):43–57

Soleimani SM, Alaqqad AR, Jumaah A, Mohammad N, Faheiman A (2021) Incorporation of recycled tire products in pavement-grade concrete: an experimental study. Crystals 11(2):161. https://doi.org/10.3390/cryst11020161

UNCCD (UN Convention to Combat Desertification) (1994) Elaboration of an international convention to combat desertification in countries experiencing serious drought and/or desertification, particularly in Africa: final text of the convention – note by the secretariat. Available online https://wedocs.unep.org/handle/20.500.11822/27569;jsessionid=794621C1042AA7332CE00CC24FC93C98. Accessed 5 July 2023

Chapter 3
Climate and Climate Change Aspects of Kuwait

Amal J. Alkandari

Abstract Climate is the average weather conditions, whereas climate change is change in climatic conditions over a minimum of 30 years' period. Kuwait lies in the climatic zone where high temperature and less rainfall are the main climatic features, exceptions are in few years (2018) when rainfall over 344.1 mm/year was recorded. In general, annual rainfall of Kuwait is less than 110 mm. Globally 0.75 °C rise in temperature has been recorded between 1850 and 2005. The Intergovernmental Panel on Climate Change (IPCC) posits an upper boundary of global sea level rise (SLR) by 2100 of 59 cm-excluding ice sheet dynamism. In Kuwait during the past two decades, atmospheric average temperature has increased, and in 2018 the highest temperature of 52.13 °C was recorded. The longest fog duration was 18 days in 2018. The highest relative humidity percentage was recorded in 2004. The records show the intensity of dust storms increased in summer seasons (May–July), dusty days are hot, and the dust remains in the atmosphere and absorbs heat. The expected climate change features in Kuwait could be prolonged drought, extreme aridity, increased evapotranspiration and water demand, extreme events (rainfall, storms), water scarcity, soil and water salinity, depletion of renewable water resources, food insecurity, land degradation and ecosystem disturbance, etc.

Keywords Weather · Temperature · Rainfall · Sea level rise · Global warming · Dust · Fog · Water source · Soil fertility · Food security · Agriculture

---

A. J. Alkandari (✉)
Desert Agriculture and Ecosystems Program, Environment and Life Sciences Research Center, Kuwait Institute for Scientific Research, Safat, Kuwait
e-mail: ajkandari@kisr.edu.kw

M. K. Suleiman, S. A. Shahid (eds.), *Terrestrial Environment and Ecosystems of Kuwait*, https://doi.org/10.1007/978-3-031-46262-7_3

3.1 Introduction

Kuwait is located in the north-western corner of the Arabian Gulf. Its total area is 17,818 km^2. It is hyper-arid country, where evapotranspiration exceeds rainfall manifolds. Summers are extremely hot and winters are mild. The desert soils are sandy in texture, infertile, and vulnerable to wind erosion. The agriculture is practiced in three areas, Abdali in the north, Wafra in the south, and Sulaibiya in the central Kuwait. Kuwait is rich in oil resources and imports most of its food to meet demand of the population. There are no surface water resources; the water demand is met through groundwater, desalination of sea water, and treated waste water (TWW).

Since the 1800s, anthropic activities released the greenhouse gases (GHGs), such as carbon dioxide (CO_2), methane (CH_4) and nitrous oxides (N_2O), etc., into the atmosphere which is known in science as primary causes of climate change (CC), as a result of burning coal, oil, and gas to produce energy. In the late 1850s, scientists began focusing on CO_2 and other gases that play a significant role in absorbing heat from solar energy. In the late 1950s, scientists approved the theory that CO_2 and other gases were responsible for CC. In 1896, Swedish chemist Svante Arrhenius (1896) was researching ways to cool the Earth by minimizing the concentration of CO_2 in the atmosphere. He presented that cut down of CO_2 levels could reduce global temperatures by about 5 °C, while doubling CO_2 levels could increase global temperatures by 5 °C.

In the early 1970s, scientists initially believed that pollutants emitted into the atmosphere could block sunlight and cool the Earth. In 1979, a declaration was issued, mission on the world's governments "to foresee and prevent potential man-made changes in climate that might have negative effects to the well-being of the humanity." Global warming gained recognition as a real phenomenon and deliberated in many international conferences. The Intergovernmental Panel on Climate Change (IPCC) was created in 1989 by the United Nations for a scientific view of CC. The IPCC's first complete report in 1990 showed severe heat waves and great hurricanes caused by rising sea surface temperatures (Parry 1990).

3.2 UN Framework Convention on Climate Change and Conference of Parties

The UNFCCC is a legally binding international treaty that sets out a framework for addressing global warming and CC. The Convention sets an overall framework for intergovernmental efforts to tackle the challenge posed by CC and emphasizes the need for international cooperation to mitigate GHG emissions and adapt to the impacts of CC in Rio "Earth Summit" which was signed by 154 states plus European Commission.

The Conference of Parties (COP) is an annual meeting where delegates from various countries discuss ways to reduce GHG discharges and their impact toward the environment. The first COP was held in 1995, and since then, several meetings have been conducted. The latest COP-27 was held in Sharm Al-Sheikh, where world leaders repeated their aim to decrease global temperature rise to 1.5 °C above pre-industrial points. The conference concluded that continued decline in worldwide GHGs emissions of 43% by 2030 relative to the 2019 level is required to limit the temperature rise to 1.5 °C. It is jointly agreed by the scientists that the GHGs have a significant impact on the food sector by affecting agriculture (Meteoblue Weather Princeton https://www.meteoblue.com/en/weather/week/princeton_united-states_5102922).

The terms "climate change" and "global warming" are often used reciprocally. Climate refers to long-term changes in average weather conditions, while weather discusses the atmospheric surroundings that occur over short periods. Universal heating is the long-term warming of Earth's surface, which is primarily caused by social actions such as fossil fuel burning. The Earth's global average temperature has risen by about 1 °C since the pre-industrial time, and this rate is mainly due to human activities (Source: NASA Global Climate Change, https://climate.nasa.gov/global-warming-vs-climate-change/).

3.3 Weather and Climate

Weather is the study of meteorological phenomena over a short period of time, typically 2–30 days, and solar flares are one of the most important factors in weather reporting. Climate refers to the distribution of weather over a historical time, which causes differences in weather conditions for an unknown period in a certain region. The temporal variation of weather changes from 1 year to another in one place causes CC. The primary driving factor for rising temperatures and causing serious impacts on the global atmosphere is GHGs discharge into the atmosphere. The CC has many adverse impacts on the Earth, including sea level rise (SLR), increasing atmospheric temperatures, and increased cloud and patterns of rainfall, floods, and the effects on plants and soil erosion due to drought or heavy rainfall (Fig. 3.1). Kuwait, for example, has experienced a shift in seasons, with winters starting in October instead of August, and the clarity of spring and autumn seasons becoming less apparent due to CC (Hardy 2003).

3.3.1 Climate Recording and Data Processing

Climatological recording and data collection are critical to understanding and addressing the most pressing weather-related problems facing the world. Climate data is used to study changes in climate over time and predict future climate trends. Data recording through weather stations is conventional method used globally. Various attributes include temperature, precipitation, pressure, wind speed,

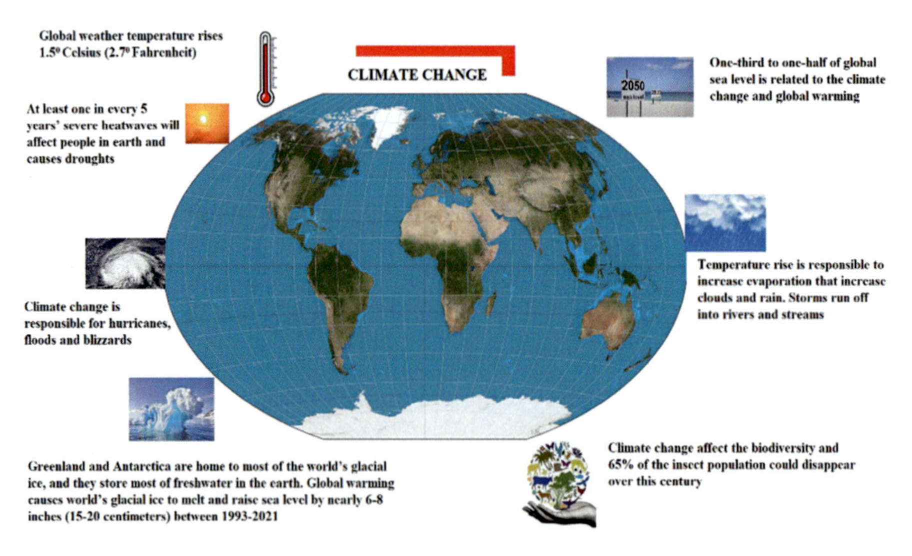

Fig. 3.1 Climate change infographic

humidity, visibility, and heat index. The World Meteorological Organization (WMO) (https://community.wmo.int/climate-data-homogenization) outlined six stages of data collection, namely, (1) data gathering, (2) data preparation, (3) data response, (4) data handing out, (5) data production, and (6) data storage. Data can be presented as graphs for decision-makers, such as government officials to respond to, manage public finances, assets such as electricity grids, and provide warnings to the public.

3.3.1.1 Trends of Temperature Change

The world has come to a consensus that global warming, resulting from GHGs emissions, poses a significant threat to the future of our Earth. The average worldwide temperature has increased by 0.2–0.4 °C, with an average of 0.05 °C per decade, and this trend is expected to continue. This rise in temperature will have an adverse effect on soil productivity and the global food industry. While some countries in colder regions such as North America have seen increased agricultural productivity as a result of global warming, hotter regions such as Asia and Africa have experienced increased evaporation, reduced rainfall, increased drought, and decreased soil and agricultural productivities. Climate data collected by NASA since 1880 has shown a steady increase in global temperatures (Fig. 3.2), highlighting the urgent need for global action to moderate the impact of worldwide warming.

The impact of the rising temperature in Kuwait can be seen on various aspects of the environment. The hot and dry weather in Kuwait can cause a significant impact on agriculture, as the soil tends to dry up quickly, requiring frequent irrigation, making difficult for farmers to grow crops. Other effects could be on air quality due to

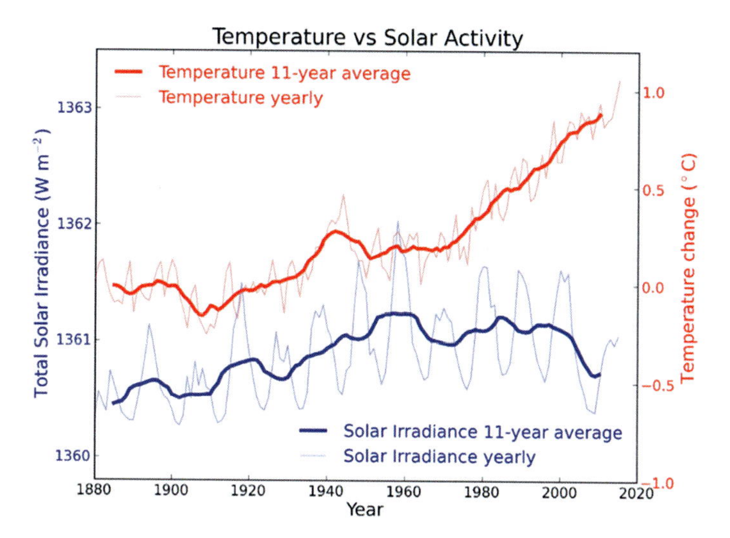

Fig. 3.2 A trend of temporal global temperature change (W refers to Watts). (Source: NASA Global Climate Change, https://climate.nasa.gov/global-warming-vs-climate-change/)

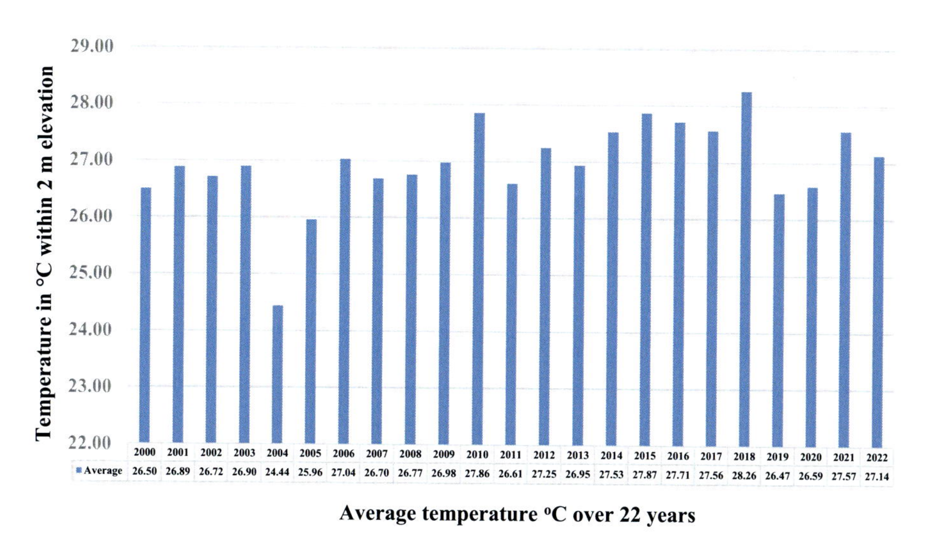

Fig. 3.3 Two decades' temperature dynamics in Kuwait. (Data source: Meteoblue Weather Department)

increased dust storms. To mitigate the impact of rising temperature in Kuwait, various measures can be taken, including promoting sustainable agriculture practices, water-saving technologies, investing in renewable energy, and implementing policies to reduce carbon emissions. Additionally, awareness campaigns can be launched to educate public about the impact of climate alterations and encourage them to take necessary measures to reduce their carbon footprint (Parry 1990). The increasing

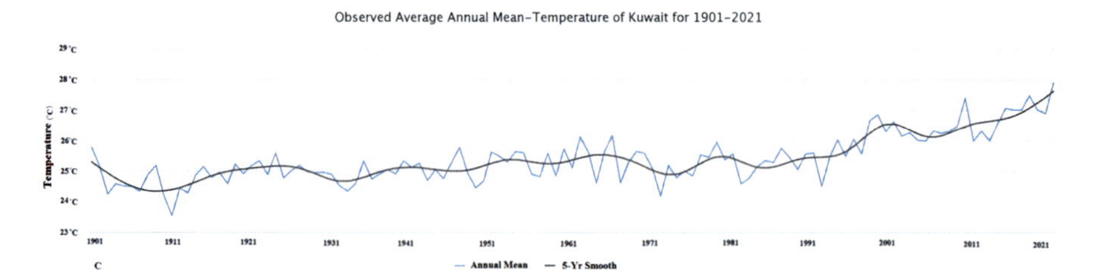

Fig. 3.4 Kuwait temperature rises. (Source: The World Bank, https://data.worldbank.org)

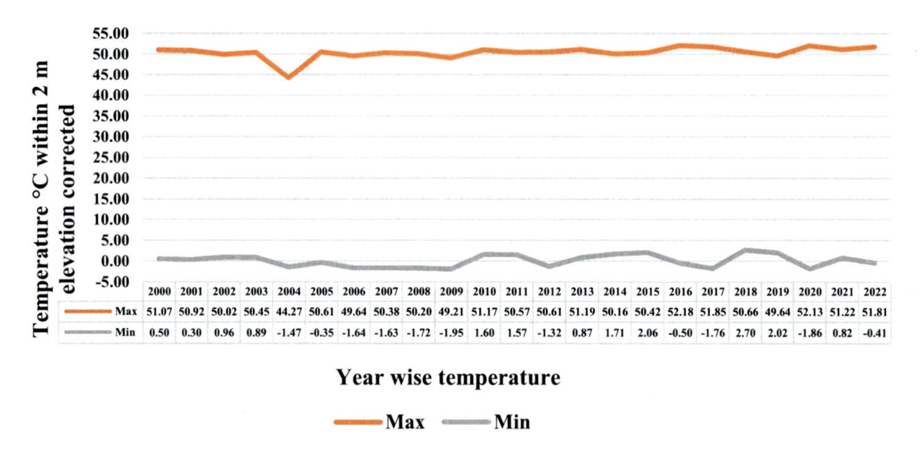

	2000	2001	2002	2003	2004	2005	2006	2007	2008	2009	2010	2011	2012	2013	2014	2015	2016	2017	2018	2019	2020	2021	2022
Max	51.07	50.92	50.02	50.45	44.27	50.61	49.64	50.38	50.20	49.21	51.17	50.57	50.61	51.19	50.16	50.42	52.18	51.85	50.66	49.64	52.13	51.22	51.81
Min	0.50	0.30	0.96	0.89	-1.47	-0.35	-1.64	-1.63	-1.72	-1.95	1.60	1.57	-1.32	0.87	1.71	2.06	-0.50	-1.76	2.70	2.02	-1.86	0.82	-0.41

Fig. 3.5 Kuwait minimum-maximum temperature dynamics during two decades. (Data source: Meteoblue Weather Department)

trend of average temperature recorded at Kuwait International Airports is shown in Figs. 3.3 and 3.4.

A review of the highest and the lowest temperature data collected over a 22-year period shows that the average lowest temperature in summer was 44.723 °C in 2004 (Fig. 3.5), which was the lowest among the 22 summers. The highest temperature was 52.13 °C in summer 2020. On the other hand, the lowest temperature data showed an average of −0.35 °C in winter 2005 (Fig. 3.5).

3.3.1.2 Sun Shine and Solar Radiation

Most of the solar energy entering the Earth's atmosphere is in the form of short waves. These waves are absorbed by the Earth's crust and warm the atmosphere above it. The troposphere contains gases that control the type of solar radiation that is allowed to enter the Earth. These gases act like a greenhouse, trapping some of

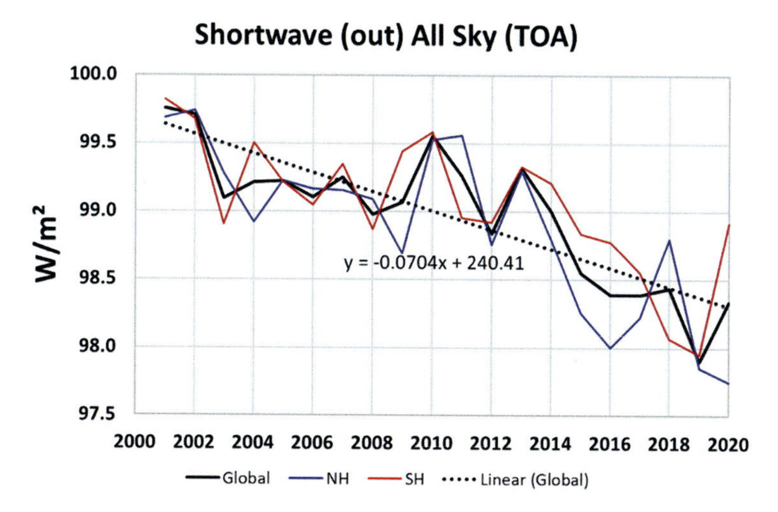

Fig. 3.6 The global shortwave reflecting to the sky recorded in the top of the atmosphere (TOA) form climate NASA (W refers to Watts, and NH (north hemisphere), SH (south hemisphere). (Source: NASA Global Climate Change, https://climate.nasa.gov/global-warming-vs-climate-change/)

the heat and causing the Earth's temperature to increase and disturbing the balance of radiation within the Earth's atmosphere. Global data shows that there has been a decrease in the amount of shortwave radiation that is being reflected out to space (Fig. 3.6). The Southern hemisphere obtains ca. 0.7 W/m^2 more than the North due to the eccentric orbit and the inclined axis of the Earth. This decrease in reflection causes the temperature of the atmosphere to increase, leading to global warming (Dübal and Vahrenholt 2021).

To understand the impact of the increase in temperature in Kuwait, it is important to analyze the data collected from various sources such as climate stations, weather balloons, buoys, and satellites. This data can help to identify the trends and changes in the weather patterns and temperatures in Kuwait over the years. In Kuwait, there is no major difference in average sunshine waves (data collected at Kuwait Airport); however, the effect of the shortwaves has increased the temperature over period of time.

3.3.1.3 Clouds

The rise in temperature increased the evaporation from water bodies (seas, lakes, oceans, soil, etc.). This, in turn, causes an increase in cloud formation resulting in warmer winters and cooler summers. The increase in cloud cover also reduces the temperature difference between day and night. The data from Kuwait international airport shows the highest cloud formation was in 2003 and 2009 (Fig. 3.7), after 2009 in general no significant changes in cloud formation observed.

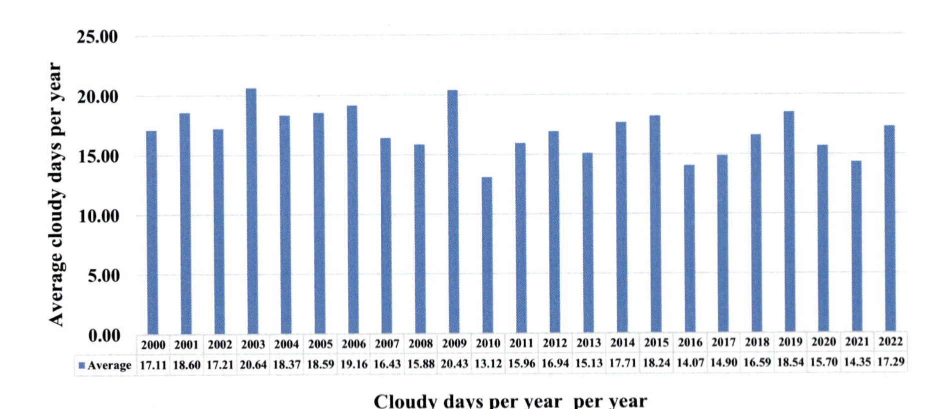

Fig. 3.7 The dynamics of cloudy days over 22 years in Kuwait. (Data source: Meteoblue Weather Department)

3.3.1.4 Rainfall

Most sub-Saharan African countries depend on rainfall for cultivation. Change in rainfall patterns and frequency can affect the growing time and availability for crops. Increased rainfall can cause soil erosion, increase land degradation, and loss of agriculture. Kuwait is a dry desert country by means of a hot and arid climate. The scarcity of rainfall in Kuwait is one of the significant challenges caused by CC, with an average annual rainfall of approximately 112 mm per year, varying from 75 to 150 mm/year. A reduction in rainfall can lead to a depletion of groundwater and may seriously affect the ability to increase food production and feed the population. Although the rainfall pattern remains mostly unchanged, there was an increase in rainfall in 2018 (Fig. 3.8).

3.3.1.5 Fog

Fog is an important atmospheric phenomenon that contributes to water input, despite its relatively low volume. Fog occurs when humidity increases in a shallow layer above the ground, especially 1 hour after sunset, when wind speeds are low, and during certain times of the year. Fog disappears when temperatures increase due to sunlight or wind speeds. Fog has direct effects on visibility, air quality, humidity, temperature, and CC. There is a positive relationship between humidity and fog formation. When humidity is high, both active and inactive hydrometeors (water droplets) are present, resulting in the formation of fog. Active hydrometeors can reduce visibility to 1 km or less, depending on the size of the water droplets. The relationship between foggy days and % humidity is shown in Figs. 3.9 and 3.10. In general the number of foggy days positively correlate with % humidity (Martial et al. 2013). In Kuwait the highest number of foggy days (18) were in 2014 and the lowest was 1 day in 2003 and there was no foggy day in 2007, and in general the fog increased in duration after 2012 (Fig. 3.9).

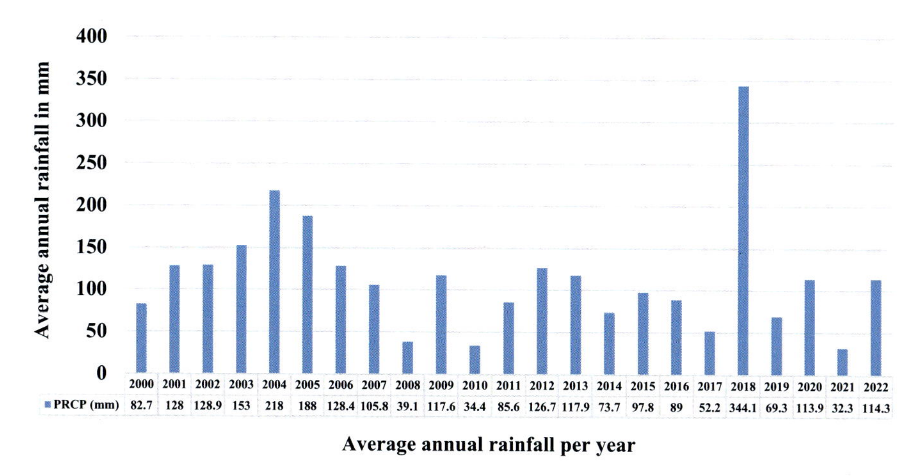

Fig. 3.8 Average annual rainfall dynamics for 22 years in Kuwait. (Data source: Meteoblue Weather Department)

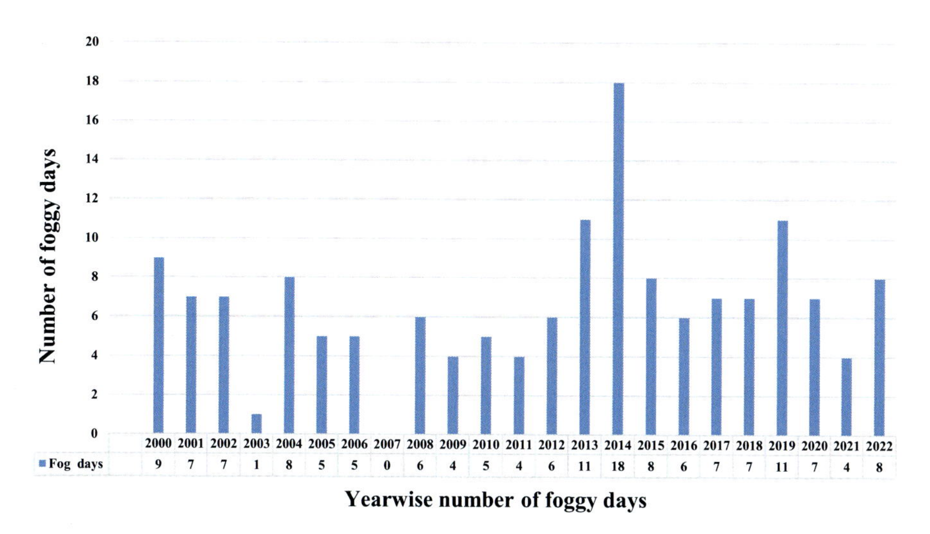

Fig. 3.9 Foggy days dynamics over 22 years in Kuwait. (Data source: Civil Aviation Meteorological department)

3.3.1.6 Relative Humidity

Relative humidity (RH) is an atmospheric phenomenon that determines the aerodynamics of evaporation demand. Changes in RH can affect the atmospheric hydrological cycle and climate aridity. The RH and climate temperature have a significant relationship in summer: as temperature increases, evaporation from water bodies increases, especially in areas near the water, while in winter, increased relative humidity causes warmer weather. Increased RH can have a positive effect on global

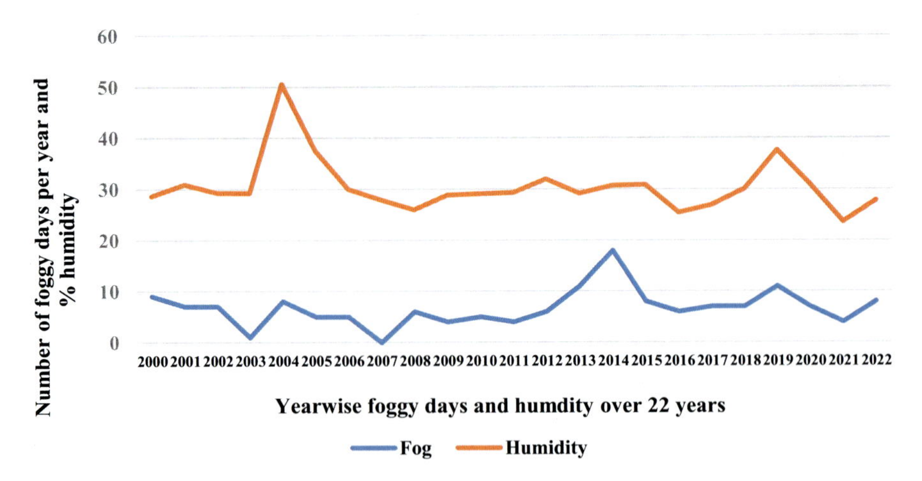

Fig. 3.10 Average foggy days and relative humidity percentage dynamics for 22 years in Kuwait. (Data source: Meteoblue Weather Department)

warming and can cause saturation, and specific humidity increases over land. Overall, the relationship between atmospheric temperature and humidity is complex. While changes in temperature and humidity can potentially cancel each other out, the specific outcomes will depend on the specific circumstances involved. Air temperature does not have a straight influence on RH. Atmospheric drivers of RH are rainfall, and land and ocean evaporation. In Kuwait, humidity rates can reach high levels in summer, making it somewhat difficult to coexist and adapt to them. The maximum average RH was recorded in 2004, when the average temperature was low (Fig. 3.11) (Vicente-Serrano et al. 2018).

3.3.1.7 Wind Speed

Wind speed is a weather phenomenon that moves the air from an area of different atmospheric pressure. The anemometer measures the direction and speed of the air current. Wind speed is measured in meters per second (m/s), miles per hour (mph), and kilometers per hour (km/h). The wind speed is measured at a standard height of 10 m (32 feet) above sea level (ASL). Local weather conditions such as hurricanes, monsoons, and cyclones are the main drivers of wind speed. The lowest wind speed that is considered safe for a walking person is from 0.4 to 1.3 m/s, while the highest wind speed is from 11 to 13.8 m/s (Grace 1988).

Wind speed plays an essential role in agriculture, as it can affect crop growth and speed of 3.5 m/s may cause microscopic harm to vegetation leaves, reducing their ability to control water. Wind can also affect ground temperature, visibility, and

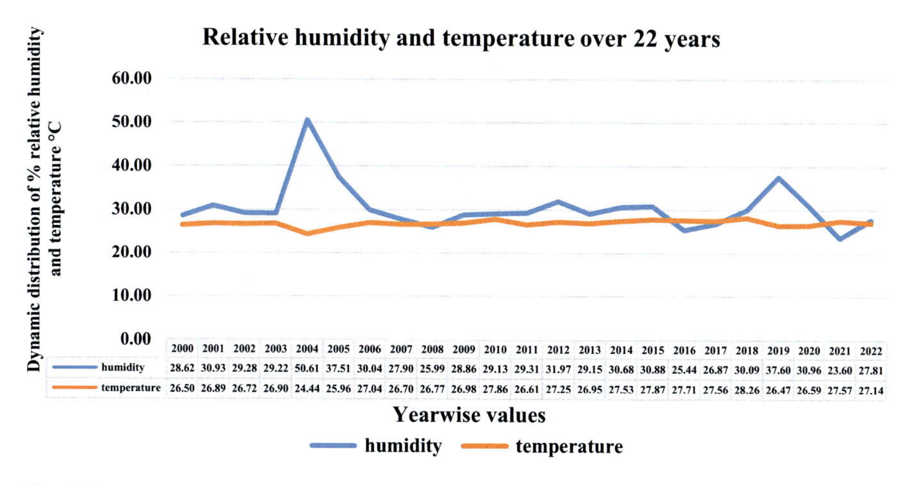

	2000	2001	2002	2003	2004	2005	2006	2007	2008	2009	2010	2011	2012	2013	2014	2015	2016	2017	2018	2019	2020	2021	2022
humidity	28.62	30.93	29.28	29.22	50.61	37.51	30.04	27.90	25.99	28.86	29.13	29.31	31.97	29.15	30.68	30.88	25.44	26.87	30.09	37.60	30.96	23.60	27.81
temperature	26.50	26.89	26.72	26.90	24.44	25.96	27.04	26.70	26.77	26.98	27.86	26.61	27.25	26.95	27.53	27.87	27.71	27.56	28.26	26.47	26.59	27.57	27.14

Fig. 3.11 Relative humidity (%) and temperature (°C) dynamics in Kuwait over 22 years. (Data source: Meteoblue Weather Department)

photosynthesis by raising dust to the atmosphere. Wind speeds exceeding 10 km/h can cause maximum damage to agriculture. In summer, hot wind speed can cause dwarfing of plants due to dehydration of plant tissues. In winter, low wind speed causes the land to absorb more sunlight, making the near surface atmosphere warmer. High wind speed can increase evapotranspiration leading to flower and fruit shedding. Additionally, high wind speed causes soil erosion and poor aeration in the root zone (Armbrust and Retta 2000).

In Kuwait, the annual average wind speed typically ranges from 3.7 to 5.5 m/s, with a mean wind power density ranging from 80 to 167 W/m^2 (W is Weibull distribution, unsurprisingly, has three parameters, shape, scale, and threshold) at a standard height of 10 m ASL. These wind speeds are generally the highest during the summer season, particularly in the northern part of the country. Wind speeds tend to be higher in open desert flat areas in the north, northern part, and southern part of the country. By measuring and analyzing wind speeds at different locations and heights, it is possible to better understand the potential for wind energy and other wind-related applications throughout the country. Within 22 years, wind speed was generally higher than 13 m/s and lower than 15 m/s, with no significant difference in speed between different years (Figs. 3.12 and 3.13) (Al-Nassar et al. 2005). Wind speed can cause dust and haziness of dust in most months of the year, especially in summer when the sand is dry and easily moved by the wind. However, wind speed has decreased in the past 10 years, which could be attributed to an increase in vegetation, particularly in urban areas.

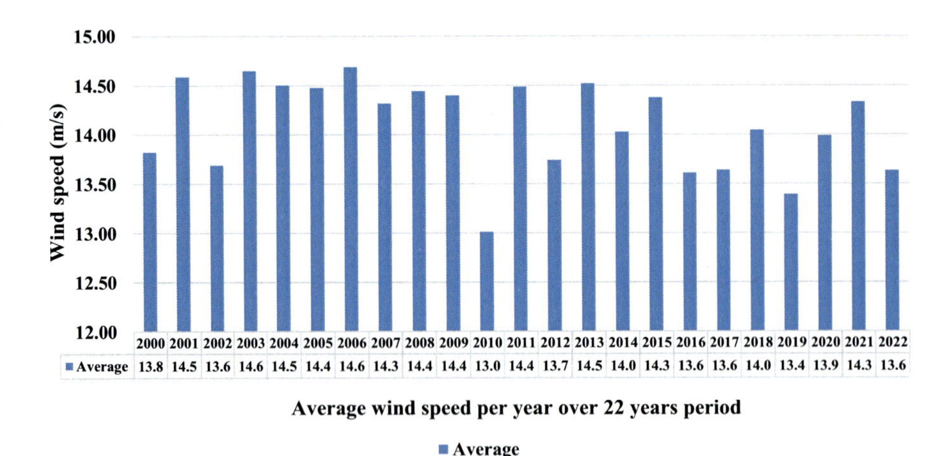

Fig. 3.12 Average maximum wind speed dynamics in Kuwait over 22 years. (Data source: Meteoblue Weather Department)

Fig. 3.13 Maximum wind speed dynamics in Kuwait over 22 years. (Data source: Meteoblue Weather Department)

3.4 Direction and Circulation (Summer and Winter in Relation to Mean Sea Level Pressure)

The CC is causing changes in the global pattern of ocean circulation. In coastal areas, there are significant differences in atmospheric pressures between warm land and cool ocean. These differences drive longshore winds that blow parallel to the coast, resulting in a change in wind direction known as the Ekman transport. This change in wind direction causes the surface water to be replaced by cold, nutrient-rich water from below (upwelling). As the atmospheric temperature increases, the land heats up more rapidly than the water, causing a rise in wind power and a growth

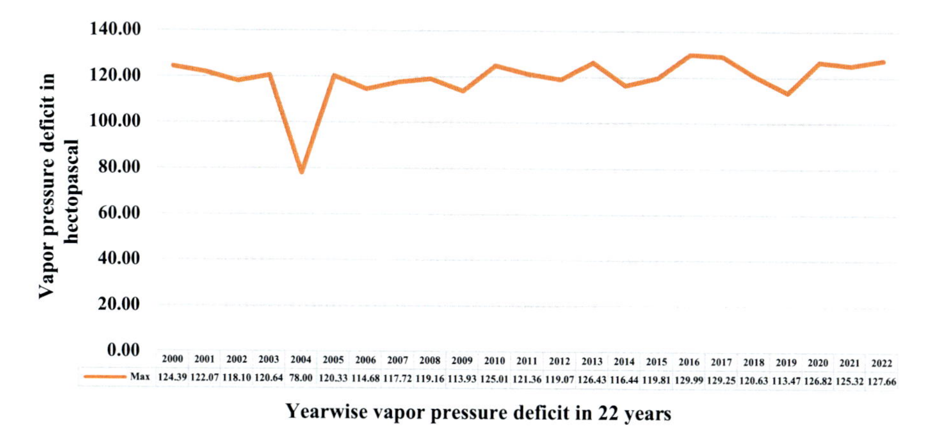

Fig. 3.14 Vapor pressure deficit (hectopascal) between dry and wet temperature over 22 years. (Data source: Meteoblue Weather Department)

in ocean upwelling. In Kuwait, the lowest vapor pressure values were recorded in 2004, when temperatures were low, and the highest value was recorded in 2021. However, overall there have been no significant differences in vapour pressure over the past 22 years (Fig. 3.14) (World Meteorological Organization, https://public.wmo.int/en/media/press-release/no-976-2001-2010-decade-of-climate-extremes).

3.5 Impact of Dust on Temperature

Kuwait is located in a region where several sources of dust exist. The western desert of Iraq, also known as the Anbar Governorate, is one of the largest sources of dust storms in the region. The Mesopotamian Flood Plain, which spans parts of Iraq and Iran, is another significant source of dust, especially during the dry season (Al-Dousari and Al-Awadhi 2012). The northeastern desert of Saudi Arabia, known as the Rub' al-Khali, is a vast desert that covers a significant portion of the Arabian Peninsula and is also a significant source of dust. The drained marshes in south Iraq, which were once a vast wetland ecosystem, have been severely degraded and contributed to the dust problem in the region. Iraq is also a significant source of dust. In addition, there are several intermediate and local sources of dust. The three intermediate sources include the drainage systems in the tri-border area between Kuwait, Saudi Arabia, and Iraq, which can generate dust during periods of low water flow; playas and drainage basins in the southern west desert of Iraq, which can become sources of dust during dry periods; and coastal sabkhas near the southern Kuwaiti border, which are areas of salt flats that can be a source of dust during windy conditions. The local sources include Bubiyan and Warba Islands, which are located off the coast of Kuwait and can generate dust due to their sandy terrain. The muddy playas, depressions, sabkhas, and intertidal zones with high tides can cause sediment to be disturbed and become a source of dust during windy conditions. Three types of dust storms are common in Kuwait: dust storms (large quantities of dust that reduce visibility to less than 1000 m due to turbulent winds), blowing dust (moderate-high winds that reduce visibility to 1.8 m), and dust haze (dust raised from the ground by a dust storm prior to the settlement).

The records from weather department at Kuwait International Airport show two patterns of dust storms. During the summer season (May–July), dusty days are hot, and the following day may become even hotter as fine dust particles can remain in the atmosphere and absorb heat. In contrast, during the winter season, cold weather and low sun waves result in decreased atmospheric temperatures and the following day making the weather even colder. These scenarios can significantly affect crops and wildlife in the area (McTainsh and Pitblado 1987; Al-Dousari 2021).

3.6 Climate Change Features and Consequences in Kuwait

Climate change is caused by the accumulation of GHGs in the near Earth's atmosphere. The GHGs are primarily emitted by human activities such as burning fossil fuels, deforestation, and industrial processes. The impacts of CC are global and affecting both developed and developing countries and all sectors of society. The CC can lead to crop failures and food shortages and displacement of populations due to SLR or extreme weather events. Therefore, CC mitigation requires a global effort to reduce GHGs emissions and build resilience to the impacts of CC (Mimura 2013). Kuwait is not exempted from CC impacts, e.g., it impacts human health, agriculture, water resources, and biodiversity, among others. The CC can also affect food production and water resources, as changes in temperature and rainfall patterns can lead to crop failures, droughts, and water scarcity. This can have ripple effects on global food security and exacerbate inequalities in access to food and water. The CC can also impact biodiversity, with rising temperatures and changing weather patterns affecting ecosystems and species distribution. This can have significant implications for the functioning and provision of ecosystem services, such as pollination and carbon sequestration. Therefore, addressing CC requires a coordinated effort across sectors to mitigate GHGs emissions, adapt to the impacts of CC, and build resilience in vulnerable communities and ecosystems.

Arid countries like Kuwait are particularly vulnerable to the impacts of CC, as they are already facing challenges related to water scarcity, extreme temperatures, and limited natural resources. The CC can exacerbate these challenges and lead to further negative impacts on the environment. The SLR can threaten coastal communities and infrastructures, as well as soil erosion, which can lead to reduced soil fertility and crop productivity. The CC can further exacerbate this issue by altering precipitation patterns and increasing the intensity of rainfall, which can lead to flash floods and erosion. The impacts of CC on biodiversity are also a significant concern in Kuwait and the Gulf region at large.

In addition, to address these challenges, Kuwait can consider various strategies, including implementing sustainable water management practices, such as desalination, water recycling, and improved irrigation techniques; investing in renewable energy sources to reduce reliance on fossil fuels and mitigate GHGs emissions; enhancing land and soil conservation measures, such as afforestation, reforestation, and soil erosion control techniques; strengthening early warning systems and disaster preparedness to respond to extreme weather events and dust storms; improving building design and urban planning to minimize heat island effects and promote energy-efficient structures; promoting research and innovation in climate adaptation strategies, such as drought-resistant crop varieties and sustainable agriculture practices.

Collaboration and international cooperation, along with effective policies and investments, are crucial for Kuwait to address the challenges posed by CC and ensure the long-term sustainability and resilience of the country.

3.6.1 Greenhouse Gasses Emission: Carbon Footprint

The increase in atmospheric concentrations of GHGs is causing the Earth's temperature to rise. However, there have been some positive developments in recent years. The use of clean energy resources such as solar and wind power has increased, and many countries have implemented policies to reduce GHGs emissions. This has led to a slight decline in global carbon dioxide emissions since 2014. In Kuwait, the National Adaptation Plan (EPA 2019) is a significant step toward addressing CC. The plan focuses on increasing resilience to CC impacts and promoting sustainable land-use practices, including reforestation and the restoration of desert areas. Reducing GHGs emissions remains critical to addressing CC and transitioning to clean energy sources such as solar and wind power is an essential step toward achieving this goal. Additionally, increasing resilience to the impacts of CC through measures such as the National Adaptation Plan is crucial to protect vulnerable ecosystems. Dynamics of CO_2 emission for the past 30 years show maximum per capita emission was in 2005 (Fig. 3.15) and progressive increase over the past three decades. The highest fossil emission by sector in Kuwait is from power industry and the lowest from noncombustion sources (Fig. 3.16).

3.6.1.1 Increase in Temperature (Global Warming)

According to Meteoblue and the World Bank Group, global warming is increasing globally and many countries are making efforts to minimize GHGs emissions. However, despite these efforts, the world has not been able to control the situation due to the increasing population and the increasing use of natural resources to boost the water and food sectors. As a result, the global temperature will continue to rise.

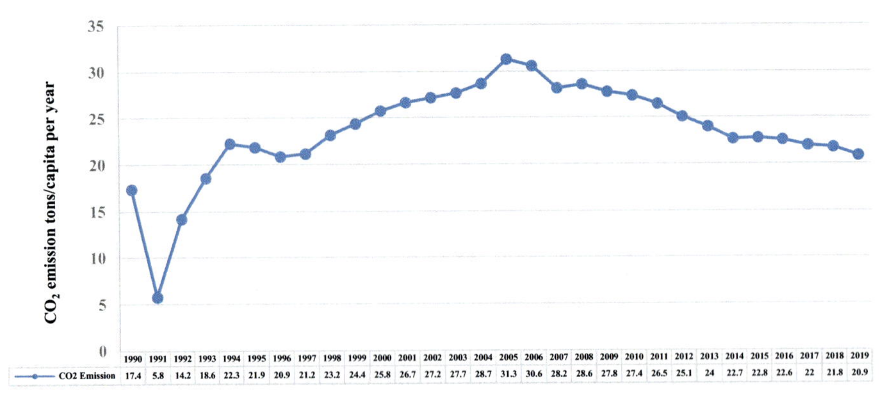

Fig. 3.15 CO_2 emission in metric tons per capita per year in Kuwait. (Source: The World Bank, https://data.worldbank.org)

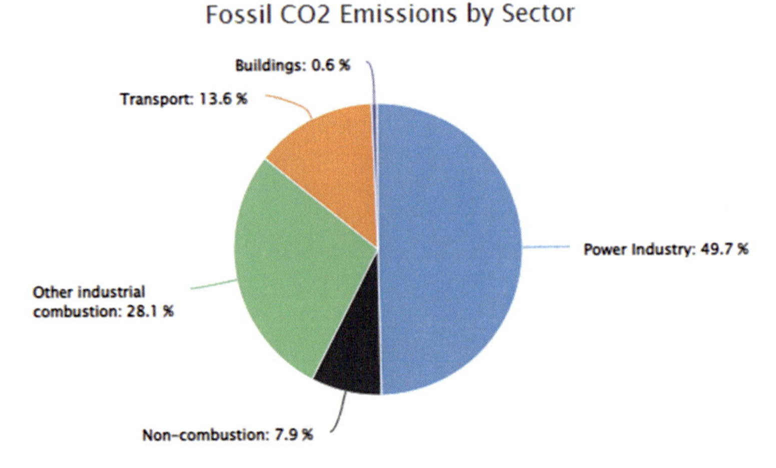

Fig. 3.16 Kuwait CO_2 emission by sector in 2016. (Source: The Worldometers, https://www.worldometers.info/co2-emissions/kuwait-co2-emissions/)

The World Bank and Meteoblue Weather Department predict that the mean temperature in Kuwait, like other countries around the world, will increase over the next decades unless global action is taken (Figs. 3.17 and 3.18).

3.6.1.2 Sea Level Rise and Impact on Shoreline

In the late nineteenth century, global typical weather change was a significant driver for SLR that has strongly impacted coastal areas worldwide. The sea level has grown over the past century due to the increasing temperature of the Earth, causing

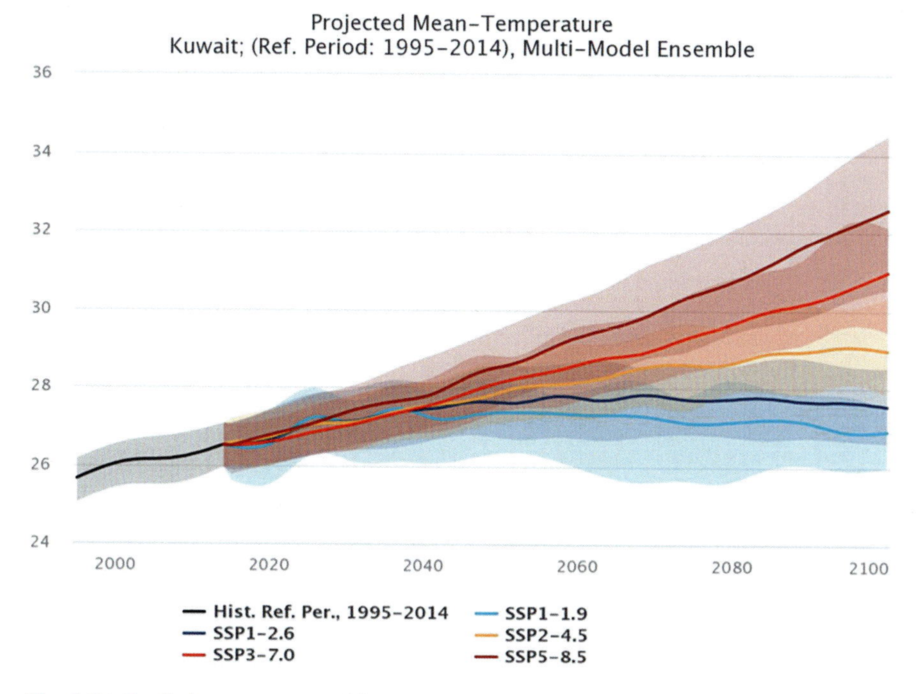

Fig. 3.17 Predicting temperature rising level by season in °C. SSPs indicate shared socioeconomic pathways. (Source: The World Bank, https://data.worldbank.org)

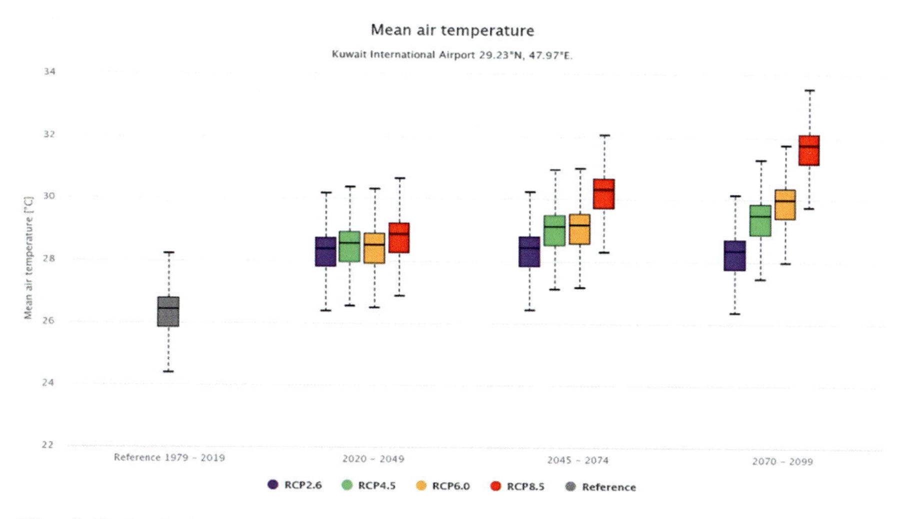

Fig. 3.18 Predicting temperature rising level in Kuwait over the twenty-first century. (Source: Meteoblue Weather Organization, https://www.meteoblue.com/en/weather/week/kuwait-city_kuwait_285787)

the glaciers and ice in both North and South poles to melt, which increases the warm water input to the bodies of water around the world. Between 1993 and 2016, the sea level globally rose about 3.2 inches (81.3 mm) above the average figure of 25, and sea levels continue to increase at a rate of 1/8 inch (3.2 mm) per year. The SLR of 1 m, 3 m, and 5 m will affect 3, 8, and 14% land area of Kuwait (Tolba and Saab 2009).

There are two ways to measure SLR, tide gauges and satellite laser altimeters. Three main factors contribute to the SLR, including geophysical phenomena such as thermal expansion of seawater, growth or ice sheets melting, and changes in water storage globally. Additionally, the size and shape of ocean basins, isostatic changes of land mass, tectonic movements, ground subsidence, changes in ocean currents, atmospheric pressure, tides, waves, tsunamis, storms, and other natural variations also play a significant role in SLR.

There are two main scenarios for predicting SLR. The first scenario involves continuous melting of ice sheets and SLR as global temperatures increase. In this scenario, valuable natural ecosystems containing rare plant communities, suitable breeding places for diverse types of coastal birds, salt marshes, mangroves, and coral reefs will be destroyed. Additionally, saltwater may move into estuaries and aquifers, sedimentation deposit patterns will change in areas near rivers and channels, and a vast area of coastal land and farms will be lost. This will result in the decrease of many sandy beaches and recreational areas, and small islands will become vulnerable to coastal erosion, necessitating the migration of people to other areas. The other scenario is if we can effectively mitigate global warming and reduce global temperatures, we can either stop or at least minimize SLR, leading to save lands and the natural ecosystem and ensuring sustainability.

In Fig. 3.19, the light blue color shows seasonal (3 months) SLR estimated from Church and White (2011) (https://psmsl.org/products/reconstructions/church.php), while the dark blue line is estimated by University of Hawaii data delivery sea level. In Kuwait, like in many other countries, SLR has been rising since 1995, and this trend continues (Fig. 3.20). This SLR could result in the loss and disappearance of many urban areas located near the sea, where the majority of the population currently resides. This presents a significant threat to these coastal communities and highlights the urgent need for measures to discourse and mitigate the special effects of CC. Failure to take action could lead to devastating consequences for the people and ecosystems of Kuwait and the wider region. Therefore, it is important for Kuwait to take proactive steps to address this issue and work toward sustainable solutions to confirm the safety and security.

3.6.1.3 Increased Drought and Crops Water Requirement

Kuwait is experiencing water scarcity due to the absence of surface waters (lakes and rivers) and harsh weather conditions, such as high temperatures and evaporation rates, which further increases the soil temperature. The availability of water is a crucial factor for sustaining life and agriculture, and it surpasses other weather

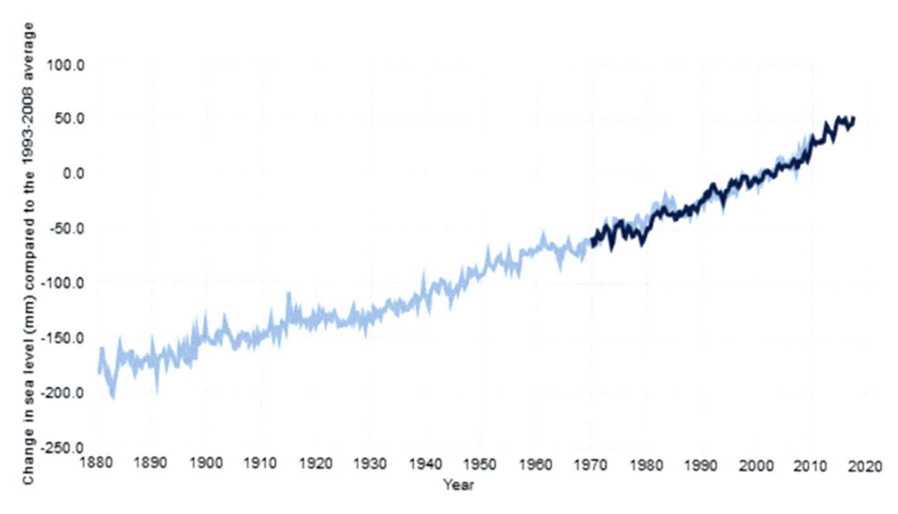

Fig. 3.19 Predicting global sea level rise from 1880 to 2020. (Source: http://arizonaenergy.org/News_17/News_Sep17/ClimateChangeGlobalSeaLevel.html)

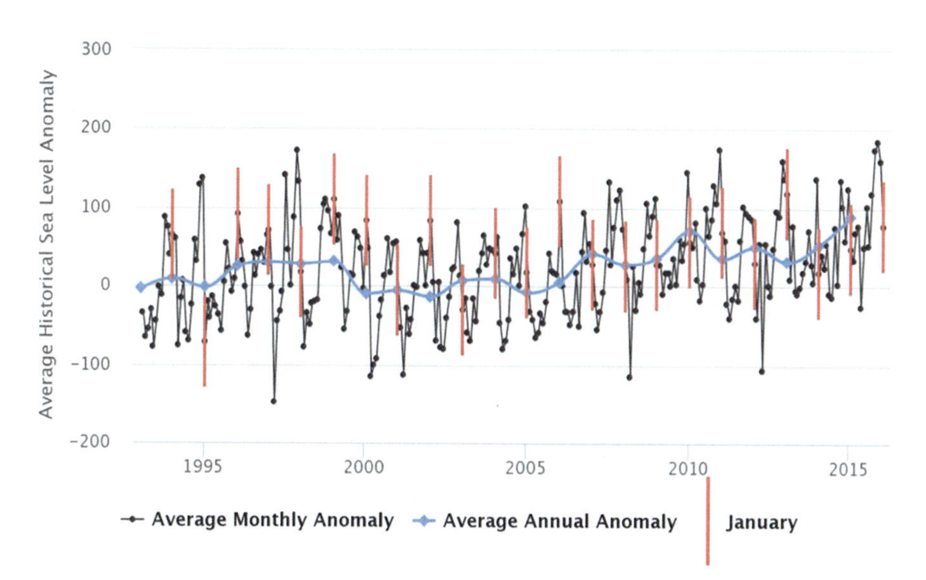

Fig. 3.20 Average sea level rise anomaly (mm) in Kuwait. (Source: The World Bank, https://data.worldbank.org)

factors. Several factors contribute to the drought in Kuwait, and the data collected from the Kuwait Weather Department and presented in Fig. 3.21 shows a high negative correlation between temperature and humidity with pressure and a moderate correlation between temperature and soil moisture. The long summer season in

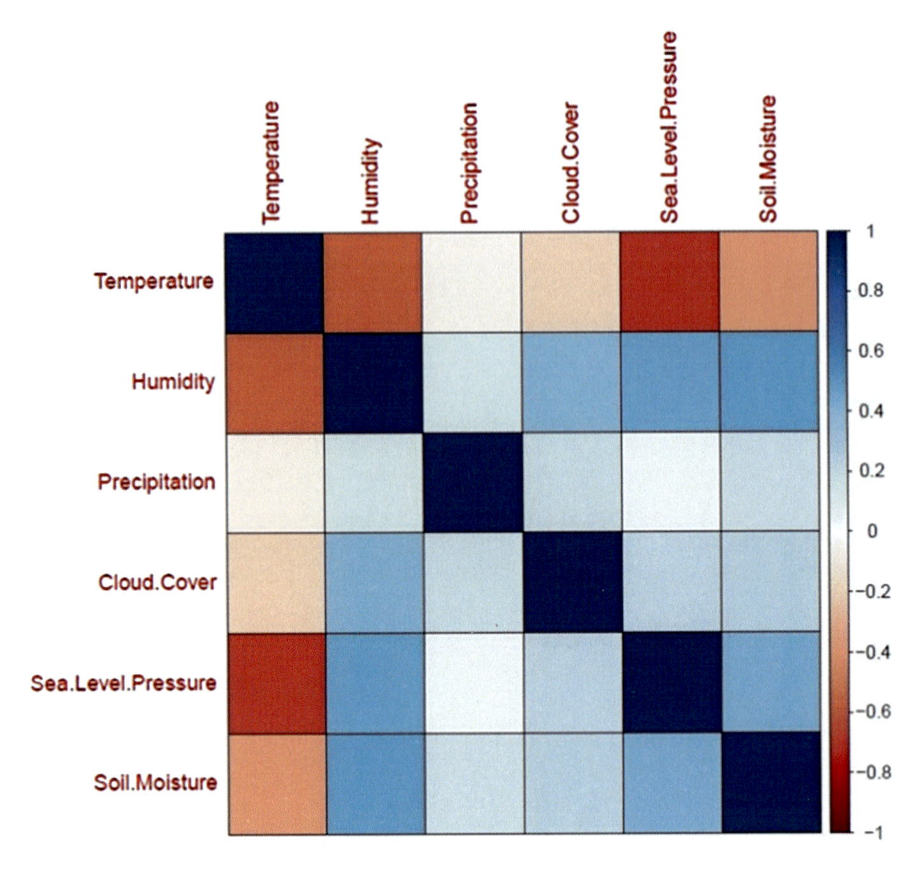

Fig. 3.21 Correlation matrix between temperature and other related variables over 22 years in Kuwait. (Data source: Meteoblue Weather Department)

Kuwait is characterized by temperatures above 50 °C, which increases the evaporation rate and further exacerbates drought conditions, ultimately affecting crop yields through soil dryness (Fig. 3.22). Temperature rise of 1.5 °C expected to shift the Mediterranean climate zone 300–500 km northward making the region as Arid (Tolba and Saab 2009).

3.6.1.4 Biodiversity Decline and Ecosystem Distribution

The CC is causing a critical shift in ecosystem distribution, leading to significant impacts on biodiversity. Harsh summer conditions contribute to soil erosion and a decline in plant cover, leaving animals to suffer in a changing environment. So far, 412 species, including 18 subspecies of birds in 65 families, have been recorded from Kuwait according to observations published by BirdLife International and Kuwait Birds (cf. Amr 2021), whereas, Pope and Zogaris (2012) listed 390 birds species and 17 subspecies in Kuwait.

Fig. 3.22 Dry soil due to prolonged drought in Al Ahmadi desert as seen in summer 2022

Bird migration, which is a natural wonder, is also affected by CC, with the timing of immigration changes due to changing temperatures. For instance, in Kuwait, flamingos and large white-headed gulls (larus) birds migrate in spring and summer. Reproduction time for animals and mammals is also changing due to CC in the last decade. Insect distribution and food collection time for hibernation have also changed.

In the oceans and seas, the growth and reproduction of fish and other organisms are linked to optimum temperatures for each species, causing a decrease or increase in some species at different times of the year. Changing sea temperature in Kuwait marine water leads to coral reef bleach. Additionally, the biodiversity of plants has changed due to CC, which indeed has a significant impact on the biodiversity decline in Kuwait.

Habitat Loss and Fragmentation

The CC can alter ecosystems and disrupt habitats, leading to habitat loss and fragmentation. Rising temperatures, changing rainfall patterns, and increased frequency of extreme weather events can result in the degradation or loss of critical habitats for various species in Kuwait. This can reduce the availability of suitable habitats for plants and animals, leading to declines in population numbers and local extinctions.

Shifts in Species Distribution

The CC can cause shifts in the geographic range and distribution of species. As temperatures rise, some species may be unable to tolerate the new conditions and may be forced to migrate or adapt to new habitats. Conversely, other species that are better suited to the changing climate may expand their range. These shifts in species distribution can disrupt ecosystems and lead to imbalances in species interactions.

Coral Reef Bleaching

Rising sea temperatures due to CC can trigger coral reef bleaching events. When corals experience prolonged exposure to high temperatures, they expel the symbiotic algae living in their tissues, causing the corals to turn white or pale. This bleaching weakens and can eventually kill the corals, leading to the loss of important reef ecosystems and the biodiversity they support.

Disruption of Ecological Interactions

The CC can disrupt ecological interactions between species. For example, changes in flowering and pollination timing can negatively affect plant-pollinator relationships. If the timing of flower blooming shifts, it may not align with the emergence of specific pollinators, leading to reduced pollination success and potential declines in plant populations.

Increased Vulnerability to Invasive Species

The CC can create more favorable conditions for invasive species to establish and thrive in Kuwait. Invasive species can outcompete native species for resources, disrupt ecological processes, and negatively impact local biodiversity. To mitigate the impact of CC on biodiversity, it is crucial to take proactive measures such as protecting and restoring critical habitats and creating connected ecological corridors to facilitate species movement and adaptation, implementing climate-smart conservation strategies that consider the anticipated impacts of CC on biodiversity, reducing GHGs emissions to mitigate the severity of CC and minimize its impact on ecosystems, conducting research and monitoring to better understand the specific vulnerabilities of Kuwait's biodiversity to CC, and promoting public awareness and education to foster a sense of stewardship and encourage actions that protect and conserve biodiversity. By addressing CC and implementing targeted conservation efforts, Kuwait can help mitigate the loss of biodiversity and ensure the long-term health and resilience of its ecosystems.

Decline in Renewable Water Resources and Use of Treated Waste Water

In 1953, the Kuwaiti government built a desalination plant to meet the growing population's water needs. The increasing population and water resource use would lead to water shortages, particularly in the groundwater wells, as most of the Kuwaiti wells contain salty water not suitable for drinking or optimum agriculture production. According to the Kuwait National Adaptation Plan 2019–2030 (EPA 2019), some of groundwater inflows from Saudi Arabia, approximately 20 Million Cubic Meter/year and the total renewable water resources per capita were 5139 m³/year in 2014 (cf. EPA 2019). Water resources play a crucial role in sustainable ecosystems and are vital for food security. Variation of fresh water consumption, desalination capacity, and population growth during 1990–2015 is shown in Fig. 3.23 (Mukhopadhyay and Akber 2018, cf. EPA 20).

The CC has significant impacts on renewable water resources, particularly in surface and groundwater resources, where inconsistency in rainfall, soil moisture, and shallow water levels can cause floods or droughts. The vulnerability of declining water resources is dependent on the increase in atmospheric temperature and water

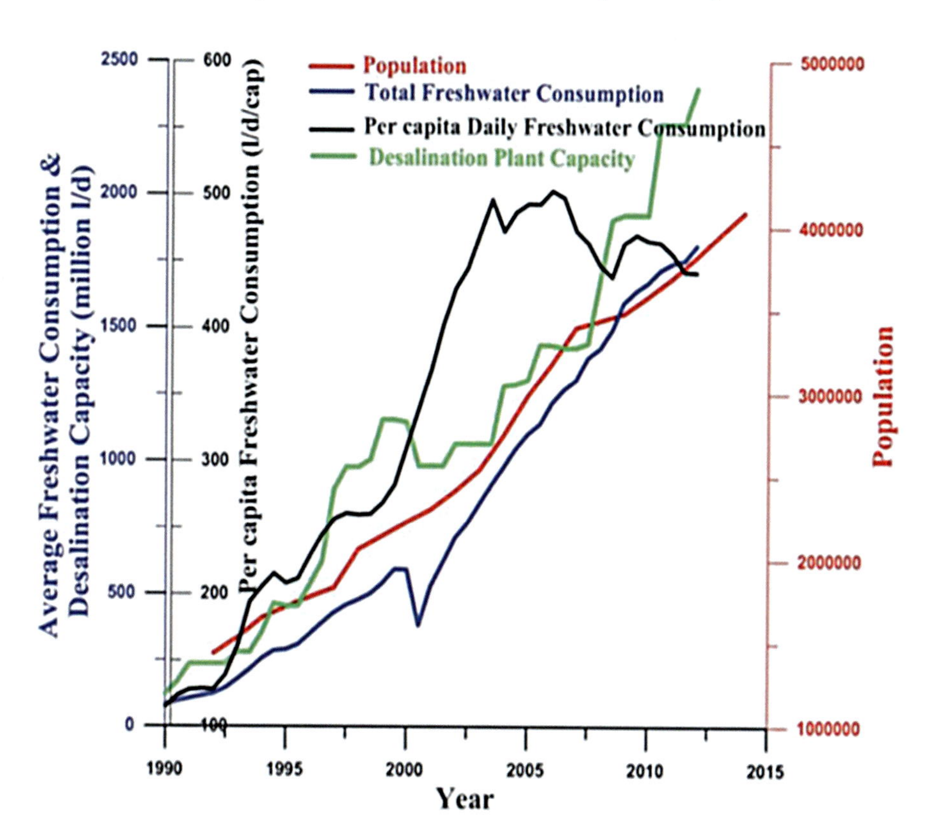

Fig. 3.23 Average fresh water consumption, desalination capacity and population growth. (Source: Mukhopadhyay and Akber 2018; cf. EPA 2019)

resource use, particularly in semiarid countries. Water resources decline will further increase with the reduction of rainfall and the increase in population. As the global temperature rises and rainfall declines, renewable water resources will become scarcer, leading to increased droughts and water resource use. Renewable water resources depend on four primary factors: water storage, water resources, water flow recharge, and water value. Weather is one of the most crucial factors for agriculture. Farmers increasingly rely on irregular rainfall patterns and renewable water resources. With declining water resources, farmers face real problems (Doll 2009).

Reduced precipitation and increased evaporation rates can lead to decreased water availability, particularly in regions that are already water-stressed. Droughts can have severe impacts on agriculture, ecosystems, and human populations, affecting food security, livelihoods, and overall water availability. Improving water resource management and governance to ensure efficient use and conservation of water is crucial.

Investment in infrastructure development should be made to store water, such as reservoirs and dams, to capture and store water during periods of high rainfall for use during dry periods. There is a need to promote water-saving practices and technologies, such as efficient irrigation systems and water recycling and reuse, enhancing watershed management and soil conservation practices to reduce runoff and promote groundwater recharge, and conducting research and monitoring to understand the local impacts of CC on water resources and develop targeted adaptation strategies. By integrating CC considerations into water resource planning and management, societies can enhance their resilience to water-related challenges and ensure the sustainable use of renewable water resources in the face of a changing climate. Finally, reusing treated sewage water and harvested rainwater in agriculture irrigation is an excellent strategy to enhance water sustainability and mitigate the impacts of CC. Reusing treated sewage water and rainwater reduces the demand for freshwater resources, thus conserving water which can be used for more productive activities.

By utilizing alternative water sources (brackish water, treated wastewater), farmers can decrease their reliance on traditional freshwater supplies, especially during periods of water scarcity or drought. Using recycled water for irrigation supports sustainable agricultural practices. It helps maintain crop productivity and reduces the stress on freshwater ecosystems, as well as minimizes the depletion of groundwater reserves. It contributes to more efficient water management, promoting long-term agricultural sustainability.

Treated waste water (TWW) contains valuable nutrients like nitrogen, phosphorus, and potassium, which can serve as fertilizers for crops. By utilizing TWW in irrigation, these nutrients can be recycled and returned to the soil, reducing the need for synthetic fertilizers. This approach supports a circular economy (CE) and reduces nutrient pollution in water bodies.

Reusing alternative water sources for irrigation helps mitigate the impacts of CC on water availability. As CC brings increased variability in rainfall patterns, utilizing TWW and rainwater can provide a more reliable water supply for agriculture, reducing vulnerability to droughts and water shortages. Reusing TWW and harvested rainwater can result in cost savings for farmers. Compared to the cost of freshwater sources, TWW is often more cost-effective, while utilizing rainwater reduces reliance on costly water supply infrastructure. This can contribute to the economic viability of agricultural operations, particularly in water-scarce regions.

Decline in Crops Yield and Impact on Food Security

Over the past decade, there has been an unusual shift in crop and plant growth patterns, which is a significant sign of CC. The growth rate of most trees in the world has changed due to the 2 °C increase in temperature since 1880. Changes in temperature during fall, winter, and spring have limited the availability of sunshine and water for plants.

Depending on the location, some countries have experienced a decline in crop yields due to increased temperatures, evaporation rates, and limited water supply, such as Kuwait. So, it is very important to insure the food sustainability which is important factor for human health and nutrition. Food sustainability is essential for human health and nutrition as it ensures access to nutritious food, protects the environment, mitigates CC, promotes food security, and offers economic benefits. By prioritizing sustainability in our food systems, we can work toward a healthier, more resilient, and equitable future for all. Conversely, some areas, such as Europe, have experienced an increase in crop yields due to rising temperatures. The CC also altered the ecological community of phytoplankton in aquatic habitats (Aragón et al. 2021).

The summer season has caused droughts in many countries around the world, making weather and climate information essential for farmers' decision-making. Providing forecasting information for farmers will help them plan their planting strategies better. Agriculture is highly susceptible to CC, with food sustainability being adversely or positively affected by it. Poor countries are particularly vulnerable to food shortages due to CC in both the short and extended period.

Limited water and rain lead to a reduction in the amount of water stored in the soil, affecting the amount of water transpired from the soil to plants. Soil temperature, which depends on air temperature and humidity around the plant, significantly affects crop yields. There is a strong relationship between rainfall, drought, and crop yields. The profitability of agriculture depends on the water requirement and weather temperature for crop yield. Water requirements not only affect agriculture but also cause serious problems for biodiversity and ecosystem disturbance.

3.7 Climate Change Adaptation Strategies for Kuwait

The CC adaptation is an essential solution to improve sustainability, and its objective is to reduce the opposing impact of climate alteration (Fig. 3.24). To encourage and improve adaptive capacity, laws need to be implemented (Aragón et al. 2021). Climate-smart agriculture (CSA) can enhance food security and productivity.

Therefore, it is crucial to adjust to CC in agriculture, and to do so, farmers have to make variations in land use or management.

Changes in land use may include using corps with higher thermal requirements in high-temperature countries or selecting different types of land that require less maintenance to enhance productivity. Changes in corps locations may also be necessary depending on the land capacity. Changes in management involve using smart technologies to control soil erosion and drainage, such as adding nanoclay in soil to hold water and nutrients (Farrar et al. 2018). The strategy should also involve controlling pests and diseases, changing farm infrastructure, and altering crop and livestock growing methods.

While agriculture has the ability to adjust to small changes in weather within economic and technological constraints, adapting to significant CC is challenging. Farmers can, therefore, choose alternative ways to enhance food productivity, such as changing irrigation requirements, using greenhouses or using different fertilization substances. There are a few examples of the alternative ways that farmers can adopt to enhance food productivity and adapt to CC. It is crucial to support and promote these sustainable farming practices through research, policy, and investments to ensure a resilient and productive agricultural sector in the face of a changing climate. In addition, healthy soils are prerequisite for sustainable food security (Shahid 2015). Here are a few potential examples:

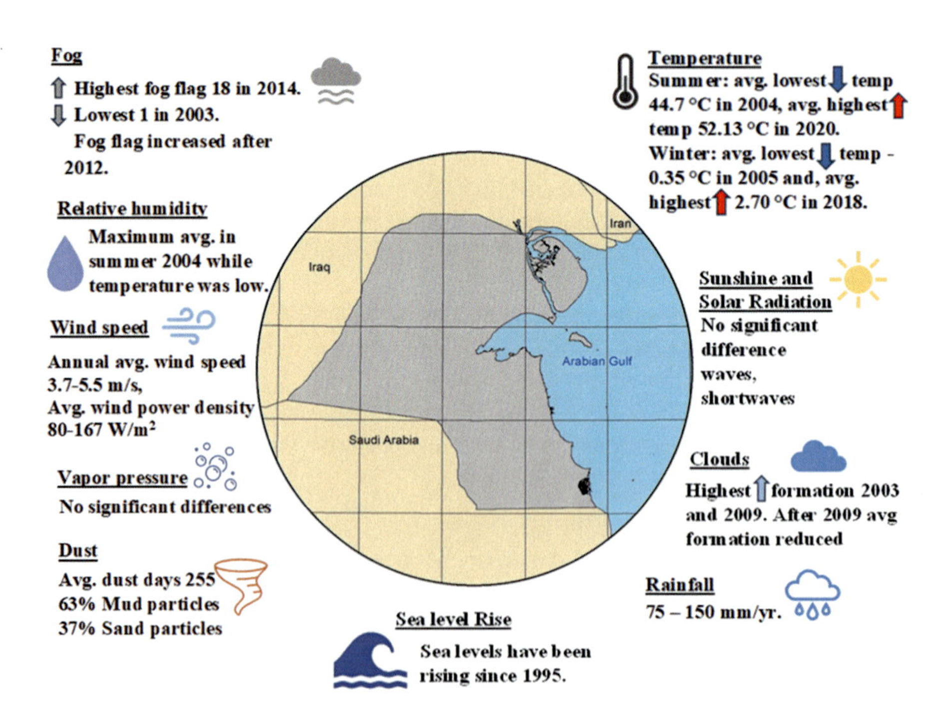

Fig. 3.24 Climate change aspects in Kuwait over the past 20 years

3.7.1 *Efficient Irrigation*

As water resources become scarcer due to changing rainfall patterns, farmers can implement more efficient irrigation practices. This includes using crops based modern irrigation systems (drip), which delivers water directly to the roots of plants, minimizing water loss through evaporation and less effects of water salinity in the plants roots (Zaman et al. 2018). Additionally, precision irrigation technologies and soil moisture sensors can help farmers optimize water usage by providing real-time data on soil moisture levels. In addition, the real-time dynamic salinity logging system established in the plants root zones can guide when to use extra water to manage root zone salinity to a safe level (Shahid et al. 2008) or using alternative crops with high salt-tolerance levels (Rao et al. 2017).

3.7.2 *Protected Cultivation*

Greenhouses and other protected cultivation methods provide controlled environments for plant growth, shielding crops from extreme weather events and temperature fluctuations (FAO 2021). These structures allow farmers to manipulate factors like temperature, humidity, and light, creating optimal conditions for crop production. This approach can help mitigate the risks associated with unpredictable weather patterns.

3.7.3 *Crop Diversification*

Farmers can adapt to changing climate conditions by diversifying their crop portfolios. Planting a variety of crops that are resilient to different environmental conditions helps mitigate the risks of crop failure. By selecting crop varieties that are better adapted to changing temperature and precipitation patterns, farmers can enhance their overall productivity and reduce vulnerability to CC using climate-smart agriculture practices (FAO 2017).

3.7.4 *Agroforestry and Conservation Agriculture*

Agroforestry involves integrating trees with crops or livestock, providing additional environmental benefits such as soil conservation, biodiversity preservation, and microclimate regulation. Conservation agriculture practices, such as minimal tillage and cover cropping, help improve soil health, retain moisture, and reduce erosion. These techniques contribute to more sustainable and climate-resilient farming systems.

3.7.5 *Improved Fertilization*

Farmers can explore alternative fertilization substances that have a lower environmental impact. For example, organic fertilizers derived from natural sources, such as compost or manure, can replace synthetic fertilizers, reducing GHGs emissions and nutrient runoff. Precision nutrient management techniques, such as site-specific fertilizer application based on soil testing, can optimize nutrient use efficiency and minimize environmental pollution. The most important in nutrients management and uses is 4R nutrients stewardship (Shahid 2018), that is, the use of fertilizers at right time, type, place, and quantity.

3.7.6 *Genetic Adaptation*

Plant breeding programs can focus on developing crop varieties with traits that are better suited to changing climatic conditions. This includes traits like drought and salt tolerance, heat resistance, disease resistance, and increased nutrient-use efficiency. By incorporating these genetic traits into crops, farmers can enhance their resilience and productivity in the face of CC.

3.8 National Adaptation Plan

The Environment Public Authority (EPA 2019) of Kuwait has developed The Kuwait National Adaptation Plan as part of its promise to preserve the local and national environment. The plan consists of medium and long-term strategies aimed at enhancing resilience to climate experiments and increasing national capability to adjust to CC. It also provides a detailed overview of the state of the environment, the sectors most affected by CC, and a CC risk assessment. To determine the socioeconomic impacts of CC risks, an index of sector vulnerability was developed using universally accepted scientific methodologies. The key sectors were identified along with their respective material risks, and stakeholders were assigned accordingly. The Government of Kuwait, through the EPA, has pledged to implement the NAP in collaboration with legislative and nonlegislative entities. A national organization device has been established to execute various initiatives aimed at safeguarding critical sectors against the adverse effects of climate alteration. Additionally, upcoming programs have been designed to address the threats faced by different sectors over time.

3.9 Strategies to Mitigate Climate Change

Although many international agreements for mitigation have not succeeded so far, however, the importance of mitigation efforts has increased to achieve the goal of stabilization. The increase in climate variability is single expected significances of CC, resulting in cold years, flood events, seasonal droughts, storms, dust storms, extreme wind speeds, and an increase in atmospheric temperature, among others. In addition, changes in food productivity must be considered, such as shifting the timing of operations such as fertilizing, controlling pests and weeds, plowing, sowing, harvesting, and more.

Countries like Kuwait need to improve strategies designed to mitigate the adverse paraphernalia of CC. This means altering existing plans to synchronize with the new set of climatic circumstances. The use of smart, effective technology such as greenhouses that provide plants with suitable climates can help farmers increase food production (FAO 2021). One of the most important mitigation alternatives is to shift from using ancient fuel energy to renewable energy, such as solar energy.

Ensuring food sustainability is indeed a crucial factor for human health and nutrition. Here are a few reasons why it is important:

3.9.1 Access to Nutritious Food

Food sustainability focuses on producing and distributing food in a way that meets the nutritional needs of the population. It aims to ensure that people have access to a diverse range of safe and nutritious food options, which is essential for maintaining good health and preventing malnutrition.

3.9.2 Environmental Conservation

Sustainable food practices take into account the long-term health of the planet. By adopting sustainable farming methods, such as organic farming, agroforestry, and crop rotation, we can minimize the use of harmful chemicals, reduce soil degradation, conserve water resources, and protect biodiversity. These efforts contribute to maintaining a healthy ecosystem, which is crucial for food production in the long run.

3.9.3 Climate Change Mitigation

Agriculture is a significant contributor to GHGs emissions, which are responsible for CC (Shahid 2010). Sustainable food systems aim to reduce the carbon footprint of agriculture by promoting practices that minimize emissions, such as efficient fertilizer use, improved waste management, and the adoption of renewable energy sources. By mitigating CC, we can protect food production systems from the adverse impacts of extreme weather events and ensure a stable food supply for future generations.

3.9.4 Food Security

Food sustainability plays a vital role in achieving food security, which means ensuring that all individuals have access to sufficient, safe, and nutritious food at all times. Rational and sustainable farming helps to cool the planet by the GHGs emission and storing carbon in soil and vegetation. Indeed, climate change in global food security is inextricably linked (Scherr and Sthapit 2009). Agriculture is 7% of GHG problem but represents 20% of the solution (IFA 2010). By promoting sustainable agriculture and reducing food waste, we can enhance the resilience of food systems, improve productivity, and create a more equitable distribution of food resources. This helps to alleviate hunger, poverty, and social inequalities.

3.9.5 Economic Benefits

Sustainable food practices can also bring economic advantages. By investing in sustainable agriculture and supporting local food systems, we can create jobs, boost rural development, and foster a thriving agricultural sector. Moreover, sustainable farming practices often result in increased productivity, cost savings, and improved market opportunities for farmers, contributing to a more sustainable and prosperous economy.

In mitigation efforts, all of these events must be taken into account and careful considerations given to minimizing their extreme effects. The CC will continue to impact sectors and regions, but efforts must be made to reach a point of sustainability. One of the most important actions is to reduce GHGs emissions through law enforcement for global climate policy. In addition, education is essential to increase alertness about the significance of the Earth and sustainability and reduce bad habits that increase GHGs emissions. Other strategies to mitigate CC include reducing disaster risks, preparing solutions for unavoidable adverse impacts, increasing food production and fresh water supply and working in research laboratories to discover new ways to mitigate CC, such as forming clouds and ice crystals. China, for example, has created artificial clouds full of dry ice or silver iodine for agriculture areas. In summary, the adversarial paraphernalia of CC can be mitigated through a combination of strategies that include education, technology, law enforcement, and research. By taking a holistic approach, countries can work toward sustainability and mitigate the extreme effects of CC.

3.10 Mitigating Climate Change: Case of Kuwait

Kuwait has the opportunity to take essential steps toward mitigating the impacts of CC and fostering a sustainable future. Here are some key actions that can be considered in Kuwait.

3.10.1 Renewable Energy Transition

Investing in renewable energy sources, such as solar and wind power, can help Kuwait reduce its reliance on fossil fuels and decrease GHGs emissions. Expanding the use of renewable energy in electricity generation, transportation, and other sectors can contribute to a more sustainable and low-carbon economy.

3.10.2 Energy Efficiency Measures

Promoting energy efficiency across various sectors, including buildings, transportation, and industry, can significantly reduce energy consumption and associated GHGs emissions. Implementing energy-efficient technologies, building standards, and transportation systems can lead to cost savings and environmental benefits.

3.10.3 Sustainable Transport

Encouraging the use of public transportation, promoting electric and hybrid vehicles, and developing infrastructure for cycling and walking can help reduce emissions from the transportation sector. Investing in efficient and sustainable transport systems can improve air quality, reduce congestion, and decrease reliance on fossil fuels.

3.10.4 *Green Building and Infrastructure*

Implementing green building practices, including energy-efficient designs, renewable energy integration, and sustainable materials, can reduce energy consumption and environmental impact. Developing green infrastructure, such as green roofs and urban parks, can contribute to climate resilience and enhance the quality of urban environments.

3.10.5 *Conservation and Restoration*

Protecting and conserving natural habitats and biodiversity is crucial for long-term sustainability. Kuwait can invest in initiatives that preserve and restore ecosystems, including wetlands, coastal areas, and marine habitats. Supporting research and conservation efforts can help safeguard the unique flora and fauna of the region.

3.10.6 *Climate Change Adaptation*

Kuwait should prioritize CC adaptation measures to minimize the impacts on vulnerable sectors such as agriculture, water resources, and infrastructure. This may involve implementing innovative irrigation techniques, enhancing water management practices, and designing climate-resilient infrastructure.

3.10.7 *Research and Development*

Investing in research endeavors focused on CC, renewable energy, sustainable agriculture, and environmental conservation can generate knowledge and technologies to address local challenges. Collaborating with regional and international partners can foster innovation and position Kuwait as a leading player in sustainability efforts.

3.10.8 *Awareness and Education*

Promoting public awareness and education on CC, sustainability, and the importance of conservation can empower individuals to make informed choices and actively participate in sustainable practices.

By taking these essential steps, Kuwait can play a significant role in mitigating CC, minimizing GHGs emissions, and ensuring a sustainable future for its flora, fauna, and agriculture systems. Such actions not only benefit the environment but also contribute to Kuwait's social and economic development in the long run.

3.11 Conclusions and Recommendations

Predicting future climate changes can be challenging, but we must follow the recent IPCC assessment. The climate change and population growth can impact global food production, leading to negative effects on food security. Addressing these impacts will require significant time and resources. Although agriculture requires some CO_2 for photosynthesis, excessive amounts can contribute to climate change and temperature increases, resulting in reduced food productivity in major exporting regions and financial losses. The most at risk areas to climate change are the semiarid temperate and subtropical areas, such as Arabia, the Maghreb, western West Africa, Horn of Africa, Southern Africa, and eastern Brazil, as well as the steamy temperate and equatorial areas, such as southeast Asia and Central America. Warming weather in these areas can cause water loss and drought, negatively impacting agriculture. Kuwait climate is a hyper-arid desert that is extremely variable, with poor-quality surface sediments that lack nutrients and organic matter. As a result, Kuwait relies heavily on food imports, with only a small fraction of food demand met by local agriculture. Despite this, Kuwait is rich in terrestrial and marine biodiversity, and climate change has resulted in changes to marine biodiversity, such as coral reef whitening. Kuwait is also extremely at risk to climate change due to its fragile conditions. As a part of the UN Decade on Ecosystem Restoration (2021–2030), Kuwait needs to adopt internationally recognized restoration practices to prevent, halt, and reverse ecosystem degradation. This effort can help to combat climate change, prevent mass extinction, and alleviate poverty, but it requires everyone to play a part.

To achieve ecological restoration, the following prerequisites must be considered:

- A strong framework to design restoration projects based on the desired goals
- Successful implementation of the project scope
- Protection of revegetated sites from external effects such as grazing and camping
- Periodic performance measurements during and after the project to monitor progress
- Replacement of dead plants to maintain plant populations
- Evaluation of the benefits of the rehabilitation project in a wider context, including its link to climate change mitigation efforts such as carbon credits
- Maintenance of post-project rehabilitation sites to ensure their continued success.

Acknowledgments I would like to express my special thanks to Miss Heba Jaber Kamal from GIS section, Science and Technology Sector of Kuwait Institute for Scientific Research for her assistance in the preparation of maps.

References

Amr ZS (2021) The state of biodiversity in Kuwait. IUCN; the State of Kuwait, Kuwait: Environmental Public Authority, Gland

Al-Dousari A (2021) Atlas of fallen dust in Kuwait. Springer Nature Switzerland AG, Cham

Al-Dousari A, Al-Awadhi J (2012) Dust fallout in northern Kuwait, major source and characteristics. Kuwait J Sci Eng 39(2):171–187

Al-Nassar W, Alhajraf S, Al-Enezi A, Al-Awadhi J (2005) Potential wind power generation in the State of Kuwait. Renew Energy 30:2149–2161

Aragón FM, Oteiza F, Rud JP (2021) Climate change and agriculture: subsistence farmers' response to extreme heat. Am Econ J Econ Policy 13(1):1–35

Armbrust DA, Retta A (2000) Wind and sandblast damage to growing vegetation. Ann Arid Zone 39(3):273–284

Arrhenius S (1896) On the influence of carbonic acid in the air upon the temperature of the ground. Philos Mag J Sci Ser 5(41):237–276

Church JA, White NJ (2011) A 20th century acceleration in global sea-level rise. Geophys Res Lett 33:L01602. https://doi.org/10.1029/2005GL024826. (Source: https://psmsl.org/products/reconstructions/church.php)

Doll P (2009) Vulnerability to the impact of climate change on renewable ground water resources: a global-scale assessment. Environ Res Lett 4(3):035006. https://doi.org/10.1088/1748-9326/4/3/035006

Dübal H, Vahrenholt F (2021) Radiative energy flux variation from 2001–2020. Atmos 12(10):1297. https://doi.org/10.3390/atmos12101297

EPA (2019) Kuwait National Adaptation Plan 2019–2030. Enhanced climate resilience to improve community livelihood and achieve. Environment Public Authority, Kuwait. https://unfccc.int/sites/default/files/resource/Kuwait-NAP-2019-2030.pdf

FAO (2017) Climate smart agriculture sourcebook. Overview-significant development, summary, pp. V – 2nd edn. ISBN 978-92-5-109988-9

FAO (2021) Protected agriculture in Kuwait. In: Unlocking the potential of protected agriculture in the GCC countries: saving water and improving nutrition. FAO, Cairo, p 216. Licence: CC BY-NC-SA 3.0 IGO.pp. 68–72

Farrar M, Gill S, Shahid SA, Babiker K (2018) Water saving innovation in urban landscaping and irrigated agriculture by using AustraBlend multimineral root zone conditioner. Asian J Sci Technol 9(2):9092–9100

Grace J (1988) Plant response to wind. Agric Ecosyst Environ 22(23):71–88. https://doi.org/10.1016/0167-8809(88)90008-4

Hardy JT (2003) Climate change causes, effects, and solutions. Wiley, England. Atmospheric sciences, ISBN: 978-0-470-85019-0, 272 pp

IFA (2010) Agricultural community addresses recommendation to negotiating governments at COP15. Fertil Agric (February 2010):8

Martial H, Jean-Charles D, Neda B, Darrel B, Laurent G, Greg R, Thierry E (2013) A comparative study of radiation fog and quasi-fog formation processes during the Paris fog field experiment 2007. Pure Appl Geophys 170(12):2283–2303

McTainsh GH, Pitblado JR (1987) Dust storms and related phenomena measured from meteorological records in Australia. Earth Surf Process Landf 12(4):415–424

Meteoblue Weather Princeton. https://www.meteoblue.com/en/weather/week/princeton_united-states_5102922
Meteoblue Weather Organization. https://www.meteoblue.com/en/weather/week/kuwait-city_kuwait_285787
Mimura N (2013) Sea-level rise caused by climate change and its implications for society. Proc Jpn Acad, Ser B, Phys Biol Ser 89(7):281–301. https://doi.org/10.2183/pjab.89.281
Mukhopadhyay AA (2018) Sustainable water management in Kuwait: current situation and possible correlation measures, Water Research Center, Kuwait Institute for Scientific Research, Kuwait. Int J Sustain Dev Plan 13(3):425–435
NASA global climate change. https://climate.nasa.gov/global-warming-vs-climate-change/
Parry ML (1990) Climate change and world agriculture, vol 14. Routledge, London. https://doi.org/10.4324/9780429345104
Pope M, Zogaris S (eds) (2012) Birds of Kuwait: a comprehensive visual guide. KUFPEC (Kuwait Foreign Petroleum Exploration Co. K.S.C.), Nicosia
Rao NK, McCann I, Shahid SA, Butt K, Al Araj B, Smail I (2017) Sustainable use of salt-degraded and abandoned farms for forage production using halophytic grasses. Crop Pasture Sci 2017(68):483–492. https://doi.org/10.1071/CP16197
Scherr SJ, Sthapit S (2009) State of the world 2009: into a warming world. The Worldwatch Institute, Washington
Shahid SA (2010) Climate change-its impacts and biosaline agriculture. Farming Outlook 9(2):24–28
Shahid SA (2015) The need for healthy soils for sustainable food security: where do we stand? Farming Outlook 14(3):3–9
Shahid SA (2018) Salt-affected soils—4 R nutrient stewardship. Farming Outlook 17(2):19–22
Shahid SA, Dakheel A, Mufti KA, Shabbir G (2008) Automated in-situ soil salinity logging in irrigated agriculture. Eur J Sci Res 26(2):288–297
The World Bank. https://data.worldbank.org
The Worldometers. https://www.worldometers.info/co2-emissions/kuwait-co2-emissions
Tolba MK, Saab NW (2009) Arab environment climate change – impact on climate change on Arab countries. Report of the Arab Forum for Environment and Development (AFED), technical
Vicente-Serrano SM, Nieto R, Gimeno L, Azorin-Molina C, Drumond A, El Kenawy A, Dominguez-Castro F, Tomas-Burguera M, Peña-Gallardo M (2018) Recent changes of relative humidity: regional connections with land and ocean processes. Earth Syst Dynam 9:915–937. http://arizonaenergy.org/News_17/News_Sep17/ClimateChangeGlobalSeaLevel.html
World Meteorology Organization. https://community.wmo.int/climate-data-homogenization
World Meteorological Organization. https://public.wmo.int/en/media/press-release/no-976-2001-2010-decade-of-climate-extremes
Zaman M, Shahid SA, Heng L (2018) Irrigation systems and zones of salinity development. In: Zaman M, Shahid SA, Heng L (eds) Guidelines for salinity assessment, mitigation and adaptation using nuclear and related techniques. Springer, pp 91–111

Chapter 4
Ecological Footprint and Biocapacity of Kuwait and Proposed Eco-resources Management Strategies: A Review

Shabbir Ahmad Shahid and Majda Khalil Suleiman

Abstract Kuwait is a desert and hyper-arid country with low rainfall, high temperature, and scarcity of arable land and water resources. Kuwait produces food locally to a limited extent, and imports food to meet population demand. Thus, the demand on nature "Ecological Footprint-EF" is higher than the "Biocapacity (BC)." On the demand side, the EF tracks the use of six categories of productive land areas, that is, cropland, grazing area, fishing grounds, built-up land, forest area and carbon demand on land, and same categories are measured from supply point of view to represent BC except carbon footprint. The review of EF and BC of Kuwait revealed its average EF in 1999 was 2.89 global hectares (gha)/capita/year against BC of 0.79 gha/capita/year (a deficit of 2.1 gha/capita/year). Over the past two decades (1999–2018), the EF significantly fluctuated and reached to 7.9 gha/capita/year (2018) against BC of 0.5 gha/capita/year (2018) with a deficit of 7.4 gha/capita/year. Kuwait is eco-plus in fishing grounds and produces sufficient fish to meet the local demand and has the surplus. In terms of built-up area, Kuwait meets the population demand adequately. All Gulf Cooperation Council (GCC) countries are eco-resources deficit to various levels, and the deficit is met through the import of food and fodder products. In this chapter the EF and BC of Kuwait, Gulf Cooperation Council countries, and the world are presented, compared, and discussed, and various options are proposed to offset the deficit in eco-resources through various ways.

Keywords Kuwait · Eco-resources · Global hectare · Cropland · Grazing land · Fishing grounds · Build up areas · Forests · Carbon footprint

S. A. Shahid (✉) · M. K. Suleiman
Desert Agriculture and Ecosystems Program, Environment and Life Sciences Research Center, Kuwait Institute for Scientific Research, Safat, Kuwait
e-mail: sshahid@kisr.edu.kw; mkhalil@kisr.edu.kw

M. K. Suleiman, S. A. Shahid (eds.), *Terrestrial Environment and Ecosystems of Kuwait*, https://doi.org/10.1007/978-3-031-46262-7_4

4.1 Introduction

Kuwait is a desert and hyper-arid country with low rainfall and high temperature. The soils are sandy and infertile due mainly to poor organic matter and clay contents (Shahid and Omar 1999; Omar and Shahid 2013). All these factors lead Kuwait to depend on other countries to import food to meet population demand, in addition to local food production mainly the vegetables and forage production. Kuwait is capital rich but food-insecure country within national boundaries. Therefore, the demand of Kuwait on nature "ecological footprint-EF" is higher than the "Biocapacity (BC)." In this chapter efforts have been made to report the EF and BC of Kuwait and reviewed the past trend of fluctuation since 1961. The EF and BC are measured in universal unit "global hectare-gha." On the demand side, the EF tracks the use of six categories of productive land areas including cropland, grazing land, fishing grounds, built-up land, forest area, and carbon demand on land; on the supply side, the BC represents the productivity of ecological resources except carbon demand (GFN 2020).

The footprint accounts are based on international data sources including FAOSTAT, UN Comtrade, International Energy Agency (IEA) (Kitzes et al. 2009). These data platforms receive the data from national statistical offices that are responsible for the accuracy of provided data. The Global Footprint Network works with national governments around the globe and invites to review the data in its National Footprint Accounts for accuracy and completeness, to assure the data is valid and reliable. This verified data is then used by government for various purposes. The review of EF and BC of Kuwait, Gulf Cooperation Council (GCC) countries and the World has revealed that the human demand on nature (EF) is higher than the production capacity (BC) and the overuse of resources may lead to degrade the resources and decrease productive capacity. A review of EF and BC of GCC countries was published by Shahid and Ahmed (2014) showing a general deficit of BC in all GCC countries. In addition, the paper presents various options to increase local agriculture production for food security and recommendations made to develop agreement with other countries to lease land for crop production in a *win-win* situation scenario.

This review of EF and BC of Kuwait revealed its average EF in 1999 was 2.89 gha/capita/year against BC of 0.79 gha/capita/year (a deficit of 2.1 gha/capita/year). Over the past two decades (1999–2018), the EF significantly fluctuated and reached to 7.9 gha/capita/year (2018) against BC of 0.5 gha/capita/year (2018) with a deficit of 7.4 gha/capita/year. The deficit is significant, and within national resources and harsh climatic conditions, it is not possible to fill the gap; therefore, this difference has to be met, for example, in the case of crops and fodders through import from countries where BC is more than EF. In this chapter the EF and BC of Kuwait are presented and discussed, and various options are proposed to offset the deficit through various ways. A brief comparison of EF and BC of Kuwait, GCC countries, and the World is also presented. The authors took the permission from the

GFN to extract the relevant information for this chapter and credit the source in acknowledgement and citations where appropriate.

It is to be noted that, in this chapter, we depended mainly on the data reported by the Global Footprint Network's National Footprint and Biocapacity Accounts, which are all based on UN-Statistics (FAOSTATs) and we have not gone through the underlying metrics behind these statistics to investigate to what extent accurately reflect the situation of Kuwait. Further investigation of the underlying metrics is beyond the scope of this chapter; however, such an investigation requires significant resources and time to complete such an important task.

The overall objectives of this chapter are:

- Report EF and BC of Kuwait for the last two decades (1999–2018).
- Compare average national EF and BC (gha/capita/year) with GCC countries and the world average.
- Identify main concerns of hiking EF causing significant deficit in natural capital (EF-BC).
- Identify national efforts to reduce the gap between EF and BC.

4.2 Ecological Footprint and Biocapacity of Kuwait

The accounting of the nature capital can be considered in two sides, the demand side, that humans place on bioproductive areas (EF), and, the supply side, biocapacity (BC), the nature's availability to provide the resources and ecosystem services that are annually consumed by humans (Kitzes et al. 2009). The measurement unit of EF and BC is global hectares (gha), which is the common unit to make the results comparable in the global context.

The data about the EF and BC of Kuwait, GCC countries, and the world is downloaded from the National Footprint Accounts 2020 edition, provided by the York University Footprint Initiative and Global Footprint Network (https://data.footprintnetwork.org). The data is downloaded as Excel files for further analyses and presentations in the form of figures and tables in this chapter. More specifically these two important terms are defined below.

4.2.1 Ecological Footprint (EF)

The EF calculated by Global Footprint Network (Wackernagel et al. 2002; Ewing et al. 2010) is an accounting system tracking the amount of biologically productive land and water areas that are required by a country or the world to produce the natural resources it consumes and to absorb the emissions it generates, using prevailing technology and management strategies. In general, these assessments of EF in national reporting system may not be located but could be located elsewhere and imported in the form of countries resources (Mancini et al. 2016; Wackernagel and Beyer 2019; EEA 2020).

4.2.2 Biocapacity (BC)

The BC is the amount of productive area that is available to generate these resources and to absorb the waste (Ewing et al. 2010). Simply the BC represents the productivity of its ecological assets (GFN 2020). The aggregated human demand (EF) and nature's supply (BC) can be used to calculate resources deficit at the national and global scales.

4.3 Description of Ecological Footprint Measurement Categories

There are six categories measured in order to quantify the EF (demand) and BC (supply side). The sum of these demands is called people's EF and the ability of ecosystems to renew biomass is called BC. Both BC and EF can be tracked and compared against each other, based on two simple principles: (1) one can add up all the competing demands on productive surfaces, i.e., the surfaces that contain the planet's Biocapacity (Table 4.1, GFN 2020); (2) by scaling these areas proportional to their biological productivity, they become commensurable (Keßler et al. 2020). According to the glossary of GFN (GFN 2020) the five area types for BC supporting the six EF demand types are the following.

To learn more into the methodologies for the calculation of EF and BC at national and global scales, the reader is referred to Ewing et al. (2010). The inclusion of detailed calculation procedures is beyond the scope of this chapter. However, the limitations in the calculations of EF are listed below.

4.4 Limitations of Ecological Footprint Calculation

In addition to the limitations and uncertainties of EF and BC measurement methodologies used in national footprint accounts (GFN 2020), the European Environment Agency (EEA 2020) took the initiative and described the limitations of EF calculation (Keßler et al. 2020), which the national systems should consider to review their policies to improve BC against EF. These limitations are listed and briefly given below with slight modifications with respect to Kuwait conditions, more details can be seen elsewhere (EEA 2020).

Table 4.1 Six categories defining ecological footprint and biocapacity (GFN 2020)

Cropland	Forests	Grazing land
Cropland is the most bioproductive of all the land-use types and consists of areas used to produce food and fiber for human consumption, feed for livestock, oil crops, and rubber. Due to lack of globally consistent data sets, current cropland footprint (CLP) calculations do not yet take into account the extent to which farming techniques or unsustainable agricultural practices may cause long-term degradation of soil. The CLP includes crop products allocated to livestock and aquaculture feed mixes, and those used for fibers and materials.	Provide two services: the forest product footprint (FPF), which is calculated based on the amount of lumber, pulp, timber products, and fuel wood consumed by a country on a yearly basis. It also accommodates the carbon footprint (CF), which represents the carbon dioxide emissions from burning fossil fuels. It is represented by the area necessary to sequester these carbon emissions. The CF component of the EF is calculated as the amount of forest land needed to absorb these carbon dioxide emissions.	Grazing land is used to raise livestock for meat, dairy, leather, and wool products. The grazing land footprint (GLF) is calculated by comparing the amount of livestock feed available in a country with the amount of feed required for all livestock in that year, with the remainder of feed demand assumed to come from grazing land.
Fishing grounds	**Built-up land**	**Carbon demand**
The fishing grounds footprint (FGF) is calculated based on the estimates of the maximum sustainable catch for a variety of fish species. These sustainable catch estimates are converted into an equivalent mass of primary production based on the various species' levels. This estimate of maximum harvestable primary production is then divided amongst the continental shelf areas of the world. Fish caught and used in aquaculture feed mixes are included.	The built-up land footprint (BLF) is calculated based on the area of land covered by human infrastructure – transportation, housing, industrial structures, and reservoirs for hydropower. Built-up land may occupy what would previously have been cropland.	The carbon footprint (CF) component is calculated as the amount of forest land needed to absorb these carbon dioxide emissions. Currently, the CF is the largest portion of humanity's footprint. The CF is one part of a full EF analysis where the greenhouse gas (GHG) emissions are translated into global hectares necessary to absorb these emissions (Kitzes et al. 2009). The CF also includes embodied carbon in imported goods.

4.4.1 Non-ecological Aspects of Sustainability

It depicts that even both the EF and BC are in balance, poor management of resources may lead to depletion of resources and ecosystem services.

4.4.2 Depletion of Nonrenewable Resources

The national footprint accounts do not track the amount of nonrenewable resource stocks, such as oil, natural gas, coal, or metal deposits. The footprint associated with these materials is based on the regenerative capacity used or compromised by their extraction and, in the case of fossil fuels, the area required to assimilate the wastes they generate, such as forest. In the case of Kuwait, the benefits of petroleum and gas products are not accounted in terms of BC. This is discussed briefly in Sect. 4.4.5.

4.4.3 Inherently Unsustainable Activities

Some industries release gasses and pollutants in the soil environment, such as the use of sewage sludge in agriculture fields add heavy metals in soil. And the release of radioactive materials and persistent synthetic compounds are not used in footprint calculation. These activities do not show BC for using them. However, the loss of BC due to these pollutants can be witnessed.

4.4.4 Ecological Degradation

The measurement of ecological degradation is another area to be seriously considered in footprint accounts. In most of the water-scarce countries where fresh water is not sufficient to irrigate the crops, the farmers are using marginal quality groundwater high in salinity to irrigate crops, which is impacting farm lands' present productivity, and reduce future productivity too. Currently world is losing 2000 ha farmland daily due to salinity. This loss should be quantified in BC measurement, although, such detailed studies are difficult to accomplish at the national levels.

4.4.5 Resilience of Ecosystems

The vulnerability or resilience of ecosystem is not identified in footprint accounts. Therefore, the footprint is merely an outcome measure documenting how much of the biosphere is being used compared with how productive it is.

The authors strongly argued that the oil and natural gas production is considered as pressure on the environment through GHGs emission, although petroleum products are used in many industries to produce many goods used as resource or perhaps considered as BC. This area is in its infancy as how these resources can be considered as a part of BC, rather than considering in EF. Currently these resources are considered under EF, thus making the gap between EF and BC wider for oil producing countries, although these countries earned significant financial resources through export.

4.5 Assumptions on Which EF and BC Measurement is Based

The following information is summarized with slight modifications from Wackernagel et al. (2002) who have set up six fundamental assumptions to account EF and BC of the globe or at the national level:

- The most of the resources consumed by human and activities and the wastes can be tracked.
- Most of these resource and waste flows can be quantified in terms of the biologically productive area necessary to maintain them. Resource and waste flows that cannot be measured in terms of biologically productive area are excluded from the assessment, leading to a systematic underestimate of the total demand these flows place on ecosystems.
- By scaling each area in proportion to its bioproductivity, different types of areas can be converted into the common unit of average bioproductivity (gha). This unit is used to express both EF and BC.
- Because a gha of demand represents a particular use that excludes any other use tracked by the footprint and all global hectares in any single year represent the same amount of bioproductivity, they can be summed. Together, they represent the aggregate demand or EF. In the same way, each hectare of productive area can be scaled according to its bioproductivity and then added up to calculate BC.
- As both EF and BC are expressed in gha, human demand as measured by EF accounts can be directly compared to global, regional, national, or local BC.
- Area demanded can exceed the area available. If demand on a particular ecosystem exceeds that ecosystem's regenerative capacity, the ecological assets are being diminished. For example, people can temporarily demand resources from forests or fisheries faster than they can be renewed, but the consequences are smaller stocks in that ecosystem. When the human demand exceeds available BC, this is referred to as overshoot.

4.6 Two Decades Biocapacity and Ecological Footprint of Kuwait (Gha/Capita/Year)

Table 4.2 presents the trend of total demand of Kuwait on nature (EF) and bioproductivity (BC) resources over 1999–2018 period. The total EF of Kuwait increased from 5.63×10^6 gha (1999) to 32.68×10^6 gha/capita/year in 2018, a manifold increase in 2018 compared to 1999. Conversely, the BC was increased to a limited extent from 1.55×10^6 gha (1999) to 2.07×10^6 gha (2018).

Table 4.2 Two decades (1999–2018) dynamics of land uses, total Biocapacity (BC) and Ecological Footprint (EF) of Kuwait

Year	Records	Built-up land	Carbon	Cropland	Fishing grounds	Forest products	Grazing land	Total
		Global hectares						
1999	BC	263,637	0	38,106	1,215,170	8976	22,888	1,548,777
1999	EF	263,637	2,564,926	1,539,233	123,359	359,396	785,463	5,636,015
2000	BC	198,288	0	41,025	1,217,083	9232	22,924	1,488,553
2000	EF	198,288	4,783,077	1,565,270	151,708	363,087	566,558	7,627,989
2001	BC	160,150	0	40,273	1,217,635	9494	22,935	1,450,486
2001	EF	160,150	8,165,482	1,478,775	151,118	351,601	563,938	10,871,064
2002	BC	181,663	0	44,962	1,218,270	9806	22,947	1,477,648
2002	EF	181,663	10,191,608	1,328,357	122,857	359,692	678,398	12,862,576
2003	BC	223,706	0	54,737	1,213,213	10,048	22,851	1,524,555
2003	EF	223,706	11,178,612	1,451,632	153,285	380,345	884,943	14,272,523
2004	BC	233,012	0	52,273	1,210,697	10,259	22,804	1,529,045
2004	EF	233,012	14,380,422	1,626,567	153,818	578,780	839,112	17,811,711
2005	BC	263,893	0	57,383	1,206,813	10,508	22,731	1,561,328
2005	EF	263,893	18,608,696	1,923,760	153,030	592,505	998,272	22,540,156
2006	BC	326,527	0	67,908	1,203,687	10,785	22,672	1,631,579
2006	EF	326,527	19,163,744	2,177,756	185,732	598,935	952,693	23,405,387
2007	BC	348,420	0	73,595	1,201,186	11,044	22,625	1,656,870
2007	EF	348,420	20,185,656	2,125,998	214,987	745,445	1,026,528	24,647,034
2008	BC	346,367	0	68,975	1,203,515	11,281	22,669	1,652,806
2008	EF	346,367	22,486,339	2,191,284	211,523	668,341	992,624	26,896,478
2009	BC	364,533	0	70,624	1,203,312	11,522	22,665	1,672,655
2009	EF	364,533	23,384,382	2,289,239	260,430	485,626	983,857	27,768,067
2010	BC	420,128	0	79,222	1,200,244	11,779	22,607	1,733,981
2010	EF	420,128	22,537,971	2,121,915	225,347	572,774	955,516	26,833,650
2011	BC	671,609	0	123,338	1,198,308	11,734	22,571	2,027,560
2011	EF	671,609	22,777,139	2,231,686	210,785	727,524	794,679	27,413,422
2012	BC	656,148	0	110,539	1,193,534	11,654	22,481	1,994,355
2012	EF	656,148	23,434,455	2,075,289	279,864	701,763	779,928	27,927,448
2013	BC	709,239	0	124,815	1,194,161	11,642	22,492	2,062,349
2013	EF	709,239	21,363,444	2,386,478	255,186	724,875	1,192,661	26,631,883
2014	BC	807,361	0	124,179	1,196,178	11,628	22,530	2,161,876
2014	EF	807,361	21,848,836	2,876,484	286,634	790,357	1,300,668	27,910,341
2015	BC	720,123	0	88,709	1,198,207	11,595	22,569	2,041,202
2015	EF	720,123	26,420,640	2,557,775	313,974	773,631	1,390,539	32,176,682
2016	BC	660,621	0	82,417	1,200,885	11,571	22,619	1,978,113
2016	EF	660,621	27,388,293	2,626,895	301,624	765,977	1,261,831	33,005,240
2017	BC	680,741	0	91,135	1,203,712	11,544	22,672	2,009,804
2017	EF	680,741	27,058,793	2,862,004	335,298	955,465	1,325,620	33,217,921
2018	BC	745,928	0	89,002	1,205,103	11,524	22,699	2,074,255
2018	EF	745,928	26,817,468	2,773,407	355,276	844,016	1,147,537	32,683,632

Source: GFN 2022 (https://data.footprintnetwork.org)

4.7 The Ecological Footprint and Biocapacity/Capita/Year (Gha) of Kuwait

The EF and BC (gha/capita/year) of six components from 1999 to 2018 is presented in Table 4.3. This clearly shows a significant increase on demand side (EF) compared to supply side (BC). The results in Table 4.3 are further extrapolated for clarification in the form of Figs. 4.1, 4.2 and 4.3. Figure 4.1 shows the trend of EF gha/capita/year (1999–2018), and Fig. 4.2 shows the trend of BC gha/capita/year from 1999 to 2018. Figure 4.1 shows a consistent increase of EF from 1999 (2.89 gha/capita/year) to a maximum of 10.13 gha/capita/year in 2008; later EF decreased slightly, which seems to be due to better environmental conditions, especially there was better rain in the last few years, especially in 2018 the rainfall was at a maximum (344 mm/annum) which improved vegetation cover and increased the carbon sink.

Figure 4.2 shows the two decades trend of BC dynamics. It appears that BC was decreased slightly but consistently from 1999 (0.79 gha/capita/year) to 2018 (0.5 gha/capita/year), with a deficit of 0.29 gha/capita/year in 2018.

Relative distribution of EF and BC in 1999 and 2018 is shown in Fig. 4.3. This clearly shows the increasing and decreasing trend of EF and BC over the two decades. Figure 4.3 shows the significant difference between the EF and BC; this suggests poor eco-resources of Kuwait; however, these eco-resources deficiencies are compensated by using income from the oil business through international trade to import goods to meet the demand of population and animal sector.

In summary, it can be stated that the trend has clearly shown a significant increase (173%) in EF from 1999 (2.89 gha/capita/year) to 2018 (7.9 gha/capita/year). On the other hand, the BC is decreased from 0.79 gha/capita/year (1999) to 0.5 gha/capita/year (2018), although a small change but a decrease of 37% over two decades (Figs. 4.2 and 4.3).

The decrease in BC and significant increase in EF of Kuwait may be due to the impact of climate change and resources constraints (poor soil resources, loss of vegetation due to Gulf war, water scarcity, prolonged drought, and desertification) limiting the BC to the lowest level and running ecological deficits is an increasing risk in Kuwait. Generally, these eco-resources do not appear in the national financial accounts of the country; however, the natural resources are fundamental assets that keep the country environment a balance. It is hoped that the natural desert capital will increase after the mega desert rehabilitation plan is implemented, and the deserts are rehabilitated through revegetation plan (Alenezi et al. 2021; Madouh et al. 2021). This will increase soil carbon sequestration and release some pressure due to carbon footprint in the country.

For more clarification and to view the EF and BC from a *bird-eye-view* context, a relative distribution of EF and BC gha/capita/year over the past two decades (1999–2018) is shown in Figs. 4.4 and 4.5, illustrating the significant gap between these two nature measurement components, with a final deficit (EF-BC) of 7.4 gha/

Table 4.3 Two decades (1999–2018) trend of biocapacity and ecological footprint of Kuwait in gha/capita/year (data quality: 3A, isoa2)

Year	Records	Built-up land	Carbon	Crop land	Fishing grounds	Forest products	Grazing land	Total
		Global hectares/capita/year						
1999	BC	0.14	0.00	0.02	0.62	0.00	0.01	0.79
1999	EF	0.14	1.31	0.79	0.06	0.18	0.40	2.89
2000	BC	0.10	0.00	0.02	0.60	0.00	0.01	0.73
2000	EF	0.10	2.34	0.77	0.07	0.18	0.28	3.73
2001	BC	0.08	0.00	0.02	0.58	0.00	0.01	0.69
2001	EF	0.08	3.88	0.70	0.07	0.17	0.27	5.17
2002	BC	0.09	0.00	0.02	0.57	0.00	0.01	0.69
2002	EF	0.09	4.77	0.62	0.06	0.17	0.32	6.02
2003	BC	0.10	0.00	0.03	0.56	0.00	0.01	0.71
2003	EF	0.10	5.17	0.67	0.07	0.18	0.41	6.60
2004	BC	0.11	0.00	0.02	0.55	0.00	0.01	0.69
2004	EF	0.11	6.54	0.74	0.07	0.26	0.38	8.09
2005	BC	0.12	0.00	0.03	0.53	0.00	0.01	0.69
2005	EF	0.12	8.20	0.85	0.07	0.26	0.44	9.93
2006	BC	0.14	0.00	0.03	0.51	0.00	0.01	0.69
2006	EF	0.14	8.07	0.92	0.08	0.25	0.40	9.86
2007	BC	0.14	0.00	0.03	0.48	0.00	0.01	0.66
2007	EF	0.14	8.06	0.85	0.09	0.30	0.41	9.84
2008	BC	0.13	0.00	0.03	0.45	0.00	0.01	0.62
2008	EF	0.13	8.47	0.83	0.08	0.25	0.37	10.13
2009	BC	0.13	0.00	0.03	0.43	0.00	0.01	0.59
2009	EF	0.13	8.29	0.81	0.09	0.17	0.35	9.84
2010	BC	0.14	0.00	0.03	0.40	0.00	0.01	0.58
2010	EF	0.14	7.53	0.71	0.08	0.19	0.32	8.97
2011	BC	0.21	0.00	0.04	0.38	0.00	0.01	0.64
2011	EF	0.21	7.19	0.70	0.07	0.23	0.25	8.65
2012	BC	0.20	0.00	0.03	0.36	0.00	0.01	0.60
2012	EF	0.20	7.00	0.62	0.08	0.21	0.23	8.34
2013	BC	0.20	0.00	0.04	0.34	0.00	0.01	0.58
2013	EF	0.20	6.06	0.68	0.07	0.21	0.34	7.55
2014	BC	0.22	0.00	0.03	0.32	0.00	0.01	0.59
2014	EF	0.22	5.92	0.78	0.08	0.21	0.35	7.56
2015	BC	0.19	0.00	0.02	0.31	0.00	0.01	0.53
2015	EF	0.19	6.89	0.67	0.08	0.20	0.36	8.39
2016	BC	0.17	0.00	0.02	0.30	0.00	0.01	0.50
2016	EF	0.17	6.92	0.66	0.08	0.19	0.32	8.34
2017	BC	0.17	0.00	0.02	0.30	0.00	0.01	0.50
2017	EF	0.17	6.67	0.71	0.08	0.24	0.33	8.19
2018	BC	0.18	0.00	0.02	0.29	0.00	0.01	0.50
2018	EF	0.18	6.48	0.67	0.09	0.20	0.28	7.90

Data source: GFN 2022 (https://data.footprintnetwork.org)

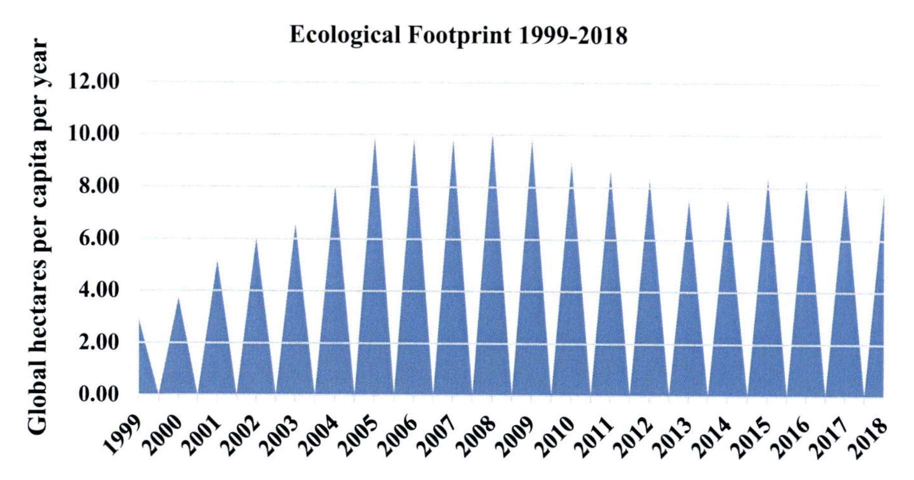

Fig. 4.1 Dynamics of ecological footprint in global hectares capita^{-1} year^{-1} from 1999 to 2018. (Data source: GFN 2022 (https://data.footprintnetwork.org))

capita/year recorded in the year 2018. Detailed breakdown of various components of both the EF and BC is shown in Table 4.3.

4.8 The Major Concerns of High EF of Kuwait

It is important to analyze the causes of high EF of Kuwait compared to BC. Analyzing the results reported by GFN (2020), we observed that on the resources production side (biocapacity), five areas are described, i.e., (i) cropland, (ii) forests, (iii) grazing land, (iv) fishing grounds, and (vi) built-up land) and are supporting the six footprint demand sides, (i) cropland footprint, (ii) forest product footprint, (iii) carbon footprint, (iv) grazing land footprint, (v) fishing grounds footprint, and, (vi) the built-up land footprint. The carbon (oil/gas) is not considered in the BC side. The total eco-resources of components (2018) of Kuwait are in deficit (carbon, cropland, forestry products, and grazing land); however, they are either equal (EF = BC) in the case of built up area, or eco-plus in fishing grounds (BC 0.29 gha/capita/year versus 0.09 EF gha/capita/year). However, the climate change in future may affect ocean in terms of building up of water heat adversely affecting across marine ecosystems. Other factors of future fishing decline could be due to unsustainable fishing (overfishing), coastal development, and pollution. Potential rise in seawater temperature due to climate change may affect spawning period of fish and shrimp and relocation to other suitable areas, disturbing the fish industry in Kuwait and the region (Omar and Roy 2010). The above discussion suggests there is immediate need to improve the productive capacity of three nature measuring components (cropland, forestry products, and grazing land) of Kuwait that should also release pressure on carbon footprint, however, to a limited extent.

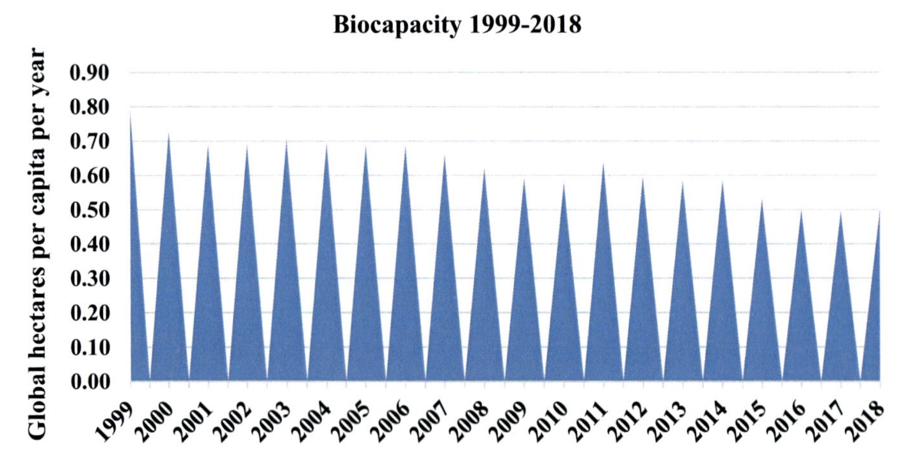

Fig. 4.2 Dynamics of biocapacity in global hectares/capita/year from 1999 to 2018. (Data source: GFN 2022 (https://data.footprintnetwork.org))

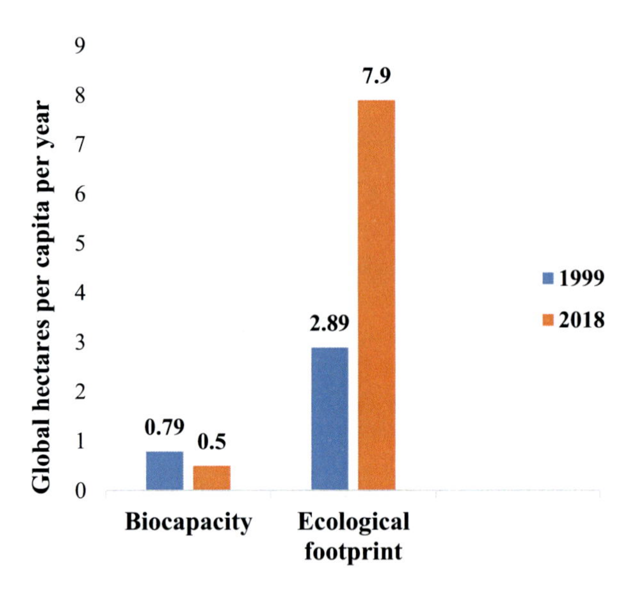

Fig. 4.3 Dynamics of biocapacity and ecological footprint of Kuwait per capita per year in the years 1999 and 2018. (Data source: GFN 2022 (https://data.footprintnetwork.org))

The recent results of Kuwait EF and BC (GFN 2020) depicts the main concern is about carbon footprint as the major source of negative impact on nature capital. The latest results of year 2018 show the carbon footprint as a result of CO_2 emissions associated with the use of fossil fuel included in consumed products, which accounts for 6.48 gha/capita/year on the total EF of 7.90 gha/capita/year. Therefore, if we ignore the carbon footprint of Kuwait, the deficit (EF-BC) of eco-resources is 1.42 gha/capita/year in contrast to 7.4 gha/capita/year including carbon footprint in the total EF of Kuwait. The deficit of 1.42 gha/capita/year for Kuwait is very closed to

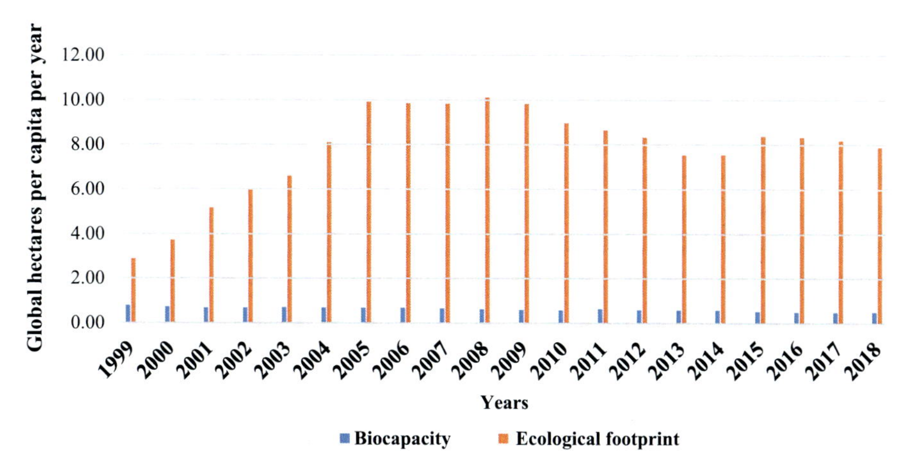

Fig. 4.4 Biocapacity versus ecological footprint (global hectares/capita/year) of Kuwait (1999–2018). (Data source: GFN 2022 (https://data.footprintnetwork.org))

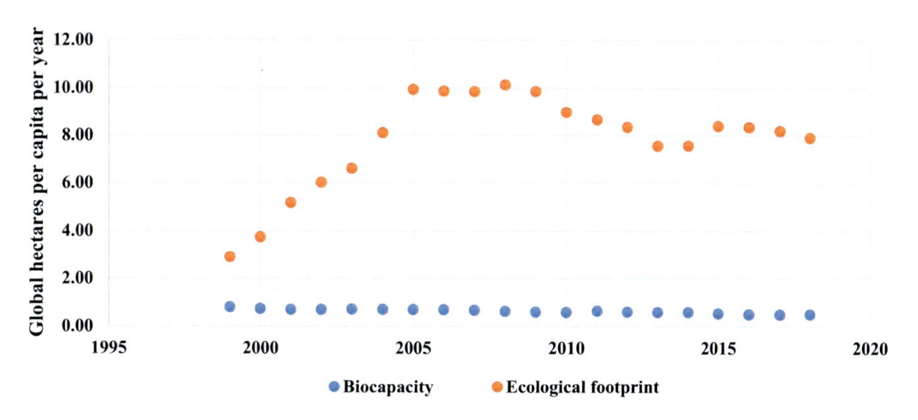

Fig. 4.5 Dynamics of biocapacity and ecological footprint of Kuwait gha/capita/year over two decades. (Data source: GFN 2022 (https://data.footprintnetwork.org))

the world deficit of 1.26 gha/capita/year (considering the world total EF of 21.2 billion gha and BC 12.1 billion gha and population of 7.2 billion).

The imbalance between the global level of EF and BC is due to the fact the in the past five decades the world has significantly transformed to global trade, high consumption and burgeoning population growth, and enormous expansion of urbanization. Until 1970, humanity's EF was smaller than the Earth's rate of regeneration. To feed and fuel and our twenty-first-century lifestyles, we are overusing the Earth's biocapacity by at least 56% (WWF 2020).

4.9 Global Trend of Ecological Footprint and Biocapacity over the Past Six Decades

On the global level, the planet is in eco-deficit of 9.1 billion gha in 2018 (Table 4.4; Fig. 4.6) compared to a surplus BC of 2.69 billion gha in 1961. In 1970, the resources on the EF and BC sides were nearly equal, when we required one Earth planet. The two decades' trend indicates that there is slight increase of BC compared to sharp increase of EF. This gives us alarm that if we continue exploiting the resources beyond their capacity, we may be living on the resources of the future generations and will require over two planets to survive, and unfortunately there is no virtual planet to import, as we import virtual water (water embedded in food products) in the form of imported food, as if this water was used locally to produce the same food within the national boundaries.

4.10 Comparison of Ecological Footprint and Biocapacity for GCC Countries and the World (2018)

The data about total EF and BC for the GCC countries and the world is downloaded from GFN website and presented in Table 4.5. A comparative presentation of EF and BC of GCC countries is shown in Fig. 4.7. Figure 4.7 shows the level of human demand on nature (EF) is the highest in Saudi Arabia in the region, and the decreasing trend is Saudi Arabia > UAE > Qatar > Kuwait > Oman > Bahrain. The trend in BC is Saudi Arabia > UAE > Oman > Qatar > Kuwait > Bahrain. The significant difference in EF and BC among the GCC countries is due to the total area of the country and the eco-resources available and their management in each country. The total demand of GCC countries on the nature (EF) is 1.70% (360.56 million gha) of the total world EF (21,176 million gha), whereas the GCC countries contribute 0.26% (31.01 million gha) to the total biocapacity (12,077 million gha) of the world.

In contrast to low BC contribution of GCC countries to the world BC, about 50% of the world's oil reserves are located in the Gulf region. These rich oil resources are considered to contribute to carbon footprint on the demand side, and the GFN considers biocapacity "zero" from oil resources, although oil products are used in many industries producing beneficial goods. *This is the main reason why GCC countries have high EF and low BC and the gap (EF-BC) is wider compared to the other non-oil producing countries*. We therefore, suggest GFN to consider the value of oil and gas resources toward contributing to biocapacity and to come up with some formulas and factors to convert oil produced to global hectares.

Table 4.4 Total world trend of biocapacity and ecological footprint in Billions gha (1961–2018)

Year	Biocapacity	Ecological footprint
	Billion global hectares	
1961	9.74	7.04
1962	9.78	7.26
1963	9.78	7.54
1964	9.83	7.85
1965	9.85	8.13
1966	9.95	8.48
1967	9.98	8.70
1968	10.03	9.05
1969	10.04	9.43
1970	10.08	9.99
1971	10.14	10.31
1972	10.10	10.59
1973	10.21	11.15
1974	10.15	11.052
1975	10.18	11.00
1976	10.24	11.52
1977	10.25	11.76
1978	10.36	12.04
1979	10.34	12.36
1980	10.34	12.12
1981	10.44	11.96
1982	10.53	11.91
1983	10.50	11.90
1984	10.65	12.37
1985	10.73	12.57
1986	10.75	12.79
1987	10.79	13.15
1988	10.72	13.44
1989	10.85	13.78
1990	10.97	13.99
1991	10.90	13.94
1992	11.05	14.03
1993	11.02	14.01
1994	11.10	14.20
1995	11.05	14.45
1996	11.19	14.78
1997	11.24	15.01
1998	11.28	15.05
1999	11.32	15.10
2000	11.33	15.45
2001	11.39	15.62
2002	11.42	15.83

(continued)

Table 4.4 (continued)

Year	Biocapacity	Ecological footprint
	Billion global hectares	
2003	11.40	16.42
2004	11.52	17.19
2005	11.50	17.66
2006	11.54	18.13
2007	11.58	18.65
2008	11.72	18.84
2009	11.67	18.51
2010	11.77	19.56
2011	11.82	19.97
2012	11.71	19.80
2013	11.89	20.25
2014	11.93	20.23
2015	11.95	20.12
2016	12.00	20.08
2017	12.13	20.83
2018	12.08	21.18

Data source: GFN 2022 (https://data.footprintnetwork.org)

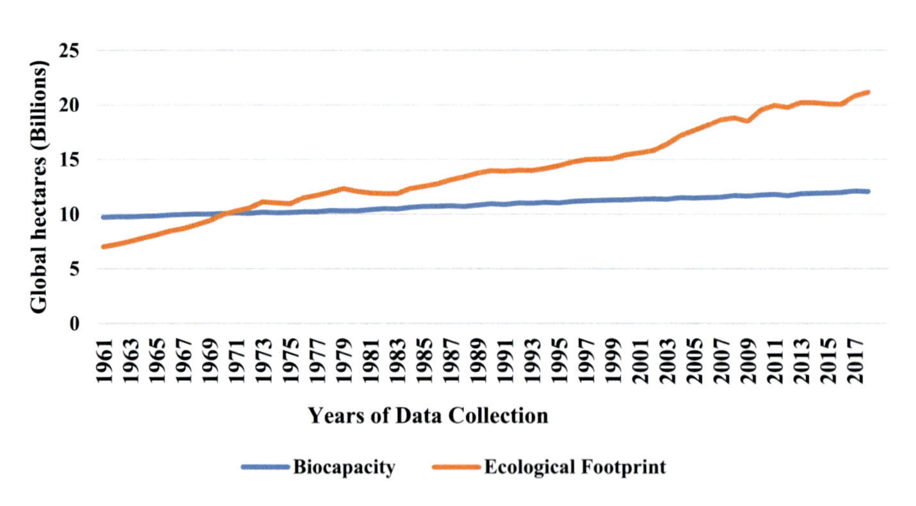

Fig. 4.6 Total world trend of biocapacity and ecological footprint in Billions gha (1961–2018). (Data source: GFN 2022 (https://data.footprintnetwork.org))

Table 4.5 Total ecological footprint and biocapacity of GCC countries and the world in 2018

GCC country	Ecological footprint (million gha)	Biocapacity (million gha)	Deficit EF-BC (million gha)
Bahrain	12.83	0.73	12.10
Kuwait	32.68	2.07	30.61
Oman	30.37	6.89	23.48
Qatar	39.70	2.56	37.14
Saudi Arabia	167.01	13.78	153.23
United Arab Emirates	77.97	4.98	72.99
Total	360.56	31.01	329.55
World	21,176	12,077	9099

Source: https://api.footprintnetwork.org/v1/data/all/2017/BCtot

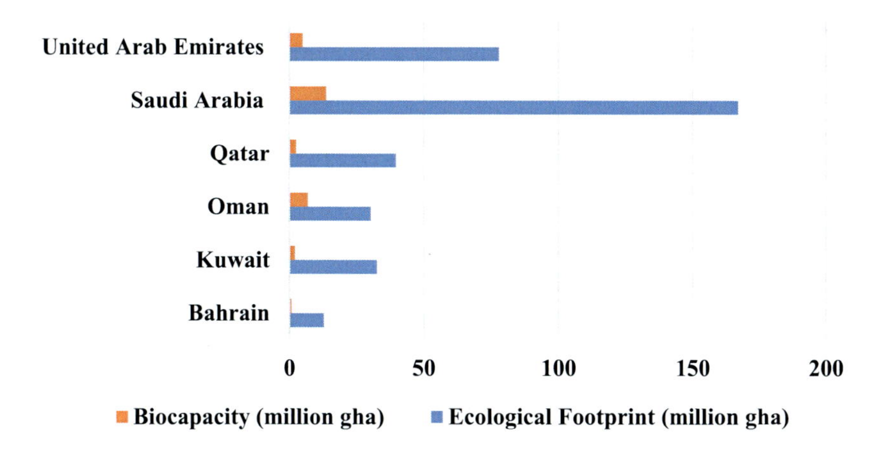

Fig. 4.7 Comparative EF and BC of GCC countries (million gha). (Data source: https://api.footprintnetwork.org/v1/data/all/2017/BCtot)

4.11 Comparison of Ecological Footprint and Biocapacity for GCC Countries and the World in Gha/Capita/Year

Table 4.6 shows GCC countries and the world gha/capita/year demand on nature (EF) and the production capacity (BC) as well as the deficit (EF-BC) capita^{-1} year^{-1}.

Table 4.6 shows variable deficit in eco-resources (EF-BC) in the GCC countries. The higher the deficit in gha/capita/year, the more resources are required from external sources to meet the demand of the population in terms of food, fodder, and other consumables other than the petroleum products.

Table 4.6 EF and BC (gha capita^{-1} year^{-1}) of GCC countries and the world

Country	EF gha/capita/year	BC gha/capita/year	Deficit gha/capita/year
Qatar	14.72	0.97	13.75
United Arab Emirates	8.95	0.53	8.42
Bahrain	8.66	0.48	8.18
Kuwait	8.03	0.48	7.55
Oman	7.29	1.46	5.83
Saudi Arabia	5.77	0.42	5.35
World	2.70	1.5	1.20

4.12 Strategies to Improve Eco-resources of Kuwait

It is clear from the earlier discussions that instead of buildup area and fish grounds, Kuwait is eco-deficit in four components of BC, that is, carbon (carbon sequestration), cropland, forest products, and grazing land. In the following section, these components are described in general and in specific, in terms of how their productivities can be improved to decrease the gap between EF and the BC.

4.12.1 Governance of Soil Resources

Governance of soil resources based on the country needs and priorities is the key to the sustainable development without compromising the ability of future generations to meet their own needs. The soil resources are the basis for the provision of ecosystem services and healthy functioning of the ecosystems. Currently due to natural (climate) and anthropogenic effects, the ecosystems are under severe pressure, including competing lands for different uses (urban development, infrastructure, agriculture, etc.). To meet the food security demand by reducing the gap between food import and local production, it requires allocation of arable area and water resources for local food production, in addition, a national plan to be in place to combat desertification and increase soil conservation practices. A national level soil survey was completed during 1995 and 1999 (KISR 1999; Omar and Shahid 2013) which provides soil information for informed decisions. A number of soil use maps (sand and gravel sources, suitability for septic tanks, sanitary landfills, seedling mortality, suitability for irrigated agriculture, etc.) were prepared (KISR 1999), constraints for each use highlighted, and recommendations made to avoid/minimize negative effects on the environment. The decision makers are suggested to use the national soil information for land-use planning and policy developments.

4.12.2 Agriculture Development in the Context of Kuwait Vision 2035

It is speculated that water and not land will be a limiting factor to increase agriculture production, an issue that will be exacerbated by the predicted trends of climate change. Indeed, agriculture sector is the greatest consumer of water, and it is not a matter of absolute use but relative use. It is therefore important to find ways as how to increase the efficiency of water for agriculture (Shetty 2006) and how to reduce the overall use.

The major emphasis of Kuwait Developmental Plan (Kuwait 2035 Vision) is on sustainable economic development through conservation and proper management of native biological and natural resources. This requires focused efforts of all institutions in Kuwait to incorporate sustainable environmental schemes in all sectors, including to increase sustainable food and feed production locally to achieve a greater degree of food security without compromising the quality of natural resources. The KISR has started a Government Initiative P-KISR-17 project titled *Pilot Study for Utilization of Modern Technologies in Sustainable Local Food Production in Support of Kuwait's Food Security*, for period of 4 years (2021–2025). This Government Initiative project is funded by The Supreme Council for Planning and Development, State of Kuwait. The results of this project will directly support accomplishment of Kuwait's Development Plan goals/ targets.

Among all 17 SDGs of the UN, SDG-2 directly deals with hunger "No-hunger," that is, "End hunger, achieve food security and improved nutrition and promote sustainable agriculture." Specifically, the Target 2.3 focuses on "by 2030 double the agricultural productivity and the incomes of small-scale food producers" and Target 2.4 focuses on "by 2030 ensure sustainable food production systems and implement resilient agricultural practices that progressively improve land and soil quality."

Another interesting area to sustain food security is the use of *Precision Agriculture*, which refers to the use of a series of technologies that allow the application of water, nutrients, and pesticides only to the places and at the time they are required, thereby optimizing the use of inputs. However, this requires investigation and investment for its implementation.

It is hoped that the results from the ambitious GI P-KISR-17 project is likely to enable Kuwait to increase 15–20% self-sufficiency levels without further deteriorating the environment and ultimately allow the State of Kuwait to meet relevant Sustainable Development Goals (SDG) set forth by United Nations (Suleiman 2020). Thus, this project represents a direct response to the government recognition of the importance of increasing local production as part of its food security strategic policy aiming at agricultural growth in an inclusive and sustainable manner while taking advantage of latest studies and innovations and optimizing economic returns to farmers. This will strengthen the food security capacity in the country and leverage investments in this field.

4.12.3 Improving Cropland Component and Biodiversity of the National Biocapacity

The national cropland and cropland productivities can be improved in various ways to reduce the current (2018) gap between EF (0.67 gha gha/capita/year) and BC (0.02 gha gha/capita/year) in cropland sector. Following are the suggestions:

- Increase agriculture area on the land currently not in use but has the capacity to be used for agriculture. Such areas can be depicted from the land suitability map for irrigated agriculture (KISR 1999; Omar and Shahid 2013).
- The increased agriculture land will require more water for irrigation, the treated sewage effluent (TSE), and the high-quality reverse osmosis (RO) water from Sulaibiya plant may be directed to the newly proposed agriculture areas. However, their restricted and/or unrestricted use must be considered and crops identified in each case.
- A need to implement climate smart agriculture (CSA) practices to increase local crop production in an environment friendly and socially sustainable way, including empowering the framers to increase farmer's resilience to climate change by reducing agriculture contribution of greenhouse gasses emission and by increasing soil carbonization through using organic soil amendments (compost, manures, biochar, etc.). As per World Bank (2011) the CSA is defined below:
  - CSA is about getting existing technologies off the shelf and into the hands of farmers and developing new technologies to meet the demands of the changing climate.
  - CSA includes proven practical techniques – mulching, intercropping, conservation agriculture, crop rotation, integrated crop-livestock management, agroforestry, improved grazing, and improved water management – but also innovative practices such as better weather forecasting, early warning systems, and risk insurance.
  - CSA seeks to increase productivity in an environmentally and socially sustainable way, strengthen farmers' resilience to climate change, and create enabling policy environment for adaptation.

Due to high aridity and hot climatic conditions in the GCC countries, there is a low biodiversity by reference to global standards; and this will further decline due to intensifying climate change; a 2 °C rise in temperature will make an extinct up to 40% of all species (Tolba and Saab 2009). While describing the biodiversity and climate change in Kuwait, Omar and Roy (2010) stressed on developing strategic plan for climate change mitigation and adaptation in Kuwait including identifying vulnerable species through field survey, monitoring and ex situ and in situ conservation, increasing protected areas, and through awareness and capacity building, as biodiversity in Kuwait is under severe threat to various natural and anthropogenic factors, where land impacted by anthropogenic activity is reaching 10.47%.

4.12.4 Carbon Footprint of GCC Countries

Global Carbon Atlas (http://www.globalcarbonatlas.org/en/CO2-emissions) reported year 2021 CO_2 emissions of various countries and the global emission (Friedlingstein et al. 2022; Andrew and Peters 2022; Peters et al. 2011). Table 4.7 presents a comparison of total ($MtCO_2$) and per capita carbon emission of GCC countries.

The carbon emission (footprint) can be reduced by using different strategies, including but not necessarily limited to the following:

- Soil carbonization to increase carbon sequestration.
- Use of organic fertilizers to improve soil carbon contents.
- Rehabilitation of degraded ecosystem to increase vegetation cover for soil protection and to reduce desertification.
- Recycling organic residues to produce organic and biofertilizers for use in soil health improvement.
- Reducing landfill with organic residues.

4.12.5 Mitigations and Adaptations

Within the perspectives of IPCC (2007), adaptation is the adjustment in natural or human systems in response to actual or expected climatic stimuli or their effects, which moderates harm or exploits beneficial opportunities. Various types of adaptation can be distinguished, including anticipatory, autonomous, and planned adaptation: (1) adaptation that takes place before the observation of CCI (anticipatory or proactive adaptation), (2) adaptation that does not constitute a conscious response to climatic stimuli but is triggered by ecological changes in natural systems and by market or welfare changes in human systems (autonomous or spontaneous adaptation), and (3) adaptation that is resulting from a deliberate policy decision, based on awareness that conditions have changed or are changing and that action is required to return to, maintain, or achieve a desired state (planned adaptation). Shahid and

Table 4.7 Population and total and per capita CO_2 emission of GCC countries

Countries	Population	Total CO_2 emission *$MtCO_2$	CO_2 emission/capita (tCO_2)
Bahrain	1,463,265	39.02	26.7
Kuwait	4,250,111	106.13	24.94
Oman	4,520,471	80.99	17.92
Qatar	2,688,235	95.67	35.59
Saudi Arabia	35,950,396	672.38	18.70
United Arab Emirates	9,365,145	204.00	21.78

*$MtCO_2$ (Million tons CO_2)

Behnassi (2014) have reported various aspects of adaptation and mitigation options for the Arab region. Examples of adaptation activities have also been defined by Klein et al. (2006).

4.12.6 Conservation Agriculture

Conservation agriculture (CA) includes low, or no tillage, practices to conserve soil moisture. The main principles (Dumanski et al. 2006) of conservation agriculture are permanent surface cover to promote minimum mechanical disturbance; healthy crop rotation; cover crops; use of balanced fertilizers, pesticides, herbicides, and fungicides; promoting precision agriculture; and promoting leguminous cops to furnish biological nitrogen in soil and adoption of agroforestry. No tillage is a practice of planting crop on untilled soil by opening a narrow slot, trench, or band of sufficient size for proper seed coverage. It has benefits: soil remains covered by crop residues from previous crops or green manure to protect soil from erosion (reduce soil degradation), reduce moisture loss (mulch), conserve organic matter, and improve nutrients (green manuring).

4.12.7 Improving Grazing Lands

Grazing lands can be improved by allocating areas for grazing and rotating the herds from one area to another to allow the plants to recover for grazing purpose. Such areas, albeit expensive, can be protected (fenced) to regenerate the plant communities after grazing. By adapting this strategy, grazing capacity "carrying capacity" can be improved over the year.

4.12.8 Combat Desertification to Improve Desert Ecosystems for Carbon Sequestration

Desertification is very common in desert environment. In Kuwait, the surface soil is sandy and loose and vulnerable to wind erosion. To combat desertification, a combination of problem diagnostic-based test, tested and approved practices are to be implemented, including but not necessarily limited to:

- Soil surface stabilization using organic soil amendments
- Protecting vegetation covers and revegetation in degraded areas
- Control on free grazing
- Creation of wind breakers

- Control on free vehicle maneuvering
- Camping on government designated areas
- Recarbonization of soils for soil carbon sequestration
- Desert soilization using plants based organic residues

4.12.9 Strengthening Extension-Research Link

A strong link between research and extension can provide the farmers newly developed technologies on rapid basis. This link should be maintained at all times, especially before the start of cropping season and continued during the growing season. In addition, the extension workers can establish demonstration plots using newly developed technologies and farmer's awareness through inviting farmers to view results from the demonstration farms and by organizing farmer days "seeing is believing concept." In addition, the extension workers can prepare user friendly guidelines in the form of flyers, brochure for distribution. Following are the important areas, which farmer may need education and training:

- Crops – improved, salt tolerant, and high yielding crop varieties.
- Soils – best soil management practices, use of organic fertilizers to improve soil tilth.
- Water – best crops based irrigation practices (modern irrigation systems – drip, sprinkler).
- Fertilizers – 4 R nutrient stewardship (right – type, source, place, time).
- Insects, pest management – integrated pest management (insecticides, biological, etc.) program.
- Easy access of farmers to the markets to sell their products.

4.12.10 Adoption of Kuwait Vision 2035: New Kuwait

The Kuwait Ministry of Foreign Affairs has set up Kuwait Vision 2035 "New Kuwait" (https://www.mofa.gov.kw/en/kuwait-state/kuwait-vision-2035/). Brief excerpt from the Kuwait Vision – 2035 is presented in this section. It uses the global indicators to track and measure Kuwait's progress with the plan and its performance, comparative to other countries, as well as setting goals and following up on the performances toward achieving the vision. Kuwait's national development plan is linked to international goals and factors by adapting them to the United Nations Sustainable Development Goals (SDGs) 2030 agenda, in order to achieve compatibility between the national development plan and the international development vision.

Among the eight Strategic Developmental Goals (8StDG) of the national development plans 2035, number 1 is focused to increase the local productivity and development of non-oil economic sectors (NOES), and number 6 outlines "Train and qualify national human resources." The seven pillars of the National Development Plan (2035 Vision) are (i) sustainable diversified economy, (ii) effective civil service, (iii) sustainable living environment, (iv) developed infrastructure, (v) high-quality healthcare, (vi) creative human capital, and (vii) global positioning. The SDG of New Kuwait 2035 requires to increase local productivity and development in non-oil economic sectors (NOES). Among NOES, agriculture sector is one of the sectors which require to be addressed to improve local agriculture and animal production sector with main emphasis on reducing water footprint of various agriculture commodities.

4.13 Conclusions and Recommendations

After reviewing different components making ecological footprint and biocapacity of Kuwait and comparing with the GCC countries, it is concluded that in all the GCC countries, there is a deficit of eco-resources defined as the difference between EF of consumption and bioproductive land (EF-BC). This leads to heavy import of food and fodder products from other countries. It is noted that in the demand side (EF), carbon footprint is high in all the countries against the zero biocapacity. This is the area to study further to correlate the oil production to contribute to BC to appreciate the oil resources contribution to the economies and helping other countries to run industries using oil and gas resources. Being the GCC countries' location in the arid region, there is impact of climate to grow crops within national boundaries to meet full demand of population. However, there are ways to improve local cop production to bridge the gap between import and local production. Strategies to improve BC, e.g., adoption of climate smart agriculture and other areas, are proposed. It is also realized that the nations who have limited eco-resources have no choice other than to manage the use of ecological assets sustainably leading the chances of economic success.

Acknowledgements We would like to thank Michelle of Global Footprint Network (GFN) for accepting our request and granting us the generous permission to use the ecological footprint data as often as possible and welcome its inclusion in the book. The suggested citation for data has been downloaded from the open data platform: York University Ecological Footprint Initiative and Global Footprint Network. National Footprint and Biocapacity Accounts, 2022 edition (https://data.footprintnetwork.org), and it has been added in the reference list and cited in the chapter.

References

Alenezi A, Quoreshi A, Bhat NR, Shahid SA, Omar SAS (2021) Ecological monitoring and evaluation of active revegetation of degraded terrestrial ecosystems SP003EC under KERP (Proposal). Kuwait Institute for Scientific Research, Kuwait

Andrew RM, Peters GP (2022) The global carbon project's fossil CO_2 emissions dataset. Last accessed 17 Oct 2022. Available at: https://zenodo.org/record/7215364. Andrew and Peters, 2022

Dumanski J, Peiretti R, Benetis J, McGarry D, Pieri C (2006) The paradigm of conservation tillage. In: Proceedings of World Association of Soil and Water Conservation, pp. 58–64

European Environment Agency-EEA (2020) Ecological footprint of European countries. https://wwweeaeuropaeu/data-and-maps/indicators/ecological-footprint-of-european-countries-2/. Last accessed 23 April 2020

Ewing B, Moore D, Goldfinger S, Oursler A, Reed A, Wackernagel M (2010) The ecological footprint atlas 2010. Global Footprint Network, Oakland

Friedlingstein P, O'Sullivan M, Jones MW, Andrew RM, Gregor L (2022) Global Carbon Budget 2022, Earth System Science Data. Available at: https://doi.org/10.5194/essd-14-4811-2022 Friedlingstein et al., 2022. Accessed 19 July2022

Global Footprint Network-GFN (2020) Glossary. https://wwwfootprintnetworkorg/resources/glossary/. Accessed 04 May 2020

GFN (2022) York University Ecological Footprint Initiative and Global Footprint Network (2022) National Footprint and Biocapacity Accounts, 2022 edition (https://data.footprintnework.org) Last accessed 27 August 2023

IPCC (2007) Climate change 2007: impacts, adaptation and vulnerability. Retrieved 10 Oct 2010 from www.ipcc.wg2.org/

Keßler S, Richard D, Zimmer S (2020) Ecological footprint – reloaded (2022). In Cooperation with Global Footprint Network: Dr. David Lin, Dr. Mathis Wackernagel. Financing/Sustainability Council of the Grand-Duchy of Luxembourg. p 56

Kitzes J, Galli A, Bagliani M, Barret J, Dige G, Ede S, Erb K, Giljum S, Haberl H, Hails C, Jolia-Ferrier L, Jungwirth S, Lenzen M, Lewis K, Loh J, Narchettini N, Messinger H, Milne K, Moles R, Monfreda C, Moran D, Nakano K, Pyhälä A, Rees W, Simmons C, Wackernagel M, Wada Y, Walsh C, Wiedmann T (2009) A research agenda for improving national ecological footprint accounts. Ecol Econ 68:1991–2007

KISR (1999) Soil survey for the State of Kuwait: reconnaissance survey, vol II & III. Kuwait Institute for Scientific Research, Kuwait

Klein RJT, Alam M, Burton M, Doughtery W, Ebi K, Fernandes M, Huber L, Rahman A, Swartz C (2006) Application of environmentally sound technologies for adaptation to climate change. Prepared for the UNFCCC Secretariat, FCCC/TP/2006/2

Madouh T, Quoreshi A, Bhat NR, Shahid SA, Omar SAS (2021) Ecological monitoring and evaluation of restoration and revegetation success of areas damaged by military fortifications (Element 1 of Claim 5000450), Sheikh Sabah Al Ahmad Nature Reserve under KERP, March 2022. Proposal. Kuwait Institute for Scientific Research, Kuwait

Mancini MS, Galli A, Niccolucci V, Lin D, Bastianoni S, Wackernagel M, Marchettini N (2016) Ecological footprint: Refining the carbon footprint calculation. Ecol Indic 61:390–403

Omar SAS, Roy WY (2010) Biodiversity and climate change in Kuwait. Int J Clim Change Strateg Manag 2(1):68–83

Omar SAS, Shahid SA (2013) Reconnaissance soil survey for the State of Kuwait. In: Shahid SA, Taha FK, Abdelfattah MA (eds) Developments in soil classification, land use planning and policy implications: Innovative thinking of soil inventory for land use planning and management of land resources. Springer, pp 85–110

Peters GP, Minx JC, Weber CL, Edenhofer O (2011) Growth in emission transfers via international trade from 1990 to 2008. Proc Natl Acad Sci 108(21):8903–8908

Shahid SA, Ahmed M (2014) Changing the face of agriculture in the Gulf Cooperation Council countries. In: Shahid SA, Ahmed M (eds) Environmental cost and face of agriculture in Gulf Cooperation Council countries-fostering agriculture in the context of climate change. Springer, pp 1–25

Shahid SA, Omar SAS (1999) Order 1 soil survey of the demonstration farm sites with proposed management. Kuwait Institute for Scientific Research, Kuwait. p viii + 144. KISR No 5463. ISBN 0 957700369

Shahid SA, Behnassi M (2014) Climate change impact in the Arab region: Review of adaptation and mitigation potential and practices. In: Behnassi M, Mteng'e MS, Ramachandra G, Shelat K (eds) Vulnerability of agriculture, water and fisheries to climate change – toward sustainable adaptation strategies. Springer, pp 15–38

Shetty S (2006) Water, food security and agricultural policy in the Middle East and North Africa Region. Working paper series, no 47, The World Bank Group, Washington, DC

Suleiman MK (2020) Pilot study for utilization of modern technologies in sustainable local food production in support of Kuwait's food security. Executive Summary. GI P-KISR-17 project proposal. Kuwait Institute for Scientific Research, Kuwait

Tolba MK, Saab NW (2009) Arab environment climate change – Impact on climate change on Arab countries. Report of the Arab Forum for Environment and Development, Technical Publications and Environment & Development Magazine, Beirut, Lebanon, p 159

Wackernagel M, Beyer B (2019) Ecological footprint – managing our biocapacity budget. New Society Publishers. ISBN 978-0865719118

Wackernagel M, Schulz NB, Deumling D, Linares AC, Jenkins M, Kapos V, Monfreda C, Loh J, Myers N, Norgaard R, Randers J (2002) Tracking the ecological overshoot of the human economy. Proc Natl Acad Sci U S A 99:9266–9271

WWF (2020) Executive summary: Living Planet Report 2020 – Bending the curve of biodiversity loss. Almond REA, Grooten M, Petersen T (eds). WWF, Gland

World Bank (2011) Policy brief: Opportunities and Challenges for Climate-Smart Agriculture in Africa, World Bank. http://climatechange.worldbank.org/sites/default/files/documents/CSA_Policy_Brief_web.pdf. Last accessed 27 Aug 2023

Chapter 5
Desertification: A Central Problem to Restoring Ecosystems

Shabbir Ahmad Shahid

Abstract Kuwait is a hot and dry country with infrequent and scanty rainfall. Therefore, deserts of Kuwait are degraded mainly due to loose sandy surface, low organic matter, and poor vegetation cover. The physical, chemical, and biological attributes of soil can support, reduce, and even completely inhibit profitable use or ecosystem services. Desertification is a central problem in Kuwait leading to declining quality of soils and their functionality to provide ecosystem services. Desertification results due to both natural and anthropogenic effects. Globally 21 billion tons of soil is eroded, resulting in a loss of 3 tons per capita per year. In addition, globally soil erosion causes 60% of crop yield loss. Three types of soil degradation processes have been identified in Kuwait (physical, chemical, biological). Among all processes soil erosion is dominant in Kuwait displacing major soil mass in the deserts, followed by salinization in the coastal areas due to seawater intrusion and subsequent evaporation causing coastal areas salinized, and saline farms due to irrigation with brackish/saline irrigation waters. Decline in biodiversity and ecological disturbances are other factors of desertification observed in Kuwait. Saltation is the main mechanism of soil particles movement, followed by creep and suspension movement. In this chapter, efforts have been made to identify land degradation indicators and threats to the deserts of Kuwait, and strategies to combat desertification are proposed.

Keywords Neutrality · Erosion · Rehabilitation · Saltation · Creep · Suspension · Desertification · Kuwait

---

S. A. Shahid (✉)
Desert Agriculture and Ecosystems Program, Environment and Life Sciences Research Center, Kuwait Institute for Scientific Research, Safat, Kuwait
e-mail: sshahid@kisr.edu.kw

M. K. Suleiman, S. A. Shahid (eds.), *Terrestrial Environment and Ecosystems of Kuwait*, https://doi.org/10.1007/978-3-031-46262-7_5

5.1 Introduction

The State of Kuwait is situated at latitudes 28°30 and 30°05 N and longitudes 46°33 and 48°35 E in the Arabian Peninsula. Its total area is 17,818 km^2. Kuwait has two main seasons, winter and summer. The summers are very hot with the daily maximum temperature averaging 45 °C in July, but temperatures as high as 51 °C are not uncommon at this time (Omar and Shahid 2013). The meteorological record (1957–2008) from the International Airport shows 119 mm mean annual rainfall and 2270 mm mean annual evapotranspiration, with rain mainly falling between November and April. The average evaporation rate ranges from 21 mm per day for July to 3 mm per day for January. The prevailing winds blow from the northwest (60% of total wind) and the southeast Kuwait falls within the geographical region that has the climate and the soil characteristics to enhance soil erosion (Higgitt 1993; Montgomery 2007). Its landscapes are dominated by loose sandy mantle vulnerable to wind erosion and hence losing productive surface layer and organic matter leading to desertification (Omar and Shahid 2013). Desertification along with climate change and loss of biodiversity has been identified as the biggest culprit preventing sustainable development in Kuwait.

Desertification refers to the degradation of land (Mainguet 1994; UNCCD 1994a, b; Adeel et al. 2005; Grainger 2013). Deserts without vegetation cover seem to be the endpoint of land degradation process. The vegetation degradation, wind and water erosion, salinization, and compaction are the main land degradation factors in Kuwait. The damage to the soil comes when wind-borne soil particles cause a cascading effect, fallback to the ground, strike the soil, and kick more particles into the air, in a chain reaction (Shahid and Abdelfattah 2008).

Many scientists view desertification differently, such as a reduction of the physical, chemical, or biological status of land, which may restrict its productive capacity (Lindskog and Tengberg 1994), or land disturbance perceived to be deleterious or undesirable (Johnson et al. 1997). In specific, the desertification is land degradation in arid, semiarid, and sub-humid areas, which are also known as drylands. Desertification, in fact, is ecosystem degradation and a complex process (GEO 4-UNEP 2006).

Land degradation (LD) means a loss or a decrease of land functions and degraded visual amenity of the landscape. The LD can be due to natural environmental conditions or due to human effects "*anthropogenic*". Both the land (wider context) and soil are essential part of the ecosystems. The healthy soils are the better for the ecosystems functions and services. Over decades, human used environment to gain benefits and in many cases the methods used to gain those benefits are unsustainable and that has led to degrade the ecosystem. Kuwait record of caring for the environment (vegetation, wildlife) especially designating the prime ecosystems as protected areas is excellent. However, there is less control on the open deserts which are prone to wind erosion and in places over grazing. The wind has a major role in drifting soil from loose surfaces and mainly from dune areas. The drifted sand cover buildings, block highways, cause dust storm and pollute the air, fill waterways and reduce their capacity, remove fertile soil and seeds, and strong winds sometimes lodges crop if not protected.

5.2 Global Perspectives of Desertification

Desertification is a common phenomenon under desert conditions such as the case of Gulf Cooperation Council (GCC) countries. Globally around 10–20% of drylands and 24% of the world's usable lands are degraded. Globally 21 billion tons of soil is eroded, resulting in a loss of 3 tons per capita per year. In addition, soil erosion causes 60% crop yield loss. Recent estimates showed the impact of climate change on land degradation, revealing a loss of about 50% of coastal wetlands over the last 100 years due to combined effects of localized human pressures, sea level rise, warming, and extreme climate events (IPCC 2023). Prolonged drought and rise in temperature in the arid and semi-arid regions are likely to trigger land degradation and crop production. The access of farmers to water for irrigation is essential in reducing or coping the drought risk and climate impacts in many regions; however, the use of brackish water may cause salinization, therefore, water quality to be considered to avoid soil salinization (IPCC 2023). Increase in temperature has many consequences, such as but not necessarily limited to desertification, reduction in precipitation, loss of biodiversity, land degradation, forest degradation, glacial retreat and related impacts, ocean acidification, sea level rise, submergence and loss of coastal ecosystems, salinization, with cascading to risks to food and water security (WGII SPM footnote 29 IPCC 2023). When the economic losses due to land degradation are viewed, it is estimated at 1.5 to 3.4 trillion Euro in 2008, which is equal to 3.3–7.5% of the global Gross Domestic Product (GDP) in 2008 (ELD 2013).

Land degradation is a global issue and a high international agenda due to many reasons including loss of productive soil which is essential for agriculture intensification to offset food demand of unprecedented population growth (Imeson 2011; Grainger 2013; IPCC 2017). The world's soil status report (FAO-ITPS 2015) indicated "the majority of the world soil resources are in only fair, poor, or very poor conditions, and that conditions are getting worse in far more cases than they are improving". This report has warned the countries to take necessary actions to mitigate the desertification by adopting best management, approved and tested soil conservation practice to minimize the desertification effects on soil resources.

5.3 UNCCD and Land Degradation Neutrality

To combat desertification, the United Nations Convention to Combat Desertification (UNCCD) was established in 1994 (UN 1994; UNCCD 1994a, b), however, the UNCCD entered into force on December 26, 1996. The vision to establish the

UNCCD was to mitigate the effects of desertification and drought through national action programs (NAP), and various countries then developed NAP to combat desertification. The UNCCD is signed by a total of 196 country parties and the European Union. Kuwait signed the UNCCD in 1995. The UNCCD is the only legally binding framework set up to address desertification and the effects of drought. The UNCCD (1994a, b) defines desertification as land degradation in arid, semi-arid, and dry sub-humid areas resulting from various factors, including climatic variations and human activities (UN 1994); however, it is intensified in the hyper-arid climatic conditions like Kuwait. The current state of global desertification reflects that it has not been controlled, instead, it is becoming even worse; and annually it is expanding at a rate of 50,000–70,000 km^2 (Ezcurra 2006; Saier 2010). The UNCCD views land degradation as a result of human-induced actions which exploit land, causing its utility, biodiversity, soil fertility, and overall health to decline. The UNCCD further elaborates that our current agricultural practices are causing soils worldwide to be eroded up to 100 times faster than natural processes replenish them.

The UNCCD stresses on the land degradation neutrality (LDN) concept, the question is how this can be achieved. The UNCCD (2023) defines LDN as "*a state whereby the amount and quality of land resources necessary to support ecosystem functions and services to enhance food security, remain stable, or increase within specified temporal and spatial scales and ecosystems*". The UNCCD recommends three actions to achieve LDN, these are, (i) avoid new degradation of land by maintaining existing healthy land; (ii) reduce existing degradation by adopting sustainable land management practices that can slow degradation while increasing biodiversity, soil health, and food production; and (iii) ramping up efforts to restore and return degraded lands to a natural or more productive state (https://www.unccd.int/land-and-life/land-degradation-neutrality/overview).

The UNCCD describes the objectives to achieve LDN, including to maintain or improve the sustainable delivery of ecosystem services and land productivity to enhance global food security, increase the resilience of land and the populations dependent on it, seek synergies with other social, economic, and environmental objectives and reinforce and promote responsible and inclusive land governance.

5.4 Land Degradation and Desertification: Kuwait Perspectives

In Kuwait human-induced land degradation is of multiple types including but not limited to salinization (agricultural farms and coastal area), soil compaction, overgrazing, loss of biodiversity and fertile soil leaving behind a very poor wildlife habitat and ecosystem, destruction of habitats and resilience in natural ecosystem, soil contamination, landfilling, mining, leaching of pesticides and fertilizers from agricultural areas and subsequent pollution of shallow groundwater.

Land degradation is one of the severe threats to the terrestrial environment and ecosystems of Kuwait making the landscape look like deserts due to loss of biodiversity, productive surface soil, and depletion of organic matter. In Kuwait human-induced land degradation is mainly due to overgrazing and recreational camping causing ground cover degradation, reducing vegetation cover, and influencing hydrological cycle and infiltration rate (Al-Dousari et al. 2000). As a result of ground cover degradation the infiltration rate decreased from a minimum of 18.46% to a maximum of 91.96% compared to non-degraded soil in the same area. In another area in Al Ahmadi infiltration rate decreased by 61.45% compared to non-degraded area (Al-Awadhi et al. 2005).

According to the soil survey of Kuwait (KISR 1999; Shahid and Omar 2022) Kuwait's landscape is classified into two soil orders based on the US Soil Classification (Soil Survey Staff 2022), these two soil orders cover 30% (Entisols) and 70% (Aridisols) of surveyed area. The Entisols are the least developed soils and are composed of loose sandy soil and are vulnerable to wind action and thus causing aeolian deposits especially on level to gently undulating plains, sand dunes, steep actively eroding slopes, and drainage depressions (Khalaf and Al Ajmi 1993). It is to be noted that Kuwait is a dry and hot country where wind erosion dominates over the water erosion, however, during the rainy season heavy rain does cause flash floods which remove significant quantity of sediments and deposit elsewhere, leaving behind the degraded land in the form of different sized rills and gullies, such a phenomenon has been observed during heavy rains in 2018 and 2023.

In order to reduce land degradation or to sustain the land resources for better services the human and climate change impact must be counteracted on a priority basis, which Kuwait is making momentous efforts to halt further land degradation by mega desert revegetation projects (Alenezi et al. 2021; Madouh et al. 2021). It is also essential to educate the community to respect the environment in general and soil and land resources in specific.

5.5 Soil Particles Movement Mechanisms in the Desert Environment of Kuwait

Bagnold (1973) distinguished three different modes of particles movement, that is creep, saltation, and suspension, while Shahid et al. (1999, 2003) and Shahid and Omar (2001) described the chain reaction of three different modes. In the particles movement mechanism, during surface *Creep* particles >500 μm are set in motion by the impact of saltating particles. They tend to roll and creep along the surface. During the rolling process, they become rounded by losing sharp edged inherited from the weathering of rocks from which these were generated. In the *Saltation* movement the particles (63–500 μm diameter) are initially rolled, a vacuum is created in the rear and in the front the air is co-pressed and the particles are lifted up in the air (Shahid and Abdelfattah 2008; Shahid and Alsumaiti 2022). On reaching the soil surface, they may rebound or become embedded when impacting the surface, or induce creep and suspension and lift the particles into the air, where they remain in atmosphere as long as air holds in the atmosphere and lodge on soil surface with rain or when wind speed is settled (Fig. 5.1). The suspension particles cause dust storm called "toze" and reach the soil surface with rain and clog the soil surface to form a surface crust. This process causes dust storms, creating many health and environment-related problems including but not necessarily limited to: dust pneumonia, plant burial, blockage of highways, air pollution, and loss of organic matter-rich top soil (Chepil 1945; Shahid and Abdelfattah 2008; Shahid and Alsumaiti 2022). However, having said this, mainly, in Kuwait dust storm is an external phenomenon as the deserts do not have enough finer particles (<63 μm) to dense dusty days. The dust enters Kuwait with wind blowing from sources in nearby countries.

The saltation (70%) followed by Creep (20%) and suspension (10%) are the mechanisms of particles movement (Burezq 2020) in the deserts of Kuwait. In another study (Ahmad and Shahid 2002) in the Wafra area the distribution was saltation (72.4%), creep (25.1%), and suspension (2.4%). This suggests, although there are differences in particles movement in different locations the trend remains the same (saltation > creep > suspension). Therefore, to minimize the sand movement there is a need to increase the soil aggregate size more than 500 μm to reduce sand movement in the deserts of Kuwait (Shahid and Abdelfattah 2008; Ahmad and Shahid 2002).

In a similar study in Abu Dhabi Emirate, United Arab Emirates (Shahid and Abdelfattah 2008; Shahid and Alsumaiti 2022) the extent of particle movement is determined as saltation > creep > suspension. In this study, particles analyses show distribution in size ranges 5–24% in creep, 70–92% in saltation, and 2–8% in suspension. Both the studies in UAE and Kuwait have confirmed that the deserts of both the countries present dominant particle size in saltation range, therefore,

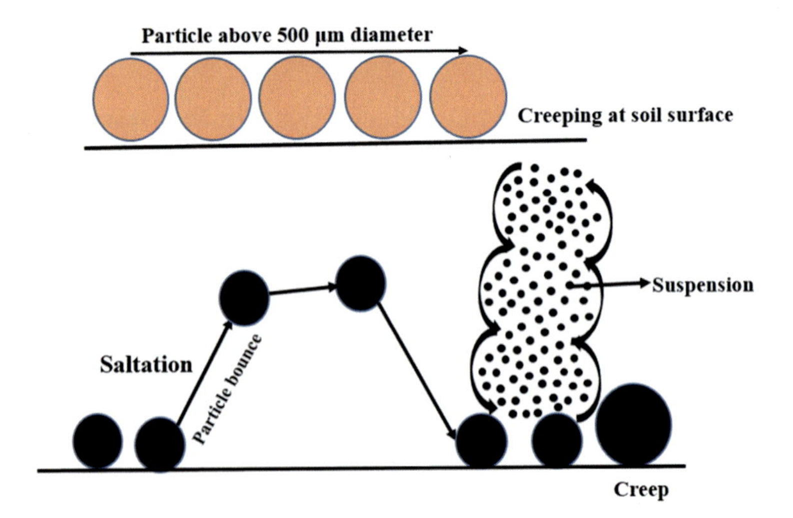

Fig. 5.1 Mechanism of particle movement in a desert environment

saltation moves the main mass of the desert environment. This suggests, to stabilize the desert soil surfaces using eco-friendly organic-based material such as carboxy-methyl-cellulose (CMC), which is commonly used in China for deserts rehabilitation and soil soilization (Yi and Zhao 2016).

5.6 High Rainfall, Flash Floods, and Water Erosion

Historic flood events in Kuwait were recorded in a number of years, e.g., 27 December 1934, 30 November 1954, 2 Feb 1993 (40 mm within 6–8 h), 11 November 1997 (105 mm within 3–4 h), January 2004, 2007, April 2008, November 2009 (Misak et al. 2013) followed by heavy rains on 9 November 2018, when 49.2 mm rain fell on Kuwait Airport on 9 November 2018. Another cause of land degradation is the flash floods caused due to unexpected high rain fall, e.g., in 2018 a rainfall of 343 mm was recorded. This has caused water erosion in low-lying areas especially in Wadis, e.g., one storm of 30–40 mm (Al-Dousari et al. 2007). This amount is the combined rain that usually occurs during October–December. The last spell of heavy rain fell on 4 December 2022, and March 2023. These floods have caused water erosion and tons of sediments flown to wadis and low-lying areas. The historical records of rainfall in Kuwait suggests that there are opportunities in area such as Jal Az Zour, Wadi Al Batin, Ahmadi ridge, to harvest rainwater, provided infrastructures are established to retain the rainwater for later use or to recharge the aquifers (Kwarteng et al. 2000). Such efforts have been made in the past (Al-Dousari et al. 2007) to harvest water in the Shuaiba industrial area.

5.7 Land Degradation Indicators and Associated Types of Land Degradation Depicted in Kuwait Deserts

The land degradation indicators can be linked to various types (physical, chemical, and biological) of soil/land degradation. The land degradation indicators observed in Kuwait's deserts are presented in Table 5.1 with respect to the type of soil/land degradation.

5.8 Determination of Loss or Gain of Soil, Surface Gravels, and Surface Soil Condition

Loss or gain of soil due to both the wind and water erosion from a specific location can be measured using the rod method. A marker rod with a known length is installed on the ground surface where loss/gain is to be measured. At the time of installation,

Table 5.1 Visual land degradation indicators depicted during field investigation and associated types of soil and land degradation

Visual indicators depicted in the field	Types of soil and land degradation					
	Wind erosion	Water erosion	Salinity/ alkalinity	Physical	Chemical	Biological
Rills	x	√	x	x	x	x
Armor layer	√	√	x	x	x	x
Nebkha	√	x	x	x	x	x
Dust storm	√	x	x	x	x	x
Bare land (salts)	x	x	√	x	√	x
Ripples on soil surface	√	x	x	x	x	x
Unstable soil surface	√	x	x	x	x	x
Hardpan (gatch)	x	x	x	√	x	x
Low organic matter	√	x	x	x	x	√
Lighter soil color	√	x	x	x	x	√
Mineral soil crust	x	√	x	x	x	x
Surface salt-crust	x	x	√	x	√	x
Soil pollution and contamination (heavy metals/oil)	x	x	√	x	√	x

the mark must be at ground level (zero). During the monitoring process, the length of the rod is measured and calculation is made to determine the loss or gain of sediment from the site (Table 5.2).

5.9 Desertification and Sustainable Management Options in Kuwait

During the overview of the desert environment in Kuwait, a number of soil surface features were observed (Table 5.3). Each feature indicates a feature of desert land degradation. Reconnaissance views of deserts were captured through flying on helicopter over the desert landscape. Visual observations from the deserts of Kuwait are also briefly described in Table 5.3. This chapter is an update of the previous work of Shahid et al. (1999, 2001). More details about the land degradation aspects of Kuwait can be found elsewhere (Shahid et al. 1999, 2003; Shahid and Omar 2001; Misak et al. 2002; Omar et al. 2005, 2006).

5.10 Drifting Sand and Their Management Strategies

Another area which needs to be addressed is to control the drifting sand in the deserts of Kuwait. Fourteen areas have been mapped which are highly vulnerable to sand encroachment (Al-Awadhi and Misak 2000; Al-Awadhi et al. 2003) dominating in the Al-Huwaimiliyah-Wafra wind corridor. Potential measures to control

Table 5.2 Field assessment methods of surface soil erosion

Soil attributes	Measurement ranges	Interpretation
Top soil loss/gain Iron rod to measure soil loss/build up in a desert environment	**Not eroded** (0 cm) **Soil eroded** 0–1.0 cm −1.1 to −2.0 cm −2.1 to −3.0 cm −3.1 to −4.0 cm −4.1 to −5.0 cm > −5.1 cm **Soil build up** +0 to 1.0 cm +1.1 to +2.0 cm +2.1 to +3.0 cm +3.1 to +4.0 cm +4.1 to +5.0 cm > +5.1 cm	Under desert environment when sites are bare (without vegetation) the soil surface is directly exposed to the action of the wind and fine particles are removed and gravels are left at the surface (armor layer). However, if sites are vegetated, the soil is protected from wind action to great extent which results in control of surface soil loss. To measure soil losses/build up, the exact locations from which sample measurements are taken should be marked, showing the field assessment to keep a record of these sites for monitoring purpose. Thus, in order to measure whether soil is lost or build-up, a rod (50 cm length) is installed at each site (20 cm belowground and 30 cm above ground). Soil build-up will show decreased aboveground length of the iron rod, whereas, the increase in aboveground length of the rod will show loss of soil. Based on these measurements weight of soil loss or build-up will be determined and monitored during the project duration. *Results interpretation*: A build-up of 1 cm soil indicates 16 kg soil deposition on an area of 1 m^2 (considering bulk density of soil as 1.6 g/cm^3). Vice versa loss of 1 cm soil indicates 16 kg soil/m^2 soils has been eroded.
Surface gravel (% cover)	0 1 to 10% 10.1 to 30% 30.1 to 50% 50.1 to 75% >75%	Surface gravel layer (Armor layer) is the concentration of gravels at surface (coarser soil particles) that would ordinarily be randomly distributed throughout the soil (in depth) (Stocking and Murnaghan 2000). The concentration of gravels indicates that finer material including organic matter has selectively been removed by wind erosion. Surface gravels are going to be estimated by two ways (i) visual assessment of % surface cover by gravels in the field, (ii) by sieving the whole soil collected from field to determine the ratio of coarse/fine (C/F). The increase in C/F will indicate erosion has increased over time, if C/F is decreased, it indicates site is protected and soil build up occurred during assessment period. The aggregates buildup in 2–1 mm fraction will show soil structure build up.

(continued)

Table 5.2 (continued)

Soil attributes	Measurement ranges	Interpretation
Surface soil condition Surface soil condition varies depending upon its exposure to environmental conditions, soil near to sea will be crusted with salts, eroded soil will present surface gravels, impact of water creates soil mineral crust, nutrients rich soil presents biological crust, heavy eroded location exposes subsurface (truncated) like gatch (hard set subsurface feature), impact of heavy water flow creates small (rills) and large channels (gullies)	Loose Soil crust Salt crust Rain crust Biological crust Gravelly loose Gravelly crusted Truncated	*Loose surface* – Soil material is loose and cohesive sample cannot be obtained (recently deposited soil material) *Crusted surface* (a thin surface in mm thickness) *Crust* – a thin soil surface in millimeters thickness *Soil crust* – bonded soil particles into thin layer *Salt crust* – layer of crystallized salts *Rain drop impact crust* (dispersed, puddled, dried, and settled in water) *Biological crust* (algal, cyanobacteria, and microphytic crusts) *Gravelly surface* Gravels of various sizes left on surface after fine material eroded by wind *Truncated surface* Exposed subsoil due to erosion Exposed hardpan “gatch” *Water eroded surface* Signs of small-sized drain channels (rills) and medium to large-sized “gullies”

Table 5.3 Visual observations of land degradation/desertification in the terrestrial environment of Kuwait

Desertification indicators	Visual observations from the deserts of Kuwait
Build-up around shrubs/plants – soil accumulates around the plants or tree trunks. This shows soil build-up in one area and erosion in another area from where the soil has been picked up. This shows the sediment is recycled in the terrestrial desert-type landscape, revealing soil loss and built-up is a simultaneous process. The shrubs and trees are the barriers to the moving sand in the desert environment. Soil mounds around plants are called "nebkha".	
Armor layer – the wind in the desert environment removes loose sandy material and leaving behind the heavy gravels. Overtime the surface gravels are increased and cover the surface to various levels. The gravel layer is called "armor layer" and shows signs of high-level land degradation. These gravels then protect further loss of sand from the area and serve as buffering agents.	
Loose and fragile – the desert environment presents diversified features including loose sandy soil, crusted surface, water erosion, sand accumulation around the plants, salty landscape, and even special fractured features revealing fragile ecosystem. The reason being high temperature and low rainfall leading to soft-rock *(lithified calcareous sand)* to fracture and exposing internal surface for further weathering. The fracture features are the niches for native plants to grow as evident in the image.	
Desert dumping – dumping the unwanted waste materials in the deserts that occupy the area for further use. The material may become a source of drifting sediment with the effect of wind or it protects the wind effects in case the material is strong enough and not picked up by the strong wind. Such features are not common, but occur in few areas. Dumping changed the visual look of the desert landscape.	

(continued)

Table 5.3 (continued)

Desertification indicators	Visual observations from the deserts of Kuwait
Build-up against barriers – erecting a series of fences in the deserts also plays important role in minimizing the movement of drifting sand. If the surface of the accumulated sand is planted with native vegetation adopted to local environment, it becomes a permanent blockade to drifting sand. The protected areas in the deserts are in general fenced, thus minimizing the wind effects. Fence lines are the barriers to moving sand.	
Blockage of highways – drifting sand if not properly controlled by erecting combined trees and shrubs-based shelter belts, the drifting sand can then accumulate on the roads causing serious accidents and slowing the traffic. Such features require regular clearing of roads for traffic and are costly. Another solution to minimize the sand movement around the roads area is sand surface stabilization using organic-based binders such as carboxy methyl cellulose (Yi and Zhao 2016)	
Desert crusting and Hardsetting – this is another feature observed in the desert where soil surface is being impacted by the rain drops, caused depressions and the rain drop splash action has created crusted surface. These crusts resist the effect of wind to dislodge the sediment and thus protect the subsurface from loss. The crusted surface is strong restricting even the seeds germination even if the soil seed bank exists, this is true as no plants are seen in the crusted area.	
Loss of vegetation – the desert area which is under severe threats of wind especially in the wind corridor where wind is blowing continuously and not providing sufficient time to develop soil to retain seeds and to grow plants. Such areas are usually bare and devoid of vegetation cover. However, if water is made available to grow trees and shrubs, this area may be rehabilitated to combat further degradation. The trees growing in the degraded land is an evidence of life.	

(continued)

Table 5.3 (continued)

Desertification indicators	Visual observations from the deserts of Kuwait
Waterlogging – water ponding in deserts is another indicator of land degradation. This is caused due to the rise of water table in the coastal area due to seawater intrusion, in inland due to the existence of hard pan at shallow depths and the rainwater ponds, accumulation of flowing rainwater in depressions, over-irrigation in agriculture farms developed on shallow depth soils where hardpan is near the soil surface.	
Salinization – salts accumulation in the desert environment has been observed in three places, (i) the landscape near the coast where seawater intrusion and subsequent evaporation has caused the salts accumulation; (ii) inland sabkha (salts in depression areas); and (iii) agriculture farms irrigated with saline/brackish water. The latter can cause harmful effects to plants and crop yield can be reduced significantly. Salinization in Kuwait is predicted as an early warning of land degradation (Shahid et al. 1998). Overall 38.54% of terrestrial desert landscape is affected by salinity (Burezq et al. 2022; Shahid 2022). This area does not include salinized agriculture farms.	
Water Erosion – high rainfall in the number of years caused the flash floods resulting into mass movement of soil sediment from the highlands such as Jal Az Zour and Wadi Al Batin in Kuwait. The effect of water can be seen in the form of small- and medium-sized rills and gullies (water flow channels). The water not only removes soil sediment but also removes plants while moving on the landscape with high force. Finally, the sediment ends up in depressions where it creates ponds in the deserts.	
Water channels – such features are common in the deserts due to the heavy flow of water after high-intensity rainfall. The flowing water detaches the soil particles and flow through small channels “rills” which over time are increased into different-sized “gullies” and become a permanent feature for water flow after heavy rain. If these rills/gullies are not taken care of, their sizes over time are increased and become permanent feature of water flow. If the water flow is directed to an area where infrastructures are developed for water harvesting (dams), the water can be accumulated for either use or for aquifer recharge.	

(continued)

Table 5.3 (continued)

Desertification indicators	Visual observations from the deserts of Kuwait
Off-road vehicles movement – this can cause two effects, (i) loss of vegetation in the area where vehicles are moving, (ii) hard setting and compaction. The compaction then results into significant reduction in infiltration rates thus causing water to lodge at surface leading to waterlogging. Once the area is waterlogged, the native vegetation cannot survive, where over time *Australis phragmies* (reeds) grows as dense vegetation.	
Soil extraction – mining of soil is another activity leading to disruption of soils and exposure to wind erosion. The mined material is used for industrial purposes including but not necessarily limited to construction material (limestone for cement production), gravels for use in concrete, gypsum for construction material (gypsum slabs, setting cement) and to reclaim saline-sodic and sodic soils, etc. the lithified sand, calcareous material is used as road beds.	
Plants roots exposure – the wind and water erosion moves soil sediments and can expose the plant's roots to the surface. After measuring the length of roots from the soil surface and knowing the bulk density of the soil, the quantity of soil removed can be calculated. Considering the bulk density of desert soil 1.6 g cm^{-3}, soil loss to a depth 1 cm in an area of 1 m^2 will be 16 kg.	
Desert camping – November to march is the camping season in Kuwait. During this time, family erect camps in the deserts and spend evenings there especially during the weekends. Such activities lead to loss of vegetation in the camping area as well as makes hard setting/crusting in deserts, where after heavy rains water ponding can be seen in deserts. The camping area has been designated by the government to avoid wider spread of landscape disturbance.	

(continued)

Table 5.3 (continued)

Desertification indicators	Visual observations from the deserts of Kuwait
Oil contamination – oil contamination during the Gulf War significantly affected the terrestrial environment and ecosystems of Kuwait. The contamination was of different levels and types in the form of soil surface (oil soot, oil crust, oil lakes), and subsurface (oil trenches) contamination. After over 30 years of Gulf War, still in many places soils are contaminated with hydrocarbons, and are under remediation process.	

wind erosion include but not necessarily limited to sand stabilization using organic-based binder through soilization (Yi and Zhao 2016) process, erecting windbreakers of drought-tolerant trees, construction of bund walls, and fences. Sand soilization (turning sand into soil) is possible by using plant-based cellulose to aggregate loose sand to attain eco-mechanical attributes of natural well-structured soil to withstand wind effects, retain water, nutrients, and air. Sand soilization is a remarkable transformation based on the revelation of the eco-mechanical attributes of soil (Yi and Zhao 2016; Yi et al. 2016).

In earlier works, Misak et al. (2009, 2013) proposed three scenarios to manage the hazards of drifting sand.

Scenario 1: Establishment of two green belts consisting of 10 rows of *Prosopis juliflora, Ziziphous spina-christi*, and *Tamarix aphylla* trees. The first belt of 25 km length at Huwaimiliya, the other (130 km long) at Ras Al-Sabiyah-AlSalmi. In addition, sand stabilization (using ecomat, coir, plant residues) between the two belts is also part of this scenario.

Scenario 2: Afforestation of 615 km^2 area by growing *Prosopis juliflora, Ziziphous spina-christi,* and *Tamarix aphylla* trees along the Huwaimiliya-Al Wafra corridor, provided the trees are irrigated with treated sewage water for at least 1 year for their establishment. In addition, conservation/revegetation of native shrubs (*Haloxylon salicornicum, Rhanterium epapposum*) is also proposed to trap drifting sand.

Scenario 3: Establishment of green belts in the northern and central parts of Kuwait, revegetation of southern Kuwait by *Rhanterium epapposum*, stabilization of sandy bodies using various materials (green residue, mulching).

5.11 Large-Scale Desert Rehabilitation

The anthropogenic effects during the Gulf War have degraded the deserts of Kuwait to a great extent. This has caused damages in many ways including but not necessarily limited to the loss of vegetation, disruption of desert soil surface and triggering wind erosion, oil contamination, etc. Currently, two large-scale projects are being implemented under the Kuwait Environmental Remediation Program-KERP (Alenezi et al. 2021; Madouh et al. 2021). These rehabilitation projects are being implemented separately in different places, where revegetation will be accomplished using native vegetation adapted to Kuwait environmental conditions. The areas which will be rehabilitated are briefly given below.

KERP SP003EC (Alenezi et al. 2021)

Four areas will be revegetated:

1. Al-Huwaimilya Nature Reserve Area
2. Wadi Al Batin Nature Reserve Area
3. Umm Qudair Nature Reserve Area
4. Al Khuwaisat Nature Reserve Area

KERP SP004EC (Madouh et al. 2021)

The project is implemented at Sabah Al Ahmed Nature Reserve (SAANR), both at the terrestrial and coastal sites.

Both the projects are under the implementation phase (2021–2026), and the results of the ecosystem restoration due to large-scale vegetation are not available. However, the introduction and scope of different tasks (soil investigation, vegetation survey, wildlife survey, roots and soil microbiology, dust and sand collection, of both the projects) are outlined in a recently published KERP projects manual (Alkandari and Shahid 2023).

In addition to these two projects, Kuwait Oil Company (KOC) is implementing the biggest in the world environmental project to revegetate 42 square kilometers of land in the north and south of the country, by cultivating nearly 10 million desert trees of 12 varieties, shrubs and fungal herbs suitable for Kuwait's natural environment (www.arabtimesonline.com).

5.12 Conclusions and Recommendations

From the above results, it is concluded that the deserts of Kuwait are under serious threats of land degradation, where desertification is recognized as a central point inhibiting the speed of desert recovery and ecosystem restoration. This is due to its location in the arid region where evaporation exceeds manifolds to precipitation, mainly due to high temperature, poor vegetation covers, and low rainfall. Further, the desert soils are loose and sandy in nature, poor in organic matter, thus highly vulnerable to both the wind and water erosion. The wind recycles the surface sediments over the whole year and causes worse environmental conditions like dust storms, accumulation of sand on highways, around building, etc. The analyses of soil particles movement in the deserts of Kuwait revealed "saltation" mode of movement is the dominant one followed by creep and suspension movement. This suggests to improve surface soil structure development and soil aggregation to increase size not easily picked by wind action. The wind erosion can be controlled or at least minimized by adopting various strategic measures, including surface soilization using the environment-friendly organic-based material, erecting shelter belts "windbreakers" in the deserts and across the highways, installing infrastructure to harvest water into depression areas for use or to recharge the depleted aquifers. A number of diversified land degradation indicators have been observed in the deserts of Kuwait requiring multi-discipline mitigating actions including physical, chemical, hydrological, and biological methods. Finally, Kuwait should make all efforts to achieve the objectives of land degradation neutrality set up by UNCCD to sustain the desert environment to keep the soils healthy for food security and the delivery of healthy ecosystem services in the long run.

References

Adeel Z, Safriel U, Niemeijer D, White R, de Kalbermatten G, Glantz M (2005) Ecosystems and human well-being: desertification synthesis: a report of the millennium ecosystem assessment. World Resources Institute, Washington, DC

Ahmad M, Shahid SA (2002) Aeolian soil particles movement mechanisms in the deserts of Kuwait. Technical Report KISR No 6563, Project No EC001G, Kuwait Institute for Scientific Research Kuwait, p 24

Al-Awadhi JM, Misak RF (2000) Field assessment of Aeolian sand processes and sand control measures in Kuwait. Kuwait J S Eng 27(1):158–176

Al-Awadhi JM, Misak RF, Omar SAS (2003) Causes and consequences of desertification in Kuwait. A case study of land degradation. Bull Eng Geol Environ 62(2):107–115

Al-Awadhi JM, Omar SAS, Misak RF (2005) Land degradation indicators in Kuwait. Land Degrad Dev 16(2):163–176

Al-Dousari AM, Misak RF, Shahid SA (2000) Soil compaction and sealing in Al-Salmi area, Western Kuwait. Land Degrad Dev 11(5):401–418

Al-Dousari AM, Misak RF, Al Gamily H, Neelamani N (2007) Integrated system for flood management in Shuaiba area and its vicinities. KISR ECO055C final report, KISR 8910

Alenezi A, Quoreshi A, Bhat NR, Shahid SA, Omar SAS (2021) Ecological monitoring and evaluation of active revegetation of degraded terrestrial ecosystems SP003EC under KERP, Kuwait Institute for Scientific Research Kuwait publication (Proposal)

Alkandari A, Shahid SA (2023) Field assessment guiding manual for Kuwait Environmental Remediation Program (KERP) project sites. Kuwait Institute for Scientific Research Kuwait Publication, pp 71 KISR 17999

Bagnold RA (1973) The physics of blown sand and desert dunes, 5th edn. Chapman and Hall, London

Burezq H (2020) Combating wind erosion through soil stabilization under simulated wind flow condition – case of Kuwait. Int Soil Water Conserv Res 8(2):154–163

Burezq H, Shahid SA, Baron HJ (2022) Salts in the terrestrial environment of Kuwait and proposed management. In: FAO. 2022. Halt soil salinization, boost soil productivity – Proceedings of the Global Symposium on Salt-affected Soils. 20–22 October 2021. Rome. https://doi.org/10.4060/cb9565en, pp 267–268

Chepil WS (1945) Dynamics of wind erosion. I. Nature of movement of soil by wind. Soil Sci 60:305–320

ELD (2013) Economic of land degradation. https://www.eld-initiative.org/en/about/the-eld-initiative. Accessed 11 Apr 2023

Ezcurra E (2006) Global deserts outlook. United Nations Environment Programme, Nairobi

FAO-ITPS (2015) Status of the world's soil resources. Main report. Food and Agriculture Organization of the United Nations and Intergovernmental Panel on Soils, Rome Italy

GEO 4.-UNEP (2006) Dent D, Chapter Land: in Global Environment Outlook 4 'UNEP 81_114'

Grainger A (2013) The threatening desert: In: Controlling desertification. EarthScan Publications Ltd, London

Higgitt D (1993) Soil erosion and soil problems. Prog Phys Geogr: Earth Environ 17:461–472

Imeson A (2011) Desertification, its causes and why it matters. In: Desertification, land degradation and sustainability, p 5–41

Intergovernmental Panel on Climate Change (IPCC) (2017) IPCC special 538 report on global warming of 1.5 °C expert review of the first order draft and start of the third lead author meeting

Intergovernmental Panel on Climate Change (IPCC) (2023) Longer report: synthesis report of the IPCC sixth assessment report (AR6)

Johnson DL, Ambrose SH, Bassett TJ, Bowmen ML, Isaacson JS, Crummey DE, Johnson DN, Lamb P, Saul AM, Winter-Nelson AE (1997) Meanings of environmental terms. J Environ Qual 26:81–589

Khalaf FI, Al-Ajmi D (1993) Aeolian process and sand encroachment problems in Kuwait. Geomorphology 6:111–134

KISR (1999) Soil survey for the State of Kuwait: reconnaissance survey, vol II & III. Kuwait Institute for Scientific Research, Kuwait

Kwarteng A, Viswanathan M, AlSenafy M, Rashid T (2000) Formation of groundwater in northern Kuwait. J Arid Environ 46:137–155

Lindskog A, Tengberg A (1994) Drylands, sustainable use of rangelands into the twenty-first century. Technical report. International Fund for Agricultural Development (IFAD) Rome

Madouh T, Quoreshi A, Bhat NR, Shahid SA, Omar SAS (2021) Ecological monitoring and evaluation of restoration and revegetation success of areas damaged by military fortifications (Element 1 of Claim 5000450), Sheikh Sabah Al Ahmad Nature Reserve under KERP, March 2022. Kuwait Institute for Scientific Research (Proposal)

Mainguet M (1994) Desertification: natural background and human mismanagement, 2nd edn. Springer Science & Business Media, Berlin

Misak RF, Al-Awadhi JM, Omar SAS, Shahid SA (2002) Soil degradation in Kabd area, southwestern Kuwait City. Land Degrad Dev 13:403–415

Misak RF, Al Sudairawi M, Al-Dousari A, Al Gamilly H (2009) Long term national program for managing the hazard of shifting sands in terrestrial environment of Kuwait. EUD, Kuwait Institute for Scientific Research, Kuwait (Proposal)

Misak RF, Khalaf KI, Omar SAS (2013) Managing the hazards of drought and shifting sands in dry lands: the case study of Kuwait. In: Shahid SA, Taha FK, Abdelfattah MA (eds) Development in soil classification, land use planning and policy implications: innovative thinking of soil inventory for land use planning and management of land resources. Springer Science Business Media, Dordrecht, pp 707–729

Montgomery DR (2007) Soil erosion and agricultural sustainability. Proc Natl Acad Sci U S A 104(33):13268–13272

Omar SAS, Shahid SA (2013) Reconnaissance soil survey for the State of Kuwait. In: Shahid SA, Taha FK, Abdelfattah MA (eds) Developments in soil classification, land use planning and policy implications: innovative thinking of soil inventory for land use planning and management of land resources. Springer Science Business Media, Dordrecht, pp 85–110

Omar SAS, Bhat NR, Shahid SA, Assem A (2005) Land and vegetation degradation in war-affected areas in the National Park/nature Reserve of Kuwait: a case study of umm Ar Rimam. J Arid Environ 62:475–490

Omar SAS, Bhat NR, Shahid SA, Assem A (2006) Land and vegetation degradation caused by military activities: a case study of the Sabah Al-Ahmad nature Reserve of Kuwait. Eur J Sci Res 14(1):146–158

Saier MH Jr (2010) Desertification and migration. Water Air Soil Pollu 205(S1):31–32

Shahid SA (2022) Innovative thinking and use of salt-affected soils in irrigated agriculture. In: FAO. 2022. Halt soil salinization, boost soil productivity – Proceedings of the Global Symposium on Salt-affected Soils, 20–22 October 2021. Rome, pp 292–293 https://doi.org/10.4060/cb9565en

Shahid SA, Abdelfattah MA (2008) Soils of Abu Dhabi emirate. In: Perry R (ed) Terrestrial environment of Abu Dhabi Emirate. Environment Agency, Abu Dhabi, pp 71–91

Shahid SA, Alsumaiti T (2022) Soil conservation practices and efforts made to combat desertification in The United Arab Emirates. In: Li R, Napier TL, El-Swaify SA, Sabir M, Rienzi E (eds) Global degradation of soil and water resources-regional assessment and strategies. Springer-Science Press Beijing Joint Publication, pp 325–341

Shahid SA, Omar SAS (2001) Causes and impacts of land degradation in the arid environment of Kuwait. In: Bridges EM et al (eds) Response to land degradation, pp 76–77

Shahid SA, Omar SAS (2022) Kuwait soil taxonomy. Springer, New York, p 149

Shahid SA, Omar SAS, Grealish G, King P, El-Gawad MA, Al-Mesabahi K (1998) Salinization as an early warning of land degradation in Kuwait. Prob Desert Dev 5:8–12

Shahid SA, Omar SAS, Al-Ghawas S (1999) Indicators of desertification in Kuwait and their possible management. Desertif Control Bull 34:61–66

Shahid SA, Omar SAS, Al-Ghawas S (2001) Evaluation of aeolian soil movement mechanisms as a function of particle size analysis. Kuwait Inst Sci Res Kuwait Ann Rep 5851:37–40

Shahid SA, Omar SAS, Misak R, Abo-Rezq H (2003) Land resources stresses and degradation in the arid environment of Kuwait. In: Alsharhan AS, Wood WW, Goudie AS, Fowler A, Abdellatif EM (eds) Desertification in the third Millenium. Swets & Zeitlinger Publishers, Lisse, pp 351–360

Soil Survey Staff (2022) Keys to soil taxonomy, 13th edn. USDA-NRCS

Stocking M, Murnaghan N (2000) Land degradation guidelines for field assessment. Overseas Development Group University of East Anglia, Norwich

UN (1994) Convention of desertification. Information Programme on Sustainable Development, New York

UNCCD (1994a) United Nations Convention to Combat Desertification, Adopted 17 June 1994 in Paris, France, United Nations, New York. http://www.fao.org/desertification/article_html/ed/1.htm.2008-12-11

UNCCD (1994b) Elaboration of an international convention to combat desertification in countries experiencing serious drought and/or desertification, particularly in Africa, U.N. Doc. A/AC.241/27, 33 I.L.M.1328

UNCCD (2023). https://www.unccd.int/land-and-life/land-degradation-neutrality/overview. Accessed 19 July 2023

Yi Z, Zhao C (2016) Desert “Soilization”: An eco-mechanical solution to desertification. Engineering 2:270–273

Yi ZJ, Zhao CH, Gu JY, Yang QG, Li Y, Peng K (2016) Why can soil maintain its endless eco-cycle? The relationship between the mechanical properties and ecological attributes of soil. Sci China Phys Mech Astron 59(10):104621-2. https://doi.org/10.1007/s11433-016-0217-2

Part II
Water Resources, Salinization Aspects and Modeling

Chapter 6
Current Status, Challenges, and Future Management Strategies for Water Resources of Kuwait

Khalid Hadi

Abstract Water resources have been the basic need for mankind. Arid region like Kuwait is reported with insufficient quantity of precipitation for groundwater recharge and to serve as the surface water resource. Hence, the available water resources to meet the water needs are fresh to brackish groundwater, treated wastewater, and desalination plants. The climate change, increase in population, urbanization, and industrialization have triggered the need for establishing the strategic plans in conserving and managing the available water resources. The chapter focuses on the available sources of water in Kuwait, their quality and utilization, tentative solutions, and the road map ahead. In certain sites of the well fields, groundwater quality has been deteriorating over time. The assessment of increased water utilization in the recent years along with its supply and projected demand will pave the way toward attaining the UN sustainable development goals. The limited availability of freshwater and lack of institutional monitoring and management are the main challenges and they are discussed with probable solutions. The research highlights the importance of integrated water resources management strategies for the long-term viability of Kuwait's water supply that would lead to a circular water economy.

Keywords Water sources · Efficient utilization · Challenges · Desalination · Strategic solution · Groundwater · Treated wastewater

6.1 Introduction

Water is a critical resource for modern humanity and ecosystem services. The Quran states that God created all life from water and has complete control over everything. "Even if you take the ablutions in a flowing river, do not waste the water", the

---

K. Hadi (✉)
Water Research Centre, Kuwait Institute for Scientific Research, Safat, Kuwait
e-mail: khadi@kisr.edu.kw

M. K. Suleiman, S. A. Shahid (eds.), *Terrestrial Environment and Ecosystems of Kuwait*, https://doi.org/10.1007/978-3-031-46262-7_6

Prophet Mohammed (Peace Be Upon Him) said, emphasizing the point (Smith and Al-Maskati 2007). In this scenario, the UN thrusts on Sustainable Development Goal (SDG) 6 "Ensure access to water and sanitation for all" to provide safety "water sanitation and hygiene". Globalization, institutional strengthening, and public participation are all aimed at bringing about these outcomes by the year 2030. The success of all 17 SDGs depends critically on SDG-6. Energy, food, and water are the three most important natural resources for sustaining life on Earth. Their insecurity stifles social stability and economic growth. The SDG-6 emphasizes the need of ensuring access to clean water and transforming society into a water-smart one, where water is a precious resource and managed to minimize waste and maximize productivity. Clean water and sanitation are critical to the success of Agenda 2030 (Romano and Akhmouch 2019; Makarigakis and Jimenez-Cisneros 2019).

Kuwait is located in the northern Arabian Peninsula and covers 17,818 square kilometers area. Summer highs regularly exceed 50 °C. Annual evapotranspiration (2266 mm) exceeds rainfall (120 mm) (Al-Senafy and Abraham 2004). Rainfall barely recharges aquifers (Amer et al. 1990; Al-Sulaimi et al. 1996). In Kuwait only desalinated saltwater, a minor amount of scarce natural groundwater, and treated sewage effluent (TSE) are viable sources of usable water (Abdel-Jawad et al. 1999). Groundwater production made up 22.3% of the total water expenditure in 2006, while desalination accounted for 54.6%, and wastewater reuse accounted for 23.0% (MEW 2007; MPW 2007). It is anticipated that Kuwait being one among the 11 water-scarce countries will need 14.8 Mm^3 day^{-1} of desalination water to keep up with the urban domestic water consumption (0.265 m^3 $capita^{-1}$ day^{-1}). Water consumption per person in Kuwait is the highest in the world. The master plan of Kuwait (Al-Kuwait Al-Youm 2008) forecasts a population of 5,368,000 by 2030, requiring more desalination facilities to produce 2.8 million m^3 of drinking water day^{-1}. Nevertheless, the brine discarded from treatment plants to the sea increases seawater salinity (Purnama et al. 2005).

6.2 Groundwater Studies in Kuwait Based on Scopus Database

The literature survey concerning the studies conducted on the groundwater of Kuwait listed in Scopus database entailed 65 more relevant articles with predominantly 38 keywords (Fig. 6.1a). The network diagram was attempted to understand the interrelationship between the keywords. Each keyword is represented by a circle and the size of the circle is proportional to the frequency of usage. The most commonly used research keywords represented were with regard to groundwater, groundwater resources, aquifer, and hydrogeology, followed by water management, water quality, hydrochemistry, recharge, and groundwater pollution. Further, the intensity of lines relating to each circle indicates the frequency association, thus indicating the association of most of the groundwater research in Kuwait spins around hydrogeology, aquifer, hydrogeochemistry, water quality, water

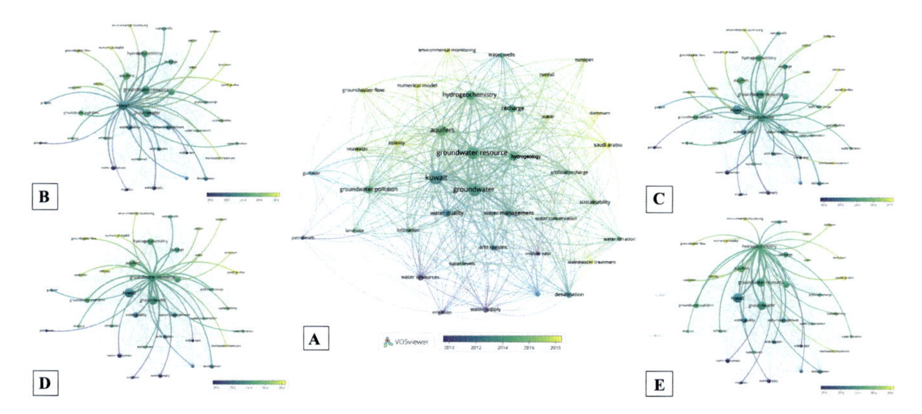

Fig. 6.1 Key word linkages of the previous water studies on water research in Kuwait. The size of the circles is proportional to the frequency of occurrence and the colors are related to the chronology of research conducted

management, and recharge. Further, the variation in color grades reflects the chronology of the publications. Though the initial publications in groundwater research started since the early 1980s, the frequency of publications notably increased after 2005 and the scaling reflects lesser frequency of publications before 2010. The recent studies on groundwater are more toward groundwater flow, numerical modeling, environmental monitoring, isotopes, and rainfall. These studies are also related to the Dammam aquifer and the adjoining Saudi Arabia region due to the transboundary nature of the aquifer.

The Keyword inter-linkage (KI) with respect to groundwater studies in Kuwait reflected (Fig. 6.1b) more frequent association of aquifers, groundwater resources, groundwater, hydrogeology, water quality, and management. The recent focus is more on environmental monitoring, water treatment, numerical modeling, isotopes, and addressing the salinity issues in water. A similar association of networking with keywords groundwater, groundwater resources, and hydrogeochemistry (Fig. 6.1c–e) indicated a similar trend as that of the KI of Kuwait groundwater studies. Inferences from KI revealed that the earlier studies focused on pollution and resource assessment, subsequently studies focused on treatment, recharge, pollution, and sustainability. In the recent days the technological advancements have helped in productive modeling of groundwater resources, through numerical flow modeling and geochemical modeling of water resources. Apart from the software application, the innovative desalination and wastewater treatment technologies by developing new membranes and application of isotopes (stable and radioactive) along with noble gases in groundwater studies have added another dimension to water research in Kuwait (Kumar et al. 2021; Rashid et al. 2022).

The integrated water management system still faces issues in achieving the targets. These issues are data collection, because the region still has significant gaps in measuring the current system. Such gaps obstruct evidence-based policymaking significantly. A gap of strong research on water management strategies for developing policies for the sustainability of water resources to adopt climate change and urban development exists. Apart from the data gaps the current status of the water resources in Kuwait, owing to its contribution from different sectors (groundwater, desalinated water, and treated wastewater) should be integrated and planned to determine the sectorial challenges. This step will help in contributing to address the challenges, and to derive future developmental strategies and sustainable management plans. Thus the current chapter focuses on the current status of the water resources in Kuwait, utilization, challenges, solutions, and future directions.

6.3 Current Status of Water Resources in Kuwait

In dry regions, such as Kuwait, preserving and protecting the quality of the groundwater supply for as long as feasible is essential for sustainable development. Groundwater, desalinated water, and recycled wastewater supply meet the majority of Kuwait's water needs. The water well fields, desalination locations, and wastewater treatment plants are distributed throughout Kuwait (Fig. 6.2).

6.3.1 Groundwater Resources

To irrigate crops, farmers in Abdally, and Wafra, pump water from deep underground aquifer. The freshwater resource is provided to the farmers at heavily discounted rates with no restrictions, and the people tend it to use beyond the requirement. Because the farmers are not restricted with regard to groundwater pumping, this leads to a decrease in groundwater levels and degradation of groundwater quality, as well as the abandoning of numerous farming regions (Al-Rashed et al. 1998). On the contrary, in several residential districts of Kuwait City and its suburbs, groundwater levels have risen dramatically over the past 20 years due to excessive irrigation of rapidly developing gardens (both private and public), water leak lines, and sewer systems. Between 1984 and 1987, the Ministry of Electricity and Water and Renewable Energy (MEWRE) commissioned the Kuwait Institute for Scientific Research (KISR) to perform an in-depth investigation on the nature and scope of the issue and to provide potential solutions (Hamdan 1987; Senay et al. 1987). As the demand for groundwater has increased over the past few years, every GCC country has noticed a reduction in groundwater levels. As a result, in Kuwait, water levels in both the Kuwait Group aquifer and the Dammam formation have been falling rapidly in recent years. It was estimated by Mukhopadhyay et al. (1994) that if extraction continues at its current rate, the central region of Kuwait will witness a maximum fall of roughly 50 m. But certain areas of the city have drainage issues during storms since groundwater lies within a meter of the surface (Al-Rashed et al. 1998) due to the presence of a gatch layer (an impermeable layer) (Riedel and Simon 1973).

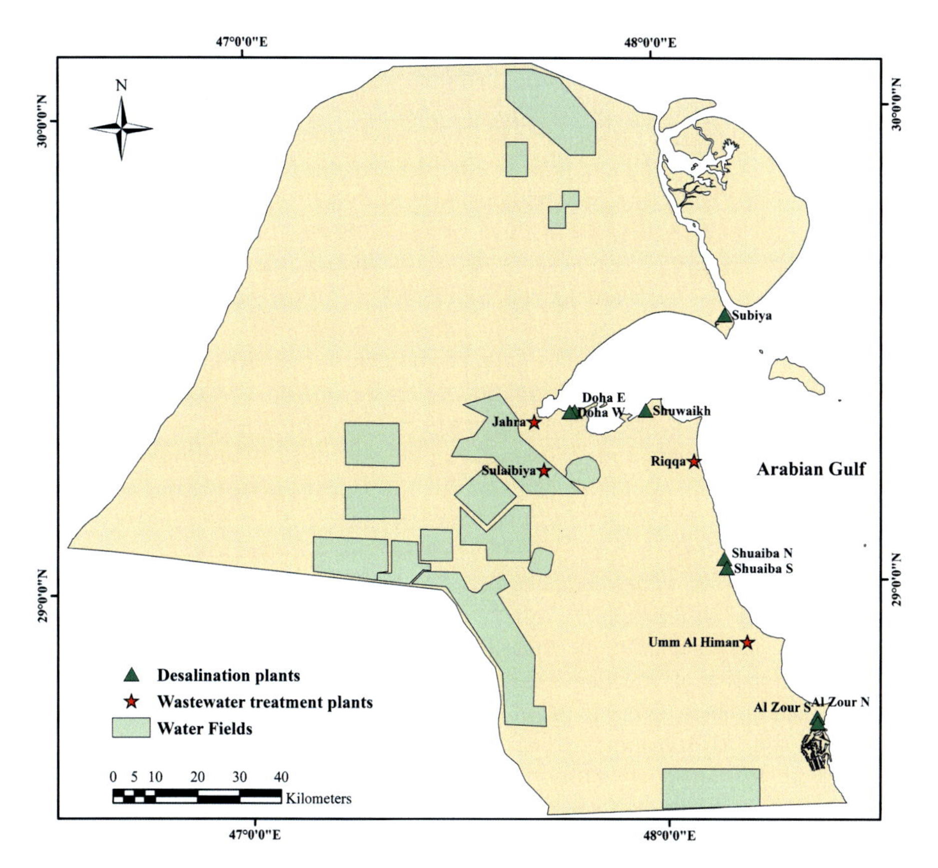

Fig. 6.2 Location of major well fields in Kuwait along with the locations of desalination and the wastewater treatment plants

Earlier researchers have studied Kuwait's water resources, and the majority of it focuses on assessing and characterizing the resources (Bergstrom and Aten 1965; Senay 1977; Mukhopadhyay et al. 1996, 2004; Al-Sulaimi et al. 1997; Al-Rashed and Sherif 2000; Fadlelmawla et al. 2007; Al-Ruwaih and Almedeij 2007). While the water resources of the country were outlined by Fadlelmawla and Al-Otaibi (2005), they addressed the country's existing and future production and demand, as well as criticized some of the country's current management techniques and by offering alternatives to manage. Al-Otaibi and Mukhopadhyay (2005) provided an overview of some of the potential methods, with an emphasis on lowering water consumption. Despite documented mismanagement of the water resources of Kuwait, steps have not been taken for their future management (Fadlelmawla and Al-Otaibi 2005; Al-Otaibi and Abdel-Jawad 2007), which may lead to a freshwater deficit in Kuwait (Darwish 2001).

6.3.1.1 Groundwater Aquifers

Groundwater in Kuwait is divided into two types based on total dissolved solids (TDS): brackish water (3000–10,000 mg/L) and saline (10,000–35,000 mg/L) (Abdullah et al. 2021). Applications besides drinking have increased for the brackish groundwater in households (Mukhopadhyay and Akber 2018). The type of water that is mostly produced from Sulaibiya, Shagaya, Wafra, Um-Qusir, and Atraf is considered as brackish (Abdullah et al. 2021). In addition, the country's central and southwestern parts are characterized with brackish groundwater type where the salinity is higher than 5000 mg/l (Mukhopadhyay and Akber 2018). Kuwait has two major aquifers, the Kuwait group aquifer (Post Eocene age) uncomfortably overlies the Dammam Formation (Eocene age) of the Hassa Group (Abdullah et al. 2021). The Kuwait Group is composed of mainly unconsolidated to semi-consolidated clastic sediments, whereas the Dammam Formation consists of chertified and silicified from the top parts (Al-Alati and Gad 2018; Bhandary et al. 2018). The Kuwait Group aquifer increases its thickness from 150 m (Southwest) to about 400 m (Northeast) (Al-Ruwaih 2017). The Kuwait Group is further divided into three different formations, which are Dibbdiba, Lower Fars, and Ghar Formations in an ascending order (Al-Ruwaih 2017). In the early 1950s, freshwater lenses (Quaternary age) were discovered in the north of Kuwait groundwater, with a total dissolved solids (TDS) value of less than 1000 mg/L (Al-Alati and Gad 2018). The presence of this fresh groundwater is found in two depression areas namely Raudhatain and Umm Al-Aish (Al-Alati and Gad 2018). It is interesting to note that these freshwater lenses were floating on top of the highly saline groundwater (TDS > 100,000 mg/L), separated by their differences in density (Al-Alati and Gad 2018). Two major aquifers, the Dammam and Umm er Radhuma, are known to be transboundary aquifers shared between three countries namely Saudi Arabia, Kuwait, and Iraq with a total area of 246,000 km^2 (MEW 2021).

6.3.1.2 Water Wells

Significant work has previously been conducted to provide a foundational understanding of Kuwait's groundwater resources (Parsons Corporation 1963; Burdon 1966; Hantush 1971; Thomas 1966; Burdon and Al-Sharhan 1968). The Kuwait groundwater is mostly brackish to saline; however, the exceptions are groundwater lenses in the north Kuwait region (Fig. 6.3a). Large quantities of brackish groundwater are withdrawn for use in industrial and agricultural expansion, building, and water mixing with purified water. Kuwait's brackish water has an average TDS concentration of around 4000 mg/L. For the purpose of supplying water to Kuwait City, the Kuwaiti Ministry of Electricity, Water and Renewable Energy (MEWRE) built the Sulaibiya water well field in 1954 (Fig. 6.4). As a result, numerous well fields were built in various parts of Kuwait, drawing water from the Kuwait Group aquifers and Dammam Formation, to satisfy the growing water needs of the country. Shigaya's well fields began pumping out brackish groundwater for non-drinking

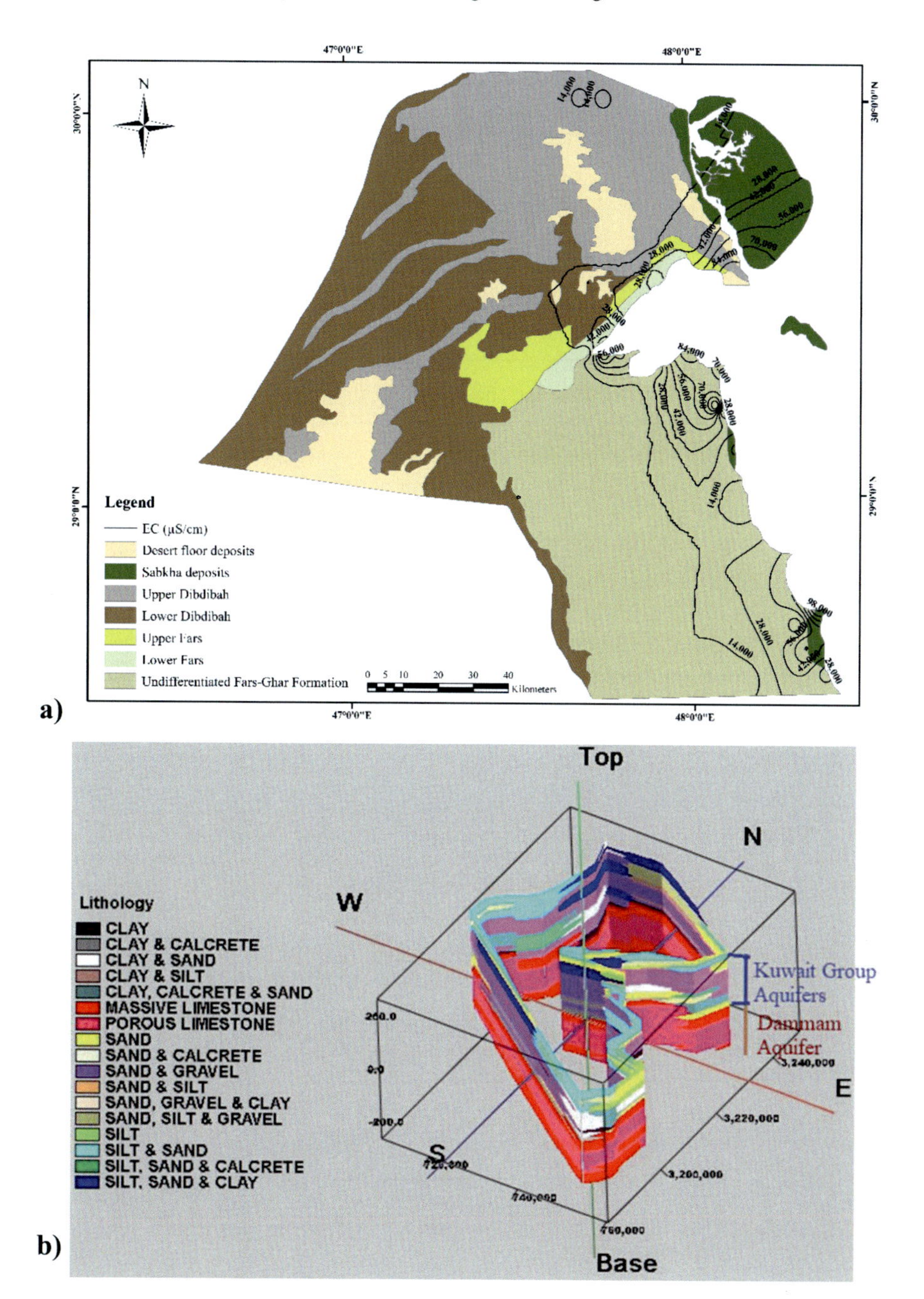

Fig. 6.3 The groundwater quality based on electrical conductivity (depicted as contour lines) on the geology map: (**a**) The lithological cross-section of Shigaya well fields; (**b**) developed litho-log data depicting the Kuwait group of aquifer and Dammam Formation

purposes in 1972. The Umm Gudair well field was established in 1986. In 1990, it was enlarged alongside the 1987-created Wafra well field. The fields' brackish groundwater was utilized to mix with desalinated water for irrigation and other non-drinking purposes.

In the 1960s, researchers found fresh, potable groundwater (TDS 500 mg/L) at Raudhatain and Umm Al-Aish. Exploration of the northeastern region increased after the discovery of freshwater lenses in order to locate further lenses and, by extension, additional supplies of water, primarily for agricultural purposes. Many private agricultural farms were also developed alongside the construction and extension of existing agricultural facilities, and these farms have begun either permanently or temporarily extracting brackish water from the Kuwait Group aquifer. Most of these farms can be found in the southern region of Wafra and the northern region of Abdaly. Many private household abstraction wells in Kuwait also tap into the brackish water of these two aquifers, adding to the already extensive network of agricultural abstraction points. It is unclear how much water is being pumped from private wells in Kuwait.

Figure 6.4 provides the details of the well fields where water is being extracted, the year of operation, the sum of all wells, and the anticipated production capacity. In Kuwait, the Dammam Formation aquifer is the primary source of groundwater for the Abdaliyah, Sulaibiya, Wafra, and Shigaya (B, C, and D) well fields. However, the Kuwait Group aquifer became increasingly important for groundwater production beginning in the 1970s and especially throughout the 1980s. A small number of dual-completion wells targeting the Dammam Formation and the Kuwait Group

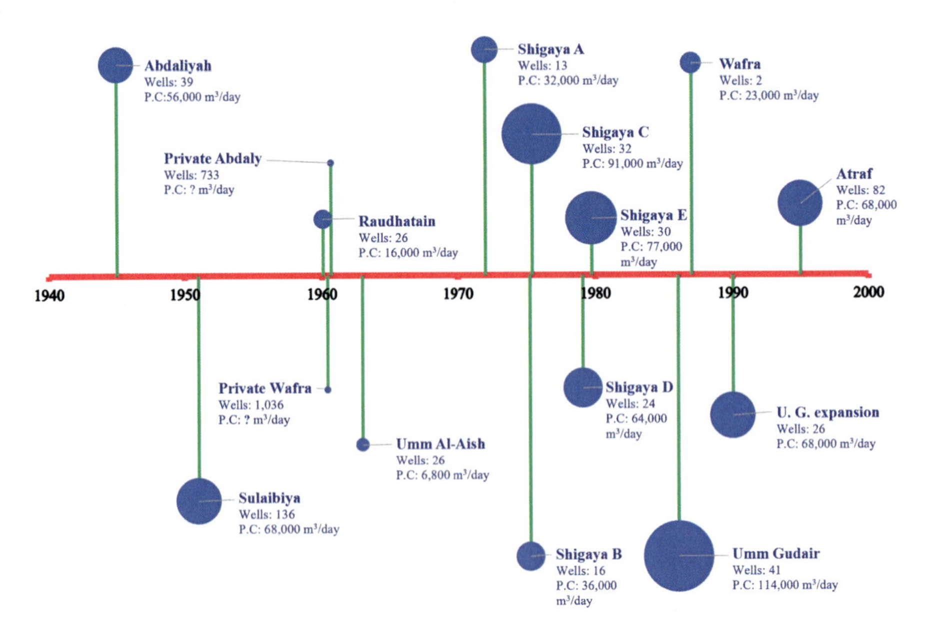

Fig. 6.4 Timeline of water well field development in Kuwait. The size of the circle is proportional to the production capacity (P.C). The symbol "?" indicates paucity of data

aquifers were drilled in Shigaya in the early 1970s (A). Shigaya (E) saw the development of 29 dual-completion wells in the early 1980s, and in 1986, the Umm Gudair well field was built to produce from 59 dual-completion wells and 8 Kuwait Group wells. During 1988–1989 eight wells from the Kabd field were available for utilization. In the mid-1990s, the Atraf field was developed as a replacement for the aging Sulaibiya field. The wells served as the foundation upon which the hydrogeological database was built. In 1989, it was estimated that each region produced around 200,000 m^3 day^{-1} on average from wells on the Wafra and Abdali farms (Al-Sulaimi et al. 1994, 1996). Groundwater is being extracted at a rate of between 300,000 and 400,000 m^3 day^{-1} in each of these regions. It has been observed that the potentiometric head of the Dammam Formation aquifer has decreased during the 1970s as a result of the extraction of brackish water. When the TDS began to rise at an alarming pace in the generated water from the Raudhatain freshwater fields between 1963 and 1967, productivity was cut to around 1 mega imperial gallon per day (MIGD) (4545 m^3 day^{-1}). Raudhatain Bottling Company produces up to 0.5 MIGPD (2275 m^3 day^{-1}) of water from the Raudhatain field. Kuwait's conventional freshwater resources are about 6 million m^3 (Hamoda 2001) and fresh groundwater aquifers (TDS <1000 mg/L) were mined for drinking water. Groundwater supplies are now a "national strategic reserve", although 137,000 m^3 are bottled daily (Fadlelmawla and Al-Otaibi 2005).

Quality of Major Groundwater Fields

Salt water ingression into the aquifer is a common source of quality decline. Kuwait has two main mechanisms that contribute to its declining quality. There is saline groundwater around all the brackish groundwater fields, and mixes with the fresh water and deteriorates the quality (Aliewi et al. 2021). As water levels continue to drop as a result of extraction, the saline liquid will begin to flow laterally. As the salinity of the groundwater in Kuwait is higher than that of the seawater in certain areas (>75,000 μS/cm) (Bhandary et al. 2018, Sabarathinam et al. 2020), seawater intrusion alone not considered as the main source of salinization. Since the 1960s, irrigation has progressively grown, reaching 200,000 m^3 day^{-1} in 1989 (Al-Sulaimi et al. 1996). After 30 years of irrigation, Al-Sulaimi et al. (1994) found that the groundwater TDS in both agricultural regions had risen from around 3000 mg/L to around 8000 mg/L. As a result of declining groundwater quality, several farms in Abdally and Wafra have been abandoned. Farmers in many locations have been compelled to over-irrigate due to rising TDS levels in the groundwater utilized (Al-Rashed and Al-Senafy 1995). Due to this fact, groundwater quality continues to decline and groundwater levels continue to fall (Fig. 6.3a). If the current rate of withdrawal from the Kuwait Group aquifer in the Wafra farm region continues, the quality of the water could significantly deteriorate (Akber et al. 1999). So, aquifer exploitation must be managed to prevent a decline in produced water quality and productivity due to a reduction in the driving head (Al-Otaibi and Mukhopadhyay 2005).

6.3.2 Desalination

Kuwait relies on seawater desalination to meet its rising population's freshwater needs. Conventional desalination technologies in Kuwait include thermal and membrane-based Reverse Osmosis (RO), Multi-Effect Distillation (MED), and Multistage Flash Distillation (MSF). The total installed capacity in 2020 is 683 MIGD and the rejected brine is about 1.020 Billion Imperial Gallons per Day (BIGD). The MSF is old and installed at 66.8% of Kuwait's average fresh water supply. The first desalination plant was built in Kuwait City in 1953 (AlAli 2008). The desalination in Kuwait plant installed at Shuwaikh during 1980 was of the submerged type, with the Multi-Stage Flash (MSF) installed being the first commercial used globally (Finan and Kazimi 2013). The large commercial seawater reverse osmosis (SWRO) desalination was first commissioned in Kuwait in 2010 by the Ministry of Electricity and Water and Renewable Energy (MEWRE) along with the cooperation of KISR (Finan and Kazimi 2013).

Desalination plants in Kuwait have been rapidly expanding in the recent years (Fig. 6.5) and are noticeable around the country, such as Doha (East and West), Shuwaikh, Sibiya, Shuaiba (North and South), and Zour (North and South) (Al-Zubari et al. 2017). Through the desalination process, a total of 84.8% was supplied for municipal water demands (Al-Zubari et al. 2017). It is well known that not only the cost of expanding desalination plants that is increasing, but also the energy consumption from the country's oil and gas to power up these plants. The production capacity varies from one plant to another. Based on the Ministry of Electricity and Water (MEW 2021), it was stated that the Doha West desalination plant has the highest production capacity reaching up to 170.4 MIGD, since 1951. Kuwait built the first large-scale desalination facility during the 1960s (Al-Wazzan and Al-Modhaf 2001).

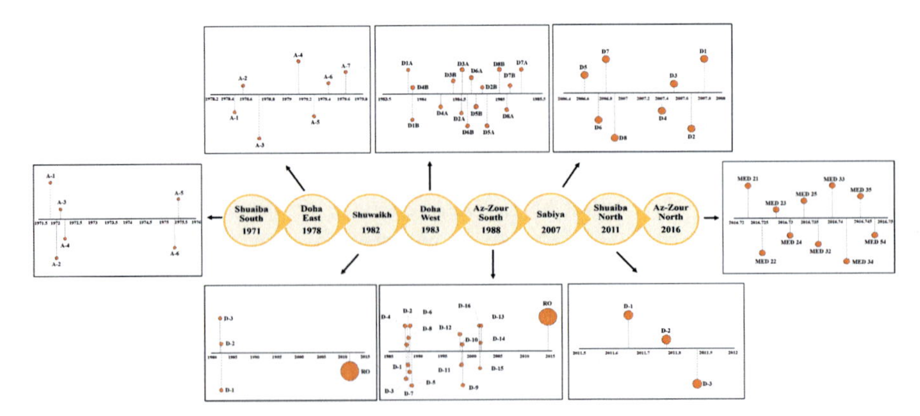

Fig. 6.5 Timeline of desalination plants in Kuwait. The size of the circle is proportional to the production capacity (P.C). R.O. (Based on the desalination water system was installed in 2019 at the Doha east station)

Kuwait developed the reverse osmosis (RO) technique for the Arabian Gulf seawater in 1987 (MEW 2007). The Al-Zour North's saltwater desalination facility has supplied 29.3% of the nation's freshwater since 2007 (567,000 m^3 day^{-1}). The RO technology is more advanced and increasing worldwide, and in Kuwait. The RO contributes 17.5% to Kuwait's average freshwater production. Conventional desalination methods in Kuwait are still expensive and energy-intensive. Kuwait's co-generation plants use only heavy crude oil to generate 90% of the nation's freshwater. Desalination plants used 462 million Giga Joules (GJ) of energy, 54% of national fuel use; desalination uses 0.7 kg/kWh and emits 15.7 kg/m^3 CO_2, respectively (Darwish et al. 2007). Conventional systems have water recovery, scaling, fouling, corrosion, and waste concentrate production restrictions (brine). Conventional technologies in Kuwait are limited by their usage of fossil energy, which emits greenhouse gases that harm the environment. Kuwait has the fourth-largest saltwater desalination capacity and produces 6% of worldwide desalinated water. The co-generation plants, for instance, provide 90% of the nation's freshwater and use only heavy crude oil to generate energy. Further, waste brine management stresses desalination companies the most. Global brine production is 142 million m^3 day^{-1}. The Middle East and North Africa generate 70.3% of worldwide brine, over 100 million m^3 day^{-1} (Jones et al. 2019).

6.3.3 *Wastewater*

Kuwait is considered the first country to construct wastewater treatment plants (WWTP) since 1950 (Aleisa and Al-Zubari 2017). It is worth noting that in the year 2005, Kuwait was the first country to launch the largest WWTP globally, namely Al-Sulaibiya WWTP, using Reverse Osmosis (RO) technology for potable water production (Aleisa and Al-Zubari 2017). The agriculture, landscaping, recharge, and other non-restricted non-portable uses of tertiary-treated wastewater (Alhumoud et al. 2003) may be an appropriate solution to reduce the strain on groundwater and desalination plants brought on by the rising demand for freshwater, since the 1970s (Fig. 6.6).

Kuwait has a well-established wastewater collecting and treatment system. There are 4700 kilometers of gravity drains and 1600 kilometers of pressure mains that collect wastewater from metropolitan areas in Kuwait (Al-Essa 2000). In total, there are 17 primary and 52 secondary pumping stations that transfer the wastewater throughout the system (Al-Ruwaih and Almedeij 2007). After the Ardiya municipal wastewater treatment plant (MWTP) closed in 2004 (MEW 2007), Kuwait's Ministry of Public Works (MPW) has managed 4 big MWTPs since 2005. Since December 2005, the Sulaibiya MWTP, the world's largest WTP by production capacity, has used ultrafiltration (UF) and RO to remove microorganisms and TDS (Gagne 2006). It treats 425,000 m^3 of municipal wastewater per day and produces 319,000 m^3 of treated water reflecting high recovery. Kuwait recycled 62% of its 76 million m^3 of municipal wastewater in 2001 (UNEP 2001).

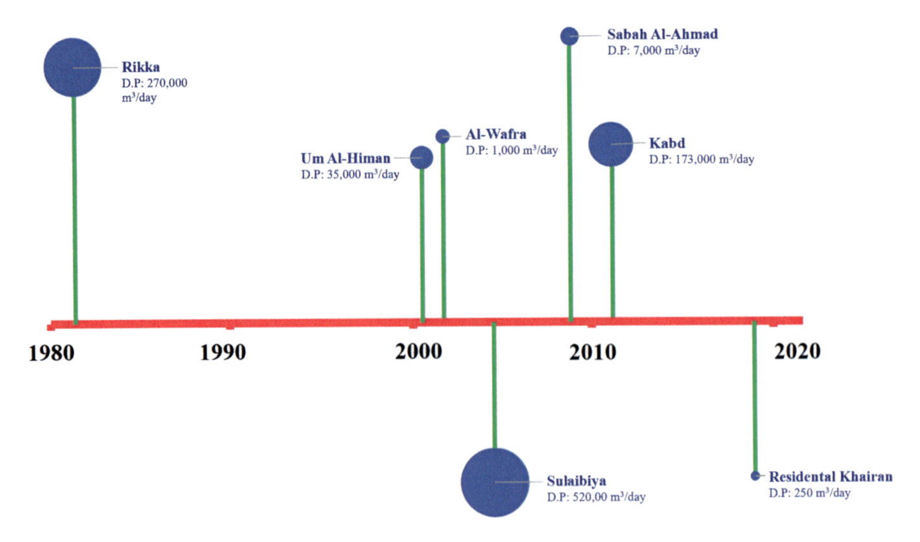

Fig. 6.6 Timeline of wastewater treatment plants in Kuwait. The size of the circle is proportional to the production capacity and D.P indicates daily production

Municipal wastewater treatment plants in Kuwait produce sludge from wastewater and may produce methane (CH_4) in huge digesters, acts as a clean fuel and raw material, with several uses (Al-Damkhi et al. 2009). The WWTP in Kuwait upgraded the treatment efficiency from secondary to tertiary treatment (Aleisa and Al-Zubari 2017). The Sulaibiya WWTP treats about 64% of the country's sewage water and is separated into biological treatment and reclamation plants (Aleisa and Alshayji, 2019). Around 75% of wastewater is treated by RO techniques and only 58% is reused. The untreated sewage accounting for nearly about 30% is dumped into the ocean (Al-Shammari and Shahalam 2006). About 19% of the effluent is being put to use in agricultural endeavors (Ismail 2015). A statistical study on the Ardiya catchment, which serves as the wastewater source for Al-Sulaibiya WWTP shows that between 2015 and 2020, there is a minimal increase in water demand and a minor rise in wastewater generation. Due to the COVID-19 pandemic travel prohibition and subsequent population decline in 2020, there was a modest rise in water demand and associated wastewater generation (Ahmed et al. 2022). Wastewater treatment could be utilized to produce methane (CH_4) from anaerobic treatment, as well as many minerals nutrients, and biosolids that could be utilized for land application in agriculture, provided the Kuwait Environmental Public Authority (KEPA) permits such uses. The resource recovery component of wastewater treatment may result in enhanced economy for wastewater treatment and reuse.

6.4 Water Utilization

The utilization of water resources in the region could be governed by understanding the water demand and supply, future water needs, and considering the virtual water footprint.

6.4.1 Water Demand and Supply

The anticipation of future water needs is a major factor in water infrastructure planning (Gleick et al. 2003). The diversity of the population that the water system must cater to is a significant obstacle in regional water demand modeling. The gross domestic consumption in the GCC region shows Kuwait as the fourth largest per capita consumer (Fig. 6.7). This heterogeneity must be taken into account whenever a regional Water Demand Models (WDMs) with a high level of detail is needed. The two-stage process involves first grouping customers with similar consumption patterns into larger groups, and then creating a WDM for each of those groups. The WDMs deliver beneficial data for policymakers and planners in the water distribution system (WDS) sector by shedding light on consumers' actual water use patterns and future needs (Avni et al. 2015).

Mukhopadhyay et al. (2001) studied the potential of artificial neural networking as a tool for predicting future values, specifically with regard to residential water use, in Kuwait. According to the data, the average water use per person in residential villas in Kuwait ranged from 180 to 2018 liters per day. This is one of the highest per capita water consumption rates in the world (Fig. 6.7). The authors inferred that the high consumption of water in the relatively higher-income class of Kuwait is mainly due to the surplus availability at a low price. The study also inferred that the summer's high temperatures likely have a role in the season's higher consumption rates. As domestic use accounts for >80% of all consumed water in Kuwait, thus the

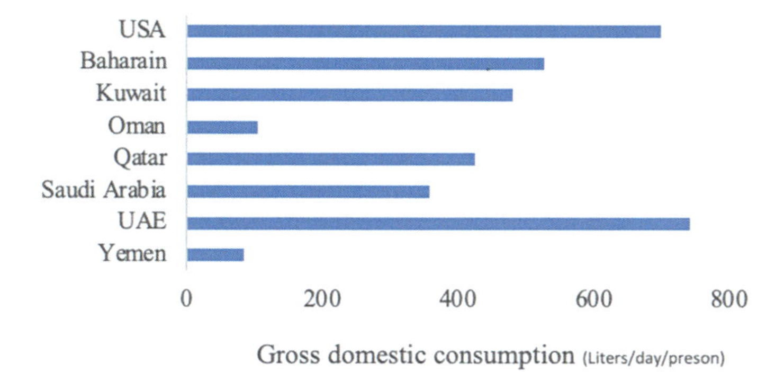

Fig. 6.7 Gross domestic consumption of the water in Kuwait and comparison to adjoining countries

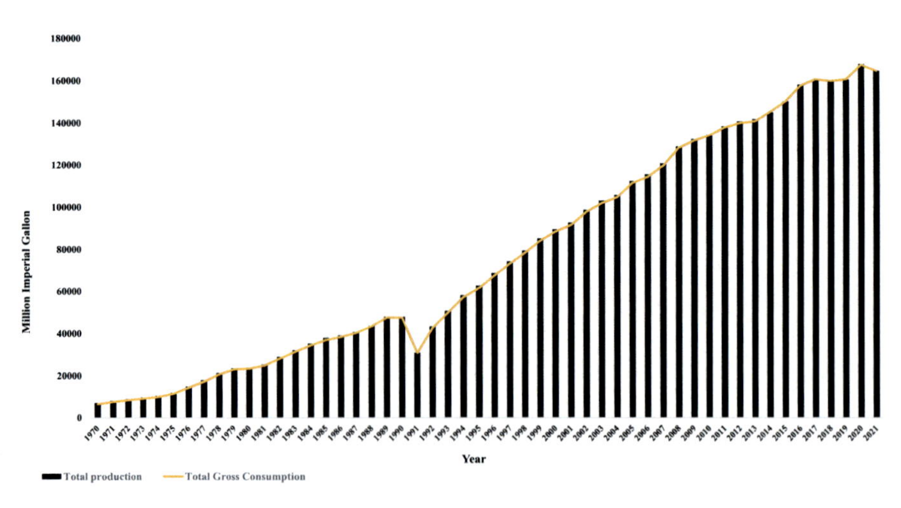

Fig. 6.8 The gross consumption and production of freshwater in Kuwait since 1970

rate of growth in demand for residential water is significant. The gross consumption for the past decades has been increasing (Fig. 6.8) proportionate to the production (Al-Otaibi and Mukhopadhyay 2005).

Due to the extreme weather patterns that characterize Kuwait, the annual use of freshwater is unevenly spread over several months of the year. In spite of the significance of examining the effects of washing machines and sprinklers on freshwater use, to date, no research has been done (Burney et al. 1993; Burney et al. 2001).

6.4.2 *Projection*

Since the mid-1980s, brackish water delivery to homes for irrigation has been curtailed due to the large fall in water levels in the groundwater fields of Kuwait, leading to the increased freshwater use. However, the widespread adoption of this solution is hampered by the cultural taboo against recycling wastewater. Kuwait City's groundwater level was affected by increasing urbanization and increased water demand in residential areas (Hamdan and Mukhopadhyay 1991). The residential regions of Kuwait were criticized to have massive amounts of underground water extracted (El-Nahhas et al. 2000). Nevertheless, a lack of data is still a significant barrier to effective water resource management. Fadlelmawla and Al-Otaibi (2005) stated that MEWRE and Ministry of Public Works (MPW) groundwater future plans show poor management and excessive use. Abusam (2008) recommended that the Kuwaiti agricultural master plan for 1990–2015 should encompass wastewater irrigation and landscaping. In 2006, 576,000 m^3 day^{-1} of high-quality treated effluents were produced, yet 35% was wasted and dumped into the Arabian Gulf (MEW 2007).

Due to widespread use, the level of brackish and fresh groundwater has dropped by 3–50 m. The TDS increased from 3000 to 8000 mg/L (Al-Sulaimi et al. 1996; Al-Rashed et al. 1998). Water shortages endanger long-term, fair economic development (Hamoda 2001). Al-Otaibi and Mukhopadhyay (2005) stated that as a result of extensive exploitation of groundwater reserves, the country will need to invest a shocking US$7 billion in desalination plants by 2025. A dedicated stormwater drainage system would allow for the collection, treatment, and storage of runoff for later use (Murtaugh 2006). Further, there hasn't been any in-depth research conducted on the increasing demand for brackish water in Kuwait or its potential future growth. The production of desalinated and TWW include power and thus treated water is used for domestic, industrial, and agricultural purpose. This linkage between water, energy, and food has led to the water-energy-food nexus (Siderius et al. 2019). The agricultural sector utilizes 19% of TWW and 54% of groundwater (Al-Murad et al. 2017). Hence the increase in the agricultural land and shifting to more water-loving crops will lead to an increase in irrigational water demand. The predicted calculation for freshwater requirement in Kuwait is estimated to range from 722 million m^3 $year^{-1}$ to 3036 m^3 $year^{-1}$ (Ibrahim et al. 2020).

6.4.3 *Virtual Water and Water Footprint*

Arab countries depend on international food trade due to freshwater and arable land scarcity, and hot climatic conditions. Experts increasingly link worldwide trade to freshwater scarcity, known as virtual water. Agricultural and industrial products contain "virtual" water. Tony Allan advocated the virtual water trade, and increased food imports, which saved domestic water and reduced friction between adjacent countries in the GCC region (Allan 2002, 2011; Mohammed and Darwish 2017). Surface and groundwater (blue), rainwater (green), and the amount of water essential to digest the contaminants to within the permissible level (grey) all contribute to what is known as water footprint (WF) for the production of crops (Qasemipour and Abbasi 2019).

International trade consumes over one-third of world water (Chen and Chen 2013). In 2010, Iran, Turkey, and Israel imported 273 km^3 $year^{-1}$ and exported 480 km^3 $year^{-1}$. South and Eastern Mediterranean net virtual water import trade averaged 49,123 Mm^3 between 1998 and 2002. Cereals include 67%, vegetable oil 27%, and sugar 10%. Qatar has a 1554 m^3 $year^{-1}$ $capita^{-1}$ water footprint (Mohammed and Darwish 2017), somewhat greater than the worldwide average (1240 m^3 $year^{-1}$ $capita^{-1}$), however, significantly less than the USA (2480 m^3 $year^{-1}$ $capita^{-1}$) and other European countries like Spain and Italy (2300–2400 m^3 $year^{-1}$ $capita^{-1}$) (Hoekstra and Chapagain 2007). The 1058 m^3 $year^{-1}$ $capita^{-1}$ is much greater than the MENA average of 601 m^3 $year^{-1}capita^{-1}$ (2010). Over 1995–1999, 695 Gm^3 $year^{-1}$ of virtual water travelled between countries (Hoekstra and Hung 2005). Virtual water estimation enabled decision-makers better use of water resources and make strategic water reallocation and conservation decisions (Mohammed and Darwish 2017).

The water footprint of a country is accounted by internal and external processes. The internal water footprint is based on the consumption of products in the countries utilizing countries' water resources and the external water consumption of imported products within the country, mainly concerning with products produced by utilizing the water resources of other countries (Hoekstra and Chapagain 2007). To obtain an indicator reflecting the consumption of water, the concept of water footprint (WF) was developed in 2002 (Hoekstra and Hung 2002). The WF is similar to the 1990s ecological footprint idea (Rees 1992; Wackernagel et al. 1997). Allan proposed importing virtual water (not real water) to partially solve Middle East water scarcity (Allan 1994). The internal water footprint (IWF) is calculated as follows:

IWF = Agricultural water usage + industrial utilization of water + domestic utilization − water virtually exported to other countries.

The external water footprint (EWF) is calculated based on the difference between the water import and export. The WF considered all categories in the development of the product (blue, green, and grey) 7450 Gm^3 $year^{-1}$, is the world WF. India has the greatest footprint at 987 Gm^3 $year^{-1}$. Due to green water consumption, agriculture's overall water use (6390 Gm^3 $year^{-1}$) is even higher than previously projected (use of soil water). Irrigation losses of 1590 Gm^3 $year^{-1}$ (Chapagain and Hoekstra 2004) raise the overall agricultural water usage to 7980 Gm^3 $year^{-1}$.

6.5 Challenges and Solutions

A study in Iran used water self-sufficiency (WSS) and WF as indices to assess water resource management (Karandish and Hoekstra 2017). The WSS index measures how well a place can provide for its own water needs in order to grow food without importing it from elsewhere (Hoekstra and Hung 2002). This metric is employed not only in global assessments of water resources (Liu et al. 2017), but also at the regional level, where it facilitates better water management (Zhuo et al. 2016). Al-Otaibi and Abdel-Jawad (2007) propose a "water security" or "strategic storage capacity" (SSC) system to meet Kuwait's water demand across all sectors. They estimated that 24% of Kuwait's annual water usage is in this SSC and suggested using artificial recharge technology to store it in a suitable groundwater aquifer. A study estimated that about 1573 Mm^3 of water could be stored in the northern Kuwait reservoir, which would suffice the domestic needs of the country for 3 years (Murtaugh 2006). The main challenges in the management of water resources in Kuwait are the demand for freshwater, and the increase in population and agricultural needs, with poor water management and infrastructure (Aliewi et al. 2017; Tariq et al. 2022). The loss of water through transportation and networking either due to leakage or illegal connections referred to as unaccounted-for water (UfW) and it is generally a non-revenue water loss, which should also be considered to address the challenges in the management of water resources in Kuwait (Tariq et al. 2022). Further, Abusam and Shahalam (2013) reported irrigating with treated water has also led to a drop in crop productivity.

6.5.1 Water Security

The extensive extraction of groundwater resources can threaten both environmental and socioeconomic sustainability. In order to assess such threats a number of useful indices are defined, such as the water dependency index (WDI), along with water scarcity (WS) and water self-sufficiency (WSS) indices (Qasemipour and Abbasi 2019). Kuwait and other GCC nations' water management issues are largely institutional to emphasize the fact that "No one owns water". The subsidized drinking water supplied in the country is 1.356 Mm^3 produced at a cost of 247 million Kuwait Dinar (MEW 2007). The 2007 Tarsheed campaign promotes "smart usage" of water and power (Al-Damkhi et al. 2009).

6.5.2 Conservation Policy and Pricing

Overconsumption of water happens if the price is below the marginal cost because additional water is delivered at a cost that is higher than the value. Wastewater disposal and treatment costs, as well as the cost of repairing the external environment after it has been damaged by untreated wastewater, are included in the monetary price tag associated with water consumption. Profitability necessitates setting the water rate at a level where revenues are sufficient to cover operation and maintenance costs, depreciation costs, and yield a reasonable rate of return on invested capital. The idea of social equity is promoted to ensure that low-income communities have access to adequate freshwater resources and adequate waste disposal services. It supports water rate subsidies for the low-income, who due to financial constraints, use less clean water than is recommended for human health yet cannot afford the market rate.

Development of policy stressed three basic criteria to focus on: (a) to conserve the natural resources, (b) to increase the supply through innovative treatment methods and identification of new groundwater resources, and (c) to decrease the need for utility (Fadlelmawla 2009). Since family income is the second most important factor in Kuwait's consumption rate, higher-income families should benefit more from the flat-rate subsidy model than the lower-income ones (Mukhopadhyay et al. 2001). Block-rate pricing should replace the flat price, and rates should be adjusted to encourage water efficiency.

6.5.2.1 Rationalizing Tariff and Policy

The current approach to water resource management is obsolete and must be replaced with one that considers water supply and demand as a whole (Ismail and Al-Maskati 2002). As water is a basic need, the pricing of water should be affordable by people of all income categories (Kayaga et al. 2003). Lifeline tariffs should be within 3–5% of household income in order to make the commodity available across all income classes of society (Smith and Al-Maskati 2007). The economic, financial viability, social equality, and simplicity principles are the four pillars upon which the pricing of water supplies is built. Individuals should be able to respond to price signals and change their consumption provided the tariff structure is understandable.

The fixed fee has been established as US$0.58 per m^3, but the supply and production of water were estimated at US$ 3.0 per m^3 (Darwish and Al-Najem 2005). This can be inferred that the government of Kuwait provides a subsidy of domestic water accounting for a value of 1 billion US$ $year^{-1}$ according to the rates of 2002. It was also inferred that family income is identified to play a key role in the volume of water consumption. Based on output and billing rates, Milutinovic (2006) estimated yearly net benefits of US$ 420 million could be attained.

There are two primary categories of water tariff structures: volumetric and non-volumetric.

1. Base tariff is a rate that stays the same no matter how much water a customer uses over the course of a given period of time. As a consumer system, it is unrivaled.
2. The customer is charged based on the volume of water utilized.
3. Tariffs based on individual slabs have two variations: The Reduced Tariff Slab (RTS) and the Increased Tariff Slab (ITS). There are different pricing structures for different volume ranges of water used in this model. By volume, larger users are charged more under ITS and less under RTS.
4. The demand for water and the expected seasonal variation should be accounted when determining the water price. The ITS of this variety is used exclusively during the warm season (AWWA-WEF A 2005).

6.5.3 *Reducing Desalinization Cost*

The water research center of Kuwait Institute for Scientific Research (WRC-KISR) is developing numerous new desalination technologies to address the above challenges. Kuwait's 2035 water sustainability goal is also the program's eighth strategic plan. Desalination research is focusing on renewable energy alternatives to fossil energy. Low fossil fuel usage boosts GDP and helps conventional technologies to meet energy demand. Desalination uses solar thermal, photovoltaic (PV), wind, and geothermal energy. Forward osmosis (FO) desalination technique based on natural osmosis could be coupled with solar energy in the future. In 2012, WRC at KISR installed, commissioned, and tested a global pilot scale research using FO-based seawater desalination technology. The WRC is now merging FO technology with solar energy for Kuwaiti saltwater desalination. Membrane distillation (MD) desalination research is integrating waste heat. Hybrid technologies that combine RO, FO, and MD are currently the most cost- and energy-effective. Scientists in the Water Desalination Technology (WDT) program at KISR have recently been awarded patents in the United States for their ground-breaking innovations in hybrid seawater desalination technologies that aim to maximize water recovery while reducing energy use and brine waste (US 10,308,524 B1, US10,940,439 B1). The WDT scientists are granted a US patent related to the extraction of commercially valuable minerals from the waste brine (US 10,280,095B1). This invention is a key breakthrough towards attaining the KISR goal to achieve Zero Liquid Discharge (ZLD) desalination approach in Kuwait. The development of innovative membranes with novel techniques for polymeric membrane synthesis is the other key area of research. The WDT researchers are granted with two US patents in this regard (US 10,124,297 B1, US 11254691 B1). At present, WRC research is mainly focused on the following aspects to tackle the future water challenges in the state of Kuwait:

- Innovative Membrane and Thermal Desalination Technologies
- Turbid Seawater Treatment and Desalination Processes
- Salts and Minerals Extraction Technologies
- Minimal Liquid Discharge (MLD) Technologies
- Zero Liquid Discharge (ZLD) Processes
- Desalination Technologies using Renewable Energy
- Innovative Desalting Processes using Waste Heat
- Innovative Desalination Technologies for Emergency using the Renewable Energy Resources
- Development of Innovative Membranes Competing with the Existing Membranes in the Market

6.5.4 Artificial Recharge

The *state-of-the-art* of recharging the aquifer in Kuwait was implemented by the water resources development and management program, with the assistance of MEWRE. Studying aquifer properties is necessary to assess whether artificial storage is feasible or not. The recharging water might come from either recycled gray water or collected rainwater. The idea is to conserve the rainwater from the brief and strong rainy period for later use. Although water harvesting is typically done on a household or community size, however, larger watershed or sub-basin scales are possible in Kuwait provided infrastructures are developed to contain the rainwater. To determine the feasibility of water harvesting in Al-Raudhatain, Murtaugh (2006) has employed models to calculate the surface runoff. Contradictory findings have been published on the topic of surface runoff generation in this region, as instant percolation of the groundwater was reported after extreme short-term rain events (Grealish et al. 1998) and the intense surface flow of shorter time period (Al-Sulaimi et al. 1997). Despite being followed by subsurface runoff, isotopic evidence for quick infiltration has been offered by Fadlelmawla et al. (2007) in favor of the latter (Fadlelmawla 2009).

6.6 Future Directions for the Management of Water Resources

The introduction of new technologies like Internet of Things (IoT) could help to monitor the water resources effectively. Further, the implementation of effective regulations along with the reuse of water focusing on a circular economy would also help in the conservation of resource.

6.6.1 *IoT-Based Monitoring and Management*

Digitalization monitors water flows and identifies water amounts for reuse, supporting the water sector's circular economy paradigm. Artificial intelligence (AI) technologies have increasingly being used to turn data into useable information to improve urban water cycle infrastructure efficiency and enhance maintenance and operational choices (Al Aani et al. 2019; Li et al. 2021, Elzain et al. 2021, 2022). Digitalization requires resources, employee experience, and website security (Marcon et al. 2019). Digitalization allows the city's water cycle to adopt modern technologies. Monitoring infrastructure, operations, and consumption generates large data sets required for complex analysis (Iansiti and Lakhani 2020). Data can be used to reduce energy use, increase output, improve facility efficiency, and extend asset life, all of which contribute to the system's long-term viability (Hernández-Chover et al. 2022).

The automated tools are used to monitor the production, supply, distribution, utilization, and manage them by using different sensors and meters from which data can be sent to the server or the central repository through different methods. Such a system ensures high quality and on-demand delivery (Public Utilities Board Singapore 2016).

Smart water focuses on real-time monitoring,

1. For preventive maintenance
2. For network planning and operation
3. For water consumption

Challenges are:

1. Availability of sensors and IoT-based acquisition and analytical tools
2. Analytical efficiency to handle the robust data and obtain information
3. Development of IT skills of the workers is needed as Smart Water Grid
4. Public messaging must effectively address issues to achieve technology acceptance
5. Smart Water Grid technologies need more study and testing to attain their full potential

Our current economic paradigm requires a never-ending supply of raw materials, most of which are non-renewable, to generate everyday goods and services. The limited lifespan of these items and services, significant and difficult waste creation, and increased consumption of new commodities make this economic model unsustainable (Castellet-Viciano et al. 2022). Intensive exploitation of groundwater extracts for domestic utilization could be prevented by treating the wastewater and thus preventing the depletion of natural resources (Levoso et al. 2020; Hernández-Chover et al. 2022).

6.6.2 Circular Economy

The European Commission's 2020 Circular Economy Action Plan and European Green Pact emphasize digitalization (Friends of the Earth Europe 2020). These programs aim to create a sustainable and green digital sector. Digitalization may help industries achieve sustainability goals and plays key role in the circular economy (CE) and in the transition phase from the linear economic model. The CE is dependent on the transformation and the development cycles that maximize production efficiency, extend the usable life of items, materials, resources, and reduce waste (Zając and Avdiushchenko 2020). The circular economy's 4 "Rs"—reduce, reuse, repair, and recycle—solve long-term problems including urban water cycle management (Hernández-Chover et al. 2022; Rybkowska and Schneider 2011). The CE can help manage water sustainably (Tahir et al. 2018; Sauve et al. 2021). The water sector benefits from closed-loop supply chains, value retention, waste minimization, and resource efficiency. The CE can also help manage water supply since it overlaps with numerous spatial scales, governance levels, implementation methods, and economic sectors (Morseletto et al. 2022).

The unsustainable "take-make-waste" economic paradigm prompted the CE's creation. Korhonen et al. (2018), and Schöggl et al. (2020) view the CE as a system-level solution to consumption and production issues. The CE has only lately been conceptualized and analyzed scientifically (Korhonen et al. 2018; Desing et al. 2020). Several factors need fixing this issue. First, numerous UN Sustainable Development Goals (SDGs) involve water and aquatic habitats. Second, water crises jeopardize peace (WEF 2020). Third, scientific and popular CE definitions emphasize materials and energy, water is essential to most human activities

(Salminen et al. 2022). The basis for water-smart design CE threats to the water environment has to be recognized, analyzed, and immediately attended to. Industrial agriculture and recharge purposes have been focused with treated wastewater but it has to be used with caution considering the microbial and chemical dangers (Voulvoulis 2018; Aleisa and Alshayji 2019; Salminen et al. 2022).

The wastewater from municipal sources mainly constituted by grey water and only 50% of the grey water is recycled and reused, natural methods can also be adopted for grey water treatment (Samayamanthula et al. 2019). About 30% of the water could be saved by domestic consumption by producing 90 Mm^3 $year^{-1}$ from TWW (Abusam 2008). Industrial wastewater and sanitary wastewater are released directly into the Arabian Gulf (UNEP 2001). Secondary firms in the Shuwaikh Industrial Area (SIA) will generate wastewater of about one Mm^3 day^{-1}, adding to the 5000 m^3 from the Sabhan industrial area (UNEP 2001). Untreated wastewater will poison the oceans. Recycling wasted water is still rare in Kuwait (Abusam 2008), despite its benefits (Eriksson et al. 2002; Müllegger et al. 2003) and cost (Friedler 2008; Yu 2008).

6.7 Conclusions and Recommendations

Intensifying and expanding wastewater recycling, reevaluating desalination technology, assessing subsurface water resources, planning ahead for critical storage of reservoir, use of renewable energy to power for wastewater treatment and desalination processes, and, most importantly, creating awareness on water resources are key players in future water management. Optimizing municipal wastewater treatment, which is equivalent to 148 US$ million $year^{-1}$, can save Kuwait's marine environment, limited water resources, and reduce the desalination capacity. The recharge wells were recommended by earlier studies to manage Kuwait's groundwater sustainably. In this regard, the Raudhatain aquifer was also recommended for artificial recharge with desalinated water and RO-treated wastewater. Importing and storing water for emergencies require foreign sources alike the earlier proposals during 2001, Turkey Peace Pipeline (TPP), and supply of 293 million m^3 of water to Kuwait from Iran. Wastewater treatment could be utilized to produce methane from anaerobic treatment, and many minerals, and biosolids could be utilized for land application in agriculture after testing its impact under local conditions. The resource recovery component of wastewater treatment may result in enhanced economy for wastewater treatment and reuse.

Integrated water resource management for Kuwait can be achieved by adopting three concepts (a) to develop an institutional setup; (b) governance and legislative measure; and (c) planning and development. The institutional setup can consist of Ministry of the Electricity and Water and Renewable Energy (MEWRE), Public Authority for Agriculture Affairs and Fish Resources (PAAFR), Kuwait Environmental Public Authority (KEPA), Ministry of Public Works (MPW), and KISR. The MEWRE will be considered as the custodian for the water well fields, desalination plants and the wastewater treatment plants, implementing the policy legislations, governance, and conservation. The PAAFR monitors the wells in the agricultural region, crop water requirements, treated water reuse, and virtual water import and export. The MPW to concentrates on distributing, networking and operations, unaccounted-for water, and leakages during transmission through pipelines. KEPA will device action plans for environmental safety accounting for the desalination rejects and residual sludge. KISR to monitor, through sampling, analysis, and interpretation apart from developments of new innovative treatment technologies and techniques for augmenting surface and subsurface water storage. The data developed by KISR will be shared with MEWRE, PAAFR, KEPA, and MPW.

Governance and legislations will focus on decision on subsidies, water tariffs, protection and conservation of water resources and legal action on defaulters. Planning and development will be based on the status of the current utilities, major sectors to be focused on are, IoT-based monitoring of the water resources through linked relational database management system, water conservation to be introduced in the academic curriculum, installation of conservation devices, annual water audit consisting of water balance related to cost, development of additional water sources though runoff harvesting, exploration of submarine aquifers and water linkages with GCC countries.

Further, water-smart CEs can be focused effectively to utilize abstracted water and eliminate losses. Water may be recycled in the technosphere by recovering energy and chemicals from used water. Water-smart CEs exploit secondary materials and produce energy without risking water-related ecosystems or human health. Authorities in charge of water supply have been focusing on boosting supplies of desalinated water and brackish groundwater for decades, despite the fact that these sources are both extremely costly and scarce. In addition, not much has been done to make the best use of, or even protect, the existing stock of resources.

References

Abdel-Jawad M, Ebrahim S, Al-Tabtabaei M, Al-Shammari S (1999) Advanced technologies for municipal wastewater purification: technical and economic assessment. Desalination 124(1–3):251–261

Abdullah MJ, Zhang Z, Matsubae K (2021) Potential for food self-sufficiency improvements through indoor and vertical farming in the Gulf cooperation council: challenges and opportunities from the case of Kuwait. Sustainability 13(22):12553

Abusam A (2008) Reuse of greywater in Kuwait. Int J Environ Stud 65(1):103–108

Abusam A, Shahalam AB (2013) Wastewater reuse in Kuwait: opportunities and constraints. WIT Trans Ecol Environ 179:745–754

Ahmed ME, Al-Haddad A, Bualbanat A (2022) Possible impacts of covid-19 pandemic on domestic wastewater characteristics in Kuwait. J Environ Eng Landsc Manag 30(3):393–411. https://doi.org/10.3846/jeelm.2022.17634

Akber A, Sherif M, Ghoneim H, Alsenafy MN (1999) Evaluation of current ground water conditions in Al-Wafra and Al-Abdally farms areas. KISR Report # 5582

Al Aani S, Bonny T, Hasan SW, Hilal N (2019) Can machine language and artificial intelligence revolutionize process automation for water treatment and desalination? Desalination 458:84–96

Al-Alati HN, Gad MI (2018) Study the seasonal fluctuations of groundwater characteristics in Al-Raudhatain and umm Al-Aish depressions, North Kuwait. J Eng Res Appl 8(1):43–56

AlAli EH (2008) Groundwater history and trends in Kuwait. WIT Trans Ecol Environ 112:153–164

Al-Damkhi AM, Al-Fares RA, Al-Khalifa KA, Abdul-Wahab SA (2009) Water issues in Kuwait: a future sustainable vision. Int J Environ Stud 66(5):619–636

Aleisa E, Alshayji K (2019) Analysis on reclamation and reuse of wastewater in Kuwait. J Eng Res 7(1):1–13

Aleisa E, Al-Zubari W (2017) Wastewater reuse in the countries of the Gulf cooperation council (GCC): the lost opportunity. Environ Monit Assess 189:1–15

Al-Essa W (2000) Wastewater management in Kuwait. In: Al-Sulaimi J, Asano T (eds) Proceedings of the workshop on wastewater reclamation and reuse. Arab School for Science and Technology, Kuwait Foundation for the Advancement of Science, Kuwait, p 4

Alhumoud JM, Behbehani HS, Abdullah TH (2003) Wastewater reuse practices in Kuwait. Environmentalist 23:117–126

Aliewi A, El-Sayed E, Akbar A, Hadi K, Al-Rashed M (2017) Evaluation of desalination and other strategic management options using multi-criteria decision analysis in Kuwait. Desalination 413:40–51

Aliewi A, Bhandary H, Al-Qallaf H, Sabarathinam C, Al-Kandari J (2021) Assessment of the groundwater yield and sustainability of the transboundary Dibdibba aquifer using numerical modelling approach. Groundw Sustain Dev 15:100678. https://doi.org/10.1016/j.gsd.2021.100678

Al-Kuwait Al-Youm (2008) The approval of Kuwait's third master plan (KMP3). The Amiri Decree No. (255/2008), issue No. 888, pp 1–56, Ministry of Information, Official Gazette of the State of Kuwait

Allan JA (1994) Overall perspectives on countries and regions. In: Rogers P, Lydon P (eds) Water in the Arab world: perspectives and prognoses. Harvard University Press, Cambridge, pp 65–100

Allan T (2002) The Middle East water question: Hydropolitics and the global economy. Bloomsbury Publishing

Allan T (2011) Virtual water: tackling the threat to our planet's most precious resource. B. Tauris & Co. Ltd, London

Al-Murad M, Uddin S, Rashid T, Al-Qallaf H, Bushehri A (2017) Waterlogging in arid agriculture areas due to improper groundwater management—an example from Kuwait. Sustainability 9(11):2131

Al-Otaibi A, Abdel-Jawad M (2007) Water security for Kuwait. Desalination 214(1–3):299–305

Al-Otaibi M, Mukhopadhyay A (2005) Options for managing water resources in Kuwait. Arab J Sci Eng 30(2C):55

Al-Rashed MF, Al-Senafy MN (1995) Groundwater impact on inorganic contamination of soil in Wafra farms. Kuwait Institute for Scientific Research Report # 4643

Al-Rashed MF, Sherif MM (2000) Water resources in the GCC countries: an overview. Water Resour Manag 14:59–75

Al-Rashed M, Al-Senafy MN, Viswanathan MN, Al-Sumait A (1998) Groundwater utilization in Kuwait: some problems and solutions. Int J Water Resour Dev 14(1):91–105

Al-Ruwaih FM (2017) Hydrogeology and groundwater geochemistry of the clastic aquifer and its assessment for irrigation, Southwest Kuwait. Aquifers-Matrix Fluids:107–133. https://doi.org/10.5772/intechopen.71577

Al-Ruwaih FM, Almedeij J (2007) The future sustainability of water supply in Kuwait. Water Int 32(4):604–617

Al-Senafy M, Abraham J (2004) Vulnerability of groundwater resources from agricultural activities in southern Kuwait. Agric Water Manag 64(1):1–15

Al-Shammari SB, Shahalam AM (2006) Effluent from an advanced wastewater treatment plant—an alternate source of non-potable water for Kuwait. Desalination 196(1–3):215–220
Al-Sulaimi J, Viswanathan MN, Szekely F, Al-Senafy MN (1994) Geohydrological studies of Al-Wafra and Al-Abdally farm areas, Kuwait Institute for Scientific Research Report # 4404 (Unpublished)
Al-Sulaimi J, Viswanathan MN, Naji M, Sumait A (1996) Impact of irrigation on brackish ground water lenses in northern Kuwait. Agric Water Manag 31(1–2):75–90
Al-Sulaimi J, Khalaf FJ, Mukhopadhyay A (1997) Geomorphological analysis of paleo drainage systems and their environmental implications in the desert of Kuwait. Environ Geol 29:94–111
Al-Wazzan Y, Al-Modaf F (2001) Seawater desalination in Kuwait using multistage flash evaporation technology—historical overview. Desalination 134(1–3):257–267
Al-Zubari W, Al-Turbak A, Zahid W, Al-Ruwis K, Al-Tkhais A, Al-Muataz I, Abdelwahab A, Muraad AA, Al-Harbi M, Al-Sulaymani Z (2017) An overview of the GCC unified water strategy (2016–2035). Desalin Water Treat 81:1–18. https://doi.org/10.5004/dwt.2017.20864
Amer A, Barrat JM, Mukhopadhyay A (1990) Assessment of groundwater resources in Kuwait using a finite difference model. Int J Water Resour Dev 6(2):104–114
Avni N, Fishbain B, Shamir U (2015) Water consumption patterns as a basis for water demand modeling. Water Resour Res 51(10):8165–8181
AWWA-WEF A (2005) Standard methods for the examination of water and wastewater. Edición 21:5–10
Bergstrom RE, Aten RE (1965) Natural recharge and localization of fresh ground water in Kuwait. J Hydrol 2(3):213–231
Bhandary H, Sabarathinam C, Al-Khalid A (2018) Occurrence of hypersaline groundwater along the coastal aquifers of Kuwait. Desalination 436:15–27
Burdon DJ (1966) Report to the State of Kuwait on investigation of the Dammam limestone aquifer in Kuwait. United Nations Food and Agriculture Organization (FAO), Rome
Burdon DJ, Al-Sharhan A (1968) The problem of the palaeokarstic Dammam limestone aquifer in Kuwait. J Hydrol 6(4):385–404
Burney NA, Al-Mutairi N, Ramadhan M, Al-Jazzaf M (1993) Assessment of electricity and water sector in Kuwait. Kuwait Institute for Scientific Research Report # TE 001c (Unpublished)
Burney N, Mukhopadhyay A, Al-Mussallam N, Akber A, Al-Awadi E (2001) Forecasting of freshwater demand in Kuwait. Arab J Sci Eng 26(2):99–113
Castellet-Viciano L, Hernández-Chover V, Hernández-Sancho F (2022) The benefits of circular economy strategies in urban water facilities. Sci Total Environ 844:157172
Chapagain AK, Hoekstra AY (2004) Water footprints of nations. Value of Water Research Report *Series* No. 16. UNESCO-IHE, Delft
Chen ZM, Chen GQ (2013) Virtual water accounting for the globalized world economy: national water footprint and international virtual water trade. Ecol Indic 28:42–149
Darwish MA (2001) On electric power and desalted water production in Kuwait. Desalination 138(1–3, 183):–190
Darwish MA, Al-Najem N (2005) The water problem in Kuwait. Desalination 177(1–3):167–177
Darwish MA, Al Otaibi S, Al Shayji K (2007) Suggested modifications of power-desalting plants in Kuwait. Desalination 216(1–3):222–231
Desing H, Brunner D, Takacs F, Nahrath S, Frankenberger K, Hischier R (2020) A circular economy within the planetary boundaries: towards a resource-based, systemic approach. Resour Conserv Recycl 155:104673
El-Nahhas F, Sherif M, Abdullah W, Hadi K, Ghoneim, H (2000) Long-term operation, monitoring and assessment of the drainage system at Shamiyah and Kaifan, Part 1: main report, Kuwait Institute for Scientific Research Report # 5840, (Unpublished)
Elzain HE, Chung SY, Park KH, Senapathi V, Sekar S, Sabarathinam C, Hassan M (2021) ANFIS-MOA models for the assessment of groundwater contamination vulnerability in a nitrate contaminated area. J Environ Manag 286:112162

Elzain HE, Chung SY, Senapathi V, Sekar S, Lee SY, Roy PD, Hassan A, Sabarathinam C (2022) Comparative study of machine learning models for evaluating groundwater vulnerability to nitrate contamination. Ecotoxicol Environ Saf 229:113061
Eriksson E, Auffarth K, Henze M, Ledin A (2002) Characteristics of grey wastewater. Urban Water 4(1):85–104
Fadlelmawla A (2009) Towards sustainable water policy in Kuwait: reforms of the current practices and the required investments, institutional and legislative measures. Water Resour Manag 23:1969–1987
Fadlelmawla A, Al-Otaibi M (2005) Analysis of the water resources status in Kuwait. Water Resour Manag 19:555–570
Fadlelmawla A, Hadi K, Zouari K, Kulkarni K (2007) Hydrochemical investigations of recharge and subsequent salinization processes at Al-Raudhatain depression in Kuwait. Hydrol Sci J 53(1):204–223
Finan A, Kazimi MS (2013) Potential benefits of innovative desalination technology development in Kuwait. Kuwait Center for Natural Resources and the Environment. Massachusetts Institute of Technology, Cambridge. Report No.2139
Friedler E (2008) The water saving potential and the socio-economic feasibility of greywater reuse within the urban sector–Israel as a case study. Int J Environ Stud 65(1):57–69
Friends of the Earth Europe (2020, April 30) Principles for transformation: how the European green deal can achieve system change. Available online: http://www.foeeurope.org/Principles-for-transformation. Accessed 25 July 2021
Gagne D (2006) The world's largest membrane-based reuse project. Technical Paper, Water & Process Technologies, General Electric, Terrace
Gleick PH, Haasz D, Henges-Jeck C, Srinivasan V, Wolf G, Cushing KK, Mann A (2003) Waste not, want not: the potential for urban water conservation in California. Pacific Institute for Studies in Development, Environment, and Security, Oakland, p 165
Grealish G, Omar S, Quinn M (1998) Affected area soil survey—assessing damage magnitude and recovery of the terrestrial ecosystem—follow-up of natural and induced desert recovery. Kuwait Institute for Scientific Research, Report No FA 015C. KISR, Kuwait
Hamdan L (1987) Study of subsurface water rise in the residential areas of Kuwait. Volume I–Main Report, Kuwait Institute for Scientific Research, Volume 1-Main Report No 2227. KISR, Kuwait
Hamdan L, Mukhopadhyay A (1991) Numerical simulation of subsurface-water rise in Kuwait City. Groundwater 29(1):93–104
Hamoda MF (2001) Desalination and water resource management in Kuwait. Desalination 138(1–3):385–393
Hantush MS (1971) Memorandum on the Shagaya groundwater project. Groundwater Department, Ministry of Electricity and Water, Kuwait
Hernández-Chover V, Castellet-Viciano L, Bellver-Domingo Á, Hernández-Sancho F (2022) The potential of digitalization to promote a circular economy in the water sector. Water 14(22):3722
Hoekstra AY, Chapagain AK (2007) Water footprints of nations: water use by people as a function of their consumption pattern. Water Resour Manag 21:35–48
Hoekstra AY, Hung PQ (2002) A quantification of virtual water flows between nations in relation to international crop trade. Water Resour 49:203–209
Hoekstra AY, Hung PQ (2005) Globalisation of water resources: international virtual water flows in relation to crop trade. Glob Environ Chang 15(1):45–56
Iansiti M, Lakhani KR (2020) Competing in the age of AI: strategy and leadership when algorithms apnd networks run the world. Harvard Business Press, Cambridge
Ibrahim T, Omar Y, Maghraby F (2020) Water demand forecasting using machine learning and time series algorithms. In: International conference on Emerging Smart Computing and Informatics (ESCI), March 12–14, 2020, Pune, India, p 325–329. IEEE. https://doi.org/10.1109/ESCI48226.2020.9167651
Ismail H (2015) Kuwait: food and water security. Future Directions International

Ismail M, Al-Maskati H (2002) Water distribution management towards waste conservation. In: Proceedings of the international conference on water resources management in arid regions (WaRMAR), Kuwait Institute for Scientific Research, March 23–27, 2002, Kuwait

Jones E, Qadir M, van Vliet MT, Smakhtin V, Kang SM (2019) The state of desalination and brine production: a global outlook. Sci Total Environ 657:1343–1356

Karandish F, Hoekstra AY (2017) Informing national food and water security policy through water footprint assessment: the case of Iran. Water 9(11):831

Kayaga S, Calvert J, Sansom K (2003) Paying for water services: effects of household characteristics. Util Policy 11(3):123–132

Korhonen J, Honkasalo A, Seppälä J (2018) Circular economy: the concept and its limitations. Ecol Econ 143:37–46

Kumar R, Ahmed M, Bhadrachari G, Al-Muqahwi S, Thomas JP (2021) Thin-film nanocomposite membrane comprised of a novel phosphonic acid derivative of titanium dioxide for efficient boron removal. J Environ Chem Eng 9(4):105722

Levoso AS, Gasol CM, Martínez-Blanco J, Durany XG, Lehmann M, Gaya RF (2020) Methodological framework for the implementation of circular economy in urban systems. J Clean Prod 248:119227

Li L, Rong S, Wang R, Yu S (2021) Recent advances in artificial intelligence and machine learning for nonlinear relationship analysis and process control in drinking water treatment: a review. Chem Eng J 405:126673

Liu X, Klemeš JJ, Varbanov PS, Čuček L, Qian Y (2017) Virtual carbon and water flows embodied in international trade: a review on consumption-based analysis. J Clean Prod 146:20–28

Makarigakis AK, Jimenez-Cisneros BE (2019) UNESCO's contribution to face global water challenges. Water 11(2):388

Marcon É, Marcon A, Le Dain MA, Ayala NF, Frank AG, Matthieu J (2019) Barriers for the digitalization of servitization. Proc CIRP 83:254–259

Milutinovic M (2006) Water demand management in Kuwait (Master dissertation). Massachusetts Institute of Technology

Ministry of Electricity and Water (MEW) (2007) Statistical Year Book: Water 2006. MEW

Ministry of Electricity and Water (MEW) (2021) Statistical yearbook water in Kuwait. Ministry of Electricity & Water & Renewable Energy

Ministry of Public Works (MPW) (2007, November 10–13) Sanitary Engineering Department, MPW 1st Conference on Public Works, Sheraton, Kuwait

Mohammed S, Darwish M (2017) Water footprint and virtual water trade in Qatar. Desalin Water Treat 66:117–132

Morseletto P, Mooren CE, Munaretto S (2022) Circular economy of water: definition, strategies and challenges. In: Circular economy and sustainability, pp 1–15

Mukhopadhyay A, Akber A (2018) Sustainable water management in Kuwait: current situation and possible correctional measures. Water Stud 13:425

Mukhopadhyay A, Al-Sulaimi J, Barrat JM (1994) Numerical modeling of groundwater resource management options in Kuwait. Groundwater 32(6):917–927

Mukhopadhyay A, Al-Sulaimi J, Al-Awadi E, Al-Ruwaih F (1996) An overview of the tertiary geology and hydrogeology of the northern part of the Arabian gulf region with special reference to Kuwait. Earth Sci Rev 40(3–4):259–295

Mukhopadhyay A, Akber A, Al-Awadi E (2001) Analysis of freshwater consumption patterns in the private residences of Kuwait. Urban Water 3(1–2):53–62

Mukhopadhyay A, Al-Awadi E, Oskui R, Hadi K, Al-Ruwaih F, Turner M, Akber A (2004) Laboratory investigations of compatibility of the Kuwait group aquifer, Kuwait, with possible injection waters. J Hydrol 285(1–4):158–176

Müllegger E, Langergraber G, Jung H, Starkl M, Laber J (2003, April 7–11) Potentials for greywater treatment and reuse in rural areas. In: Ecosan—closing the loop—Proceedings of the 2nd international symposium on ecological sanitation, Germany,p 799–802

Murtaugh KA (2006) Analysis of sustainable water supply options for Kuwait (Master dissertation). Massachusetts Institute of Technology

Parsons Corporation (1963) Groundwater resources of Kuwait, vols I–III. Ministry of Electricity and Water, Kuwait

Public Utilities Board Singapore (2016) Managing the water distribution network with a smart water grid. Smart Water 1(1):4

Purnama A, Al-Barwani HH, Smith R (2005) Calculating the environmental cost of seawater desalination in the Arabian marginal seas. Desalination 185(1–3):79–86

Qasemipour E, Abbasi A (2019) Assessment of agricultural water resources sustainability in arid regions using virtual water concept: case of South Khorasan Province, Iran. Water 11(3):449

Rashid T, Sabarathinam C, Al-Qallaf H, Bhandary H, Al-Jumaa M, Shishter A, Al-Salman B (2022) Evolution of hydrogeochemistry in groundwater production fields of Kuwait–Inferences from long-term data. Chemosphere 307:135734

Rees WE (1992) Ecological footprints and appropriated carrying capacity: what urban economics leaves out. Environ Urban 4(2):121–130

Riedel G, Simon AB (1973) Geotechnical properties of Kuwaiti "gatch" and their improvement. Eng Geol 7(2):155–165

Romano O, Akhmouch A (2019) Water governance in cities: current trends and future challenges. Water 11(3):500

Rybkowska A, Schneider M (2011) Housing conditions in Europe in 2009. Eurostat Statist Focus 4(1):1–12

Sabarathinam C, Bhandary H, Al-Khalid A (2020) Tracing the evolution of acidic hypersaline coastal groundwater in Kuwait. Arab J Geosci 13:1146

Salminen J, Määttä K, Haimi H, Maidell M, Karjalainen A, Noro K, Pohjola J (2022) Water-smart circular economy–conceptualisation, transitional policy instruments and stakeholder perception. J Clean Prod 334:130065

Samayamanthula DR, Sabarathinam C, Bhandary H (2019) Treatment and effective utilization of greywater. Appl Water Sci 9:1–12

Sauve S, Lamontagne S, Dupras J, Stahel W (2021) Circular economy of water: tackling quantity, quality and footprint of water. Environ Dev 39:100651

Schöggl JP, Stumpf L, Baumgartner RJ (2020) The narrative of sustainability and circular economy-a longitudinal review of two decades of research. Resour Conserv Recycl 163:105073

Senay Y (1977) Groundwater resources and artificial recharge in Rawdhatain water field. Ministry of Electricity and Water, Kuwait, p 35

Senay Y, Hamdan L, Yaqubi A (1987) Study of subsurface water rise in the residential areas of Kuwait. Kuwait Institute for Scientific Research, Report Number 2227, KISR, Kuwait

Siderius C, Conway D, Yassine M, Murken L, Lostis PL (2019) Characterizing the water-energy-food nexus in Kuwait and the Gulf region. LSE Middle East Centre Paper Series (28). Middle East Centre, LSE, London

Smith M, Al-Maskati H (2007) The effect of tariff on water demand management: implications for Bahrain. Water Sci Technol Water Supply 7(4):119–126

Tahir S, Steichen T, Shouler M (2018) Water and circular economy: a white paper. Ellen MacArthur Foundation, Arup

Tariq MAUR, Alotaibi R, Weththasinghe KK (2022) A detailed perspective of water resource management in a dry and water scared country, Kuwait, under the conventional and modern approaches. Front Environ Science 2418

Thomas HE (1966) Water resources program of Kuwait. Mimeographed Report. Ministry of Electricity and Water, Kuwait

UNEP, United Nation Environment Programme (2001) Overview of the socio-economic aspects related to the management of municipal wastewater in West Asia (including all countries bordering the Red Sea and the Gulf of Aden). UNEP/PERSGA/ROPME Workshop on Municipal Wastewater Management in West Asia, 10–12 November, Manama, Bahrain. Available online at: http://www.gpa.unep.org/documents/socio-economic_west_asia_english.pdf. Accessed 25 Nov 2008

Voulvoulis N (2018) Water reuse from a circular economy perspective and potential risks from an unregulated approach. Curr Opin Environ Sci Health 2:32–45
Wackernagel M, Onisto L, Callejas Linares A (1997) Ecological footprints of nations: how much nature do they use? How much nature do they have? Centre for Sustainability Studies, Mexico
WEF, World Economic Forum (2020) The global risk report—insight report, 15th edn
Yu X (2008) Use of low quality water: an integrated approach to urban storm water management (USM) in the greater metropolitan region of Sydney (GMRS). Int J Environ Stud 65(1):119–137
Zając P, Avdiushchenko A (2020) The impact of converting waste into resources on the regional economy, evidence from Poland. Ecol Model 437:109299
Zhuo L, Mekonnen MM, Hoekstra AY (2016) Consumptive water footprint and virtual water trade scenarios for China—with a focus on crop production, consumption and trade. Environ Int 94:211–223

Chapter 7
Groundwater Salinization in Kuwait: A Major Threat to Indigenous Ecosystems

Dalal Sadeqi, Amjad Sami Aliewi, Habib Al-Qallaf, and Tareq Rashed

Abstract The process of salinization of both the vadose (including soil subzone) and the saturated zones of aquifers in Kuwait is addressed. For the upper soil subzones closer to the ground surface where evaporation takes place, an increase in soil salinity will reduce plant growth and decline crop yield, impair soil health, affect permeability, and reduce the survival rate of the indigenous fauna. Moreover, the increase in salinity in the saturated zone will considerably reduce the sustainable yield of the aquifers and will increase the salinity levels in the fresh and brackish groundwater aquifers. In this study, the vertical salinity profiles of the utilized aquifers at several locations in the entire state of Kuwait are analyzed to explain the process of salinity stratification. Sustainable yield and recharge of the aquifers in Kuwait are also analyzed. The study investigated the response of the aquifers to a pumping test where the water level was compared to the salinity values. The study concluded that due to the low natural rate of recharge, over-exploitation of the groundwater, and the increase in the pumping rate, the salinity of the aquifers increased. Therefore, integrated management of groundwater in Kuwait is essential to preserve the holistic process involved in the well-being of the eco-hydrology of the aquifer.

Keywords Aquifer · Brackish groundwater · Dammam aquifer · Kuwait group · Salinity vertical profiles · Salinity stratification · Environment · Desert

---

D. Sadeqi (✉) · A. S. Aliewi · H. Al-Qallaf · T. Rashed
Water Research Center, Kuwait Institute for Scientific Research, Safat, Kuwait
e-mail: dsadeqi@kisr.edu.kw; aaliewi@kisr.edu.kw; hqallaf@kisr.edu.kw; trashed@kisr.edu.kw

M. K. Suleiman, S. A. Shahid (eds.), *Terrestrial Environment and Ecosystems of Kuwait*, https://doi.org/10.1007/978-3-031-46262-7_7

7.1 Introduction

Salinization deteriorates soil quality, crop/grass productivity, and soil ecosystem services. Globally, 424 and 833 million hectares of topsoil (0–30 cm) and subsoil (30–100 cm) respectively, are salt-affected (FAO 2021). The US Salinity Lab Staff (1954) defines saline soil with electrical conductivity of extract from saturated soil paste (ECe) is equal or greater than 4 dS/m and the exchangeable sodium percentage (ESP) is less than 15. However, recently, Soil Science Division Staff (2017) sets the limit of ECe 2 dS/m for a very slightly saline soil category. Soil salinity develops as a result of the accumulation of soluble salts. Soil salinity effects crops growth and the yield by reducing water availability to plants (Bresler et al. 1982; Hillel 1994; Ramoliya and Pandey 2003; Shahid et al. 2018a; Shahid 2022). Subsurface water moves to soil surface through capillary action and with subsequent evaporation the surface salinity is increased. The buildup of soil salinity reduces the choices of crops. The soil salinity can be dryland or secondary. The dryland salinity is developed through rising groundwater and subsequent evaporation of the soil water (salts dissolved in the embedded rocks), whereas the secondary salinity refers to the salinization of soil due to human activities such as farmland salinity due to the use of salty water for irrigation (Shahid et al. 2018a). The primary salinity (due to the dissolution of soil material) is in-built salinity in the soil parent material. In Kuwait, based on a soil survey of Kuwait (KISR 1999) and the soil data analyzed by Burezq et al. (2022), overall 38.54% terrestrial desert landscape is affected by primary salinity (ECe > 2 dS/m). This area does not include salinized agriculture farms due to a lack of farm-level salinity survey. However, farmland salinity has been indicated both in Abdaly and Wafra farms (Shahid et al. 2002; Al-Rashed and Al-Senafy 2004). Soil salinity in Kuwait has been identified as early warning of land degradation (Shahid et al. 1998).

The climatic, topographic, and geologic factors (saline upconing and saltwater intrusion (Ghassemi et al. 1995) can also affect soil salinization. Lithology is an important geological factor in the soil salinization process since weathering and dissolution of minerals and rocks is a direct source to bring saline materials to the soil zone to cause dryland salinity. Also, low topography (closed depressions) helps accumulate salts in the soil zone (Kovda 1954) such as the development of inland sabkhas in Kuwait (Omar et al. 2002). Furthermore, anthropogenic factors worsen the soil salinization process, such as due to poorly managed irrigation, non-effective drainage schemes, and heavy application of fertilizers and agrochemicals (Shokri-Kuehni et al. 2020). Thomas and Middleton (1993) and Ceuppens and Wopereis (1999) reported that salinization of the soil due to unsuitable irrigation practices has affected about 60 million hectares (i.e., 24% of all irrigated land worldwide). Other estimates (FAO-AQUASTAT 2013; Qadir et al. 2014; Shahid et al. 2018b) presented 20% (62 million hectares) of the total global irrigated area (310 million hectares) affected by salinity.

The utilization of poor-quality salty water for irrigation increases the soil salinity in agriculture farms and urban landscapes in Kuwait. Irrigation with brackish/saline water and treated wastewater (TWW) not treated to an advanced level will cause soil salinity and ion toxicity (Sparks 1995). In support to this observation, Kotb et al. (2000) reported that intensive irrigation without an effective drainage scheme is the reason behind soil salinization in the Nile Delta. Shallow unconfined aquifers do exist in Kuwait and in several countries. They often affect the functioning of ecosystems. These aquifers are affected by water evaporation from land surfaces (Lee et al. 2005, Muwamba et al. 2018; Shokri and Salvucci 2011). In the case of shallow aquifers (less than 2 m) (Al Senafy 2011), water and dissolved salts moved by capillary action from the water table to the soil surface (Shokri and Salvucci 2011; Tsypkin and Shargatov 2018) and with subsequent evaporation deposit near the ground surface, leading to a reduction in plant growth, impairing soil health and reducing crop yields (Dabach 2021). The excess salts in soil cause physiological drought in crops leading to complete plants failure. When aquifers have deep water tables, i.e., greater than 2 m below the ground surface, capillaries do not extend to the soil surface. Therefore, the vaporization process takes place below the soil zone and vapor diffuses through the overlying dry layer (Milly 1986; Ross et al. 1991; Saravanapavan and Salvucci 2000; Shokri and Or 2011; Shokri and Salvucci 2011; Smits et al. 2012; Kamai and Assouline 2018). In general, the overexploitation of shallow aquifers (recycling for irrigation) causes a drop in water levels and an increase in salinity.

Studying the spatial and vertical dynamics in groundwater salinity can be useful to determine the sustainable yield of brackish aquifers and the recharge to brackish aquifers from rainfall (Aliewi et al. 2020a, 2021). The sustainable yield of shallow aquifers is susceptible to salinization (Werner et al. 2013; Jiao and Post 2019) especially when there is an evidence of sea level rise (SLR) and seawater intrusion (Werner and Simmons 2009; Passeri et al. 2015) to nearby coastal land due to excessive groundwater pumping, sewage disposal, and agricultural activities (Abidi et al. 2017).

Later in this chapter, focus is placed on the utilized aquifers of Kuwait, as well as discussions made on the vertical salinity profiles, sustainable yield in brackish aquifers, and aquifer recharge (Messerschmid and Aliewi 2022) especially for shallow aquifers (Aliewi et al. 2022). The reason to emphasize on the concepts of sustainable yield and aquifer recharge is due to the uncertainties to measure them for brackish aquifers (Hogan et al. 2004; Huet et al. 2016; Walker et al. 2019). The difficulty to measure sustainable yield and aquifer recharge is related to the fact that an accurate estimation of the water balance of an area (e.g., rainfall, runoff, and evapotranspiration parameters) and salinity variation especially in the vertical dimension is rarely possible (Aliewi et al. 2022). McKenzie (2001) and Aliewi et al. (2022) formed a good bond between chlorides (Cl) and carbonates (CO_3) in rainfall and Cl in recharge water from sedimentary aquifers respectively.

Previous studies investigated the link between the groundwater and the salinity of the soil. Groundwater is known to be closely related to the salinity of the soil (Seeboonuang 2012). Pearson's correlation coefficient was performed, and the study concluded that a high positive correlation of $r = 0.87$ exists between the shallow groundwater and the topsoil salinity (Abliz et al. 2016). Weert and Gun (2012) studied the effects of large-scale irrigation water on the groundwater table, which will lead to evaporation of the water table and consequently increase the mineralization of the soil. A study conducted in northeastern Thailand investigated the physical and chemical properties of the groundwater in relation to the soil salinity, the results indicated that the high salinity of the topsoil is closely linked to shallow groundwater (Seeboonuang 2012). Jesus et al. (2015) linked the capillary rise of the groundwater with the soil salinity as the main process for soil salinization. This chapter emphasizes the vertical salinity profiles to estimate sustainable yield and aquifer recharges for brackish aquifers.

7.2 Utilized Aquifers in Kuwait

In Kuwait, the utilized shallow aquifers are affected with shallow marine to sabkha conditions during the Paleocene to the Eocene ages. The lithology of these aquifers was uncomfortably deposited on the underlying Cretaceous rocks. Two aquifer systems are utilized in Kuwait: Kuwait Group aquifers and Dammam aquifers (Table 7.1). Dammam Formations are of Eocene age and are carbonate in nature. The potential of Eocene age aquifer to produce sufficient volumes of groundwater for domestic and irrigation demands is discussed by Aliewi (2022). The lower Eocene represents the sabkha environment formed by evaporites (mostly anhydrite-$CaSO_4$). The Dammam aquifer is divided into three parts (upper, middle, and lower) (Table 7.1). In general, the three parts of the Dammam aquifer are dolomite, massive chalky, laminated dolomite, and nummulitic limestone. The sediment of the Dammam aquifer is deposited in the geological paleo-environments.

Table 7.1 Geologic classification and lithology of aquifers in Kuwait

Age	Group	Formation	Aquifer	Lithology
Recent	Kuwait Group	Recent	–	Surface deposits; beach sand, sand, gravels, silt, etc.
Pleistocene		Dibdibba	AQ-1	Coarse sand and gravels, calcretized at lower parts
Miocene		Fars	–	Evaporites, fine sand, and clay with fossiliferous limestone
		Ghar	AQ-2	Sand and gravels occasionally calcretized with clay intercalations thicker towards the bottom
Eocene	Hasa Group	Dammam	AQ-3	Upper: chalky dolomicrite chertified at top
				Middle: laminated limestone and dolomicrite with lignetic seams
				Lower: nummulitic limestone with shale at the bottom
Paleocene		Rus	–	Anhydrite with dolomitic limestone
		Um Radhuma	–	Limestone, dolomite, anhydrite

Source: Al-Senafy (2001)

In the north-east of Kuwait, dense brines are identified in the Dammam aquifer. The main recharge to the regional Dammam aquifer is in the Kingdom of Saudi Arabia. A chert zone is separating the Dammam aquifer from the Kuwait Group aquifers. The Kuwait Group consists of unconsolidated, poorly sorted sand mixed with gravel, sandstone, clay, and silts (Table 7.1). The thickness of Kuwait Group aquifers (KGA) differs from the north to the south. In the south, the KGA are thin ranging from about 150 m at the Saudi Arabian border to less than 30 m at the western Arabian Gulf in the east. The thickness of KGA in north-easterly direction is in excess of 400 m. The regional flow defines Kuwait's position as a regional groundwater discharge zone, with discharge being induced by the stagnant and dense brines in the north and the Arabian Gulf in the south-east. The vertical flow gradient induced by the dense brines may be a significant contributing factor for upward leakage and hence recharge to the water-bearing horizons overlying the Dammam aquifers.

The Kuwait Group aquifers are:

- *Dibdibba Aquifer*: it is unconfined to semi-confined. It is overlaid by recent deposits.
- *Fars Aquitard*: it is semi-confined. It acts as an aquitard between the Dibdibba aquifer and the Ghar aquifer. It is only found in Northern Kuwait.
- *Ghar Formations Aquifer*: it is confined. It is with clay intercalations thicker towards the bottom.

7.3 Methodology

The methodology used in this chapter is based on groundwater sampling campaign all over Kuwait to determine electrolytes concentration (electrical conductivity-EC, cations, and anions). In particular, the vertical salinity profiles of several wells were determined to assess salinity variations in the horizontal plane (xy dimensions) and in the vertical dimension (z dimension). Mathematical calculations were carried out for some hydrogeological processes such as sustainable yield of brackish aquifers and recharge to brackish aquifers. Figure 7.1 shows the location of the wells that were used in this study for the sampling campaign in order to establish the salinity vertical profiles. As seen, the map (Fig. 7.1) is georeferenced with x and y coordinates to illustrate the GPS coordinates of the group of wells selected for this study.

7.4 Results and Discussion

One of the most important outputs of this investigation is the vertical salinity profiles. These salinity profiles can be used to assess a number of processes such as vertical recharge to aquifers from the ground surface and ecosystems and the

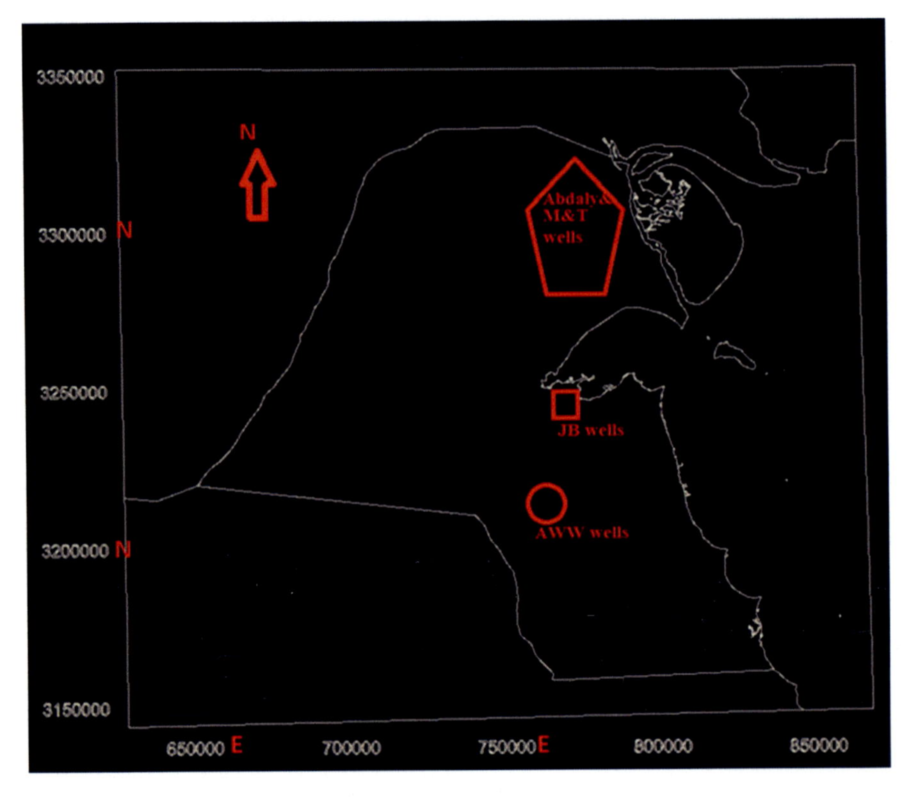

Fig. 7.1 Locations map showing the cluster of wells used for the sampling campaign to generate salinity vertical profiles

mixing processes occurring as a result of transboundary fluxes to Kuwait from outside especially Saudi Arabia. This will be discussed in detail for the utilized aquifers in Kuwait as seen in the following subsections.

7.4.1 Vertical Salinity Profile

The vertical salinity profile was determined by logging the electrical conductivity (EC) of 16 wells tapping both the Kuwait Group aquifer and the Dammam Formation. Wells number AWW-27, AWW-43, AWW-52, AWW-72, AWW-82, AWW-148, AWW-746, and AWW-747 are located in Abdalyeh water field. These wells penetrate the Upper Dammam aquifer and are considered abandoned. Wells number JB-A, JB-B, JB-C, JB-D, and JB-E are located in the Jaber Al Ahmed residential area, all of which tap through the upper Kuwait Group aquifer. A constant pumping with recovery test was also performed on these wells to establish their efficiency and operation conditions to observe how salinity changes during the

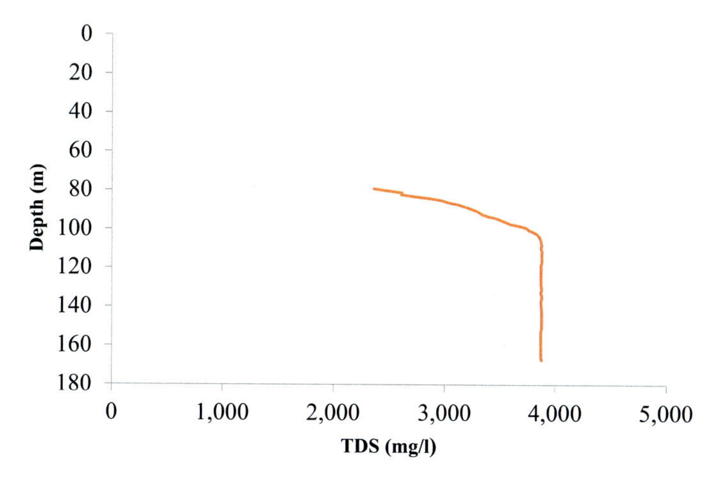

Fig. 7.2 Vertical salinity profile (mg/l) for a well (AWW-52) tapping the upper Eocene aquifer at Kabd region in Kuwait

pumping and recovery phases. Also, the salinity profile of the Dibdibba aquifer of the Kuwait Group aquifer was examined by three wells located in Abdaly farms in northern Kuwait (well T1A, well M1U, and a third pilot well).

7.4.1.1 Upper Dammam Aquifer (Upper Eocene)

Figure 7.2 illustrates a well tapping the Upper Eocene aquifer (well number AWW-52) at Abdalyeh region located in the middle of Kuwait. Its static water level is at about 80 meters below ground level (mbgl). Figure 7.2 indicates the transition zone has a width of 27 m between brackish water of total dissolved solids (TDS) 2361 mg/l and more brackish water (TDS = 3872 mg/l). This indicates a few observations: (i) the effects of recharge on the upper segment of the aquifer is effectively reflecting the vertical recharge process mainly from rainfall and also from pipe networks leakage; (ii) the relatively extensive transition zone width of 27 m at the location of this well reflects the higher degree of heterogeneity of the upper segment of the Upper Eocene and; (iii) in fact, that segment of this aquifer in Kuwait is reported (Al-Senafy 2001) to be karst dolomitic limestone. The quality of the groundwater then remains stagnant from a depth of 106.15 to 167.15 mbgl.

The vertical salinity profile of well number AWW-27 (Fig. 7.3) indicates no stratification in the TDS values, signifying that either the recharge from the ground surface is minimal or the effects of the lateral transboundary flux are limited.

The vertical salinity profile of well number AWW-43 (Fig. 7.4) illustrates the TDS values remain unchanged from depth 67.5 to 114.5 mbgl. A salinity transition zone (width = 13 m, a wide zone) is observed from depth 114.5 to 127.5 mbgl. The TDS values are slightly decreasing after a depth of 127.5 mbgl, which indicates the possibility of transboundary groundwater influx with lower TDS than the indigenous groundwater.

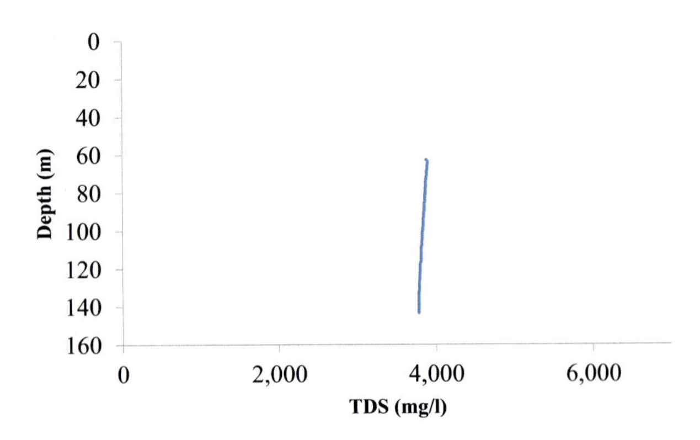

Fig. 7.3 Vertical salinity profile (mg/l) for a well (AWW-27) tapping the upper Eocene aquifer at Kabd region in Kuwait

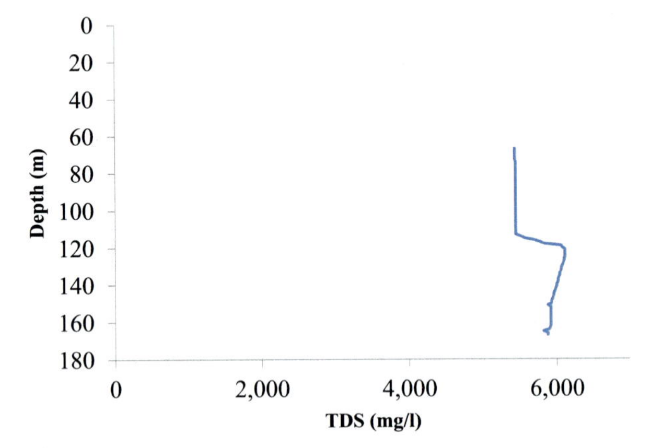

Fig. 7.4 Vertical salinity profile (mg/l) for a well (AWW-43) tapping the upper Eocene aquifer at Kabd region in Kuwait

The vertical salinity profile of well number AWW-72 (Fig. 7.5) illustrates two transition zones. The first transition zone continues with a width of 6 m (average width), leading to a 205 mg/l increase in the TDS from the top of this transition zone to its bottom, which reflects the effects of aquifer recharge from the upper zones. The second transition zone has a width of 9 m (slightly wide), with an increase in its TDS of 378 mg/l between the top of this transition zone and its bottom. This indicates that the influx of transboundary groundwater with a higher TDS value could possibly affect the overall quality of the native groundwater accommodated in heterogeneous/anisotropic porous media.

The vertical salinity profile of well number AWW-82 (Fig. 7.6) indicates two transition zones. The first transition zone continues with a width of 1 m (narrow width) and has a TDS difference of 602 mg/l from the top of the transition zone to its bottom. This reflects the effects of aquifer recharge from the upper zones. The second transition zone has a depth of 14 m (slightly wide), and the TDS increased by 294 mg/l from the top of the transition zone to its bottom indicating the increase may be due to the effects of transboundary groundwater.

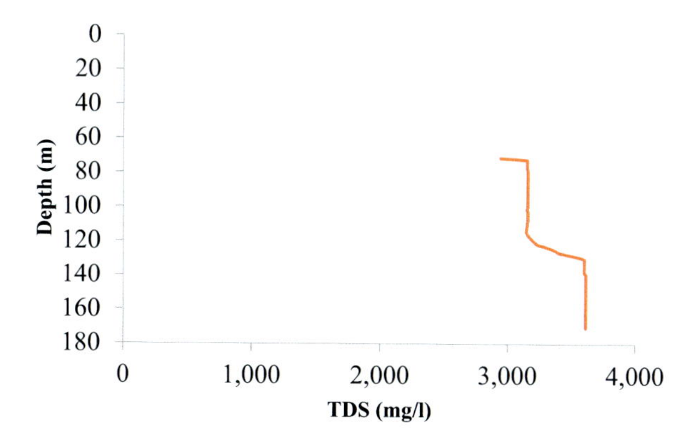

Fig. 7.5 Vertical salinity profile (mg/l) of a well (AWW-72) tapping the upper Eocene aquifer at Kabd region in Kuwait

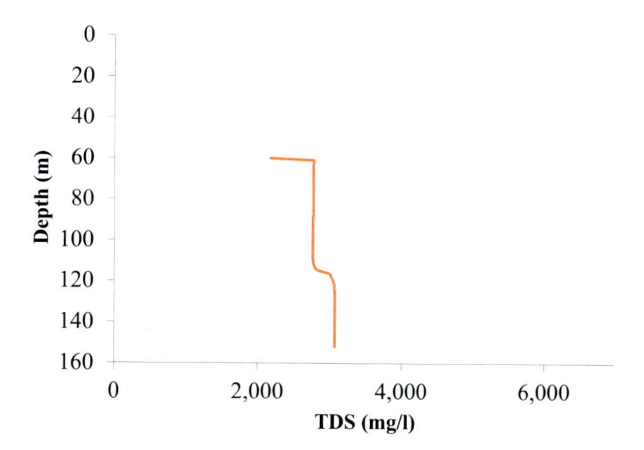

Fig. 7.6 Vertical salinity profile (mg/l) of a well (AWW-82) tapping the upper Eocene aquifer at Kabd region in Kuwait

The vertical salinity profile of wells number AWW-148 and AWW-746 (Figs. 7.7 and 7.8) exhibits no stratification in the TDS value, indicating a lack of recharge from the upper zones of the aquifers and no possible transboundary groundwater influx.

The vertical salinity profile of well number AWW-747 (Fig. 7.9) displays two transition zones, one of which has a width of 1 m (narrow width), and displays a TDS increase of 557 mg/l from the top of the transition zone to its bottom, therefore, indicating the effects of groundwater recharge from the upper zones of the aquifer. The second transition zone occurs at a depth of 139.17 m and continues for a width of 2 m (narrow width), where the TDS values changed from 2490 to 2925 mg/l, which could be the result of transboundary flux with higher TDS values.

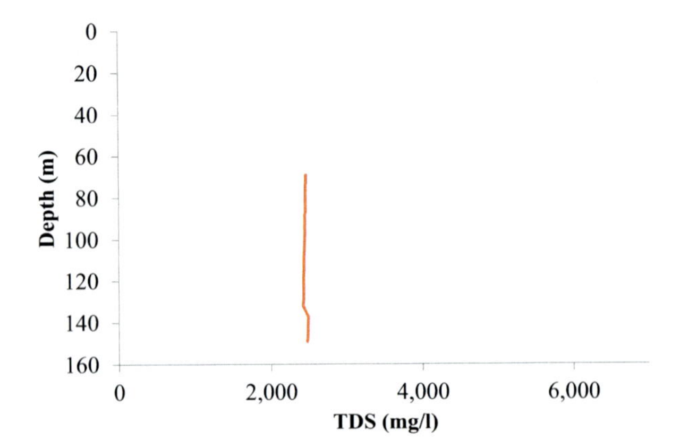

Fig. 7.7 Vertical salinity profile (mg/l) of a well (AWW-148) tapping the upper Eocene aquifer at Kabd region in Kuwait

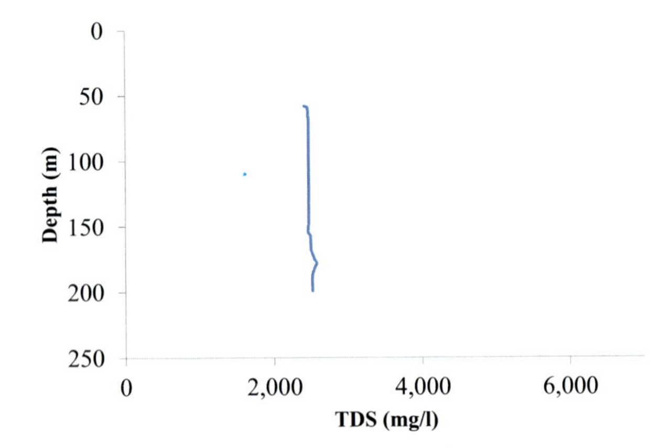

Fig. 7.8 Vertical salinity profile (mg/l) of a well (AWW-746) tapping the upper Eocene aquifer at Kabd region in Kuwait

7.4.1.2 Upper Kuwait Group Aquifer at Jaber Al-Ahmad-Al-Qirawan Region

In Fig. 7.10, the vertical salinity profiles for wells JB-A and JB-B reflect a significant effect of ground surface recharge as there is a distinct stratification in the TDS vertical profile. However, for wells JB-C and JB-D the change in the TDS values is not drastic indicating that the effects of the surface recharge on the groundwater quality are not significant, and the stratification of the TDS vertical profile is insufficient.

The vertical salinity profiles for each of the wells are summarized to highlight the major differences in the reaction of the subsoil to the natural recharge of groundwater (Table 7.2). It is evident that the transition zone for each of the wells is narrow.

The wells number JB-A, JB-B, JB-C, and JB-D are located in a residential area where the top zone of the aquifer could be affected by different anthropogenic activities such as agriculture, or by freshwater leakage from the freshwater supply network. To assess the conditions and performance of the wells a constant rate pumping

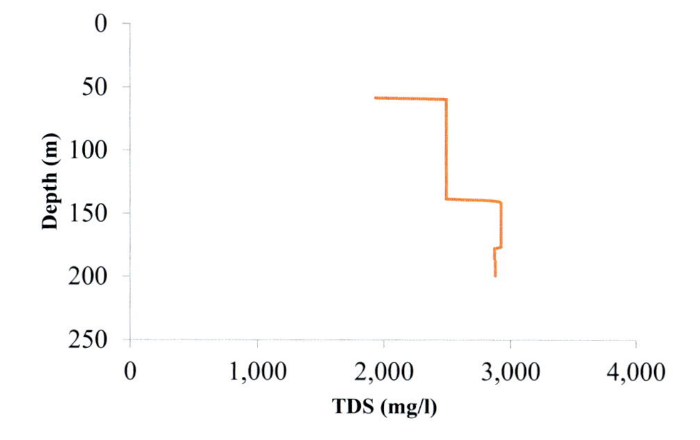

Fig. 7.9 Vertical salinity profile (mg/l) for a well (AWW-747) tapping the upper Eocene aquifer at Kabd region in Kuwait

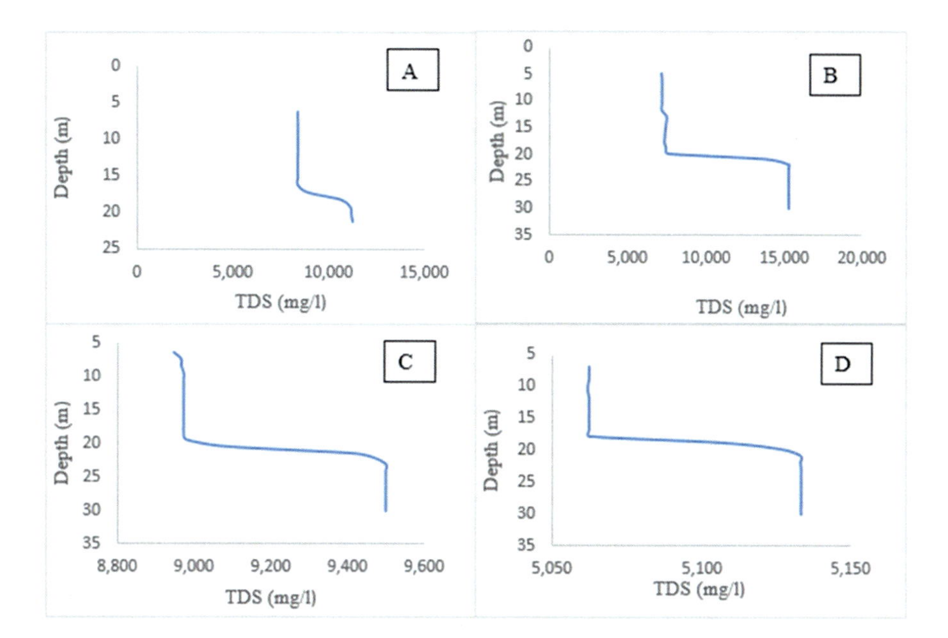

Fig. 7.10 Vertical salinity profiles (mg/l) of wells: A) JB-A, B) JB-B, C) JB-C, D) JB-D

Table 7.2 Vertical salinity profile of the wells tapping the Upper Kuwait Group Aquifer at Jaber Al-Ahmad-Al-Qirawan Region

Well number	TDS range	Transition zone width
JB-A	8358–10,656 mg/l	4 m
JB-B	7526–15,232 mg/l	2 m
JB-C	8979–9466 mg/l	2.6 m
JB-D	5062–5133 mg/l	3 m

test was performed on three wells located in Jaber Al Ahmad residential area. The test was performed on well number JB-C for approximately 3 days (Fig. 7.11). The water level has declined by 1.2 m resulting in a 30% increase of TDS values (9309–12,074 mg/l). It was also noticed that when the operation of the pump was stopped, the TDS value was increased by 16.5%. However, the final TDS value of 10,074 mg/l is 8% higher than the initial quality of the groundwater in the well.

Figure 7.12 shows the drawdown in the water level of well JB-D (Fig. 7.12) reached 1.1 m after continuous pumping for 5 days. As a result, the TDS level has increased from 2598.4 mg/l to approximately 8600 mg/l, which is a 331% increase.

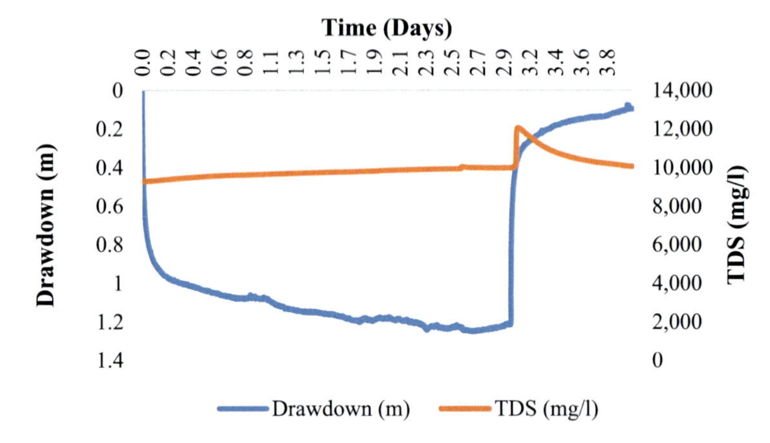

Fig. 7.11 Constant rate pumping test on well number JB-C tapping the upper Kuwait Group Aquifer at Jaber Al-Ahmad-Al-Qirawan Region

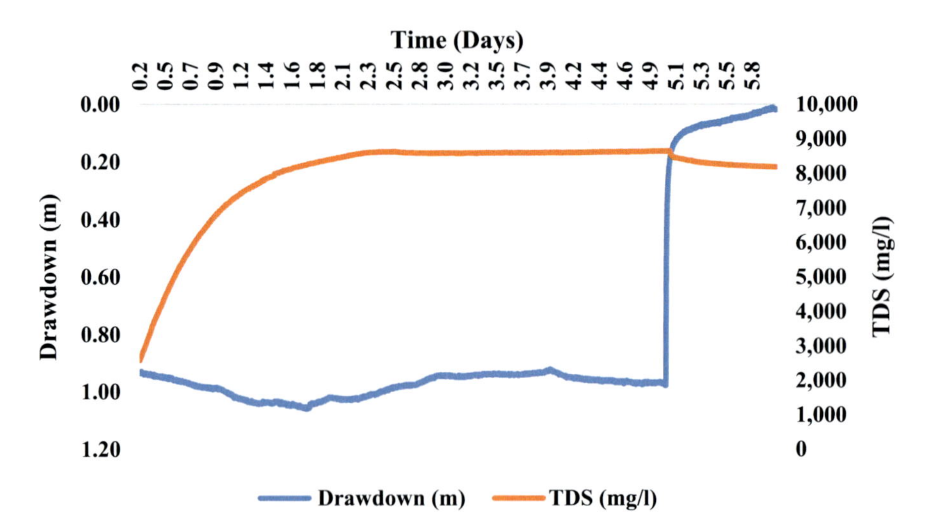

Fig. 7.12 Constant rate pumping test on well number JB-D tapping the upper Kuwait Group Aquifer at Jaber Al-Ahmad-Al-Qirawan Region

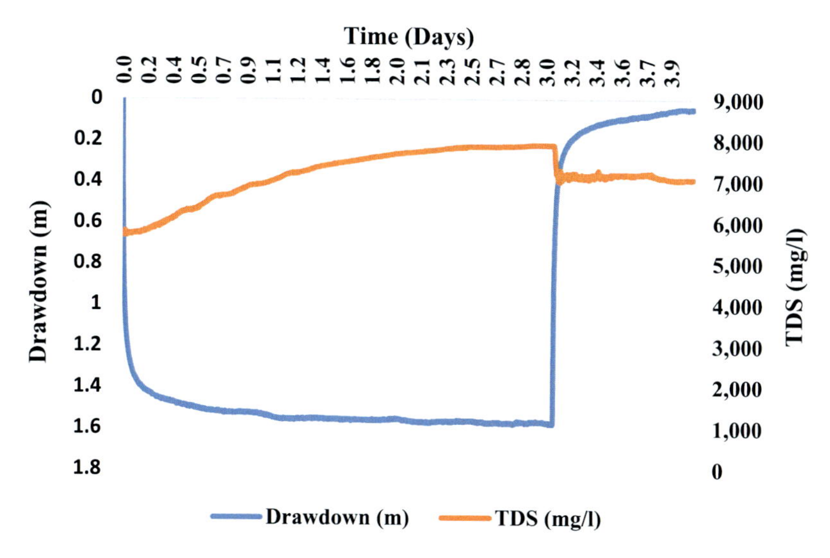

Fig. 7.13 Constant rate pumping test on well number JB-E tapping the upper Kuwait Group Aquifer at Jaber Al-Ahmad-Al-Qirawan Region

Figure 7.12 also reflects the effects of stopping the operation of the pump on the overall quality of the groundwater, as the TDS values decreased by 5%.

As shown in Fig. 7.13, after conducting the pumping test on well JB-E, the water level reached its maximum at 1.6 m after 3 days. Therefore, the quality of the groundwater declined as the TDS value increased from 5744 to 7885 mg/l. When the operation of the pump has stopped the TDS value decreased by 11%, and it reached to 7056 mg/l, which is still considerably higher than the initial quality of the groundwater in the well.

7.4.1.3 Dibdibba Aquifer (Upper Kuwait Group) at Abdaly Farms

The vertical salinity profile of well T1A located upstream of Abdaly farms (Fig. 7.14) infers the transition zone width of 1 m which is an extremely narrow transition zone. The difference in TDS between the top of transition zone and its bottom is 938 mg/l, implying that the recharge from the upper zones of the aquifer is relatively limited, and the stratification of the TDS vertical profile is insufficient.

The vertical salinity profile of a well (Fig. 7.15) located on a pilot farm in Abdaly illustrates two transition zones. The first transition zone has a width of 3 m (narrow width) and the TDS value increased by 5145 mg/l from the top of the transition zone to the bottom at 6727 mg/l. The quality of the groundwater changed due to recharge from the upper zones of the aquifer. The second transition zone has a width of 5 m (narrow width), and it exhibits a TDS is 7203 mg/l at the top of the transition zone to its bottom at 11,760 mg/l, which is the result of the transboundary groundwater influx characterized by a higher TDS value than the native water.

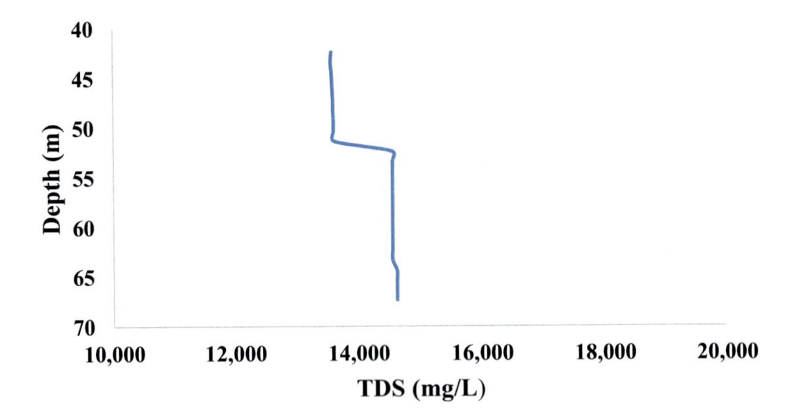

Fig. 7.14 Vertical salinity profile (mg/l) for a well (T1A) tapping the Dibdibba Aquifer (Upper Kuwait Group Aquifer) at Abdaly Farms

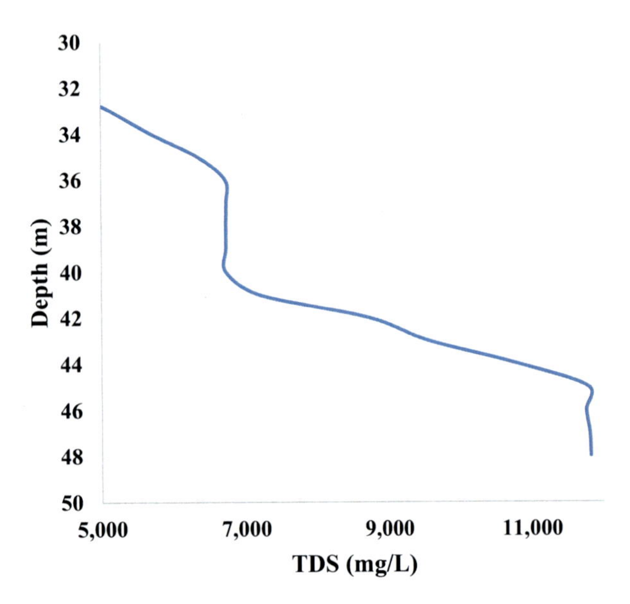

Fig. 7.15 Vertical salinity profile (mg/l) for a well tapping the Dibdibba Aquifer (Upper Kuwait Group Aquifer) at Abdaly Farms

The vertical salinity profile of well number M1U located downstream of Abdaly farms (Fig. 7.16) shows the width of the transition zone equals to 9.5 m (slightly wide), and the TDS increased by 5404 mg/l from the top of the transition zone to its bottom, which indicates that recharge from the upper zones of the aquifer significantly effects the overall quality of the groundwater in the specified location.

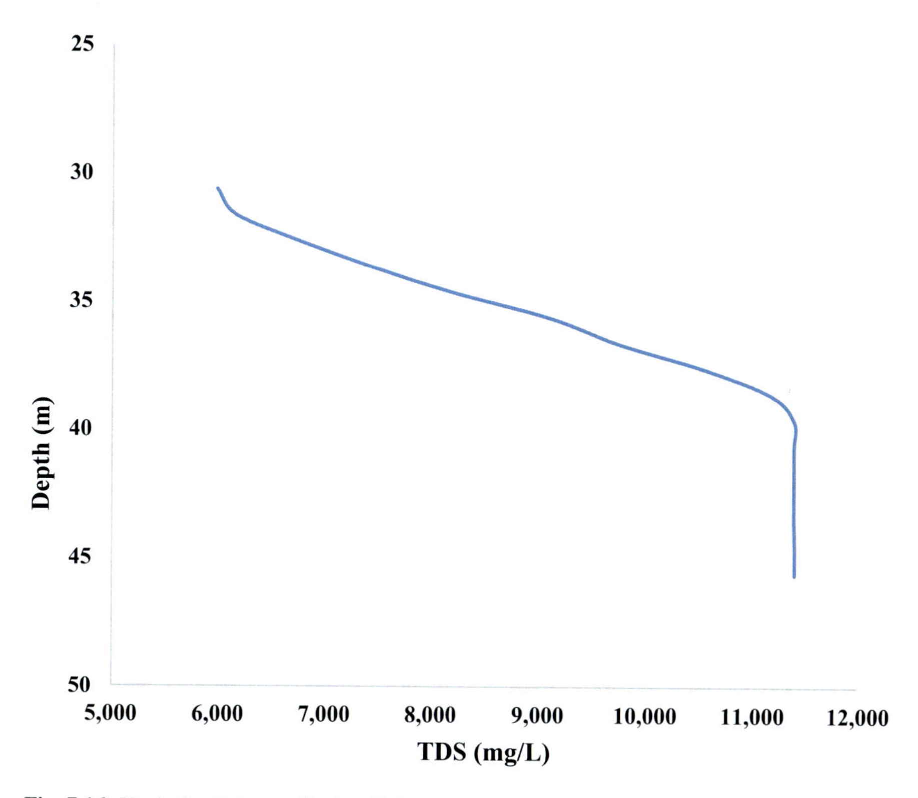

Fig. 7.16 Vertical salinity profile (mg/l) for a well (M1U) tapping the Dibdibba Aquifer (Upper Kuwait Group Aquifer) at Abdaly Farms

7.4.2 *Sustainable Yield of Brackish Groundwater Reserve in Abdaly Farming Area*

The sustainable yield of aquifers storage (i.e., potential of utilizable brackish groundwater volumes) was estimated from Dibdibba aquifer in northern Kuwait using the Eq. 7.1 of Fetter (2001):

$$V_w = A.b.Sy \tag{7.1}$$

where:
V_W = the total groundwater Volume (m^3) (sustainable yield)
A = the surface areal extent of the aquifer (m^2)
b = average saturated thickness (m) for unconfined aquifer
Sy = the average effective porosity

During the calculations, the following were considered:

- The areas of TDS greater than 10,000 mg/l were considered not suitable for extraction of brackish groundwater.
- The areas with transmissivity values less than 10 m^2/day were considered not suitable for further extraction of brackish groundwater.
- The areas where the aquifer thickness or the saturated thickness is less than 1 m were eliminated.

Based on the above criterion, the areas of suitable portions of Dibdibba with suitable salinity level, transmissivity, and thickness were determined using Arc GIS. This investigation is based on the efforts made by Al-Qallaf et al. (2021) and Aliewi et al. (2021). They estimated the volume of the brackish groundwater aquifer below the Abdaly farming area during 2010.

This thickness was estimated from the study of the vertical variation in groundwater quality. The effective porosity of the wells in the northern Kuwait was taken from pumping tests at a value of about 0.12 (Aliewi et al. 2020b, c, 2013).

The following data were used to calculate the sustainable yield (utilizable groundwater reserve in the Abdaly area in 2010) using Eq. 7.1:

- The total area of the water table aquifer is 300 km^2
- The average saturation thickness of groundwater is 21.5 m
- Effective porosity is 0.12

Accordingly, the sustainable yield of brackish groundwater aquifer below Abdaly was estimated as 0.8 billion cubic meters (bcm). During the period from 1989 to 2010, the groundwater level had declined about 3.23 m in average.

Bhandary et al. (2023) estimated the average saturation thickness of the brackish groundwater aquifer (Dibdibba Aquifer) from the vertical variation in groundwater TDS measured in 2021 in Abdaly monitoring wells to be 12.3 m. By applying the Eq. 7.1, the sustainable yield of the Dibdibba aquifer is 0.5 bcm. By comparing this value with the one estimated by Al-Qallaf et al. (2021), it is obvious that the volume of the brackish groundwater reserve has reduced from 0.8 bcm in 2010 to 0.5 bcm in 2021. This is a reduction of 37.5% of sustainable yield in 12 years due to saline water upconing as a result of over pumping (Al-Rashed and Aliewi 2018; Aliewi et al. 2020a). Table 7.3 shows the difference in the thickness of brackish water in the monitoring wells between 2009 and 2021. It shows the groundwater in four wells (M1-U, M1-L, M5-L, and T3-A) changed from partially brackish to totally saline, and the thickness of the brackish groundwater reduced in all the wells except the well T2-A. Table 7.3 shows that the maximum reduction in the thickness of brackish water is 31 m, which is a significant change. Provided the pumping to continue at the same rate, the groundwater at Abdaly will become completely saline within a period of 16 years.

Table 7.3 Changes in brackish groundwater thickness (m) between 2009 and 2021

Well No	2009	2021	Change in thickness (m)
M1-U	4	0	4
M1-L	3	0	3
M2-U	11	7	4
M2-L	40	35	5
M3-U	7	5	2
M3-L	17	9	8
M4-U	19	4	15
M4-L	0	0	0
M5-U	7	4	3
M5-L	28	0	28
M6-U	10	4	6
M6-L	20	19	1
M7-U	12	11	1
M7-L	33	7	26
T1-A	35	–	–
T1-B	–	22	–
T2-A	28	28	0
T2-B	35	11	24
T3-A	31	0	31
T4-A	28	12	16
T4-B	31	6	25

7.4.3 Estimation of Recharge to Aquifer from Changes in Aquifer Salinity

As explained earlier, the vertical salinity profiles can be used to determine the aquifer recharge to brackish aquifers. This concept was investigated thoroughly by Aliewi et al. (2021). The investigation is based on the fact that chloride (Cl) in rainfall water dilutes the underlying relatively deeper saline waters making it possible to determine the thickness of the unsaturated zone that was affected by recharging water with less Cl concentration. This recharging water can come from different sources such as irrigation and wastewater returns or other sources. This investigation considers the background concentration to be the long-term concentration of Cl from other sources. In their investigation, Aliewi et al. (2021) developed a new equation (Eq. 7.2) based on vertical salinity profiles:

$$R_{CC} = A.b.Sy\left[\frac{\mathrm{Cl}_0 - \mathrm{Cl}_m}{\mathrm{Cl}_m - \mathrm{Cl}_P}\right] \tag{7.2}$$

where:
Cl_0: background chloride concentration in the brackish water aquifer (mg/l) representing all other sources except precipitation
Cl_m: measured chloride (mg/l) in the water samples of the aquifer after rainfall events
Cl_p: chloride (mg/l) in rainfall
R_{cc}: recharging flux [(Million Cubic Meter (Mm^3) per unit time)]
A: the surface area of the aquifer under recharge stresses (m^2)
b: saturated thickness of the aquifer only affected by infiltrating water (m) measured from the vertical salinity profile
Sy: specific yield of the unconfined aquifer under recharge stresses

Equation 7.2 was applied to Kuwait aquifers using the data for the extreme rainfall events that took place in the first two weeks of November 2018 (Table 7.4).

The results are shown in Fig. 7.17. The results of rainfall events in November 2018 show that the average aquifer recharge is 9.3% from rainfall during the two weeks of November 2018.

Aliewi et al. (2021) reported that wherever the water level is closer to the ground surface such as I-1 and I-2 zones (Fig. 7.17), then the aquifer recharge from rainfall is high at a percentage of 23.5%. Also, they reported that for deep water levels such as B and D zones, the aquifer recharge from rainfall is less than 1%.

7.4.4 *Effects of Groundwater Misuse on the Ecosystem*

Misuse of groundwater has been witnessed in the agricultural areas of Kuwait, where the farmers have exploited the available groundwater by digging unauthorized wells and by over pumping. Irrigated water used for farming has a total dissolved solids ranging from 2500 to 10,000 mg/l. Due to the high temperature and high evaporation rates the salinity of soil increased, where the soil affected by salinity covers an area of 209,000 hectares, and the soil salinity in Abdaly farms ranges from ECe 5 to 25 dS/m (Choukr-Allah et al. 2023). As a result, the cultivated area in Abdaly was decreased (Misak and Hussain 2023). While other farms experienced a rise in the water table due to the unregulated irrigation. Eventually, the root zone will become oxygen deprived due to waterlogging, and in turn deprive the growth of plants (Sophocleous 2003) (Fig. 7.18a). In arid lands such as Kuwait with high temperature and evaporation rates, the water in the topsoil or in the shallow groundwater leaves behind a layer of salts which has proven to be lethal for plant growth (Sophocleous 2003) (Fig. 7.18b).

Drilling of wells to retrieve groundwater has negative repercussions on the surrounding wells (Fig. 7.19). The waste resulting from borehole drilling mainly consists of crushed rock and remnants of mud, which in turn has negative spatial effects on sediment macrofauna community structure (Bakke et al. 2013). A groundwater ecosystem which is directly linked to the soil offers an immense value such as purification of water, active biodegradation, nutrient recycling, and mitigation of floods,

Table 7.4 Recharge estimation to Kuwait shallow aquifers (1–15 Nov 2018) using the Modified Chloride Mass Balance (CMB) equation (Eq. 7.2)

Rainfall Station→	A	B	C	D[a]	E	F	G	H	I1[a]	I2	Total
A (km^2)	1529	3105	2960	2087	763	2501	457	485	1458	1099	16,444
Rainfall (Mm^3)	263	199	299	292[a]	159	540	74	88	204[a]	154	1776
Cl_p (mg/l)	16	16	16	16	16	16	16	16	16	16	
Cl_m (mg/l)	3960	5000	1910	939[a]	3081	5180	18,900	4110	30[a]	2315	
Cl_p/Cl_m	0.004	0.003	0.008	0.017[a]	0.021	0.003	0.0008	0.0133	0.53[a]	0.02	
b (m)[b]	3	3	3	1	2	3	1.5	2	3	3	
Sy (%)	1.7	1.7	1.7	1	1	1	1	1	1.7	1.7	
Groundwater (Mm^3)	76	155	148	21	15	75	7	10	73	55	635
Cl_0 (mg/l)	5325	5001	2250	1007	10,271	6300	23,640	13,700	4148	3830	
R_{CC} (Mm^3)	26.5	0	26.6	1.5[a]	35.8	16.3	1.7	22.7	172.4[a]	36.2	165.1
% of recharge	10.1	0.0	8.9	0.3[a]	22.5	3.0	2.3	25.8	85.0[a]	23.5	9.3 (average)
$(Cl_0 - Cl_m)/(Cl_m - Cl_p)$	0.35	0	0.18	0.074[a]	2.35	0.22	0.25	2.34	2.37[a]	0.66	

[a]Rainfall stations excluded from analysis

[b]The saturated thickness affected by direct rainfall recharge in the vertical salinity profiles

After Aliewi et al. (2022)

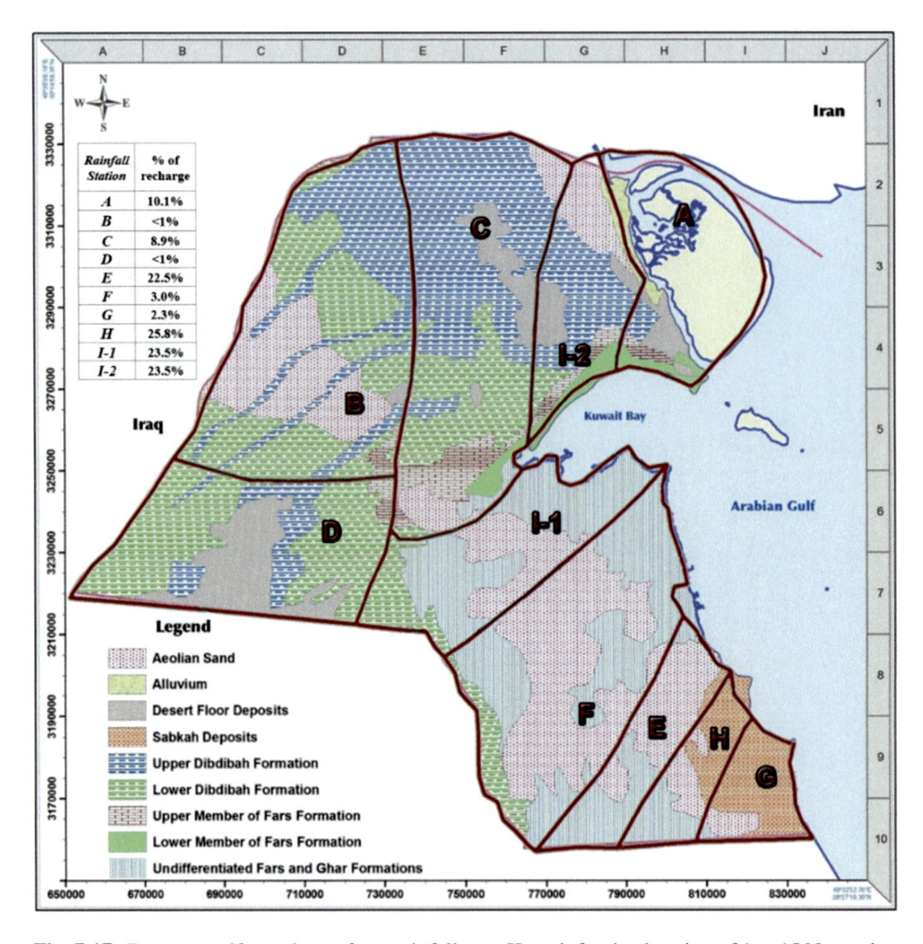

Rainfall Station	% of recharge
A	10.1%
B	<1%
C	8.9%
D	<1%
E	22.5%
F	3.0%
G	2.3%
H	25.8%
I-1	23.5%
I-2	23.5%

Fig. 7.17 Percent aquifer recharge from rainfall over Kuwait for the duration of 1 to 15 November 2018. (Source: Aliewi et al. 2022)

which are directly connected to the presence and activity of microorganisms and metazoan (Griebler and Avramov 2015). Therefore, the protection and management of groundwater will require a holistic understanding of the system at different spatial and temporal scales, also an assessment of the resistance and resilience of that system to anthropogenic impacts. Additionally, ecosystems will also affect some hydrological variables, such as aquifer recharge.

Fig. 7.18 (**a**) Dewatering of shallow aquifers and (**b**) the effects of the shallow water level on plant growth

7.5 Conclusions and Recommendations

Salinity variations in the utilized aquifers of Kuwait were investigated in several locations. It has been observed that soil salinity reduces plant growth and soil health leading to bare lands. The increase in soil salinity is caused by the encrustation process as a result of evaporation in the upper soil zone. Also, the salinized soil reduces the biodiversity of soil flora and fauna. This study has illustrated that the increase in salinity due to over pumping in the saturated zone of the Dibdibba aquifer underlying the Abdaly farms will considerably reduce its sustainable yield. In fact, the sustainable yield of the Dibdibba aquifer at Abdaly farms was reduced to 37.5% between 2010 and 2021 converting fresh groundwater lenses and brackish groundwater bodies to saline. The vertical salinity profiles of the utilized aquifers at several locations in the entire state of Kuwait were generated and analyzed to explain the aquifer recharge process. It has been shown that in regions where the water level is closer to the ground surface, the percentage of aquifer recharge from rainfall is high at 23.5%. Also, for deep water levels, the aquifer recharge from rainfall is lower than 1%. This study recommends to control abstractions from wells at Kuwait farms to protect the precious groundwater resources from further degradation and to keep the ecosystem healthy to provide ecosystem services.

Fig. 7.19 Borehole drilling site

References

Abidi JH, Farhat B, Mammou AB, Oueslati N (2017) Characterization of recharge mechanisms and sources of groundwater salinization in Ras Jbel coastal aquifer (Northeast Tunisia) using hydrogeochemical tools, environmental isotopes, GIS, and statistics. J Chem Article ID 8610894:20. https://doi.org/10.1155/2017/8610894

Abliz A, Tiyip T, Ghulam A, Halik Ü, Ding JL, Sawut M, Zhang F, Nurmemet I, Abliz A (2016) Effects of shallow groundwater table and salinity on soil salt dynamics in the Keriya Oasis, Northwestern China. Environ Earth Sci 75(3):260, 1–15. https://doi.org/10.1007/s12665-015-4794-8

Aliewi AS (2022) Assessment of the productivity potential of the Eocene formation to yield uncontaminated groundwater. In: Senapathi V, Sekar S, Viswanathan PM, Sabarathinam C (eds) Groundwater contamination in coastal aquifers. Elsevier, pp 141–154. https://doi.org/10.1016/B978-0-12-824387-9.00004-9. https://www.sciencedirect.com/science/article/pii/B9780128243879000049
Aliewi AS, Al-Odwani A, Qallaf H, Rashid T, El- Mansour M, Al Mufleh S (2013) Design of dewatering schemes using analytical and numerical methods at residential areas in Kuwait. Int Water Technol J 3(4):217–231
Aliewi AS, Al-Enezi H, Al-Maheimid I, Kandari J, Al-Haddad A, Al-Qallaf H, Rashid T, Sadeqi D (2020a) Sustainability of brackish groundwater utilization from the Eocene aquifer for oil exploration operations in central Kuwait. Environ Dev Sustain 22:4639–4653. https://doi.org/10.1007/s10668-019-00401-9
Aliewi AS, Bhandary H, Al-Kandari J, Sabarathinam C, Al-Qallaf H, Dashti F, Al-Otaibi F (2020b) Assessment of the potential Dibdibba as a shared regional aquifer. Water Research Centre. Kuwait Institute for Scientific Research, Kuwait. Final report of project number WM066C. 211 p
Aliewi AS, Al-Qallaf H, Rashid T, Al-Odwani A (2020c) Hydraulic evaluation of a dewatering scheme in shallow aquifers in Kuwait. Q J Eng Geol Hydrogeol 53(1):125. https://doi.org/10.1144/qjegh2019-044
Aliewi AS, Al-Kandari J, Al-Khalid A, Bhandary H, Al-Qallaf H (2021) Modelling the effect of high level of total dissolved solids (TDS) for the sustainable utilization of brackish groundwater from saline aquifers in Kuwait. Environ Dev Sustain 23(2):2204–2223. https://doi.org/10.1007/s10668-020-00670-9
Aliewi AS, Bhandary H, Sabarathinam C, Al-Qallaf H (2022) A new modified chloride mass balance approach based on aquifer hydraulic properties and other sources of chloride to assess rainfall recharge in brackish aquifers. Hydrol Process J 36(3). https://doi.org/10.1002/hyp.14513
Al-Qallaf H, Aliewi AS, Ramadan A, Abdulhadi A, Jose J, Shishter A, Al-Salman B, Al-Jumaa M, Dashti F, Al-Otaibi F (2021) Prediction of the impact of climatic changes on the fresh/usable groundwater accumulation at the Northern watershed of Kuwait, Report No. KISR 17002. Kuwait Institute for Scientific Research, Kuwait City
Al-Rashed M, Aliewi AS (2018) Water resources sustainability in Kuwait against United Nations sustainable development goals. In: Azar E, Raouf MA (eds) Sustainability in the Gulf: challenges and opportunities, p 312
Al-Rashed M, Al-Senafy M (2004) Assessment of ground water salinization and soil degradation in Abdaly farms, Kuwait. Agric Marine Sci 9(1):17–19
Al Senafy M (2011) Management of water table rise at Burgan oil field, Kuwait. Emirates J Eng Res 16(2):27–38
Al-Senafy M (2001) Geohydrology of fresh groundwater lenses in arid environment, Kuwait. In: Proceedings of the groundwater quality conference, 2001. Sheffield, pp 167–168
Bakke T, Klungsøyr J, Sanni S (2013) Environmental impacts of produced water and drilling waste discharges from the Norwegian offshore petroleum industry. Mar Environ Res 92:154–169
Bhandary H, Akber A, Chidambaram S, Aliewi A, Al-Qallaf H, Rashid T, Dhanuradha SVV, Sugumaran K, Jayaramu Y, Alambi R, AlSabti B, Sadeqi D, Al-Otaibi F (2023) Effect of reverse osmosis units' brine reject from agricultural farms in Kuwait on groundwater quality and levels (WM077C). Final Report, KISR No.17938, Kuwait
Bresler E, McNeal BL, Carter DL (1982) Saline and sodic soils: principles, dynamics, modeling. Springer
Burezq H, Shahid SA, Baron HJ (2022) Salts in the terrestrial environment of Kuwait and proposed management. In: FAO (ed) Halt soil salinization, boost soil productivity – Proceedings of the global symposium on salt-affected soils, 20–22 October 2021. Rome, pp 267–268. https://doi.org/10.4060/cb9565en
Ceuppens J, Wopereis MCS (1999) Impact of non-drained irrigated rice cropping on soil salinization in the Senegal river delta. Geoderma 92:125–140

Choukr-Allah R, Mouridi ZE, Benbessis Y, Shahid SA (2023) Salt-affected soils and their management in the Middle East and North Africa (MENA) region: a holistic approach. In: Choukr-Allah R, Ragab R (eds) Biosaline agriculture as a climate change adaptation for food security. Springer, Cham, pp 13–45

Dabach S (2021) Soil salinity and field flooding. https://ndrip.com/soil-salinity-and-field-flooding/

FAO (2021) Global map of salt-affected soil. https://www.fao.org/3/cb7247en/cb7247en.pdf

FAO-AQUASTAT (2013) Area equipped for irrigation and percentage of cultivated land. Available at http://www.fao.org/nr/water/aquastat/globalmaps/index.stm. Accessed 16 Sept 2013

Fetter CW (2001) Applied hydrogeology, 4th edn. Prentice-Hall, Inc, New Jersey

Ghassemi F, Jakeman AJ, Nix HA (1995) Salinisation of land and water resources: Human causes, extent, management and case studies. University of New South Wales Press/CAB International, Sydney/Wallingford. ISBN 0868401986, 9780868401980

Griebler C, Avramov M (2015) Groundwater ecosystem services: a review. Freshwater Sci 34(1):355–367

Hillel D (1994) Fundamentals of soil physics. Academic

Hogan JF, Phillips FM, Scanlon BR (eds) (2004) Groundwater recharge in a desert environment, the Southwestern United States. American Geophysical Union, Washington, DC

Huet M, Chesnaux R, Boucher MA, Poirier C (2016) Comparing various approaches for assessing groundwater recharge at a regional scale in the Canadian Shield. Hydrol Sci J 61(12):2267–2283. https://doi.org/10.1080/02626667.2015.1106544

Jesus J, Castro F, Niemelä A, Borges MT, Danko AS (2015) Evaluation of the impact of different soil salinization processes on organic and mineral soils. Water Air Soil Pollut 226:1–12

Jiao J, Post V (2019) Coastal hydrogeology. Cambridge University Press. https://doi.org/10.1017/9781139344142

Kamai T, Assouline S (2018) Evaporation from deep aquifers in arid regions: analytical model for combined liquid and vapor water fluxes. Water Resour Res 54(7):4805–4822. https://doi.org/10.1029/2018WR023030

KISR (1999) Soil survey for the state of Kuwait, vol 2 & 3. Kuwait Institute for Scientific Research, State of Kuwait.

Kotb THS, Watanabe T, Ogino Y, Tanji KK (2000) Soil salinization in the Nile Delta and related policy issues in Egypt. Agric Water Manag 43:239–261

Kovda V (1954) Geochemistry of the Arid Zone in USSR, Publ. House of the Acadamy of Sciences, Moskva-Leningrad, U.S.S.R

Lee J, Oliveira RS, Dawson TE, Fung I (2005) Root functioning modifies seasonal climate. Proc Natl Acad Sci Braz 102(49):17576–17581

McKenzie AA (2001) West bank aquifers-conceptual recharge estimation. Technical report No. SUSMAQ-REC#04V0.1. Sustainable Management for the West Bank and Gaza Aquifers. Palestinian Water Authority (Palestine) and Newcastle University (UK)

Messerschmid C, Aliewi AS (2022) Spatial distribution of groundwater recharge, based on regionalised soil moisture models in Wadi Natuf karst aquifers, Palestine. Hydrol Earth Syst Sci 26(4):1043–1061. https://doi.org/10.5194/hess-26-1043-2022

Milly PCD (1986) An event-based simulation model of moisture and energy fluxes at a bare soil surface. Water Resour Res 22:1680–1692

Misak R, Hussain W (2023) Groundwater in Kuwait. In: Abd el-aal Aea K, Al-Awadhi JM, Al-Dousari A (eds) The Geology of Kuwait. Regional Geology Reviews. Springer, Cham. pp. 199-214. https://doi.org/10.1007/978-3-031-16727-0_9

Muwamba A, Nkedi-Kizza P, Morgan KT (2018) Effect of water table depth on nutrient concentrations below the water table in a spodosol. Water Air Soil Pollut 229(3). https://doi.org/10.1007/s11270-018-3727-z

Omar SAS, Misak RF, Shahid SA (2002) Sabkhat and halophytes in Kuwait. In: Barth, Boer (eds) Sabkha ecosystems, pp 71–81

Passeri DL, Hagen SC, Medeiros SC, Bilskie MV, Alizad K, Wang D (2015) The dynamic effects of sea level rise on low-gradient coastal landscapes: a review. Earth's Future 3(6):159–181. https://doi.org/10.1002/2015EF000298

Qadir M, Quillerou E, Nangia V, Murtaza G, Singh M, Thomas RJ, Drechsel P, Noble AD (2014) Economics of salt-induced land degradation and restoration. Nat Res Forum 38(4):282–295. https://doi.org/10.1111/1477-8947.12054

Ramoliya PJ, Pandey AN (2003) Effect of salinization of soil on emergence, growth and survival of seedlings of *Cordia Rothii*. For Ecol Manag 176(1–3):185–194. https://doi.org/10.1016/s0378-1127(02)00271-2

Ross PJ, Williams J, Bristow K (1991) Equation for extending water-retention curves to dryness. Soil Sci Soc Am J 55:923–927

Saravanapavan T, Salvucci GD (2000) Analysis of rate-limiting processes in soil evaporation with implications for soil resistance models. Adv Water Resour 23:493–502

Seeboonruang U (2012) Relationship between groundwater properties and soil salinity at the Lower Nam Kam River Basin in Thailand. Environ Earth Sci 69(6):1803–1812

Shahid SA (2022) Innovative thinking and use of salt-affected soils in irrigated agriculture. In: FAO (ed) Halt soil salinization, boost soil productivity – Proceedings of the global symposium on salt-affected soils, 20–22 October 2021. Rome, pp 292–293 https://doi.org/10.4060/cb9565en

Shahid SA, Omar SAS, Grealish G, King P, El-Gawad MA, Al-Mesabahi K (1998) Salinization as an early warning of land degradation in Kuwait. Prob Desert Dev 5:8–12

Shahid SA, Abo-Rezq H, Omar SAS (2002) Mapping soil salinity through a reconnaissance soil survey of Kuwait and geographic information system. Annual research report, Kuwait Institute for Scientific Research, Kuwait, KSR # 6682, pp 56–59

Shahid SA, Zaman M, Heng L (2018a) Introduction to soil salinity, sodicity, and diagnostics techniquesIn. In: Zaman M, Shahid SA, Lee H (eds) Guidelines for salinity assessment, mitigation and adaptation using nuclear and related techniques. Springer, pp 1–42

Shahid SA, Zaman M, Heng L (2018b) Soil salinity: Historical perspectives and world overview of the problem. In: Zaman M, Shahid SA, Lee H (eds) Guidelines for salinity assessment, mitigation and adaptation using nuclear and related techniques. Springer, pp 43–53

Shokri N, Or D (2011) What determines drying rates at the onset of diffusion controlled stage-2 evaporation from porous media? Water Resour Res 47:W09513. https://doi.org/10.1029/2010WR010284

Shokri N, Salvucci G (2011) Evaporation from porous media in the presence of a water table. Vadose Zone J 10:1309–1318

Shokri-Kuehni SMS, Raaijmakers B, Kurz T, Or D, Helmig R, Shokri N (2020) Water table depth and soil salinization: from pore-scale processes to field-scale responses. Water Resour Res 56(2). https://doi.org/10.1029/2019WR026707

Smits KM, Ngo VV, Cihan A, Sakaki T, Illangasekare TH (2012) An evaluation of models of bare soil evaporation formulated with different land surface boundary conditions and assumptions. Water Resour Res 48(W12526):1–15. https://doi.org/10.1029/2012WR012113

Soil Science Division Staff (2017) Soil survey manual. United States Department of Agriculture Handbook No. 18

Sophocleous M (2003) Environmental implications of intensive groundwater use with special regard to streams and wetlands. In: Intensive use of groundwater: challenges and opportunities, pp 93–112

Sparks DL (1995) Environmental soil chemistry: special issue. Elsevier

Thomas DSG, Middleton NJ (1993) Salinization: New perspectives on a major desertification issue. J Arid Environ 24:95–105

Tsypkin GG, Shargatov VA (2018) Influence of capillary pressure gradient on connectivity of flow through a porous medium. Int J Heat Mass Transf 127:1053–1063

US Salinity Laboratory Staff (1954) Diagnosis and improvement of saline and alkali soils. In: USDA Agriculture Handbook 60. USDA, Washington, DC, p 160
Walker D, Parkin G, Schmitter P, Gowing J, Tilahun SA, Haile AT, Abdu Yimam Y (2019) Insights from a multi-method recharge estimation comparison study. Ground Water 57(2):245–258. https://doi.org/10.1111/gwat.12801
Weert FV, Gun JVD (2012) Saline and brackish groundwater at shallow and intermediate depths: genesis and world-wide occurrence. In: Proceedings of IAH 2012 Congress, Niagara Falls, p 9
Werner AD, Simmons CT (2009) Impact of sea-level rise on sea water intrusion in coastal aquifers. Groundwater 47(2):197–204. https://doi.org/10.1111/j.1745-6584.2008.00535.x
Werner A, Bakker M, Post EA, Vandenbohede V, Lu A, Ataie-Ashtiani BC, Barry AD (2013) Seawater intrusion processes, investigation and management: recent advances and future challenges. Adv Water Resour 51:3–26. https://doi.org/10.1016/j.advwatres.2012.03.004

Chapter 8
Predicting the Behaviour of the Salt/Fresh-Brackish Water Transition Zone During Scavenger Well Pumping: 1. Numerical Model Development and Testing

Amjad Sami Aliewi

Abstract Soil salinization is a major soil degradation process that threatens local aquifers in Kuwait as well as agricultural production, food security and environmental sustainability. Kuwait being in hyper-arid region, the shallow aquifers with salinity problem need to be controlled to protect the agricultural water resources. In aquifers, where the fresh/brackish water lenses overly the saline groundwater, pumping only fresh/brackish water may yield saline groundwater upconing and hence deteriorates the groundwater quality. In such aquifers, scavenger pumping wells could be one of the methods to exploit fresh or brackish groundwater lenses only. Numerical modelling of this exploitation is complex because the velocity of the fluids involved and the dispersion of salts are interdependent on each other. This study aims at developing a local scale finite-difference model (RASIM) that can exploit fresh/brackish water lenses through a scavenger well pumping. It uses both cylindrical and normal coordinates, recasts the algebraic approximated governing equations in symmetrical forms, and derives unique and accurate internodal averaging of the properties of aquifer geometry and permeability. Moreover, the results' accuracy is enhanced when tensor properties of dispersion are taken into consideration. RASIM was verified against a number of analytical and numerical solutions and proved competent for the exemplified cases.

Keywords Agricultural groundwater · Scavenger wells · Modelling · Density-dependent flow · Solute transport · Velocity-dependent dispersion · Verification tests

A. S. Aliewi (✉)
Water Research Centre, Kuwait Institute for Scientific Research, Safat, Kuwait
e-mail: aaliewi@kisr.edu.kw

M. K. Suleiman, S. A. Shahid (eds.), *Terrestrial Environment and Ecosystems of Kuwait*, https://doi.org/10.1007/978-3-031-46262-7_8

8.1 Introduction

In aquifers containing groundwater of different qualities (from fresh to saline), the understanding of the movement of the lower saline water body, in response to brackish water or freshwater utilization from the upper part of the aquifer through pumping, is important for any effective management and sustainable development of existing groundwater resources for agricultural purposes (Saeed et al. 2002). The presence of brackish or saline waters in the upper part of the aquifer is due to the vertical upward leakages of brine/saline water from deeper aquifers, lateral inflows of brackish/saline groundwater from regional shared aquifers, washing of salt contents in soils, mixing and dissolution of salt rocks in the formation, and the saltwater intrusion process (Aliewi et al. 2021). In some cases, the leakage from irrigation canals as freshwater returns can be large enough to form freshwater lenses floating above saline water bodies in saline aquifers. Hence, the utilization of fresh or brackish groundwater from these saline aquifers is severely vulnerable to saline water upconing and intrusion (Aliewi et al. 2001). In Kuwait, the salinity of the groundwater in the Dammam Formation increases from 2500 mg/l in the southwest to 10,000 mg/l in the central part of the country. There is an abrupt increase in salinity towards the north and east to 150,000 mg/l and higher. The water is of calcium sulphate ($CaSO_4$), sodium sulphate (Na_2SO_4) and sodium chloride (NaCl) types and is supersaturated with respect to $CaCO_3$, but it is under-saturated with respect to $CaSO_4$. In the Dammam Formation aquifer water of salinity higher than 10,000 mg/l is often associated with hydrogen sulphide (H_2S), suggesting stagnation in the aquifer in these areas.

In order to control the salinity levels in the abstracted waters from saline aquifers, so that the abstracted water quality becomes a suitable source for irrigation, scavenger wells may be used (Fig. 8.1).

Scavenger wells are methods to exploit fresh or brackish groundwater lenses overlying more saline water body as a means of protection of agricultural water resources (Alam and Olsthoorn 2014; Aliewi et al. 2019). Scavenger wells are normally used where these groundwater lenses are vital water resources, especially in the arid and semi-arid regions. Scavenger wells are screened in both “Fresh/Brackish Water Zone” and “Saline Water Zone” of the utilized aquifer (Fig. 8.1). The concept of a scavenger well relies on the fact that Transition Zone’s upconing is the result of pumping in the “Fresh/Brackish Water Zone” whereas transition zone (also called fresh-saline water interface) downconing is caused by pumping from “Saline Water Zone” alone. The two processes can be balanced by varying the pumping rates from the two zones. For this purpose, a scavenger well can be located between a group of good production wells and a source of potential pollution so that the scavenger well will pump polluted water as waste to prevent it from reaching the group of good production wells (Fader 1957; Long 1965; Dagan and Bear 1968; Zack and Candelario 1983; Fukumori et al. 1986; Zack 1988). Field investigations and modelling studies are normally used to investigate the effects of possible seawater intrusion and saline water upconing in inland (Gopinath et al. 2016) and coastal (Stein

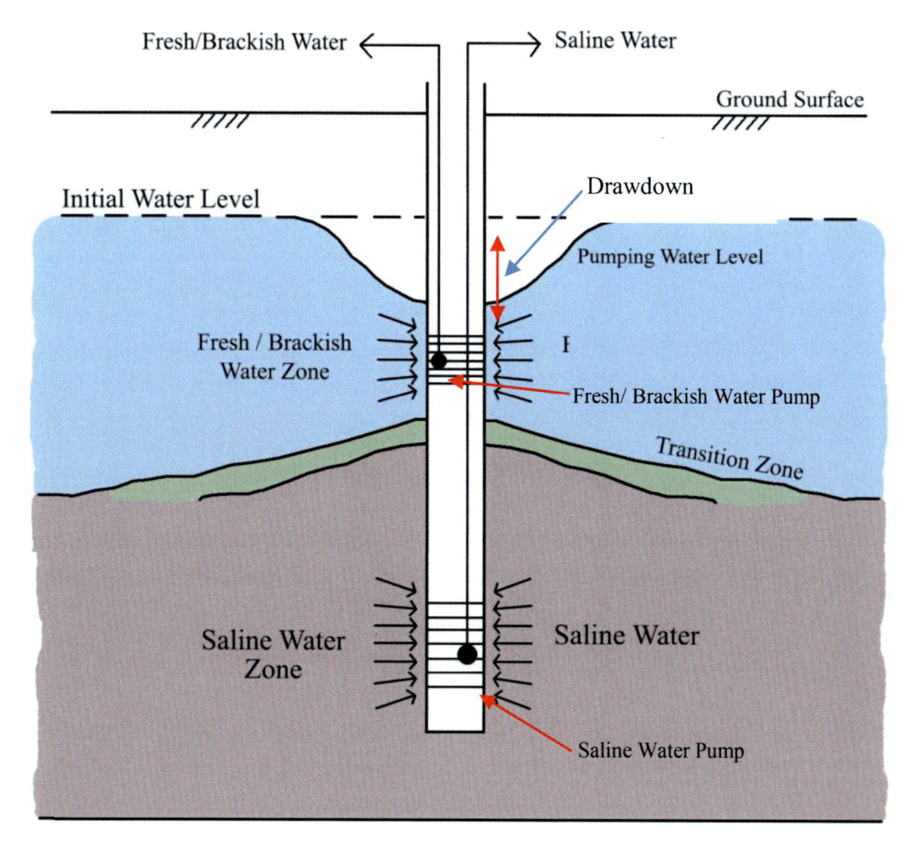

Fig. 8.1 Concept of a scavenger well pumping phenomena

et al. 2020) aquifers. Stein et al. (2020) studied the effect of long-term pumping from the saline groundwater zone on the fresh-saline water interface in a coastal aquifer in Spain. They found that in one observation well, the interface deepened some 50 m. Their modelling results showed that the target aquifer was freshening due to the continuous pumping of saline water below the interface and the overall salinity of the pumped water from above the interface was decreased by 17% near the wells. Therefore, developing a numerical model to address the salinity problems in specific locations is required. In particular, smaller (local scale) models are required to be developed to address the site-specific nature and the salinity problems. Local models can be used to investigate different operational scenarios and design suitability of the scavenger wells for specific sites in the aquifers concerned. Therefore, this research aims to develop a local numerical model (around a scavenger pumping well only) to simulate the movement of salt/fresh-brackish water transition zone during pumping stresses exemplified for Kuwait and Pakistan aquifers. The reason of selecting these two locations is the complexity of hydrogeological conditions with different sources of salinization in the aquifers. This will illustrate that the developed model is suitable for both applications.

The subject of controlling groundwater salinity through scavenger pumping for more efficient irrigation schemes in Kuwait and Pakistan in this research is divided into two parts. The first part (this chapter) is about developing and testing of a local scale numerical model (called RASIM) suitable for Kuwait and Pakistan hydrogeological conditions. The second part is about the application of the model (RASIM) and the design criteria of scavenger wells in Pakistan and Kuwait aquifers. Because the design of scavenger wells is site-specific, details about field design criteria are covered in the second paper of this research (Aliewi 2023).

In order to trust RASIM to be applied for complex hydrogeological conditions, it has to be verified for some theoretical cases such as Theis (1935) solution for transient drawdown in confined aquifers, Gelhar and Collins (1971) solution for radial groundwater flow with solute transport problem, and Henry (1964) solution for saltwater intrusion problem. In the literature, there are four famous numerical models that can deal with the transition zone approach (density-dependent flow with solute transport theory): SUTRA model (Voss 1984), Visual Modflow (McDonald and Harbaugh 1988), FEFLOW model (Diersch 2014), and SEAWAT (Langevin et al. 2007). The application of these software packages for regional aquifers yielded very useful results. However, RASIM is different from these four software packages as it is only a local numerical model that simulates the behaviour of salt/fresh-brackish water transition zone under a single scavenger well in axi-cylindrical and normal (Cartesian) coordinates. RASIM deals with groundwater movement towards a scavenger pumping well; therefore, radial flow patterns are common to most well-aquifer systems. Radial flow is used in RASIM to mean axially symmetrical flow towards the scavenger well. This requires that the flow towards the well is symmetrical radially about the z-axis of the cylindrical coordinates so that the angular coordinates will not appear in the flow equation. RASIM is developed in this research to suit axi-cylindrical and normal coordinates. Furthermore, RASIM is developed to recast the resulted algebraic approximated equations that govern the movement of the transition zone between waters of different qualities in symmetrical forms. RASIM is unique in deriving accurate internodal averaging (that can have physical and hydraulic meanings) of the properties related to aquifer geometry and permeability. RASIM takes into account strong anisotropy properties of a tensor dispersion model whose importance was addressed by a number of authors (Bennett et al. 2017; Karatzas 2017; Konikow 2011; Lee et al. 2017). These numerical benefits of RASIM will be detailed in the coming sections.

8.2 Methodology

Modelling the utilization of freshwater or brackish water from saline aquifers follows two approaches: "the sharp interface" and "the transition zone". The sharp interface approach assumes that there is an abrupt change in concentration at the interface between "fresh/brackish water" and "saline water" (Dokou and Karatzas 2012). In this approach, the width of sharp interface is assumed very small. Clearly, this assumption can create errors (Hill 1988) when field observations provide evidence that in some cases there is a wide transition zone of 16 m as an example (Terzić et al. 2010). It should be noted that sharp interface approach can yield very useful results as modelling is the art of making appropriate assumptions which are never "actually true", but necessary to understand the problem and find useful solutions. In the transition zone approach, different qualities of waters are allowed to mix. The concentrations in the transition zone change gradually from fresh/brackish water to saline water. The mixing process in this zone is mainly controlled by dispersion mechanism which is dependent on spatial variations of local groundwater velocity (Voss 1999). The governing theory of transition zone is well reported (Voss 1984; Aliewi 1993; Yin and Tsai 2019). This research addresses the transition zone approach. Most of the applications of scavenger pumping concentrate on recovering either freshwater lenses such as in the Indus Basin of Pakistan (Aslam et al. 2016; Zardari et al. 2015), or utilizing brackish groundwater bodies from saline aquifers, such as in Kuwait (Aliewi et al. 2019).

Modelling the behaviour of the transition zone during scavenger pumping requires improved, efficient and robust solutions to the algebraic approximated equations and hydraulic meaning of the averaged hydraulic properties of the aquifer on the local scale. This is carried out in this research by developing a local finite-difference numerical model (RASIM) for Kuwait and Pakistan case study aquifers. RASIM exhibits a number of numerical benefits whose details are discussed in this study. RASIM solves numerically the density-dependent groundwater flow equation coupled with the solute transport equation (Eqs. 8.1 and 8.2) (Aliewi 1993):

$$\left[\propto_{\mathrm{aq}} \frac{S_y}{b|\vec{g}|} + \left(1-\propto_{\mathrm{aq}}\right)\rho S_{OP}\right]\frac{\partial P}{\partial t} + \left(\varepsilon\frac{\partial \rho}{\partial C}\right)\frac{\partial C}{\partial t} - \nabla.\left[\left(\frac{\rho \underline{k}}{\mu}\right).\left(\nabla P - \rho\vec{g}\right)\right] = Q_{pp} \quad (8.1)$$

$$[\varepsilon\rho]\frac{\partial C}{\partial t} + \left(\varepsilon\rho\underline{V}\right).\nabla C - \nabla.\left[\left(\varepsilon\rho\underline{D}\right).\nabla C\right] = Q_p\left[C^* - C\right] \quad (8.2)$$

where

p: fluid pressure [M/(LT^2)]
b: an effective thickness of the water table [L]
α_{aq}: a parameter equals 1 for the unconfined zone of the aquifer and zero for the confined zone of the aquifer
S_y: specific yield [1] for water table (unconfined) aquifers
Q_{pp}: the fluid mass source/sink [M/L^{3}T] including sources at all boundary conditions
Q_p: the fluid mass source of any solute dissolved in the source fluid [M/(L^{3}T)]
ρ: fluid density [M/L^3], which can be approximated as:

$$\rho(\mathrm{C}) = \rho_0 + \frac{\partial \rho}{\partial C}\left(\mathrm{C} - \mathrm{C}_0\right) \quad (8.3a)$$

$\partial\rho/\partial c$: a constant value of density changes with concentration $[M_s/L^3]$
ρ_0, C_0: base values of density and concentration, respectively
S_{op}: specific pressure storativity $[M/(LT^2)]^{-1}$ expressed as:

$$S_{op} = (1-\varepsilon)\alpha_p + \varepsilon\beta \tag{8.3b}$$

β: fluid compressibility $[M/(LT^2)]^{-1}$
α_p: porous media compressibility $[M/(LT^2)]^{-1}$
ε: porosity [1]
t: time [T]
$\underline{k}$: the porous media permeability tensor $[L^2]$
μ: fluid viscosity $[M/(L_fT)]$
$\vec{g}$: gravity vector $[L/T^2]$
C: solute concentration as a mass fraction $[M_s/M]$
$\underline{V}$: the fluid velocity [L/T]
Q_pC^*: $[M/(L^3T)]$ the contribution to the dissolved species mass added by a fluid source with concentration, $C^*[M_s/M]$
$\underline{D}$: the hydrodynamic dispersion tensor $[L^2/T]$

The second rank dispersion tensor according to Scheidegger (1961) can be written as:

$$\underline{D}_{kl} = \alpha_{klmn}\frac{V_mV_n}{|\underline{V}|} \tag{8.4}$$

where
$\underline{D}_{kl}$: hydrodynamic tensor $[L^2/T]$
α_{klmn}: dispersivity tensor of porous media [L]
V_m: a general m-component of velocity [L/T]
V_n: a general n-component of velocity [L/T]
$|\underline{V}|$: magnitude value of velocity [L/T]

$\underline{D}_{kl}$ is a second-rank tensor while α_{klmn} is a fourth rank tensor. Equation 8.4 can be simplified in r-z coordinates as:

$$\underline{D} = \begin{pmatrix} D_{rr} & D_{rz} \\ D_{zr} & D_{zz} \end{pmatrix} \tag{8.5a}$$

where

$$D_{rr} = \alpha_L\frac{V_r^2}{|\underline{V}|} + \alpha_T\frac{V_z^2}{|\underline{V}|} \tag{8.5b}$$

$$D_{zz} = \alpha_L \frac{V_z^2}{|\underline{V}|} + \alpha_T \frac{V_r^2}{|\underline{V}|} \tag{8.5c}$$

$$D_{rz} = D_{zr} = \left(\alpha_L - \alpha_T\right) \frac{V_r V_z}{|\underline{V}|} \tag{8.5d}$$

$$|\underline{V}| = \sqrt{V_r^2 + V_z^2} \tag{8.5e}$$

$$\alpha_L : \text{longitudinal dispersivity of porous media} \left[\mathrm{L}\right] \tag{8.5f}$$

$$\alpha_\mathrm{T} : \text{transverse dispersivity of porous media} \left[\mathrm{L}\right] \tag{8.5g}$$

$$V_r : \text{radial velocity} \left[\mathrm{L} / \mathrm{T}\right] \tag{8.5h}$$

$$V_z : \text{vertical velocity} \left[\mathrm{L} / \mathrm{T}\right]. \tag{8.5i}$$

The anisotropic properties of dispersivity are defined in Eqs. 8.5j and 8.5k and illustrated in Fig. 8.2.

$$\alpha_L = \frac{\alpha_{L\max} \alpha_{L\min}}{\alpha_{L\min} \cos^2 \theta_{kv} + \alpha_{L\max} \sin^2 \theta_{kv}} \tag{8.5j}$$

$$\alpha_T = \frac{\alpha_{T\max} \alpha_{T\min}}{\alpha_{T\min} \sin^2 \theta_{kv} + \alpha_{L\max} \cos^2 \theta_{kv}} \tag{8.5k}$$

$\boldsymbol{\alpha}_{\mathbf{Lmax}}$: squared radius of the longitudinal dispersivity [L] ellipse in the maximum permeability direction, r_{max}

$\boldsymbol{\alpha}_{\mathbf{Lmin}}$: squared radius of the longitudinal dispersivity [L] ellipse in the minimum permeability direction, r_{min}

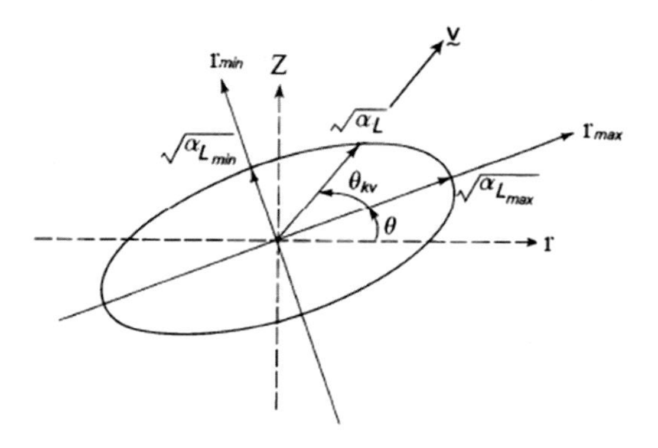

Fig. 8.2 Definition of flow-direction-dependent dispersivity in 2D. (Voss 1999)

$\boldsymbol{\alpha_{Tmax}}$: squared radius of the transverse dispersivity [L] ellipse in the maximum permeability direction, r_{max}
$\boldsymbol{\alpha_{Tmin}}$: squared radius of the transverse dispersivity ellipse [L] in the minimum permeability direction, r_{min}
$\boldsymbol{\theta_{kv}}$: angle from maximum permeability direction r_{max} to flow direction
θ : angle between r-axis to the maximum permeability direction r_{max}

The solute dispersion coefficient (Eqs. 8.2, 8.4 and 8.5) is an important property. Dispersion depends on flow velocity magnitude and direction. It is an anisotropic property in nature (Bear and Cheng 2010). The longitudinal dispersion, $\alpha_L|\underline{V}|$, is the dispersion occurring in the direction of the fluid flow while the transverse dispersion, $\alpha_T|\underline{V}|$, is the dispersion occurring perpendicular to the flow direction. Isotropic dispersion occurs when $\alpha_L = \alpha_T$, otherwise the dispersion is anisotropic. Anisotropic dispersion model depends on the anisotropy ratio (α_L/α_T) and the off-diagonal terms (D_{rz}, D_{zr} of Eq. 8.5d) in addition to flow velocity magnitude and direction. The developed model in this study is valid for applications of groundwater problems in fresh/saline aquifers (Indus Basin of Pakistan) as well as in brackish/saline aquifers (Kuwait).

8.3 Analysis and Discussions

The analysis and discussions address how RASIM can develop symmetrical forms of algebraic equations that can be solved by Incomplete Conjugate Cholesky Gradient (ICCG) preconditioning method of matrix solution for robustness and accuracy.

8.3.1 Special Numerical Treatment to the Flow Term of Eq. 8.1

In the approximation of Eqs. 8.1 and 8.2 here, only specific numerical benefits are discussed where these are relatively new with respect to the development of the numerical model (RASIM) of this study. Eq. 8.1 can be rewritten in axi-symmetrical cylindrical coordinates as (Eq. 8.6):

$$S_{tor}\frac{\partial P}{\partial t}+\varepsilon\frac{\partial \rho}{\partial C}\frac{\partial C}{\partial t}-\left[\frac{1}{r}\frac{\partial}{\partial r}\left(r\lambda_r\frac{\partial P}{\partial r}\right)+\frac{\partial}{\partial z}\left(\lambda_z\left\{\frac{\partial P}{\partial z}+\gamma\right\}\right)\right]=Q_{pp} \tag{8.6}$$

$$S_{tor}=\alpha_{aq}\frac{S_y}{b\,|\vec{g}|}+\left(1-\alpha_{aq}\right)\rho S_{op} \qquad \left[T^2/L^2\right]$$

$$\lambda_r=\frac{\rho k_r}{\mu}\,[T]\,,\ \lambda_z=\frac{\rho k_z}{\mu}\,[T]\,,\ \gamma=-\rho\,|\vec{g}|\,\left[M/\left(L^2T^2\right)\right],$$

k_r, k_z are radial and vertical intrinsic permeability [L^2].

The term 1/r of Eq. 8.6 is a singularity problem (Mitchell and Griffiths 1980) for finite difference approximation because when r is approaching zero, the result of 1/r is infinity (∞). In this research, the finite difference approximation of Eq. 8.6 is treated in a special way to gain a number of numerical benefits. This special treatment includes rewriting Eq. 8.6 first in terms of $\Omega = r^2$and the second step is to multiply the resulting equation by the actual volume of the block associated with node ij in the finite difference grid (Fig. 8.3; Eq. 8.7).

$$VOL_{ij} = \pi\left(\Omega_{i+½,j} - \Omega_{i-½,j}\right)\left(z_{i,j+½} - z_{i,j-½}\right) \tag{8.7}$$

The resulted finite difference approximation of the flow term (A_f) (which is the third term of Eq. 8.6) after multiplying it with Eq. 8.7 is represented in Eq. 8.8:

$$\begin{aligned}\tilde{A}_{f_{ij}} = 2\pi\left(z_{i,j+½} - z_{i,j-½)}\right)&\left[\lambda_{r_{i+½,j}} .2\Omega_{i+½,j}\left(\frac{P^{n+1}_{i+1,j} - P^{n+1}_{ij}}{\Omega_{i+1,j} - \Omega_{ij}}\right) - \lambda_{r_{i-½,j}} .2\Omega_{i-½,j}\left(\frac{P^{n+1}_{ij} - P^{n+1}_{i-1,j}}{\Omega_{ij} - \Omega_{i-1,j}}\right)\right] \\ + \pi\left(\Omega_{i+½,j} - \Omega_{i-½,j}\right)&\left[\lambda_{z_{i,j+½}} \cdot\left(\frac{P^{n+1}_{i,j+1} - P^{n+1}_{ij}}{z_{i+1,j} - z_{ij}} + \gamma_{i+j+½}\right) - \lambda_{z_{i,j-½}} \cdot\left(\frac{P^{n+1}_{ij} - P^{n+1}_{i,j-1}}{z_{ij} - z_{i,j-1}} + \gamma_{i,j-½}\right)\right]\end{aligned} \tag{8.8}$$

Equation 8.8 does not have any term of the form 1/Ω, and therefore, the problem of boundary singularity is avoided. The second issue that needs treatment in Eq. 8.8 is the internodal averaging of the parameters, $\Omega_{i\mp½,j}$, $\lambda_{r_{i\mp½,j}}$,$\lambda_{z_{i,j\mp½}}$, and $\gamma_{i\mp j+½}$ (which are related to aquifer geometry and permeability). The discussion here illustrates that the values of these parameters can be chosen uniquely to minimize errors due to truncation and linearization processes. In addition to that, the averaging of these parameters can be made unique to illustrate that the numerically approximated fluxes in and out at any node in the finite difference grid are equivalent to those physical fluxes obtained from applying Darcy's equation. Equation 8.8 represents the approximated flux ($\tilde{A}_{f_{ij}}$) in [M/T] dimensions *in* and *out* of a point *ij* in a mesh through its correspondence volume VOL_{ij} (Eq. 8.7; Fig. 8.3). Darcy's flux in radial direction between nodes ij and i+1, j in axi-symmetrical cylindrical coordinates (Fig. 8.3) can be calculated using the integral method as (Eq. 8.9):

$$-2\pi \int_{P_{ij}}^{P_{i+1,j}} \left[z\frac{\rho k_r}{\mu}\right] dP = \text{Radial Darcy's flux} \tag{8.9}$$

Similarly, Darcy's flux ($AD_{f_{ij}}$) in both radial and vertical (Fig. 8.3) axi-symmetrical cylindrical coordinates are (Eq. 8.10):

$$\begin{aligned}AD_{f_{ij}} = 2\pi\left(z_{i,j+½} - z_{i,j-½)}\right)&\left[\lambda_{r_{i+½,j}} .\mathrm{r}_{i+½,j}\left(\frac{P^{n+1}_{i+1,j} - P^{n+1}_{ij}}{\Omega_{i+1,j} - \Omega_{ij}}\right) - \lambda_{r_{i-½,j}} .\mathrm{r}_{i-½,j}\left(\frac{P^{n+1}_{ij} - P^{n+1}_{i-1,j}}{\Omega_{ij} - \Omega_{i-1,j}}\right)\right] \\ + \pi\left(r^2_{i+½,j} - \mathrm{r}^2_{i-½,j}\right)&\left[\lambda_{z_{i,j+½}} \cdot\left(\frac{P^{n+1}_{i,j+1} - P^{n+1}_{ij}}{z_{i+1,j} - z_{ij}} + \gamma_{i+j+½}\right) - \lambda_{z_{i,j-½}} \cdot\left(\frac{P^{n+1}_{ij} - P^{n+1}_{i,j-1}}{z_{ij} - z_{i,j-1}} + \gamma_{i,j-½}\right)\right]\end{aligned} \tag{8.10}$$

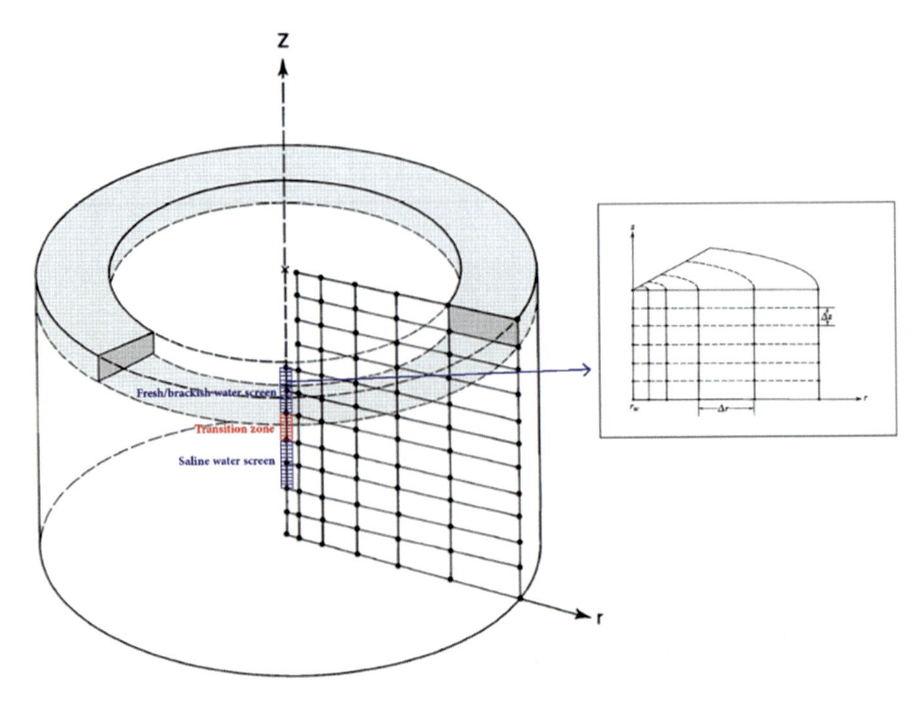

Fig. 8.3 Scavenger wells in axi-symmetrical cylindrical coordinates

It can be shown that $\tilde{A}_{f_{ij}}$ (Eq. 8.8) and $AD_{f_{ij}}$ (Eq. 8.10) are identical if,

$$r_{i\pm1/2,j} = \left[\frac{r_{i\pm1,j}^2 - r_{ij}^2}{\ln\left(\frac{r_{i\pm1,j}}{r_{ij}}\right)^2} \right]^{1/2} \tag{8.11}$$

Equation 8.11 is the harmonic mean of $r_{i\mp½,j}$ expressed logarithmically in r^2. Equation 8.11 provides a unique expression of $r_{i\mp½,j}$. However, other transformations were tried to present $r_{i\mp½,j}$ logarithmically in r such that as Eq. 8.12 but failed to match those of Darcy's physical *in* and *out* fluxes (Eq. 8.10) associated with node *ij*.

$$r_{i\pm½,j} = \frac{r_{i\pm1,j} - r_{ij}}{\ln\left(\frac{r_{i\pm1,j}}{r_{ij}}\right)} \tag{8.12}$$

Since, the harmonic mean of $r_{i\mp½,j}$ is expressed logarithmically in r^2, then it was tried to write $\lambda_{r_{i\mp½,j}}$ logarithmically in terms of r^2 as presented in Eq. 8.13 for $\lambda_{r_{i+½,j}}$ only:

$$\vec{\lambda}_{r_{i+1/2,j}} = \frac{\left[\dfrac{\ln\left(\dfrac{r_{i+1,j}}{r_{ij}}\right)^2}{r_{i+1,j}+r_{ij}}\right]^{1/2}}{\dfrac{\left[\dfrac{\ln\left(\dfrac{r_{i+1,j}+r_{ij}}{2r_{ij}}\right)^2}{\left(3r_{ij}+r_{i+1,j}\right)}\right]^{1/2}}{\lambda_{r_{ij}}}+\dfrac{\left[\dfrac{\ln\left(\dfrac{2r_{i+1,j}}{r_{i+1,j}+r_{ij}}\right)^2}{\left(3r_{i+1,j}+r_{ij}\right)}\right]^{1/2}}{\lambda_{r_{i+1,j}}}} \tag{8.13}$$

The resultant equation (Eq. 8.13) unfortunately violates the mathematical requirements of the harmonic averaging rule (Aliewi 1993) and hence it cannot be used. The important conclusion which can be drawn here is while it was possible to represent uniquely $r_{i\pm 1/2,j}$ logarithmically in r^2, in order to have continuity in conservative fluxes, it was not possible to do so for $\lambda_{r_{i\pm1/2,j}}$. However, the same continuity in fluxes can be achieved only if and only if the magnitude of $\lambda_{r_{i\pm1/2,j}}$ be averaged harmonically in logarithm r as presented in Eq. 8.14.

$$\lambda_{r_{i\pm½,j}} = \frac{\ln\left(\dfrac{r_{i\pm1,j}}{r_{ij}}\right)}{\dfrac{\ln\left(\dfrac{r_{i\pm1,j}+r_{ij}}{2r_{ij}}\right)}{\lambda_{r_{i,j}}}+\dfrac{\ln\left(\dfrac{2r_{i\pm1,j}}{r_{i\pm1,j}+r_{ij}}\right)}{\lambda_{r_{i\pm1,j}}}} \tag{8.14}$$

Other properties can be expressed in normal harmonic means (Eqs. 8.15 and 8.16).

$$\lambda_{z_{i,j\pm½}} = \frac{2\lambda_{z_{i,j}}\lambda_{z_{i,j\pm1}}}{\lambda_{z_{i,j}}+\lambda_{z_{i,j\pm1}}} \tag{8.15}$$

$$\gamma_{i,j\pm½} = \frac{2\gamma_{ij}\gamma_{i,j\pm1}}{\gamma_{ij}+\gamma_{i,j\pm1}} \tag{8.16}$$

In summary, the approximated flow term can represent the actual radial and vertical flow if $r_{i \mp ½, j}$ was rewritten logarithmically in r^2 and its approximation was multiplied by the actual volume associated with a node ij in the discretized domain. As for minimizing truncation errors when approximating $\lambda_{r_{i\pm 1/2,j}}$ and to make the approximated fluxes equivalent to those of Darcy's equation, $\lambda_{r_{i\pm 1/2,j}}$ should be expressed logarithmically in r.

8.3.2 *The Symmetry of the Finite Difference Approximation*

The special treatment presented in the above section (Eqs. 8.6–8.16) when used in the finite difference approximation of the flow term, can produce symmetrical coefficients matrix. The symmetrical forms of Eqs. 8.1 and 8.2 require many details, which are presented in Aliewi (1993). In this chapter, only the indication of how the symmetrical form can be achieved is discussed. The approximated form (Eq. 8.8) of the flow term of Eq. 8.6 can be written as:

$$\hat{A}_{f_{ij}} = a_{ij}P^{n+1}_{i+1,j} + \dot{x}_{ij}P^{n+1}_{ij} + b_{ij}P^{n+1}_{i-1,j} + D_{ij}P^{n+1}_{i,j+1} + E_{ij}P^{n+1}_{i,j-1} + F_{ij} + G_{ij} \tag{8.17}$$

All the coefficients of Eq. 8.17 are defined in Aliewi (1993). Here, only four parameters are discussed to illustrate the symmetry of matrix coefficients. The parameters are:

$$a_{ij} = 4\pi\left(z_{i,j+½} - z_{i,j-½}\right)\left(\lambda_{r_{i+½,j}}\right)\left[\frac{r^2_{i+½,j}}{r^2_{i+1,j} - r^2_{ij}}\right] \tag{8.18}$$

$$b_{ij} = 4\pi\left(z_{i,j+½} - z_{i,j-½}\right)\left(\lambda_{r_{i-½,j}}\right)\left[\frac{r^2_{i-½,j}}{r^2_{ij} - r^2_{i-1,j}}\right] \tag{8.19}$$

$$D_{ij} = \pi\left(r^2_{i+½,j} - r^2_{i-½,j}\right)\left[\frac{\lambda_{z_{i,j+½}}}{z_{i,j+1} - z_{ij}}\right] \tag{8.20}$$

$$E_{ij} = \pi\left(r^2_{i+½,j} - r^2_{i-½,j}\right)\left[\frac{\lambda_{z_{i,j-½}}}{z_{ij} - z_{i,j-1}}\right] \tag{8.21}$$

The matrix of coefficients of Eq. 8.17 can be symmetrical if and only if 1) $b_{i+1,j} \equiv a_{ij}$ 2) $E_{i,j+1} \equiv D_{ij}$. If one of the above two conditions fails, then the matrix cannot be symmetric. Verifying the values of the variable of the two conditions shows that they are identical for a rectangular finite difference grid. This is an important achievement to get robust and efficient numerical solution. In order to maintain consistency in the numerical procedure and to achieve similar aims of symmetry and minimum truncation errors, the solute mass balance (Eq. 8.2) is also multiplied by VOLij (Eq. 8.7). It can be shown using similar numerical treatment to that of the flow equation that the final approximated solute transport equations are symmetrical too as presented in Aliewi (1993) if lagging method is used to approximate the advection term and the cross-derivative components of dispersion terms. Lagging these terms refers to the use of values from the previous iteration rather than those of the current iteration in the solution cycle of the flow and the solute transport equations. To avoid approximating these terms at explicit levels, n levels, rather than the implicit level, n + 1 level, two iterations at least are required in the solution cycle for each time step. The lagging method requires using time level, n, for the first iteration and, n + 1, for the second iteration.

8.4 Model Verification and Testing

There is a need to verify the developed simulation local model, RASIM, to ensure that it performs the calculations correctly for the two case studies in Pakistan and Kuwait aquifers. This is normally confirmed by running several numerical tests for which an analytical solution is known. The verification tests used in this research cover transient drawdown in confined aquifers (Theis 1935), radial groundwater flow with solute transport (Gelhar and Collins 1971) and saltwater intrusion (Henry 1964). There are analytical solutions for these problems, therefore, RASIM numerical solutions to these problems can be verified. Since SUTRA is a well-verified simulation code (Voss 1984), it was used to numerically simulate a saline water upconing problem in a specific case study. Then RASIM was used to solve numerically the same problem with the same input data. Then the results from both codes were compared to make sure that RASIM predicts the upconing of the transition zone under pumping stresses with confidence.

8.4.1 Verification Test 1: Theis Solution for Transient Drawdown in a Confined Aquifer

The main objective of the test is to demonstrate that RASIM is capable of accurately simulating transient radial flow by showing that the solution obtained from RASIM agrees with Theis (1935) analytical solution for the same input data. The thickness of the aquifer in this test, Δz, is 1 m. The mesh contains 2 rows and 27 columns (54 nodes). The minimum and the maximum radial length increment between nodes is 2.5 m and 25 m respectively. The length of the aquifer is 500 m. The radial distance between nodes increases by a factor of 1.2915. The initial time step, Δt_0, is 1 s. This is increased by a factor 1.5 on each succeeding time step. The other input data are presented in Table 8.1. No flow was simulated across the upper and the bottom boundaries. A sink was specified at the left lateral boundary at 0.3142 kg/s for each well node. Hydrostatic pressure was specified at the right-hand lateral boundary (r = 500 m). The pressure distribution everywhere in the model is assumed to be

Table 8.1 Other input data of Theis-RASIM Test

Parameter	Value	Unit
S_{op}	1.039×10^{-6}	[ms^2/kg]
α_p	1.299×10^{-6}	[ms^2/kg]
β	4.4×10^{-6}	[ms^2/kg]
μ	0.001	[kg/ms]
ε	0.2	1
$\lvert\vec{g}\rvert$	9.81	[m/s^2]
k	2.0387×10^{-10}	[m^2]
ρ	1000	[kg/m^3]
Q_{TOT}	0.6284	[kg/s]

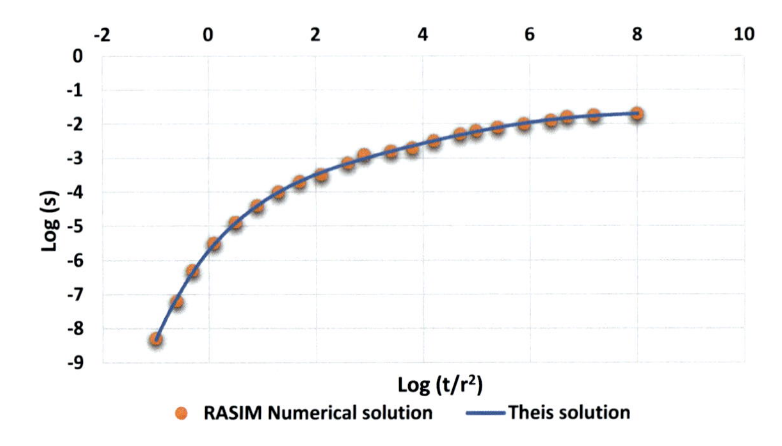

Fig. 8.4 Simulation of transient drawdown using RASIM and Theis solutions

initially hydrostatic for a constant water level. The results, obtained from RASIM were plotted in Fig. 8.4 against Theis solution at an observation well located 15.284 m from the discharge well. The match between Theis' and RASIM's results is very good with a correlation factor, $R^2 = 0.99$.

8.4.2 Verification Test 2: Gelhar Analytical Solution (Radial Flow with Solute Transport)

The analytical solution for radial flow with solute transport of Gelhar and Collins (1971) is taken as a reference solution for this test. The purpose of this test is to simulate the transient spreading of the solute front as it moves radially away from an injection well. The solution of Gelhar and Collins (1971) is presented in Eq. 8.22:

$$\frac{C - C_0}{C^* - C_0} = 0.5 \times \mathrm{erfc}\left[\frac{r^2 - r^{*2}}{2 \times \sqrt{(0.75 \times \alpha_L) r^{*3} + \left(\frac{D_m}{A_C}\right) r^{*4}}}\right] \tag{8.22}$$

where

$$r^* = \sqrt{2A_c t}$$

$$A_C = \frac{Q_{TOT}}{2\pi\varepsilon\rho B}$$

where C: solute concentration $[M_s/M]$; C^*: injected solute concentration $[M_s/M]$; C_0: base concentration $[M_s/M]$; D_m: molecular diffusion $[L^2/T]$; B: aquifer thickness [L]; erfc(y) is the complementary error function of y; other variables are defined in the previous test. The aquifer is confined, infinite, and includes a fully penetrating injection well. The fluid is injected at a rate of Q_{TOT} with a known (injected) solute concentration C^*. The concentration of the solute before injection is C_0. The fluid density is not a function of concentration. No flow takes place across any boundary except where hydrostatic pressure is specified at the far lateral boundary. The injection well is represented at the left lateral boundary as a source boundary. The mesh spacing is expanded radially from a minimum Δr of 2.5 m to a maximum value of 24.2 m by a factor of 1.06. The maximum radius is taken as 500 m. The mesh contains two rows with a 10 m vertical distance between them. The time step is constant at 4021 s. The other parameters are presented in Table 8.2.

Table 8.2 Other input data of Gelhar-RASIM Test

Parameter	Value	Unit
S_{op}	0	$[ms^2/kg]$
$C_0 = 0$	1.299×10^{-6}	$[ms^2/kg]$
$C^* = 1$	4.4×10^{-6}	$[ms^2/kg]$
μ	0.001	[kg/ms]
ε	0.2	1
$\lvert\vec{g}\rvert$	9.81	$[m/s^2]$
k	1.02×10^{-11}	$[m^2]$
ρ	1000	$[kg/m^3]$
Q_{TOT}	62.5	[kg/s]
D_m	1×10^{-10}	$[m^2/s]$
α_L	10	[m]
α_T	0	[m]

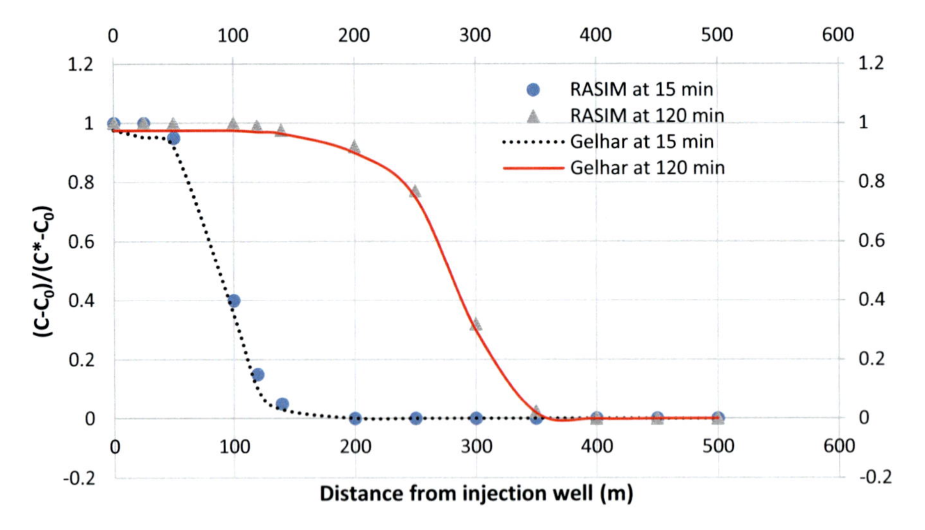

Fig. 8.5 Comparison of simulation results (RASIM-Gelhar) of transient spreading of the solute front from an injection well

One pressure solution is completed to achieve a steady state. One concentration solution is carried out at every time step. The results are obtained after 15 min (225 times steps) and 120 min (1800 times steps). The match between the results of RASIM and Gelhar solution is very good as shown in Fig. 8.5.

8.4.3 *Verification Test 3: Henry's (1964) Analytical Solution for Saline Water Intrusion*

This test examines the capability of RASIM to solve groundwater problems involving density-dependent flow and solute transport. The importance of Henry's solution is that, to date (Kalakan 2014), there is no two-dimensional analytical solution for the transient movement of a saline front including the effect of dispersion. Despite the fact that the approach developed in this research presents the mixing of different qualities of water by means of a velocity-dependent dispersion tensor, Henry's solution is still being used as a reference verification test for saline water intrusion problems.

8.4.3.1 General Description of Henry's Problem

Henry (1964) developed an analytical solution for the problem of saline water intrusion for an isotropic, homogeneous, confined aquifer under steady-state conditions. Henry solution is for 2D planar flow in normal coordinates, which RASIM can deal with in addition to axi-symmetric coordinates. In his model, he assumes that dispersion is independent of fluid flow velocity and direction. In Henry's solution, a Fourier series method is used to transform the governing equations of saline intrusion into non-linear sets of algebraic equations. Henry's problem is a cyclic flow problem. Initially, the aquifer is saturated with freshwater. The saline water moves from the sea boundary to the bottom of the freshwater body due to the differences between the two water densities. The intruding saline water mixes with the advancing freshwater from the opposite direction. This mixture is lighter than the saline water itself. The freshwater gradient forces this mixture to move seaward. The process continues until a balance between saline water entering through the lower part of the sea boundary and the saline water leaving the top portion of that boundary is achieved. Then the whole system is in equilibrium.

8.4.3.2 Input Data for Henry's Example

Time and space dimensions have been chosen to coincide with those of the dimensionless parameters of Henry. The length of the simulation run is taken as 100 min, which was enough to reach steady state. A constant Δt of 1 min is used. The finite difference mesh contains 231 nodes. Its length and thickness were 2 m and 1 m respectively. Grid spacing has been taken as 0.1 m in both the r and z directions. Dispersion is simulated by using a large constant value of molecular diffusivity and zero dispersivity. The rest of the input parameters are presented in Table 8.3.

Boundary and Initial Conditions

The left lateral boundary is the freshwater source boundary. The concentration of inflowing freshwater is C_{IN}. Both the top and the bottom boundaries are impermeable. A hydrostatic saline water pressure distribution is specified at the sea boundary

Table 8.3 Other parameters of Henry-RASIM test

Parameters	Value	Units
ε	0.35	1
k	1.020408×10^{-9}	[m^2]
C_s	0.0357	[kg (dissolved solids)/kg (saline water)]
C_{IN}	0	
ρ_s	1025	[kg/m^3]
ρ_0	1000	[kg/m^3]
$\partial\rho/\partial C$	700	[kg/m^3]
$\lvert\vec{g}\rvert$	9.8	[m/s^2]
$\alpha_L = \alpha_T = 0$	0	
D_m	6.6×10^{-6}	[m^2/s] (first case)
D_m	18.8571×10^{-6}	[m^2/s] (second case)
Q_{IN}	6.6×10^{-2}	[kg/s]

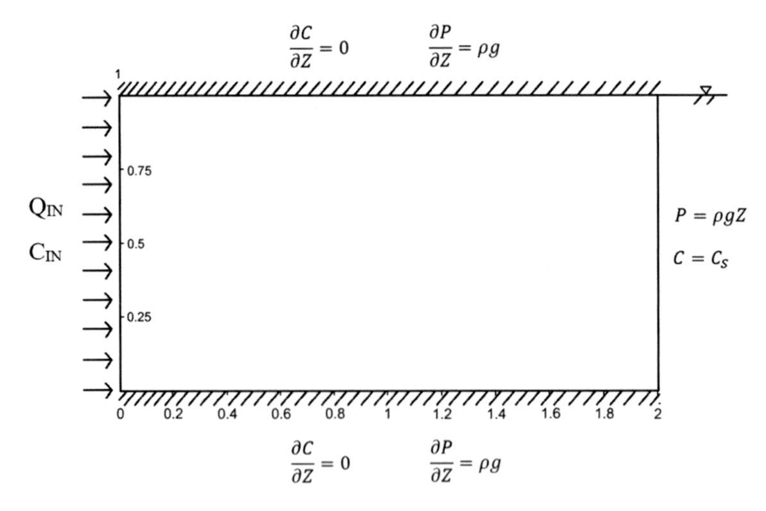

Fig. 8.6 Boundary conditions for Henry's problem

as seen in Fig. 8.6. The concentration of inflowing saline water across this boundary is C_s. The initial concentration distribution is C_{IN} everywhere except at the sea boundary. The initial pressure distribution is set by an initial simulation run with a constant concentration of C_{IN}.

8.4.3.3 Results for Henry's Example

Before presenting the results for this example, it should be noted that some authors (Simpson and Clement 2004) suggested a modification of Henry's problem to reduce the flux rate along the freshwater boundary to enhance the density-dependent effects. Henry used unrealistically high values of dispersion ($D_m = 18.8571 \times 10^{-6}$ m^2/s) to make his solution converge. This suggests that Henry model may be suitable only for wide transition zone problems. The results of using this large value of dispersion for Henry and other numerical solutions are shown in Fig. 8.7. As seen in Fig. 8.6, there are differences in the results of the numerical solutions. The only solution that is closer to Henry's over the entire domain is RASIM. It agrees favourably with the numerical models at the toe of the 0.5 isochlor near the aquifer bottom, and RASIM reasonably agrees with Henry's solution as the 0.5 isochlor approaches the sea boundary. This is because RASIM is set to simulate the inflow/outflow mechanism at the saline boundary in a realistic way. Simply, a source/sink term is added to the transport model such as any fluid leaving the system has the calculated concentration and if the fluid enters the system, it has the specified concentration (of the entering fluid) at the boundary (C_s in this example). Thus, there is no need to alter the sea boundary for Henry's problem in RASIM's approach. In fact, other researchers (Segol et al.

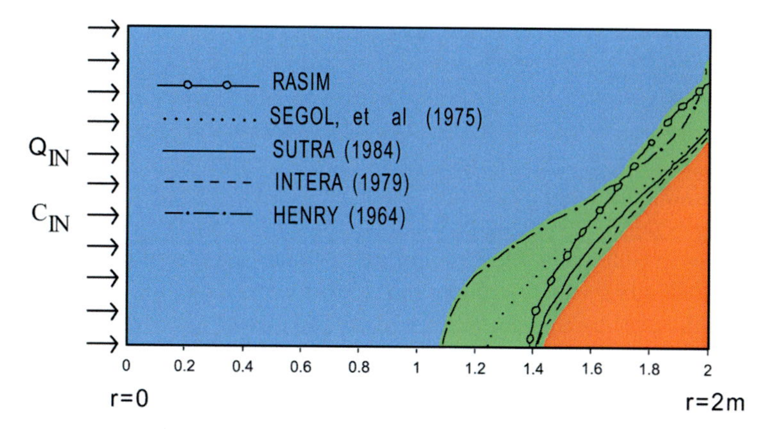

Fig. 8.7 Match of 0.5 isochlor for Henry analytical solution and several numerical solutions for $D_m = 18.8571 \times 10^{-6}$ [m²/s]

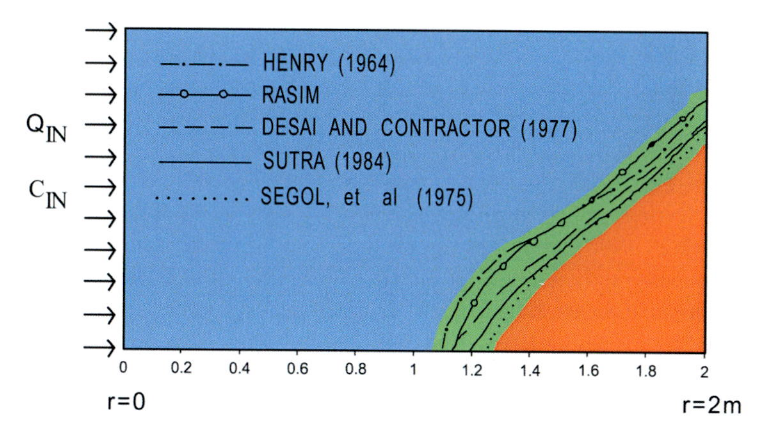

Fig. 8.8 Match of 0.5 isochlor for Henry analytical solution and several numerical solutions for $D_m = 6.6 \times 10^{-6}$ [m²/s]

1975; Desai and Contractor 1977) implemented different values of dispersion in their numerical models, such as $D = 6.6 \times 10^{-6}$ m²/s. The reason lies in the conversion of Henry's non-dimensional parameter "b" to equivalent freshwater recharge and total dispersion. The value of the total dispersion, D, as expressed in the solute transport model of RASIM equals the product of molecular diffusion, D_m, and porosity,ε. Thus, $D = \varepsilon D_m$. This means that for$\varepsilon = 0.35$ and $D = 6.6 \times 10^{-6}$ m²/s, the value of D_m should be 18.8571×10^{-6} m²/s. Now, simulating Henry's problem with this realistic value if dispersion resulted in 0.5 isochlor distribution as shown in Fig. 8.8. Figure 8.8 produces relatively narrow transition zone. Again, RASIM's results are the best to match Henry's results for the same reasons mentioned earlier.

8.4.4 *Comparing RASIM and SUTRA Results for Saline Upconing Problem*

The comparison problem (the Indus Basin Case study) was taken from GDC and BGS (1990) and Aliewi (1993) about the utilization of freshwater from saline aquifers in the Sindh Province of Pakistan. The accumulation of irrigation losses resulted in rising the water tables dramatically in the underlying aquifer causing waterlogging of crops in addition to severe salinization of uncropped land and an increase in storm water flooding. The irrigation losses have formed thin lenses of freshwater floating on the existing saline water. Thus, the problem of the case study since then has been how to get rid of waterlogging and salinity problems and how to optimize the recovery of the thin fresh groundwater lenses. Simulating the scavenger pumping in that area through using RASIM and SUTRA shows that RASIM requires much less computational effort. The results are presented in Fig. 8.9. The total depth of the aquifer in the study area is 63 m. The width of the transition zone is 6–7 m.

The match between SUTRA and RASIM is very good ($R^2 = 0.99$). Both SUTRA and RASIM were run on Newcastle University mainframe for the same data of the case study. In CPU time, SUTRA required 1060 s while RASIM needed just 615 s. RASIM has taken the advantage of the properties of the radial flow and the solute transport equations being solved to permit the use of an Incomplete Conjugate Cholesky Gradient (ICCG) preconditioning method of matrix solution for symmetric matrices. The ICCG algorithm is very fast and the casting of the equations in a symmetric form permits significant savings in computer storage. The filed data (GDC and BGS 1990) show that the measured width of the transition zone in the Indus Basin case study is around 6 m. This verification test investigates the deviation from this real measured width of the transition zone when different modelling

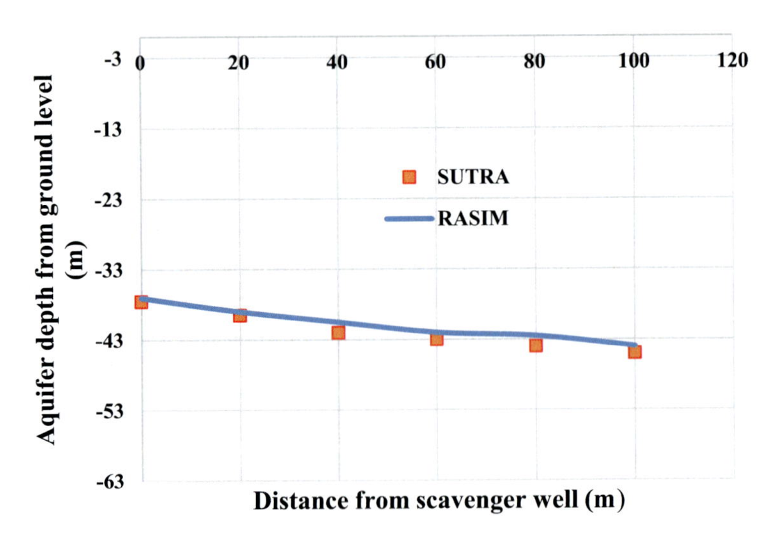

Fig. 8.9 Simulated location of 0.5 isochlor under scavenger pumping after 120 days

scenarios are carried out for isotropy and tensor properties of dispersion of saltwater. In reality both longitudinal and transverse dispersivity are dependent on flow direction.

The test here verifies what happens to the width of the transition zone if dispersivity is independent of flow direction. Also, this test investigates the effects of neglecting the dispersive flux of saltwater in r-direction or z-direction due to concentration gradients in z-direction and r-direction respectively (i.e., the off-diagonal terms of dispersion Eqs. 8.5c and 8.5d). The model RASIM was developed to take into consideration the anisotropic nature of the longitudinal and transverse dispersivity in addition to include off-diagonal terms of dispersion. The test investigated first to use anisotropy ratio of 1 and 20. Then the off-diagonal terms of dispersion (Eqs. 8.5c and 8.5d) were modelled and neglected in different modelling scenarios in RASIM. The results are presented in Table 8.4 with respect to the width of the transition zone only. When the off-diagonal terms of dispersion were neglected in RASIM, (a) the results show that the transition zone was widely dispersed everywhere and is severely dispersed at the well section (33 m width) for isotropic dispersion model, (b) using anisotropy ratio of 20 for the $\boldsymbol{\alpha}_L$ and isotropic nature for the α_T yields a width of the transition zone of 6 m, (c) using anisotropy ratio of 20 for the α_T and isotropic nature for the α_L yields a width of the transition zone of 33 m, (d) using anisotropy ratio of 20 for both α_T and α_L yields a width of the transition zone of 13 m. These observations show that the isotropy property of α_T has a stronger influence of widening the transition zone than α_L.

Remarkable improvements of the simulation results were obtained when the off-diagonal terms, D_{zr}, D_{rz}, of the dispersion model were accounted for in RASIM. Neglecting the off-diagonal terms of the dispersion model in RASIM yielded a wider transition zone with a sort of numerical instability near the well

Table 8.4 Results of modelling isotropy and tensorial properties of dispersion

Longitudinal dispersivity	Transverse dispersivity	Off-diagonal terms of dispersion	Width of transition zone (m)
Isotropic, $\boldsymbol{\alpha}_{L\max} = \boldsymbol{\alpha}_{L\min}$	Isotropic, $\boldsymbol{\alpha}_{T\max} = \boldsymbol{\alpha}_{T\min}$	Neglected	33
Anisotropic, $\boldsymbol{\alpha}_{L\max} \neq \boldsymbol{\alpha}_{L\min}$	Isotropic, $\boldsymbol{\alpha}_{T\max} = \boldsymbol{\alpha}_{T\min}$	Neglected	13
Isotropic, $\boldsymbol{\alpha}_{L\max} = \boldsymbol{\alpha}_{L\min}$	Anisotropic, $\boldsymbol{\alpha}_{T\max} \neq \boldsymbol{\alpha}_{T\min}$	Neglected	33
Anisotropic, $\boldsymbol{\alpha}_{L\max} \neq \boldsymbol{\alpha}_{L\min}$	Anisotropic, $\boldsymbol{\alpha}_{T\max} \neq \boldsymbol{\alpha}_{T\min}$	Neglected	13
Isotropic, $\boldsymbol{\alpha}_{L\max} = \boldsymbol{\alpha}_{L\min}$	Isotropic, $\boldsymbol{\alpha}_{T\max} = \boldsymbol{\alpha}_{T\min}$	Simulated	6.29
Anisotropic, $\boldsymbol{\alpha}_{L\max} \neq \boldsymbol{\alpha}_{L\min}$	Isotropic, $\boldsymbol{\alpha}_{T\max} = \boldsymbol{\alpha}_{T\min}$	Simulated	6.23
Isotropic, $\boldsymbol{\alpha}_{L\max} = \boldsymbol{\alpha}_{L\min}$	Anisotropic, $\boldsymbol{\alpha}_{T\max} \neq \boldsymbol{\alpha}_{T\min}$	Simulated	6.13
Anisotropic, $\boldsymbol{\alpha}_{L\max} = \boldsymbol{\alpha}_{L\min}$	Anisotropic, $\boldsymbol{\alpha}_{T\max} \neq \boldsymbol{\alpha}_{T\min}$	Simulated	6

boundary, while including these tensorial properties yielded half narrower transition zone and with no numerical instability. These results are also observed by Reddell and Sunada (1970). They showed that eliminating the cross-derivative terms from the simulation led to inaccurate results.

8.5 Conclusions and Recommendations

A local finite difference numerical model called RASIM has been developed to predict the behaviour of the Salt/Fresh-Brackish water transition zone during scavenger well pumping for two case studies in Kuwait and Pakistan. RASIM couples and solves density-dependent fluid flow and solute transport for anisotropic-tensoral dispersion nature. RASIM deals with aquifer-well system, therefore, cylindrical coordinates with radial symmetry were used in addition to Cartesian coordinates. This study concludes while it was possible to represent uniquely $r_{i \pm 1/2, j}$ logarithmically in r^2, in order to have continuity in conservative fluxes, it was not possible to do so for $\lambda_{r_{i\pm1/2,j}}$. However, the same continuity in fluxes can be achieved only if and only if the magnitude of $\lambda_{r_{i\pm1/2,j}}$ be averaged harmonically in logarithm r. By using such properties of accurate and unique internodal averaging properties, this research yields a symmetrical matrix of coefficients of the numerically approximated flow and solute transport equations, which make RASIM robust and efficient with minimum truncation errors. Transverse dispersivity has a stronger influence on widening the transition zone than α_L. Remarkable improvements of the simulation results were obtained when the off-diagonal terms, D_{zr}, D_{rz}, of the dispersion model were accounted for in RASIM. Neglecting the off-diagonal terms of the dispersion model in RASIM yielded a wider transition zone with a sort of numerical instability near the well boundary. RASIM has been verified and is shown to perform accurately in many cases with known analytical solutions for a range of flow and transport problems such as Theis, Gelhar and Henry solutions. In a comparison of the results obtained from RASIM with those obtained from SUTRA for a specific case study to solve a saline water upconing, it was shown that RASIM requires much less computational effort. Since RASIM is a verified and calibrated numerical model, therefore, this study recommends to apply RASIM in a number of case studies where freshwater lenses float above saline groundwater bodies such as the case of Sindh in Pakistan and where brackish water lenses float above saline groundwater bodies such as the case of Kuwait. Indeed, these recommendations were carried out in a different research studies that is presented in this book as will be seen later.

Acknowledgements The author would like to extend his appreciation to the Kuwait Foundation for the Advancement of Sciences (KFAS) for the financial support of this study under grant number PN17-25EM-03. The support of the management of the Kuwait Institute for Scientific Research (KISR) is pivotal in carrying out the various tasks of the study.

Declaration of Interest Statement There is no conflict of interest in our research results and analysis presented in this chapter.

Data Availability Statement The data that support the findings of this study are available on request from the corresponding author. The data are not publicly available due to privacy or ethical restrictions.

References

Alam N, Olsthoorn T (2014) Punjab scavenger wells for sustainable additional groundwater irrigation. Agric Water Manag 138:55–67. https://doi.org/10.1016/j.agwat.2014.03.001

Aliewi AS (1993) Numerical simulations of the behavior of fresh/saline transition zone around a scavenger well. PhD dissertation, Civil Engineering/Water Resources Engineering, Newcastle University

Aliewi AS (2023) Predicting the behaviour of the salt/fresh-brackish water transition zone during scavenger well pumping: 2. Model application in Kuwait and Pakistan. In: Suleiman MK, Shahid SA (eds) Terrestrial environment and ecosystems of Kuwait-assessment and restoration. Springer

Aliewi AS, Mackay R, Jayyousi A, Nasereddin K, Mushtaha A, Yaqubi A (2001) Numerical simulation of the movement of saltwater under skimming and scavenger pumping in the Pleistocene aquifer of Gaza and Jericho Areas, Palestine. Transp Porous Media 43:195–212

Aliewi AS, Al-Enezi H, Al-Maheimid I, Al-Kandari J, Al-Haddad A, Al-Qallaf H, Rashid T, Sadeqi D (2019) Sustainability of brackish groundwater utilization from the Eocene Aquifer for oil exploration in Central Kuwait. Environ Dev Sustain 22(5):4639–4653. https://doi.org/10.1007/s10668-019-00401-9

Aliewi A, Al-Kandari J, Al-Khalid A, Bhandary H, Al-Qallaf H (2021) Modelling the effect of high level of total dissolved solids (TDS) for the sustainable utilization of brackish groundwater from saline aquifers in Kuwait. Environ Dev Sustain 23(2):2204–2223. https://doi.org/10.1007/s10668-020-00670-9

Aslam M, Matsuno Y, Nobumasa H (2016) Performance evaluation of fresh groundwater skimming wells in the Indus basin irrigation system of Pakistan: a selected review. Mimeogr Fac Agric Kindai Univ 50:5–23

Bear JJ, Cheng HDA (2010) Introduction. In: Modeling groundwater flow and contaminant transport. Theory and applications of transport in porous media, vol 23. Springer, Dordrecht. https://doi.org/10.1007/978-1-4020-6682-5_1

Bennett JP, Haslauer CP, Cirpka OA (2017) The impact of sedimentary anisotropy on solute mixing in stacked scour-pool structures. Water Resour Res 53:2813–2832. https://doi.org/10.1002/2016WR019665

Dagan G, Bear J (1968) Solving the problem of local interface upconing in a coastal aquifer by the method of small perturbations. J Hydraul Res 6(1):15–44. https://doi.org/10.1080/00221686809500218

Desai CS, Contractor DN (1977) Finite element analysis of flow, diffusion and saltwater intrusion in porous media. In: Bathe KJ (ed) Formulation and computational algorithms in finite element analysis. MIT Press, pp 958–985

Diersch HG (2014) FEFLOW – finite element modeling of flow, mass and heat transport in porous and fractured media. Springer, Berlin/Heidelberg, vol XXXV, 996p. ISBN 978-3-642-38738-8. https://doi.org/10.1007/978-3-642-38739-5

Dokou Z, Karatzas GP (2012) Saltwater intrusion estimation in a karstified coastal system using density-dependent modelling and comparison with the sharp-interface approach. Hydrol Sci J 57(5):985–999. https://doi.org/10.1080/02626667.2012.690070
Fader SW (1957) An analysis of contour maps of 1955 water levels with a discussion of salt-water problems in south-western Louisiana. Water Resour Pamphlet 4:16–27. Published by the Department of Conservation, Louisiana Geological Survey and Louisiana Department of Public Works, Baton Rouge
Fukumori E, Wake A, Laursen E (1986) Modelling two well system in a stratified aquifer. J Water Resour Plan Manag ASCE 112(1):129–141
Gelhar LW, Collins MA (1971) General analysis of longitudinal dispersion in non-uniform flow. Water Resour Res 7(6):1511–1521
Gopinath S, Srinivasamoorthy K, Saravanan K, Suma CS, Prakash R, Senthilnathan KD, Chandarasekaran N, Yasala S, Sarma VS (2016) Modeling saline water intrusion in Nagapattinam coastal aquifers, Tamilnadu, India. Model Earth Syst Environ 2(1). https://doi.org/10.1007/s40808-015-0058-6
Groundwater Development Consultant (GDC) and British Geological Survey (BGS) (1990) Scavenger wells and pilot studies. Volumes 1 to 6
Henry HR (1964) Effects of dispersion on saltwater encroachment in coastal aquifers. USGS Paper 1613-C:C70–C84
Hill MC (1988) A comparison of coupled freshwater-saltwater sharp-interface and convective-dispersive models for saltwater intrusion in a layered aquifer system. Dev Water Sci 35:211–216. https://doi.org/10.1016/S0167-5648(08)70340-X
Kalakan C (2014) Investigation of saltwater intrusion based on the Henry problem and a field-scale problem. PhD dissertation, University of Florida
Karatzas GP (2017) Developments on modeling of groundwater flow and contaminant transport. Water Resour Manag 31:3235–3244. https://doi.org/10.1007/s11269-017-1729-z
Konikow L (2011) The secret to successful solute transport modelling. Ground Water 49(2):144–159
Langevin CD, Thorne DT Jr, Dausman AM, Sukop MC, Guo W (2007). SEAWAT Version 4: A computer program for simulation of multi-species solute and heat transport. U.S. Geological Survey Techniques and Methods Book 6, Chapter A22, 39 pp
Lee J, Rolle M, Kitanidis P (2017) Longitudinal dispersion coefficients for numerical modelling of groundwater solute transport in heterogeneous formations. J Contam Hydrol. https://doi.org/10.1016/j.jconhyd.2017.09.004
Long RA (1965) Feasibility of a scavenger-well system as a solution to the problem of vertical salt-water encroachment. State of Louisiana Department of Conservation, Geological Survey and Department of Public Works in Cooperation with United States Geological Survey, Water Resources Pamphlet No. 15
McDonald G, Harbaugh A (1988) A modular three-dimensional finite difference groundwater flow model. Techniq Water Resour Investig US Geol Surv 6:586
Mitchell AR, Griffiths DF (1980) The finite difference method in partial differential equations. Wiley International Publication
Reddell DL, Sunada DK (1970) Numerical simulation of dispersion in groundwater aquifers. Hydrological paper No. 41. Colorado State University, Fort Collins
Saeed M, Bruen M, Asghar M (2002) A review of modeling approaches to simulate saline-upconing under skimming wells. Nord Hydrol 33:165–188. https://doi.org/10.2166/nh.2002.0021
Scheidegger AE (1961) General theory of dispersion in porous media. J Geophys Res 66(10):3273–3278
Segol G, Pinder G, Gray W (1975) A Galerkin-finite element technique for calculating the transient position of the saltwater front. Water Resour Res 11(2):343–347
Simpson MJ, Clement TB (2004) Improving the worthiness of the Henry problem as a benchmark for density-dependent groundwater flow models. Water Resour Res 40:W01504

Stein S, Solo F, Yechieli Y, Shalev E, Sivan O, Kasher R, Vallejos A (2020) The effect of long-term saline groundwater pumping for desalination of the fresh-saline water interface: field observations and numerical modelling. Sci Total Environ 732(August issue). https://doi.org/10.1016/j.scitotenv.2020.139249

Terzić J, Peh Z, Marković T (2010) Hydrochemical properties of transition zone between fresh groundwater and seawater in karst environment of the Adriatic islands, Croatia. Environ Earth Sci 59(8):1629–1642. https://doi.org/10.1007/s12665-009-0146-x

Theis CV (1935) The relation between the lowering of the piezometric surface and the rate of duration of discharge of a well under using ground-water storage. Trans Am Geophys Union 16:519–524

Voss CI (1984) A finite-element simulation model for saturated-unsaturated fluid-density-dependent groundwater flow with energy transport or chemically-reactive-single-species solute transport. USGS Water Resources Investigation Report 84-4369, 409 pp. https://doi.org/10.3133/wri844369

Voss CI (1999) USGS SUTRA code: history, practical use, and application in Hawaii. Kluwer Academic, Dordrecht, 625 pp

Yin J, Tsai FTC (2019) Steady-State approximate freshwater-saltwater interface in a two-horizontal-well scavenging system. Tech Note J Hydrol Eng 24(10):06019008–06019010

Zack AL (1988) A well system to recover water from a freshwater-saltwater aquifer in Pureto Rico. USGS Water Supply Paper 2328. 15 p. https://doi.org/10.3133/wsp2328

Zack A, Candelario RM (1983) Hydraulic technique for designing scavenger-production well couples to withdraw freshwater from aquifers containing saline water. Project No. OWRT-A-075-PR (1) Final Report, US Geological Survey, San Juan, Puerto Rico Water Resources Division, 58 pp

Zardari NH, Shirazi SM, Yusop Z, Naubi I, Mangrio MA (2015) A comparison of current and design operational efficiencies of scavenger wells in lower Indus Basin of Pakistan and possibility of upconing problem. Arab J Geosci 8(10):12. https://doi.org/10.1007/s12517-015-1851-2

Chapter 9
Predicting the Behaviour of the Salt/Fresh-Brackish Water Transition Zone During Scavenger Well Pumping: 2. Model Application in Kuwait and Pakistan

Amjad Sami Aliewi

Abstract In part 1 of this study, "RASIM" was developed as a validated model for optimum operational schemes of scavenger wells' pumping in order to protect local groundwater resources from salinization. A suitable design criterion for scavenger wells requires the enhancement of the understanding of the movement of the salt/fresh-brackish groundwater transition zone around them. Therefore, this part of the study presents the application of the RASIM model with two case studies: Jaber Al-Ahmad residential area in Kuwait with the aim to maximize the recovery ratio of brackish groundwater and the agricultural areas of Sindh province in Pakistan with the aim to optimize the recovery of thin fresh groundwater lenses for irrigation purposes. For these case studies, hundreds of simulation runs were performed using RASIM to characterize the design criterion and establish optimal freshwater or brackish water recovery from the scavenger wells under diverse hydrogeological, physical and operational conditions. The results of the study showed that while it was possible to optimize the freshwater recovery ratio in the agricultural areas in the Sindh province to 80%, it was not possible to maximize the brackish water recovery ratio greater than 55% in the urban area of Jaber Al-Ahmad in Kuwait.

Keywords Scavenger wells · Salinization processes · Modelling density-driven flow · RASIM model · Design criterion · Groundwater lenses · Freshwater and brackish water recovery · Jaber Al-Ahmad Kuwait · Sindh Province Pakistan

---

A. S. Aliewi (✉)
Water Research Centre, Kuwait Institute for Scientific Research, Safat, Kuwait
e-mail: aaliewi@kisr.edu.kw

M. K. Suleiman, S. A. Shahid (eds.), *Terrestrial Environment and Ecosystems of Kuwait*, https://doi.org/10.1007/978-3-031-46262-7_9

9.1 Introduction

The verified and validated model, RASIM, developed in part 1 of this research (Aliewi 2023) is applied for two case studies (Fig. 9.1) in Kuwait and Pakistan. The Kuwait case study is for a brackish/saline aquifer while the Pakistan case study is for a fresh/saline aquifer. The lithology of the two aquifers is by some means similar of inter-bedded deposits of alluvial sand, silt and clay. However, in Kuwait, the clay content is not a continuous stratum. The brackish water in the Kuwait case study is utilized to irrigate agricultural farms, landscaping and greenery uses.

The Kuwait Group Aquifer is the target aquifer in this study. It can be categorized into three clastic subunits: Upper aquifer (principal aquifer), Middle aquitard and Lower aquifer. Lithologically, the upper Kuwait Group is represented by gravelly sand, sandy gravel, calcareous and gypsiferous sand with occasional intercalation of thin clay lenses. Although resistivity logs show variation in the resistivity of the material encountered at the locations of several wells in the study area, it has not exceeded 40-ohm meter. This low resistivity indicates porous and unconsolidated material, partially cemented, as indicated by segments of sharp increases on the resistivity logs at the locations of the wells. Relatively low resistivities of lower parts than the upper parts at the locations of some wells indicate the presence of relatively high salinity water at lower parts as compared to upper parts. The presence of relatively low salinity water may be attributed to the recharge of aquifer with water leaked from freshwater network and freshwater used to irrigate landscape areas. The brackish water in the Kuwait study area is used for greenery

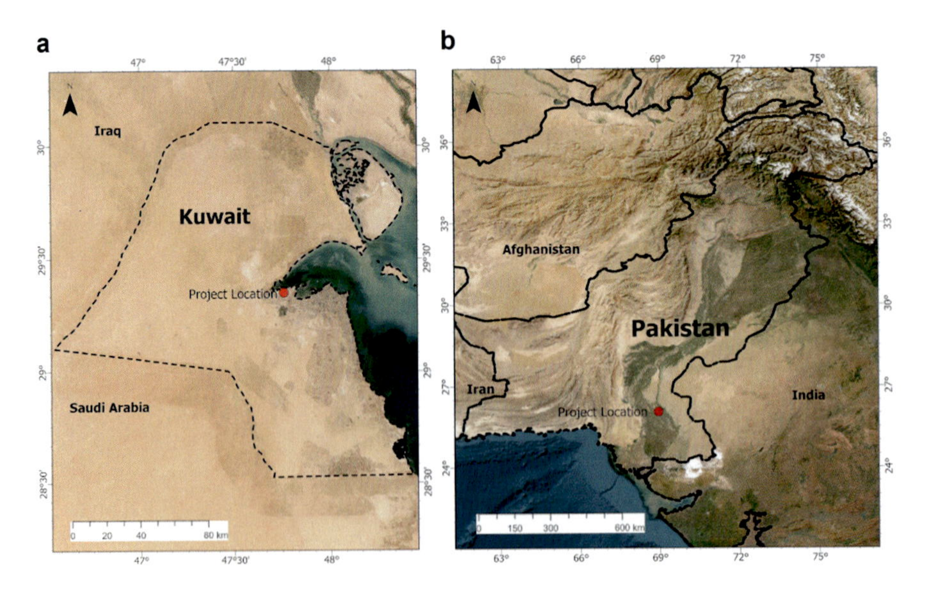

Fig. 9.1 Location map of the two case studies. (**a**) Jaber Al-Ahmad City, Kuwait. (**b**) The Sindh province of the Indus Basin of Pakistan

purposes which is important to keep the quality of the abstracted water as brackish by minimizing saline water upconing. While, the freshwater in Pakistan study area is used for agricultural water supply, landscaping, livestock, fishery and greenery purposes (PDDGS 2017). The behaviour of transition zone in fresh/saltwater aquifers can be observed more clearly than in the brackish/saltwater aquifers. Therefore, in this study, more details have been provided for the characterization of the transition zone for the Sindh province case study. As seen later, details will be provided for both case studies when the design criteria of scavenger wells are developed.

When the canal system was completed in 1932 in the Sindh agricultural areas, it was possible to irrigate with freshwater on a perennial basis. However, it is estimated (Wang 1965) that nearly one-third of the water supplied is lost from the canal distribution system forming freshwater lenses floating over saline water in the aquifer. Saeed et al. (2002) illustrated that the thickness of the fresh groundwater lenses is in the range of 30–100 m near the Indus River and main canals. Over tens of years of the accumulation of irrigation losses, the groundwater tables have risen dramatically causing waterlogging of crops in addition to severe salinization of uncropped land and an increase in stormwater flooding (Qureshi 2009; Messerschmid and Aliewi 2022). Thus, the problem since 1965 has been how to get rid of waterlogging and salinity problems, and how to develop the thin fresh groundwater lenses without increasing the salinity of the utilized groundwater resources. The Hunting Technical Services and Mott McDonald Partners (HTS and MMP 1965) recommended that freshwater lenses should be recovered by the construction of scavenger wells. Scavenger wells are methods to extract finite fresh/brackish groundwater lenses overlying saline water. The idea of pumping both fresh and saline waters using separate pumps or wells (scavenger pumping) was first shown in Pennik (1904, 1905). He used scavenger pumping just to detect the existence of saline water in the aquifer of the dune area of Amsterdam, the Netherlands. Pennink (1904, 1905) did not mention the use of scavenger pumping as a means of recovering freshwater or controlling saline water invasion. After nearly 50 years, this idea was suggested by Fader (1957). In order to recover freshwater overlying saline water in the Chicot aquifer of southwestern Louisiana (USA), Fader (1957) described a two-well system in which a scavenger well was screened in the lower part of the aquifer to draw off only saline water and a nearby supply well was screened in the freshwater section of the aquifer. In the same state Louisiana, Long (1965) tested (in the field) the feasibility of the two-well scavenging system suggested by Fader (1957) and concluded that this system could be effective to recover fresh groundwater overlying saline water under controlled conditions. Field studies (Zack and Munoz-Candelario 1984; Zack 1988) were carried out to demonstrate that screening a pumping well in both fresh and saline water will provide an effective method for extracting water from a thin freshwater layer in the coastal aquifers of Puerto Rico of the United States of America (USA). From their results, they established a hydraulic approach to estimate optimal withdrawals of freshwater based on pumping rates and the level of solute concentrations. It was demonstrated by Zack (1988) that the system was only able to produce freshwater from the aquifer when the operation of a scavenger well was coupled with a production well.

In a preparatory study, HTS and MMP (1965) recommended the use of scavenger wells to recover thin lenses of fresh groundwater in the Sindh Province of Pakistan. For the aim of recovering thin freshwater lenses, a field-testing programme has been undertaken (GDC and BGS 1990). In commenting on this fieldwork, it was concluded by Birch and Wonderen (1990) and Aliewi (1993) that scavenger wells are the most economic means to optimize the freshwater recovery in saline aquifers of the Sindh Province, Pakistan. Further investigations (Aliewi et al. 2001; Aslam et al. 2016) in the same study area about the operational efficiency of scavenger wells show that saline water upconing is negligible if the pumps are operated at the rate of 14.4 h/day.

Numerical simulations were conducted (Aliewi et al. 2020a, 2021) to test the effect of operating scavenger pumping in controlling saline water upconing in inland and coastal aquifers without mentioning the term "scavenger well". Fukumori et al. (1986) carried out a numerical study to analyse the effect of simultaneous pumping from both sides of the interface. They demonstrated that pumping water from above and below the interface at the same time gave more yield of freshwater than pumping only from above the interface. This concept of scavenger wells has been used (Wisniewski et al. 1985) to control the interface separating a dense fluid (coal tar) from a light fluid (water) in water/coal tar systems, in order to recover the denser fluid. The same principle was used to recover freshwater from contaminated groundwater with nitrates in the chalk aquifers in the United Kingdom (Stoner and Bakiewicz 1992). The development of scavenger wells in the study area was based on both field investigations and numerical modelling work. Scavenger pumping is applied in this research to extract freshwater/brackish and saline waters in the same well not from adjacent wells.

The aim of this research is to apply the RASIM model to simulate the behaviour of the salt/fresh-brackish water transition zone during scavenger well pumping to arrive at optimal design criteria for these wells under different hydrogeological, physical and operational conditions. This study will help both Kuwait and Pakistan to better utilize good quality water for irrigation services without provoking the deeper saline water to contaminate the target fresh and brackish water bodies, thus sustaining the ecosystem services on long-term basis. The study was completed in 2020.

9.2 Reference Parameters of Analysis for the Two Case Studies

The total dissolved solids (TDS) of freshwater and saline water in the Sindh case study are 550 mg/l and 29,250 mg/l, respectively. The Kuwait case study gives 3200 mg/l for brackish water and above 10,000 mg/l for saline water (Aliewi et al. 2020b). The reference hydrogeological and physical data of the two case studies are presented in Table 9.1. The aquifer system in both case studies can be treated as a single unconfined aquifer (Saeed et al. 2002; Birch and Wonderen 1990; Aliewi

Table 9.1 Actual field data for the two case studies[a]

Parameters	Value for Sindh case study (Pakistan)	Value for Jaber Al-Ahmad case study (Kuwait)
Aquifer radius, width	100 m, 63 m	100 m, 45 m
Thickness of clay layer, aquifer	8 m, 55 m	45 m aquifer
Initial K_r for clay layer, aquifer	2 m/day, 18.7 m/day	0.9 m/day, 15 m/day
Initial K_v for clay layer, aquifer	0.1 m/day, 0.9 m/day	0.12 m/day, 1.5 m/day
Initial depth to water table, recharge value	1.9 m bgl, 1.45 mm/day	2 m bgl, 15 mm/year
α_{Lmax}, α_{Lmin} ,α_{Tmax}, α_{Tmin}	5 m, 0.25 m, 0.02 m, 0.001 m	5 m, 0.25 m, 0.02 m, 0.001 m
Dispersivity ratio, anisotropy ratio	250, 20	250, 20
Porosity, specific yield	0.3, 0.06–0.12	0.35, 0.05
Initial location, width of transition zone	45 m bgl, 2 m	28 m bgl, 17 m
Salinity of freshwater, saline water	550 mg/l, 29,250 mg/l	3200 mg/l, 10,000 mg/l
Density of fresh/brackish water, saline water	1001 kg/m^3, 1029 kg/m^3	1003 kg/m^3, 1040 kg/m^3
Base value of concentration in aquifer	0 kg/kg	0 kg/kg
Density change with concentration	700 kg/m^3	700 kg/m^3
Viscosity	0.001 kg/m.s	0.001 kg/m.s
Gravity acceleration	9.806 m/s^2	9.806 m/s^2
Fluid compressibility	4.5×10^{-10} m.s^2/kg	4.5×10^{-10} m.s^2/kg
Fluid diffusivity	1.0×10^{-9} m^2/s	1.0×10^{-9} m^2/s
Solid matrix compressibility	1.0×10^{-8} m.s^2/kg	1.0×10^{-8} m.s^2/kg
Depth to top, bottom of screen	15.1 m bgl, 39.7 m bgl	7, 45 m bgl
Total well discharge	4155 m^3/day	1250 m^3/day

[a]Source of data: (GDC and BGS 1990; Aliewi et al. 2020b); *m bgl* meters below ground level

et al. 2020b). The initial concentration for the simulation study is set to the salinity of the freshwater/brackish water and saline water in the fresh and saline bodies respectively. The term “isochlor” is given for the contour line of equal salinity. The zero isochlor (contour line of equal salinity) for the Sindh case study is 550 mg/l, which is considered the freshwater limit for this case study. In the Kuwait case study, the zero isochlor of brackish water limit was 3200 mg/l. The maximum salinity of 29,250 mg/l (in Pakistan) and 10,000 mg/l (in Kuwait) define the bottom of the transition zone which is identified in the analysis of this research by the 1.0 isochlor. Intermediate isochlor magnitudes are related to their equivalent salinities. The width of the transition zone is defined as the vertical distance between the 0.0 and the 1.0 isochlors. The recovery ratio of freshwater or brackish water is defined as freshwater/brackish water abstracted through the upper screen (freshwater or brackish water pump) divided by the total abstraction of fresh/brackish and saline waters from the scavenger well. In addition, the saline upconing rate is defined as the speed at which the transition zone is proceeding towards the freshwater zone and mixes with it.

9.3 Analysis and Discussion

The section below details how the transition zone between different quality of waters moves under different hydrogeological and operational conditions including the design set up of scavenger wells.

9.3.1 The Behaviour of the Transition Zone Under Different Hydrogeological Conditions

The RASIM model was run for 300 days for Sindh Province study area utilizing the data of Table 9.1. Figure 9.2 shows that the steady state conditions were achieved after 130 days and that the rate of upconing process is high in the early stages of pumping until the transition zone reaches the bottom of the well screen at 39.7 m bgl in the saline water zone. After that, the upconing rate slows down gradually to achieve, eventually, a state of equilibrium. Figure 9.3 is the simulated salinity profile of the field conditions after reaching a steady-state upconing process at 130 days of scavenger pumping for the Sindh case study. The initial width of the transition zone was measured in the field before pumping started at 2 m. The simulated steady-state width of the transition zone was 6 m. The upconing of the transition zone over the entire domain of the modelled area after 200 days of scavenger pumping was

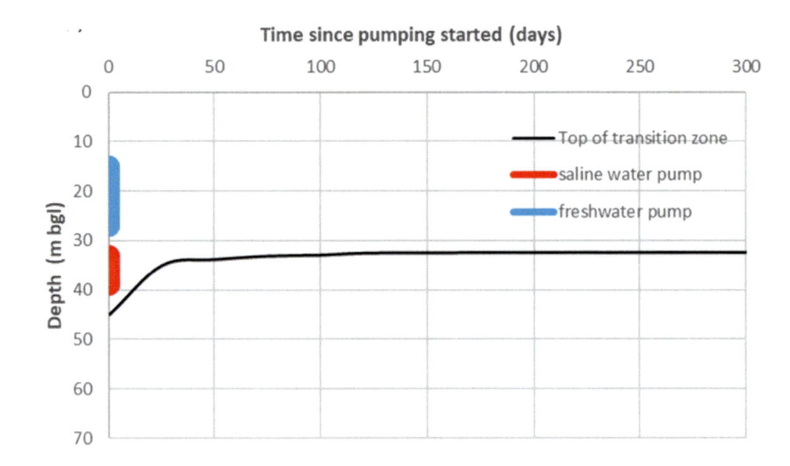

Fig. 9.2 Upconing of the top of transition zone with time. (Source: Sindh case study)

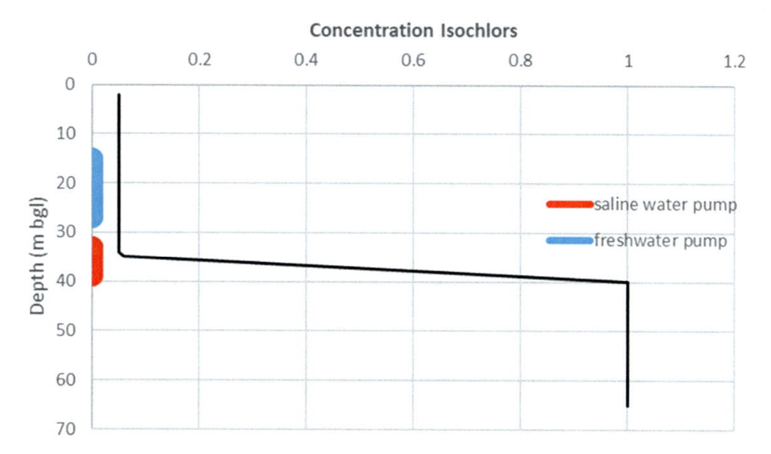

Fig. 9.3 Simulated salinity profile at scavenger well under field conditions. (Source: Sindh case study)

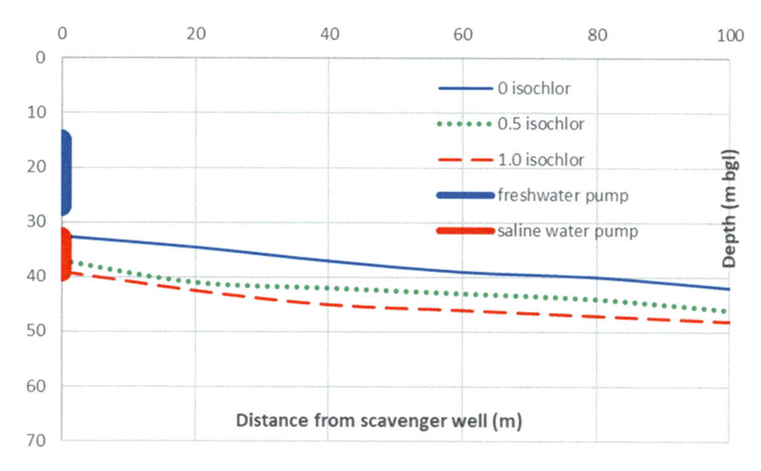

Fig. 9.4 Location of transition zone under scavenger pumping at the end of 200 days for field conditions. (Source: Sindh case study)

simulated (Fig. 9.4) using RASIM. Figure 9.4 provides evidence that the transition zone is not symmetric around the 0.5 isochlor due to the behaviour of waters during mixing by dispersion, which is controlled by the flow velocity magnitude, direction and dispersivity parameters. For the results of the standard run in the Sindh province, the freshwater recovery ratio was 71%. Freshwater and saline water screens partially penetrate the entire thickness of the aquifer. This partial penetration forces groundwater flow at the scavenger well to be almost vertical directly above and below the well screens, and almost horizontal alongside the screen sector.

It is important to investigate the effect of both longitudinal and transverse dispersivity on saline water upconing process. When higher values of transverse dispersivity (i.e., $\alpha_{Tmax} = 0.2$ m) were simulated using RASIM, the upconing process

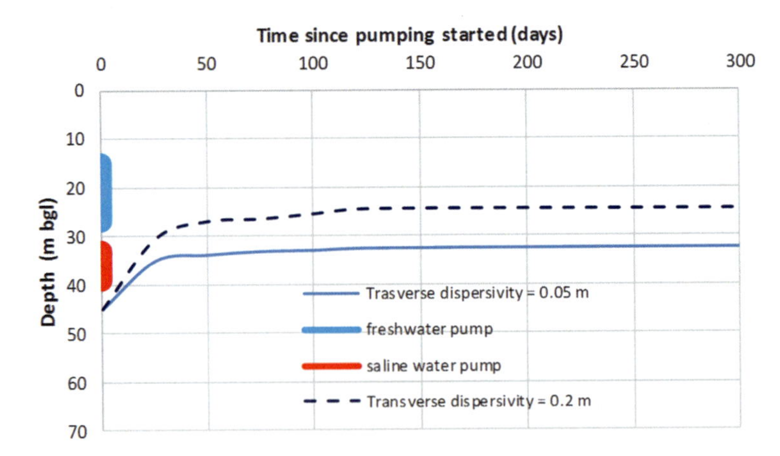

Fig. 9.5 Rate of top of transition zone upconing for different values of transverse dispersivity. (Source: Sindh case study)

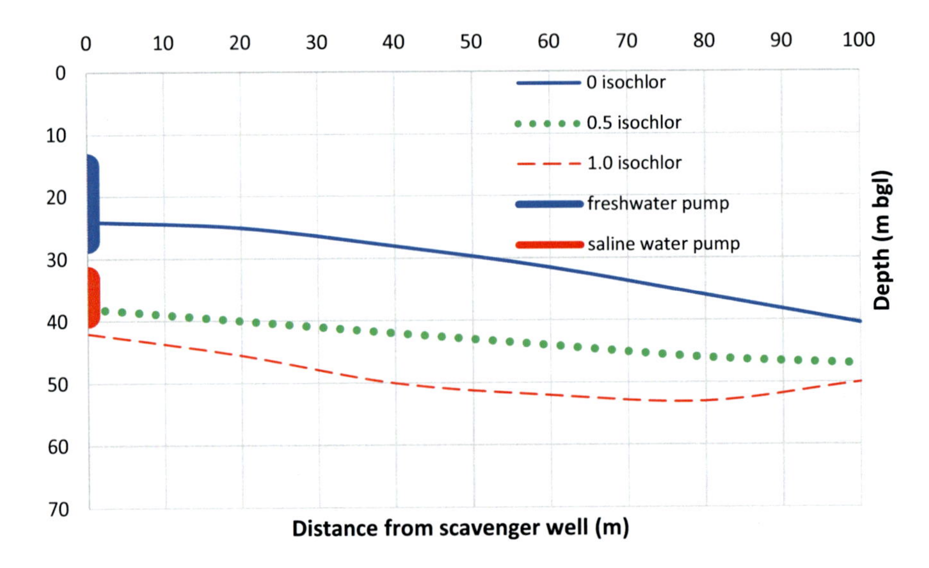

Fig. 9.6 Location of transition zone under scavenger pumping at the end of 200 days for transverse dispersivity value of 0.2 m. (Source: Sindh case study)

approaches equilibrium in a longer time (150 days) than for lower values of transverse dispersivity (i.e., $\alpha_{Tmax} = 0.02$ m) which took almost 130 days to reach equilibrium (Fig. 9.5). This means that when the system reaches equilibrium, the final position of zero isochlor for higher values of transverse dispersivity was nearer to the ground level (Fig. 9.5) than lower values of transverse dispersivity.

As seen in Fig. 9.6, the width of the transition zone at day 200 was 18 m for $\alpha_{Tmax} = 0.2$ m compared to 6 m when $\alpha_{Tmax} = 0.02$. The freshwater recovery ratios were 71% for $\alpha_{Tmax} = 0.02$ m and 36% for $\alpha_{Tmax} = 0.2$ m. The conclusion from this analysis is that the value of transverse dispersivity does characterize the transition zone of the groundwater system in this case study. When the value of α_{Lmax} is reduced from 5 m to 1 m and then increased to 10 m, no significant changes on the shape of the transition zone or the recovery ratio were observed. Therefore, the upconing process for this study is insensitive to the value of longitudinal dispersivity. These results are in agreement with the analysis conducted for aquifer systems in Hawaii, USA (Souza and Voss 1986; Voss and Souza 1986, 1987). However, Zhou et al. (2005) show that upconing process is sensitive to both transverse and longitudinal dispersivities. Also, the width of the transition zone is normally affected by aquifer recharge by rainfall (Aliewi et al. 2022).

Several simulation runs were conducted to test the sensitivity of the upconing process to initial settings of the position of the transition zone for both case studies in Sindh Province, Pakistan and Jaber Al-Ahmad City in Kuwait. The results are presented in Tables 9.2 and 9.3.

Tables 9.2 and 9.3 show that the upconing results are sensitive to the initial position of the 0.5 isochlor. In the Sindh case study, lowering the initial position of 0.5 isochlor from its current position by 5 m, resulted in improving freshwater recovery ratio by 8%, deepening the final position of the zero isochlor by 2.2 m and narrowing the transition zone by 1.1 m. On the other hand, simulating the initial position of 0.5 isochlor 5 m shallower than its current position, yielded in deteriorating the freshwater recovery ratio by 9%, rising the position of the transition zone by 2.4 m and widening the transition zone by 2.5 m. For the case study of Jaber Al-Ahmad in Kuwait, lowering the initial position of 0.5 isochlor from its current position at 28 m bgl by 7 m, 12 m and 14 m improves the recovery ratio by 14%, 24% and 28% respectively.

Table 9.2 Sensitivity to initial position of 0.5 isochlor, LS = 24.6 m

Initial position of 0.5 isochlor (m bgl)	Freshwater recovery ratio (%)	Final position of zero isochlor (m bgl)	Width of transition zone (m)
51	79	34.70	4.90
46	71	32.52	6.00
41	62	30.15	7.10
36	45	26.70	8.50

Source: Sindh case study

Table 9.3 Sensitivity to initial position of 0.5 isochlor, LS = 35 m

Initial position of 0.5 isochlor (m bgl)	Brackish water recovery ratio (%)
28	44
35	58
40	68
42	72

Source: Jaber Al-Ahmad case study

As presented in Table 9.1, the aquifer of the study area of the Sindh province is inter-bedded by silt and clay layer of 8 m. The results of extensive numerical tests show that no noticeable change in the final position of the transition zone was observed for different thicknesses of the clay layer as far as the position of the clay layer remains above the location of the screen of the freshwater pump. Because the intake of water along the length of the screen depends on the value of permeability as developed in RASIM, the recovery ratio was reduced in the simulations, which have longer extension of the clay layer in the screen section. When the clay layer was located below the bottom of the saline water screen, the results of the freshwater recovery improved significantly (by 10–15%). This is because the vertical permeability of the clay layer is much lower than that in the rest of the aquifer, the saline water could not move large distances upwards. As the clay content at the Jaber Al-Ahmad case study was not a continuous layer, its effect was not simulated.

It was also important to study the effect of changing the aquifer permeability on saline water upconing. Similar trends of the results were achieved when the permeability values were changed for the Jaber Al-Ahmed case study, so that the details here are provided for the Sindh province case study only. Increasing the horizontal permeability from 18.7 meters/day (m/d) of field conditions to 22 m/d results in lowering the final position of the top of the transition zone by 1.37 m, and improving the freshwater recovery ratio from 71% for the field condition to 75%. Reducing the horizontal permeability of the aquifer to 10 m/d results in rising the final position of the top of the transition zone by 1.5 m, and deteriorating the freshwater recovery ratio from 71% for the field conditions to 66%. In the same way, increasing the vertical permeability from 0.9 m/d of field conditions to 18.7 m/d (for isotropic conditions) results in rising the final position of the top of the transition zone by 4.62 m and deteriorating the freshwater recovery ratio from 71% for the field conditions to 55%. Reducing the vertical permeability of the aquifer to 0.37 m/d results in lowering the final position of the top of the transition zone by 1.88 m, and improving the freshwater recovery ratio from 71% for the field conditions to 79%. These analyses show that the change in vertical permeability has the opposite effect on the transition zone movement from that of the horizontal permeability. Increasing the vertical permeability results in severely worse results, especially when the vertical permeability is increased to the value of the horizontal permeability (isotropic media). For the two case studies, decreasing the salinity of saline water results in a slight rise of transition zone for less dense saline water. The width of the transition zone and the freshwater recovery ratio remain the same as the results of the standard run. The behaviour of the transition zone was insensitive to changing the values of porosity. Increasing the aquifer thickness has almost no effect on the results, but decreasing the aquifer thickness results in a slight lowering of transition which improves slightly the fresh/brackish water recovery ratio. Increasing the value of natural recharge has a negligible effect of the results of recovery ratio of fresh/brackish water or on the location of the transition.

9.3.2 *The Behaviour of the Transition Zone Under Different Design Settings*

This section details the movement of the transition zone and the different design schemes of scavenger wells in Kuwait and Pakistan.

9.3.2.1 Sindh Province Case Study

Investigating the saline water upconing mechanism under different designs and operational conditions of scavenger wells pumping is very important. A number of researchers (GDC and BGS 1990; Zardari et al. 2015; Aslam et al. 2016) pointed out that the design settings of the fresh/brackish water and saline water pumps control the upconing process. They reported the best results could be achieved for operating these pumps for a limited number of hours per day (cyclic pumping) (Aliewi and Al-Khalid 2017). This study tested different scenarios of cyclic pumping and found that the behaviour of the transition zone is the same for different factors (>0.5) of cyclic pumping. However, the width of the transition zone at the end of a recovery period of 100 days following a continuous pumping for 60 days was 24.36 m (Fig. 9.7), which is almost four times wider than the transition zone resulting from the continuous pumping. When the pumps stop operating the upconed saline water, which was at equilibrium under pumping stresses, will lose that equilibrium and fall down by gravity. This downward movement of heavier saline water will increase in the width of the transition zone considerably as shown in Fig. 9.7. The results of this study agree with those of Zhou et al. (2005). They reported that cyclic pumping may lead to a recirculating flow in the saltwater zone beneath the pumping well, widening the saltwater mound (Aliewi and Al-Khalid 2017). They also reported that the recovery process is lengthy, as it takes a long time for the

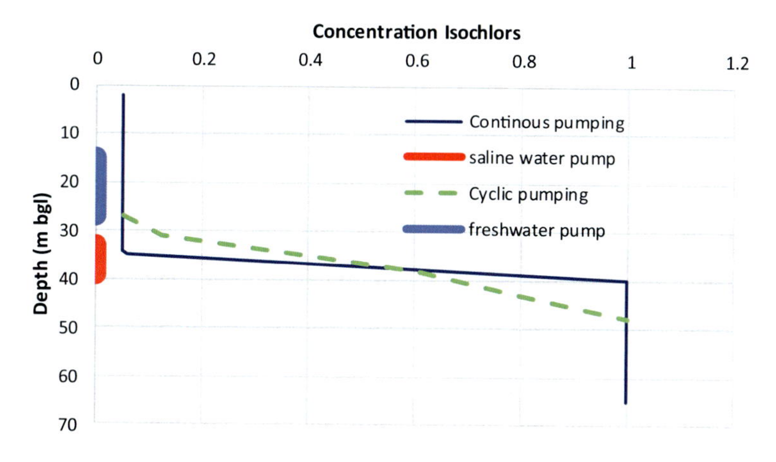

Fig. 9.7 Salinity profile for continuous and cyclic pumping

upconed saltwater to fall back near its original position prior to pumping. They also confirmed that the wider transition zone caused by hydrodynamic dispersion can never return to the initial position.

The following sensitivity analyses were conducted with respect to well dimensions, length, and location of well screens and pumping rates of the scavenger well. Altering the total well capacity of the scavenger well (total pumping from both fresh/brackish water pump and the saline water pump) between 3142 and 8040 m^3/day has no effect on the fresh/brackish water recovery ratio and it has a minor effect on the final position of the top of the transition zone. Steady state was approached at similar times for different discharge rates. Then, this research tested the effect of altering well capacity per meter of screen length (Q_{TOT}/LS) such that both the pumping rate and the screen length are changed so that their ratio remains constant at 170 m^3/day per meter of screen length. The results are presented in Table 9.4. The results show that increasing both the total pumping rate and the screen length if Q_{TOT}/LS is kept constant and the top of the screen is fixed, results in a reduction in the value of fresh/brackish water recovery ratio. This is understood from the fact that all alterations presented in Table 9.4 resulted in lowering the location of the transition facing the screens of the scavenger well towards the saline water zone. It can be noted from Table 9.4 that increasing the total pumping rate of the scavenger well and the screens of the well by 75% and 100% results in widening the transition zone 67% and 100% respectively. This is because the top of the freshwater screen remained in its location and the increase in screen length was towards the saline water zone.

The settings of the well screens for the Sindh case study were changed as presented in Table 9.5. The results from these simulations compared to the standard run of field conditions are also presented in Table 9.5. Table 9.5 shows (while keeping the same length of the screens) that raising the well screens 5 m raises the top of transition zone by 2.77 m, increases freshwater recovery to 76% and narrows the width of the transition zone to 4.72 m. Lowering the screens 5 m works exactly in the opposite direction. This means it lowers the top of the transition zone by 2.78 m, deteriorates freshwater recovery ratio to 62% and widens the transition zone to 7.03 m. These results can be explained as follows: when the screen is lowered 5 m,

Table 9.4 Results of altering Q_{TOT} and LS while keeping Q_{TOT}/LS and screen top constant

Q_{TOT} (m^3/d)	LS (m)	Altering Q_{TOT}/ LS	Freshwater recovery ratio (%)	Top of transition zone (m bgl)	Width of transition zone
3142	18.6	Decreased by 25%	73	29.00	4.88
4155	24.6	Standard run	71	32.52	6.00
6233	36.6	Increased by 50%	61	37.40	9.60
7188	42.6	Increased by 75%	55	38.99	10.01
8040	47.6	Increased by 100%	53	40.03	12.03

Table 9.5 Sensitivity of altering screen settings

Analysis	Freshwater recovery ratio (%)	Top of transition zone (m bgl)	Width of transition zone (m)
Standard run	71	32.52	6.00
Screens raised 5 m	76	29.75	4.72
Screens lowered 5 m	62	35.30	7.03

Table 9.6 Sensitivity to changing the magnitude of the screen lengths

Change	From	Fixing	Recovery ratio (%)	Top of transition zone (m bgl)	Width of transition zone (m)
Increased by 5 m **(A)**	Top (in the freshwater zone)	Bottom	76	32.35	6.24
Increased by 5 m **(B)**	Bottom	Top	69	35.09	7.21
Increased by 5 m **(C)**	Top and Bottom	Screen mid-point	71	33.52	6.73
Decreased by 5 m **(D)**	Top	Bottom	68	33.00	5.65
Decreased by 5 m **(E)**	Bottom	Top	78	29.92	4.54
Decreased by 5 m **(F)**	Top and Bottom	Screen mid-point	70	31.46	5.69

then the level of the bottom of the saline water screen is 44.7 m bgl. This means that the distance between the initial position of the top of the transition zone and the position of the bottom of the screen is 0.3 m. This distance is compared with 10.3 m for the case of raising the screen by 5 m. The outcome of this is that the transition zone moves a shorter distance to reach equilibrium when it is initially closer to the bottom of the screen.

Altering the position of the top of the freshwater screen and the bottom of saline water screen is proved to be influential on the saline water upconing process. The results are presented in Table 9.6 with the following remarks:

1. If the bottom of the screen is fixed, then increasing its length by 5 m from the top (towards the freshwater zone) increases freshwater recovery ratio by 5% but with minor effect on the final position of top of the transition zone and its width (Analysis A in Table 9.6).
2. If the top of the screen is fixed, then increasing the screen length by 5 m from the bottom towards the saline water zone has reduced freshwater recovery by 2% and lowered the top of the transition zone by 2.57 m (Analysis B in Table 9.6).

3. Increasing or decreasing the screen length while keeping the screen mid-point fixed has a minor effect on the recovery ratio (Analyses C and F in Table 9.6).
4. If the bottom of the screen is fixed, then decreasing the screen length 5 m from the top, results in a decrease of freshwater recovery ratio by 3% and has a slight change in the final position of the top of transition zone and its width (Analysis D in Table 9.6).
5. If the top of the screen is fixed, then decreasing the screen length 5 m in the saline water zone has increased freshwater recovery ratio by 7%, raised top of transition zone by 2.60 m and narrowed the transition zone by about 1m (Analysis E in Table 9.6).
6. Increasing the length of the screen from the top (while the bottom is fixed) or from the bottom (while the top is fixed) has the tendency to widen the transition zone. Decreasing the length of the screen from the top (while the bottom is fixed) or from the bottom (while the top is fixed) has the tendency to narrow the transition zone.

9.3.2.2 Jaber Al-Ahmad (Kuwait) Case Study

The data of the field conditions of the case study at Jaber Al-Ahmad area in Kuwait is presented in Table 9.1. Figure 9.8 and Table 9.7 show that, for a screen length of 30 m, different locations of the top of the screens gave different values of brackish water recovery ratios as presented earlier. It is clear that if the location of the top of the screen is closer to the static water level (at 2 m bgl) in the upper part of the aquifer then the brackish water recovery ratio will be improved.

It is clear from Fig. 9.8 that the screen operates from day 1 in both the brackish and saline water zones. However, when the top of the screen was 7 m bgl, the brackish water utilization was much larger than deepening the screen 3 m lower.

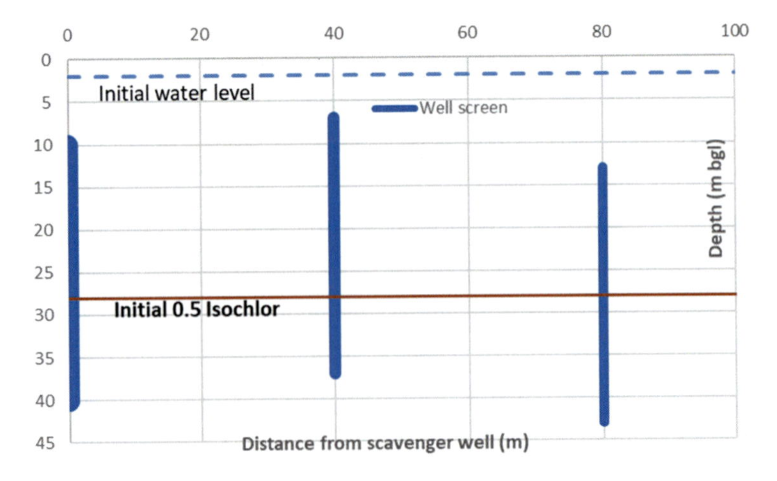

Fig. 9.8 Different settings of top of screen for the same length of nearly 30 m

Table 9.7 Simulated brackish water recovery ratios for different locations of the top of the screen (LS = 30 m, $I_{0.5}$ = 28 m bgl (field conditions))

Initial position of top of screen (m bgl)	Brackish water recovery ratio (%)
7	47
10	44
13	41

Table 9.8 Different locations and lengths of screen for scavenger pumping for field conditions at Jaber Al-Ahmad in Kuwait

Simulation	Top of screen (m bgl)	Bottom of screen (m bgl)	Screen length (m)	Location with respect to 0.5 isochlor	Brackish water recovery ratio (%)
A	5	27	22	100% in brackish zone	55
B	5	35	30	77% brackish 23% saline	50
C	5	40	35	66% brackish 34% saline	46
D	5	44	39	59 % brackish 41% saline	43
E	13	35	22	68 % brackish 32% saline	46
F	13	43	30	50 % brackish 50 % saline	40
G	13	45	32	46 % brackish 54 % saline	39

The following analyses (A–G in Table 9.8) were made to suit brackish water utilization from saline aquifers in Kuwait. It is clear from Table 9.8 that when scavenger pumping starts initially with more portions of the screen (for different screen lengths) in the brackish water zone (in saline-brackish aquifers), then better recovery ratios are obtained. However, the improvement is not that much compared to fresh-saline aquifers. For example, the best brackish water recovery ratio for the analyses in Table 9.8 was 55% when the bottom of the well screen was located initially above the 0.5 isochlor. The worst brackish water recovery ratio of 39% was in most of the well screen located in the saline water zone.

It should be noted that the analyses in Table 9.8 were conducted without keeping the ratio of total pumping rates to the length of well screen constant as was done for freshwater/saltwater aquifers of the Sindh case study. For the Jaber Al-Ahmad case study, practical considerations for the pumps available in the market were only taken.

9.3.2.3 Comparison of the Results Between the Two Case Studies

The results under different hydrogeological conditions between the two case studies worked in the same direction except that the recovery ratio of brackish water in the Jaber Al-Ahmed case study in Kuwait is less than that of freshwater in the Sindh Province case study in Pakistan. This is due to the fact that brackish water is heavier than freshwater. Another difference is the initial location of the 0.5 isochlor in the field before the pumping was started. This study shows that lowering the initial position of 0.5 isochlor from its current position resulted in improving fresh/brackish water recovery ratio by 1.6% per meter and 2% per meter for Sindh Province and Jaber Al-Ahmed Case studies respectively. The results can be attributed to the initial vertical distribution of salinity and the initial width of the transition zone. Simply the initial width of the transition zone between freshwater and saline water in the Sindh province is 2 m while it is 17 m between the brackish water zone and the saline water zone in the Jaber Al-Ahmed case study. This wide transition zone in the latter case allows more brackish water to be utilized by the pump from above the 0.5 isochlor at steady state conditions. The parameters of the design settings of the scavenger well pumps are very important in this research for the two case studies. For both case studies, if the location of the top of the fresh/brackish water screen is closer to the static water level in the upper part of the aquifer then the fresh/brackish water recovery ratio will be improved. The best recovery ratio for Sindh province was 78% when the saline water screen length was decreased from the bottom, allowing more freshwater to be captured. However, the best brackish water recovery ratio for Jaber Al-Ahmad was 55% when the bottom of the well screen was located initially above the 0.5 isochlor allowing the saline water pump to start functioning at early stages of the upconing process. The worst brackish water recovery ratio of 39% was in most of the well screen located in the saline water zone.

9.4 Optimal Design of Scavenger Wells to Maximize Fresh/Brackish Water Recovery

In the light of the above, this section provides insights into the optimal design of scavenger wells in Kuwait and Pakistan to maximize the recovery of the better quality water in saline aquifers.

9.4.1 Influential Factors of Saline Water Upconing

The aim of this part of research is to integrate the simulation results in order to optimize the design of scavenger wells for the maximum water recovery values for a controlled saline water upconing. Over 100 numerical experiments were used to determine the parameters that influence the design of scavenger wells to optimize the water recovery ratios. The following variables seem to have a negligible or a minor effect on the upconing process of saline water upconing:

- Longitudinal dispersivity
- Fluid density
- Aquifer porosity
- Initial water level
- Total pumping rate

The following parameters have the most influence on the saline water upconing:

- $\mathbf{S_T}$ is the depth to the top of the freshwater screen from the top of the aquifer [L]
- $\mathbf{I_{0.5}}$ is the initial depth to 0.5 isochlor from the top of the aquifer [L]
- $\mathbf{Q_{TOT}/LS}$ ratio, where $\mathbf{Q_{TOT}}$ is the total pumping rate of freshwater and saline water pumps. **LS** is the total length of well screens [L]
- **B** is effective aquifer thickness [L]
- $\boldsymbol{\alpha_T}$ is transverse dispersivity [L]
- **Kv** is the vertical hydraulic conductivity of the aquifer [L/T]
- **Kh** is the horizontal conductivity of the aquifer [L/T]
- Anisotropy ratio (**Kh/Kv**)

For practical reasons, it is important to decide before constructing the scavenger wells, that the thickness of fresh/brackish water lens is enough to get the desired recovery ratio. In this study, the field operations supported by simulation results in the Sindh province case study show that for a freshwater lens thickness of 28 m, the recovery ratio is 45%. If, for an example the economic situation suggests that the desired value of recovery ratio should not be less than 45%, then the lens thickness should be more than 28 m. If the simulation results do not guarantee this value of freshwater recovery, then operating a scavenger well for these conditions is not economically feasible. The situation in brackish/saltwater aquifers such as the aquifer in the Jaber Al-Ahmad case study is different. The water of the aquifer is initially mixed between different qualities in a wide transition zone of nearly 17 m. The scavenger wells pumping is operated under different design parameters so that the salinity of the pumped water does not exceed certain limits.

9.4.2 *Optimal Freshwater Recovery Ratio in the Sindh Province Case Study in Pakistan*

The simulation results show that (Fig. 9.9) the recovery ratio of freshwater was always improved if the screen of the freshwater pump was located in the upper part of the aquifer. Figure 9.9 shows that the optimal situation (80% of freshwater recovery) is to place the freshwater screen just below the clay layer in the top of the aquifer for the following conditions:

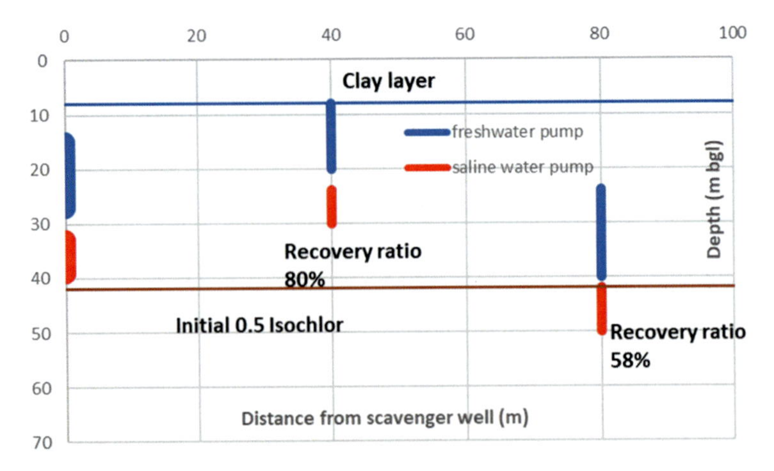

Fig. 9.9 Settings of optimal freshwater recovery ratio for Sindh province case study

- Freshwater pumping rate = 3324 m^3/day
- Saline water pumping rate = 3324 m^3/day
- Screen length in the freshwater zone = 20 m
- Screen length in the freshwater zone = 5 m
- Screen top of the freshwater screen at 8 m bgl
- Screen bottom of the saline water screen at 32.6 m bgl
- Location of 0.5 isochlor follows the field conditions at 42 m bgl

9.4.3 Optimal Brackish Water Recovery Ratio in Jaber Al-Ahmad Case Study in Kuwait

The above analysis indicates that the optimal brackish groundwater recovery ratio in Kuwait saline-brackish aquifers at Jaber Al-Ahmad is 55% for:

- Brackish water pumping rate = 688 m^3/day
- Saline water pumping rate = 562 m^3/day
- Screen length in the brackish water zone = 12 m
- Screen length in the freshwater zone = 10 m
- Screen top of the brackish water screen at 5 m bgl
- Screen bottom of the saline water screen at 27 m bgl
- Location of 0.5 isochlor follows the field conditions at 28 m bgl

It should be noted that the total pumping rate was reduced to 1000 m^3/day and then increased to 2000 m^3/day, but there was a very minor change in the results of the recovery ratio. For hydrogeological and practical conditions and considerations, 1250 m^3/day is considered the optimal pumping rate for the Jaber Al-Ahmad case study. It can be shown that the optimum average salinity of the pumped water is 6257 mg/l realizing that the 0.0 and 1.0 isochlors for Jaber Al-Ahmad are 3200 and 10,000 mg/l respectively.

9.5 Conclusions and Recommendations

The general behaviour of the transition zone under scavenger pumping works almost in the same direction for fresh/saline water and brackish/saline water aquifers under different hydrogeological conditions. The only differences are to get a wider steady state transition zone and less recovery ratio in brackish/saline water aquifers than fresh/saline water aquifers. All that was due to heavier brackish water that is distributed over a longer vertical profile in the brackish/saline water aquifers. As a general conclusion, the transition zone at equilibrium is thicker radially away from the scavenger well than it is at the well itself. In addition, the steady-state transition zone is asymmetric around the 0.5 isochlor. The narrowing/widening mechanism of the transition zone is controlled by fluid velocity, dispersivity and flow direction. The upconing mechanism is fast in the first period of simulation. Steady-state status under scavenger well pumping is normally approached in a period of 2–3 months. The following variables seem to have a negligible or a minor effect on the upconing process of saline water upconing under scavenger pumping: longitudinal dispersivity, fluid density, aquifer porosity, initial water level, aquifer recharge and total pumping rate from both freshwater and saline water pumps. The most influential parameters that control the saline water upconing under scavenger wells pumping are: depth to top of freshwater screen, initial depth to 0.5 isochlor, combination of total pumping rate and screens length, effective aquifer thickness, transverse dispersivity, vertical and horizontal hydraulic conductivities of the aquifer and their anisotropy ratio.

The optimal design criteria for scavenger wells have been established and recommended as given here. In order to maximize the recovery ratio, the top of the screen should always be located closer to the static water level in the upper part of the aquifer. The extraction should be carried out from both fresh/brackish and saline water zones simultaneously with more fresh/brackish water extraction than saline water extraction. The saline water extraction (below the initial location of the 0.5 isochlor) is placed to reduce the vertical gradient of the saline water upconing. The optimal operational scheme of scavenger wells in both case studies is 80% of freshwater recovery in the Sindh Province and 55% of brackish water recovery in the Jaber Al-Ahmed case study.

Acknowledgements The author would like to extend his appreciation to the Kuwait Foundation for the Advancement of Sciences (KFAS) for the financial support of this study under grant number PN17-25EM-03. The support of the management of the Kuwait Institute for Scientific Research (KISR) is pivotal in carrying out the various tasks of the study.

References

Aliewi AS (1993) Numerical simulations of the behavior of fresh/saline transition zone around a scavenger well. PhD dissertation, Newcastle University

Aliewi AS (2023) Predicting the behaviour of the salt/fresh-brackish water transition zone during scavenger well pumping: 1. Numerical model development and testing. In: Suleiman MK, Shahid SA (eds) Terrestrial environment and ecosystems of Kuwait-Assessment and Restoration. Springer

Aliewi AS, Al-Khalid A (2017) Numerical simulation of the effect of cyclic pumping on the dispersion of salt water in shallow aquifers. In: Proceedings of the water-society-climate conference, 2–5 October 2017, Hammamat City, vol 5, pp 9–18

Aliewi AS, Mackay R, Jayyousi A, Nasereddin K, Mushtaha A, Yaqubi A (2001) Numerical simulation of the movement of saltwater under skimming and scavenger pumping in the Pleistocene aquifer of Gaza and Jericho areas, Palestine. Transp Porous Media 43:195–212

Aliewi AS, Al-Enezi H, Al-Maheimid I, Al-Kandari J, Al-Haddad A, Al-Qallaf H, Rashid T, Sadeqi D (2020a) Sustainability of brackish groundwater utilization from the Eocene aquifer for oil exploration in central Kuwait. Environ Dev Sustain:1–15. https://doi.org/10.1007/s10668-019-00401-9

Aliewi A, Al-Qallaf H, Rashid T, Al-Odwani A (2020b) Hydraulic evaluation of a dewatering scheme in shallow aquifers in Kuwait. Q J Eng Geol Hydrogeol 53(1):125–136. https://doi.org/10.1144/qjegh2019-044

Aliewi AS, Al-Kandari J, Al-Khalid A, Bhandary H, Al-Qallaf H (2021) Modelling the effect of high level of total dissolved solids (TDS) for the sustainable utilization of brackish groundwater from saline aquifers in Kuwait. Environ Dev Sustain. https://doi.org/10.1007/s10668-020-00670-9

Aliewi A, Bhandary H, Sabarathinam C, Al-Qallaf H (2022) A new modified chloride mass balance approach based on aquifer hydraulic properties and other sources of chloride to assess rainfall recharge in brackish aquifers. Hydrol Process J. https://doi.org/10.1002/hyp.14513

Aslam M, Matsuno Y, Nobumasa H (2016) Performance evaluation of fresh groundwater skimming wells in the Indus basin irrigation system of Pakistan: a selected review. Mimeogr Fac Agric Kindai Univ 50:5–23

Birch RP, Van Wonderen JJ (1990) Exploitation of freshwater lenses in Sindh province, Pakistan. In: Transactions 4th international congress on irrigation and drainage – Rio de Janeiro Brazil, April–May 1990 Q42-R30, pp 445–460

Fader SW (1957) An analysis of contour maps of 1955 water levels with a discussion of saltwater problems in southwestern Louisiana. Water Resources Pamphlet (4), Published by the Department of Conservation, Louisiana Geological Survey and Louisiana Department of Public Works, Baton Rouge, pp 16–27

Fukumori E, Wake A, Laursen E (1986) Modelling two well system in a stratified aquifer. J Water Resour Plan Manag ASCE 112(1):129–141

Groundwater Development Consultant (GDC) and British Geological Survey (BGS) (1990) Scavenger wells and pilot studies summary

HTS and MMP (1965) Lower Indus report. WAPDA, Lahore

Long RA (1965) Feasibility of a scavenger-well system as a solution to the problem of vertical salt-water encroachment. Water Resources Pamphlet No. 15. Published by State of Louisiana Department of Conservation, Geological Survey and Public Department of Public in Cooperation with USGS

Messerschmid C, Aliewi AS (2022) Spatial distribution of groundwater recharge, based on regionalised soil moisture models in Wadi Natuf karst aquifers, Palestine. Hydrol Earth Syst Sci 26(4):1043–1061. https://doi.org/10.5194/hess-26-1043-2022

PDDGS (2017) The irrigation management strategy for irrigated agriculture of Sindh province (Pakistan). Planning and Development Department, Government of Sindh, Pakistan

Pennink JMK (1904) The water-supply system of Amsterdam in the dunes. Koninski Instituut Ingenieur Tijdschrift 1903–1904, The Hague, pp 183–238

Pennink JMK (1905) Investigations for ground-water supplies. In: Transactions American Society of Civil Engineers, International Engineering Congress, 54, Section D, Paper No. 50, pp 169–181

Qureshi A (2009) Managing salinity in the Indus Basin of Pakistan. Int J River Basin Manag 7(2):111–117. https://doi.org/10.1080/15715124.2009.9635373

Saeed M, Bruen M, Asghar M (2002) A review of modeling approaches to simulate saline-upconing under skimming wells. Nord Hydrol 33:165–188. https://doi.org/10.2166/nh.2002.0021

Souza WR, Voss CI (1986) Modelling a regional aquifer containing a narrow transition between freshwater and saltwater using a solute transport simulation: part II – analysis of a coastal aquifer system. In: Proceedings of 9th Salt Water Intrusion Meeting (SWIM), Delft 12–16 May 1986, pp 457–473

Stoner RF, Bakiewicz W (1992) Scavenger wells 1 – historical development. In: Proceedings of 12th Salt Water Intrusion Meeting (SWIM), Barcelona

Voss CI, Souza WR (1986) Modelling a regional aquifer containing a narrow transition zone between freshwater and saltwater using solute transport simulation: part I – theory and methods. In: Proceedings of 9th Saltwater Intrusion Meeting (SWIM), Delft 12–16 May 1986, pp 493–514

Voss GI, Souza WR (1987) Variable density flow and solute transport simulation of regional aquifers containing a narrow freshwater-saltwater transport zone. Water Resour Res 10:1851–1866

Wang FC (1965) Approximate theory for skimming well formation in the Indus Plain of West Pakistan. J Geophys Res 70(20):5055–5063

Wisniewski GM, Lennon GP, Villaume JF, Young CL (1985) Response of a dense fluid under pumping stress. Focus on problems solving. In: Proceedings of 17th Mid-Atlantic industrial waste conference on toxic and hazardous wastes, pp 226–237

Zack A (1988) A well system to recover water from a freshwater-saltwater aquifer in Pureto Rico. USGS, Water Supply No. 2328

Zack A, Munoz-Candelario R (1984) Hydraulic technique for designing scavenger-production well couples to withdraw freshwater from aquifers containing saline water. Project No. OWRT-A-075-PR(1), Geological Survey, San Juan, PR Water Resources Division, 58 pp

Zardari N, Sharif Z, Shirazi S, Yusop Z, Naubi I, Mangrio M (2015) A comparison of current and design operational efficiencies of scavenger wells in lower Indus Basin of Pakistan and possibility of upconing problem. Arab J Geosci 8(10):10.1007/s12517-015-1851-2

Zhou Q, Bear J, Bensabat J (2005) Saltwater upconing and decay beneath a well pumping above an interface zone. Transp Porous Media 61:337–363. https://doi.org/10.1007/s11242-005-0261-4

Part III
Terrestrial Ecosystems and Their Management

Chapter 10
Terrestrial Habitats and Ecosystems of Kuwait

Tareq A. Madouh and Ali M. Quoreshi

Abstract The habitat refers to a broad variety of resources, including biotic and abiotic elements supporting a particular species' ability to survive and adapt within an environmental context associated with attributes of a certain location. The terrestrial habitat natures of arid desert ecosystems are example of how native biota and microflora adapt to such harsh conditions. The arid desert ecosystems are the most widespread terrestrial biome on earth and are distinguished by the severe climatic conditions including prolonged drought, extreme temperatures, low and infrequent rainfall, strong winds and limited biota of flora and fauna structures. The State of Kuwait is characterized by arid desert environment, which mostly consists of flat desert plain with mild undulations. Kuwait has a typical hyper-arid desert climate with two discrete seasons. The Kuwait desert ecosystem shelters a reasonable number of a unique biota that are highly adaptable to extreme environments of various habitat structures. In this publication, focus is given to the terrestrial habitats in the desert ecosystems of Kuwait and their association with different plant species. The review depicted three different ecosystems (desert plain, sand dunes and salt marshes), and among major terrestrial habitats, smooth, rugged and active sand sheets, wadis and depressions, escarpments, barchan dunes, coastal sand dunes and nebkhas, as well as coastal and inland sabkhas are the main components.

Keywords Kuwait ecology · Arid desert ecosystem · Desert biodiversity restoration · Terrestrial desert habitats · Habitats management · Habitat conservation

T. A. Madouh (✉) · A. M. Quoreshi
Desert Agriculture and Ecosystems Program, Environment and Life Sciences Research Center, Kuwait Institute for Scientific Research, Safat, Kuwait
e-mail: tmadouh@kisr.edu.kw; aquoreshi@kisr.edu.kw

M. K. Suleiman, S. A. Shahid (eds.), *Terrestrial Environment and Ecosystems of Kuwait*, https://doi.org/10.1007/978-3-031-46262-7_10

10.1 Introduction

Most desert environments are distinguished by their limited natural resources including water scarcity, and limited food and nutrients resources (Holzapfel 2008). These limiting factors in desert ecosystems are mainly due to poor soil conditions, inadequate soil fertility, shallow and unstable sands, saline soils, deep infiltration of water into alluvial deposits, and high fluctuations of surface and groundwater, all of which play important roles in the development of characteristic desert biodiversity (McGinnies 1979). Therefore, the biota including the natural flora have evolved several adaptation mechanisms enabling them to grow and survive under the harsh and extreme environment conditions of the desert ecosystems (Madouh 2022). The types of natural vegetation in the arid desert ecosystem are mostly medium to small shrubs, dwarf bushes, perennial grasses, herbaceous plants and short-lived annuals that grow abundantly after the rain events during winter seasons. Despite the term, deserts are enormously biodiverse, harbouring innumerous biota species including various microflora communities that are specifically adapted to aridity, and severe, unpredictable environments including prolonged drought, extreme temperature fluctuations, high solar radiation and high salinity (Madouh and Quoreshi 2023).

10.2 Arid Desert Ecosystems of Kuwait

The State of Kuwait occupies a region of about 17,818 km^2 in the northeastern corner of the Arabian Peninsula. The country experiences two discrete seasons, long, dry, and hot summers, and brief winters, which are typical of a hyper-arid desert environment climate. The average maximum temperatures during the summer months (July and August) can fluctuate between 46 and 50 °C with no rainfall. Winters are chilly with occasional rainfall ranging from 110 to 150 mm/annum and average temperatures of 13 °C, sometimes as low as 3 °C (Meteorological Department 2023). The desert ecosystem of Kuwait is characterized by multiple environmental restrictions, including limited rainfall, prolonged drought, high-level of soil salinity, severe temperatures, intense sunlight, frequent dust storms, poor moisture retention, and infertile soil, all of which collectively affect the biodiversity and community structure. Despite these environmental limitations, the native desert biota including flora and fauna are well adjusted and highly adapted to these challenging climatic circumstances (Madouh 2022). Nonetheless, the natural desert ecosystem of Kuwait is currently declining due to multiple abiotic and biotic factors altering the native biodiversity (Madouh 2023). Increased human influences and uncontrolled overgrazing, together with harsh weather conditions in the desert's ecosystem, have significantly affected the loss and decline of the native biota. These circumstances have led to the development of extensive degraded rangelands throughout Kuwait's deserts (Khalaf 1989; Omar 1991; Zaman 1997; Omar et al. 2001; Misak et al. 2002; Al-Dousari et al. 2019; Madouh 2023).

The Kuwait's surface topography can be described as flat desert plain that levels gradually toward desert plateau. Both of these terrains (plains and the plateaus) dominate the major part of the country. Fewer insignificant elevations, wadis, basins, small sand dunes, inland and coastal salt marshes and small offshore islands occur within the terrain desert ecosystems (Halwagy et al. 1982; Halwagy and El-Saadawi 1992; Zaman 1997). The typical arid soils are poor in organic matter, nutrients, and very low in moisture content. These arid soils are comprised of soluble salts of Ca^{2+}, Mg^{2+}, Na^{+}, K^{+}, CO_3^{2-}, HCO_3^{-}, Cl^{-} and SO_4^{2-}. Aridisols is the dominant soil order occupying 70% and Entisols occupying 30% of the survey area of Kuwait (Shahid and Omar 2022). These soils are sandy, gravelly and poorly developed (Halwagy et al. 1982; Brown and Schoknecht 2001). The soil texture is mainly sandy, loamy sand and sandy loam, the soils are slightly alkaline in reaction (pH 8.2–8.3), and with more than 90% sand-sized particles (KISR 1999).

10.3 Terrestrial Habitats and Desert Ecosystems of Kuwait

The terrestrial desert habitats encompass diversified flora, fauna and microbial communities specifically adapted to the nature and type of habitat. According to Boulos and Al-Dosari (1994), Kuwait's native flora comprises 374 plant species from 55 families spread across a variety of habitats. Annual plants account for 68.4% of the total flora. While just 9.4% of woody shrubs and bushes and 22.2% of herbaceous perennials have been reported. The most recent taxonomic index by Al-Dosari (2021) recorded 452 species associated with 61 families of all native flora and naturalized vascular plants. In addition, the fauna of these various terrestrial habitats consists of a vast number of invertebrates, reptiles, mammals, birds, insects and one amphibian (Khalaf 1989; Amr 2021).

The distribution of native vegetation communities across the desert areas of Kuwait depends entirely on the amount of rainfall, landform and geomorphology, soil characteristics and biotic and abiotic factors. Halwagy and Halwagy (1974) and Halwagy et al. (1982) were the first to recognize Kuwait desert ecosystems and ecological classification based on the variations among habitat, landforms, soil properties, floristic composition, and dominant species. Four distinct ecosystems (sand dunes, salt marshes and salty depressions, desert plains, and desert plateaus) are identified in Kuwait. According to this initial research as well as data from the soil survey of Kuwait conducted between 1996 and 1999, Omar et al. (2007) have reported six ecosystems, (i) coastal plain and lowland, (ii) desert plain and lowland, (iii) alluvial fan, (iv) escarpment, ridges and hilly, (v) wadi and depressions, and (vi) the barchan sand dune. Therefore, the review given below is reported from the current publications regarding the ecological classification of Kuwait's desert ecosystem based on the associated plant communities, landforms and soil characteristics. Nevertheless, the review suggests three main ecological ecosystems and various habitats that exist within each particular ecosystem zone. A wide range of perennial plant species, some of which are more habitat-specific than others and are found in all different types of ecological ecosystems. In addition, numerous important annual plant species are also present in those habitats, mostly throughout the growing season. Only the perennial plants related to each ecological zone and habitat are emphasized in the classification and description given below.

10.4 Classification of Ecological Ecosystems and Habitats

In the following section, various types of ecosystems encountered in Kuwait along with plant species associated with the ecosystems are described.

10.4.1 Desert Plain Ecosystem

The desert plain ecosystem covers the majority of Kuwait and stretches over from the country's south-westerly borders to its northeastern region. In the desert plain ecosystem, there are a range of soil types and three different types of aeolian sand accumulations, i.e., sand drifts, sand sheets and wadi fill (Halwagy and Halwagy 1974; Khalaf et al. 1984; Khalaf 1989; Omar et al. 2007). Additionally, smooth, rugged and active sand sheets are the three distinct groups identified across the desert plain ecosystem (Khalaf 1989). All of these habitat types support various biota populations and diversity (Fig. 10.1). However, *Rhanterium epapposum, Haloxylon salicornicum, Panicum turgidum* and *Cyperus conglomeratus* are the most

Fig. 10.1 The desert plain ecosystem with the distribution of *Rhanterium epapposum* community

predominant population species in this ecosystem (Halwagy and Halwagy 1974; Omar et al. 2007).

10.4.2 Smooth, Rugged and Active Sand Sheets Habitats

The smooth sand sheets habitats cover a large area of the desert plains ecosystem particularly in the north and southern part of Kuwait's desert (Fig. 10.2). It is a commonly levelled and flat surface with coarse sand particles and granules. The rugged sand sheets are similar in nature to the smooth sand sheets habitat but with rough surface, diverse, denser vegetation and are formed due to sand accumulation around the vegetation making nebkha feature (Fig. 10.3). The active sand sheet habitats are thin smooth layers of mobile sand due to the prevailing north-westerly wind depositing over other surface sediments and are made up of a mixture of residues (Fig. 10.4) (Khalaf 1989; Al-Awadhi et al. 2005). Khalaf et al. (2013) reported thick, rough sand sheets that once dominated the southern part of the country are now thin, smooth, and covered in a thin layer of gravels lag.

Several other perennial plants species are associated with these habitats of desert plain ecosystem such as *Astragalus spinosus, Aeluropus lagopoides, Centropodia forsskalii, Convolvulus oxyphyllus, Fagonia bruguieri, Fagonia glutinosa, Fagonia indica, Gynandriris sisyrinchium, Helianthemum lippii, Moltkiopsis ciliata,*

Fig. 10.2 Smooth sand sheet habitat dominated with *Haloxylon salicornicum*

Fig. 10.3 Rugged sand sheet habitat dominated with *Cyperus conglomeratus* showing nebkha feature

Polycarpaea repens, Pennisetum divisum and *Stipagrostis plumosa* (Halwagy and Halwagy 1974; Halwagy et al. 1982; Omar et al. 2007; Abdullah and Al-Dosari 2022).

10.4.3 Wadis and Depressions Habitats

One of the most noticeable desert landforms are wadis and depressions showing physiographic inconsistencies causing parallel alterations in plant species distribution (Kassas and Girgis 1964). Several wadis occur across the deserts of Kuwait including Al-Awjah, Wadi Al-Batin and Al-Seer. Among all wadis, Wadi Al-Batin is considered the country's largest wadi, which delineates the northwest boundary with Iraq. Its breadth is 8–11 km, and depth of 50 m or so below the adjacent plateau's elevation (Misak et al. 2002). Um Ar-Rimam is one of the more significant desert depressions and it is situated adjacent to the northern region of Kuwait Bay at Sabah Al-Ahmad Nature Reserve (SAANR). The desert depressions commonly provide favourable conditions for diverse native desert plants. Shallow depressions (Khubrat), are also found in the desert ecosystem of Kuwait, which are formed during the heavy rains as the water accumulates forming water ponds. The Kuwait desert contains a number of shallow depressions, including Rawdatain, Um Al-Aish,

Fig. 10.4 Dynamics of active sand sheets habitat actions

Khubrat Al-Awazem, Um Sidra, Khairan, Khubrat Hoshan along with other minor Khubrat.

A wide range of plant species diversity is associated with the wadi and depressions habitats. Some of the perennial plant species observed in Wadi Al-Batin include *A. spinosus*, *Astragalus sieberi, H. salicornicum* and *Zilla spinosa*. In Um Ar-Rimam as well as the other depressions habitats, the perennial species *Calligonum polygonoides, Convolvulus pilosellifolius, G. sisyrinchium, H. lippii* and *Ochradenus baccatus* are seen (Fig. 10.5) as reported by (Halwagy and El-Saadawi 1992; Omar et al. 2007; Abdullah and Al-Dosari 2022). Halwagy and El-Saadawi (1992) presented a detailed description of the floristic composition of Um Ar-Rimam depression and identified the following perennials: *C. forsskalii, Citrullus colocynthis, Cynodon dactylon, Fagonia glutinosa, Lycium shawii, M. ciliata, Onobrychis ptolemaica, P. divisum, P. repens, R. epapposum* and *S. plumosa* (Fig. 10.6a, b).

10.4.4 Escarpments Habitats

An *escarpment* is a steep slope or long cliff that forms as a result of faulting or erosion and separates two relatively level areas of different elevations. Escarpments frequently produce distinctive and varied desert landscapes. The Jal Az-Zor

Fig. 10.5 Variety of desert plant species occupying the Um Ar Rimam depression

a) A. spinosus b) *P. divisum*

Fig. 10.6 *A. spinosus* in Wadi Al-Batin habitat (**a**) and *P. divisum* in Um Ar-Rimam habitat (**b**)

Fig. 10.7 Jal Az Zor escarpment landscape

escarpment runs parallel to Kuwait Bay's northern shoreline (Fig. 10.7). Al-Liyah, Ahmadi and As-Sabriyah are among the numerous ridges that are found in Kuwait's desert. The Jal-Az-Zor escarpment is approximately 60 km long with the highest point of 145 m (above sea level).

The primary wadi system is parallel to the trend of the escarpments and ridges (Al-Sarawi 1995). These habitats are commonly dominated by perennials such as *Anabasis setifera, A. spinosus, Halothamnus iraqensis* and *L. shawii* (Omar et al. 2007). A mixed stand of sparse annual vegetation occurs during the rainy season of the winter months (Fig. 10.8a, b).

10.4.5 Sand Dunes Ecosystem

In the following section, different dunes recorded in Kuwait are described along with associated diversity of plant species.

10.4.5.1 Barchan Dunes, Coastal Sand Dunes and Nebkhas Habitats

The formation of sand dunes including barchan dunes, coastal sand dunes and nebkhas habitats is largely influenced by the wind speed and direction in the arid regions. The barchan dunes are usually large in size, crescent shape, vary in size, and are connected to mobile sand movements (Fig. 10.9) (Al-Sarawi 1995; Khalaf

Fig. 10.8 The vegetation cover at Jal Az Zor Escarpment habitat (**a**) and Al-Liyah ridge habitat (**b**)

Fig. 10.9 The barchan sand dune formation without vegetation

1989). They are generated by a dominating wind direction, but if the wind direction changes, they can change their shape and move over the desert terrain. Two major strips of barchan sand dunes are found in Kuwait's desert, one near Um Niqa and the other at Al-Huwaymiliah. These dunes are typically devoid of vegetation, and they extend in a northwest-southeast direction (Khalaf 1989). The term "nebkha" refers to the immobile dunes that form a ring around bushes and are dispersed over the northeast to southeast coast. Nebkhas are a key factor in preventing land degradation and have a significant influence in the dynamics of the vegetation species variety (Fig. 10.10). The coastal dunes are typically pyramid-shaped accumulation of sand created by deposition around discrete vegetation clusters (Al-Sarawi 1995). These coastal sand dunes are commonly located along the southern and northern coastal strips, and are stabilized sand dunes (Fig. 10.11).

The perennial vegetation associated with the barchan dunes include *H. salicornicum, M. ciliata* and *P. repens* (Omar et al. 2007; Abdullah and Al-Dosari 2022). The coastal sand dunes are occupied by the perennials *Atriplex leucoclada, Seidlitzia rosmarinus, H. iraqensis, L. shawii, Nitraria retusa, P. divisum* and *Zygophyllum qatarense* (Halwagy and Halwagy 1974). The desert nebkhas are formed around *A. spinosus, C. colocynthis, C. conglomeratus, H. salicornicum, L. shawii, R. epapposum* and *P. turgidum, Halocnemum strobilaceum, N. retusa* and *Tamarix aucheriana* plants species form coastal nebkhas (Al-Dousari et al. 2008).

Fig. 10.10 Desert nebkhas formed around *H. salicornicum*

Fig. 10.11 The formation of coastal sand dune dominated by *L. shawii*

10.4.6 Salt Marshes Ecosystem

In this section, various features of coastal and inland sabkhas are presented and discussed. Sabkhas are salt-scalds (rich in salts) formed at the coastal areas due to seawater intrusion and subsequent evaporation. The sabkhas are either devoid of vegetation due to very high salt contents not supporting the plants or salt-loving plants "halophytes" are growing (Omar et al. 2002). The inland sabkhas are developed due to water accumulation in low-lying areas and evaporation leaving the salts at surface, these sabkhas are commonly underlain by dense layer "gatch" impeding downward water movement.

10.4.6.1 Coastal Sabkhas and Inland Sabkhas Habitats

Salt marshes are one of the Kuwait's most prevalent terrain formations, known locally as a "Sabkha". These salt marshes are salt flats or shallow depressions covering around 10% of the country's total area. On the basis of their incidence and location, they are divided into coastal sabkha and inland sabkha. Among the most noticeable surface topographical features in the Kuwaiti coastal zone is coastal sabhka, which often occurs around sea level or at the level of subsurface water (Kleo and Al-Otaibi 2011). It is seen to be present along the southern coastal plain (Julayah_Az-Zour strip) as well as the northern coastal plain, which encompasses Khuwaisat, Khor As Sabiyah, Doha, Ad-Dhubaiyah, Al-Khiran, Bubiyan Island, Warba Island, and portions of Faylaka Island. Due to tidal action, the area tends to be inundated with seawater, making the soil hypersaline and forming marine-based sabkha. Inland sabkha, on the other hand, occasionally flooded due to rain or when runoff water builds up in the depressions. Some of these salt marshes are considered as natural reserve specifically Khuwaisat (Al-Jahra Natural Reserve) which is Kuwait's the largest salt marsh occupying an area of 18.00 km^2 and completely protected to conserve the native and adapted biodiversity (Fig. 10.12). Nevertheless, Al-Hurban and Gharib (2004) indicated that under the influence of the predominant north-westerly wind, both coastal and inland sabkhas are vulnerable to mobile sand encroachment. Both types of sabkhas can be categorized as belonging to the gravelly to slightly gravelly muddy sand texture classification.

Coastal sabkhas are generally dominated by the halophytic perennials (*A. lagopoides, Aeluropus littoralis, Cornulaca monacantha, Cressa cretica, H. strobilaceum, Juncus rigidus, N. retusa, S. rosmarinus, T. aucheriana* and *Z. qatarense*) (Fig. 10.12) as reported by various scientists (Halwagy and Halwagy 1977; Omar et al. 2002; El-Ghareeb et al. 2006; Abdullah and Al-Dosari 2022). In Khuwaisat salt marshes, *Phragmites australis* is the dominant plant species along with several other halophytic plants such as *A. littoralis, C. cretica, J. rigidus, N. retusa* and *Tamarix aucheriana* (Fig. 10.13) (Halwagy and Halwagy 1977; Al-Yamani 2021).

Fig. 10.12 Khuwaisat salt marsh habitat with adapted salt-tolerant plants

Fig. 10.13 The halophyte *Tamarix aucheriana* in the coastal sabkha

10.5 Management and Conservation of Terrestrial Habitats

Terrestrial habitats in the desert ecosystem support diverse flora and fauna. The diversity and structure of the vegetation are crucial in preserving the ecosystem's biodiversity (Bidak et al. 2015). The fragile desert ecosystem is persistently subjected to a variety of abiotic and biotic environmental stresses. The severity of drought conditions, high temperatures and increased salinity as well as uncontrolled human activity may possibly have a significant impact on the habitat diversity and vegetation structure. Despite the abiotic factor stresses affecting the desert ecosystem of Kuwait, human activities are considered to be the main contributor to habitat degradation and vegetation decline. Unrestricted rangeland grazing, seasonal recreational activities, quarries and landfills, and the military occupation by Iraq, all contributed considerably to the degradation of the country's terrestrial ecosystems (Misak et al. 2002; Al-Awadhi et al. 2003; Kalander et al. 2021).

Considering the above threats, there is a need to develop habitat and ecosystem-based management and conservation plan to protect biodiversity. It is also essential to create restoration plans for severely damaged areas to halt additional habitat loss. Actions include determining the primary cause of habitat degradation and giving priority to those species that are rare or at high danger of extinction. Additional plans should include in-situ and ex-situ conservation strategies for improved management. Seed collection and seed bank storage are highly important for conservation of native plant materials using living resources. More research and modelling studies to enhance conservation efforts and monitor habitats performance are suggested to be conducted.

10.6 Conclusions and Recommendations

The distinctive terrestrial desert habitats of Kuwait's arid desert ecosystem are home to a diverse range of living organisms, including flora, fauna and microbial communities that are well adapted to aridity. The desert ecosystem of Kuwait encompasses several topography and terrain systems that commonly determine the desert habitat types to support the characteristic vegetation communities and diversity. According to the literature and previous studies, there does not seem to be a distinct difference between habitat structure and ecosystem function. The majority of the literature has not focused on quantifying and mapping the ecosystem and habitat services, as our review demonstrates the difference between the dynamics of ecosystems and habitats role. However, the review presented in this chapter describes three main ecological ecosystems and various habitats that are found inside each individual ecosystem zone. These ecological ecosystems are desert plain, sand dunes and salt marshes and the habitats include smooth, rugged and active sand sheets, wadis and depressions, escarpments, barchan dunes, coastal sand dunes and nebkhas, coastal and inland sabkhas. Further investigation is urgently required to categorize the biological ecosystems according to their functionality and the habitats they support. The ecological service assessments in our opinion are essential components to gain more traction in the fields of science, policy and decision-making steps.

Acknowledgements The Kuwait Institute for Scientific Research (KISR) provided the resources and help needed to acquire the information for this review chapter, which the author gratefully acknowledges. The Kuwait Foundation for the Advancement of Sciences (KFAS) is acknowledged for supporting the scientific research. The authors would like to express their gratitude to Mrs. Jasmine P. Sali for formatting, editing, and preparing the figures as well as Eng. Harby Tawfeeq for his technical input. Both are colleagues at the Desert Agriculture and Ecosystems Program of Environment and Life Sciences Research Center, Kuwait Institute for Scientific Research Kuwait.

References

Abdullah MT, Al-Dosari ME (2022) Vegetation of the state of Kuwait. IUCN/Environment Public Authority, Gland/Kuwait City

Al-Awadhi JM, Misak RF, Omar SS (2003) Causes and consequences of desertification in Kuwait: a case study of land degradation. Bull Eng Geol Environ 62:107–115

Al-Awadhi JM, Al-Helal A, Al-Enezi A (2005) Sand drift potential in the desert of Kuwait. J Arid Environ 63(2):425–438. https://doi.org/10.1016/j.jaridenv.2005.03.011

Al-Dosari M (2021) Revision of the flora of Kuwait, nomenclature and revisiting the taxonomy of *Arnebia* (Boraginaceae). Dissertation, Arabian Gulf University, Bahrain

Al-Dousari AM, Ahmed MO, Al-Senafy M, Al-Mutairi M (2008) Characteristics of nabkhas in relation to dominant perennial plant species in Kuwait. Kuwait J Sci 35(1A):129

Al-Dousari AM, Alsaleh A, Ahmed M, Misak R, Al-Dousari N, Al-Shatti F, Elrawi M, Willaim T (2019) Off-road vehicle tracks and grazing points in relation to soil compaction and land degradation. Earth Syst Environ 3(3):471–482

Al-Hurban A, Gharib I (2004) Geomorphological and sedimentological characteristics of coastal and inland sabkhas, Southern Kuwait. J Arid Environ 58(1):59–85. https://doi.org/10.1016/S0140-1963(03)00128-9

Al-Sarawi MA (1995) Surface geomorphology of Kuwait. GeoJournal 35:493–503. https://doi.org/10.1007/BF00824363

Al-Yamani YF (2021) Fathoming the northwestern Arabian gulf: oceanography and marine biology, 1st edn. Kuwait Institute for Scientific Research, Kuwait City

Amr ZS (2021) The state of biodiversity in Kuwait. IUCN/Environmental Public Authority, Gland/Kuwait City

Bidak LM, Kamal SA, Halmy MW, Heneidy SZ (2015) Goods and services provided by native plants in desert ecosystems: examples from the northwestern coastal desert of Egypt. Glob Ecol Conserv 3:433–447. https://doi.org/10.1016/j.gecco.2015.02.001

Boulos L, Al-Dosari M (1994) Checklist of the flora of Kuwait. Journal of the University of Kuwait (Science) 21:203–217

Brown G, Schoknecht N (2001) Off-road vehicles and vegetation patterning in a degraded desert ecosystem in Kuwait. J Arid Environ 49(2):413–427. https://doi.org/10.1006/jare.2000.0772

El-Ghareeb RM, El-Sheikh MA, Testi A (2006) Diversity of plant communities in coastal salt marshes habitat in Kuwait. Rendiconti Lincei 17:311–331

Halwagy M, El-Saadawi W (1992) Drought and changes in the bryoflora and angiosperm flora in Kuwait in the years 1974–1990. Acta Bot Neerlandica 41(2):183–195. https://doi.org/10.1111/j.1438-8677.1992.tb00497.x

Halwagy R, Halwagy M (1974) Ecological studies on the desert of Kuwait. II. The vegetation. J Univ Kuwait (Sci) 1:87–95

Halwagy R, Halwagy M (1977) Ecological studies on the desert of Kuwait. III. The vegetation of the coastal salt marshes. J Univ Kuwait (Sci) 4:33–74

Halwagy R, Moustafa AF, Kamel SM (1982) On the ecology of the desert vegetation in Kuwait. J Arid Environ 5(2):95–107. https://doi.org/10.1016/S0140-1963(18)31543-X

Holzapfel C (2008) Deserts. In: Jørgensen SE, Fath BD (eds) Encyclopedia of ecology. Academic, London, pp 879–898. https://doi.org/10.1016/B978-008045405-4.00326-8

Kalander E, Abdullah MM, Al-Bakri J (2021) The impact of different types of hydrocarbon disturbance on the resiliency of native desert vegetation in a war-affected area: a case study from the State of Kuwait. Plants 10(9):1945. https://doi.org/10.3390/plants10091945

Kassas M, Girgis WA (1964) Habitat and plant communities in the Egyptian desert. V. The limestone plateau. J Ecol 52(1):107–119

Khalaf FI (1989) Desertification and aeolian processes in the Kuwait desert. J Arid Environ 6(2):125–145. https://doi.org/10.1016/S0140-1963(18)31020-6

Khalaf FI, Gharib IM, Al-Hashash MZ (1984) Types and characteristics of the recent surface deposits of Kuwait, Arabian Gulf. J Arid Environ 7(1):9–16. https://doi.org/10.1016/S0140-1963(18)31399-5

Khalaf FI, Al-Awadhi J, Misak RF (2013) Land-use planning for controlling land degradation in Kuwait. In: Shahid SA, Taha FK, Abdelfattah MA (eds) Developments in soil classification, land use planning and policy implications. Springer, Dordrecht, pp 669–689

KISR (1999) Soil survey for the state of Kuwait. AACM International, Adelaide

Kleo AA, Al-Otaibi O (2011) The sustainable development of Kuwaiti Sabkhas. Domes 20(1):27–49. https://doi.org/10.1111/j.1949-3606.2011.00064.x

Madouh TA (2022) Eco-physiological responses of native desert plant species to drought and nutritional levels: case of Kuwait. Front Environ Sci 10:Article 785517. https://doi.org/10.3389/fenvs.2022.785517

Madouh TA (2023) The influence of induced drought stress on germination of *Cenchrus ciliaris* L. and *Cenchrus setigerus* Vahl.: implications for rangeland restoration in the arid desert environment of Kuwait. Research in Ecology 5(1):1–11. https://doi.org/10.30564/re.v5i1.5426

Madouh TA, Quoreshi AM (2023) The function of arbuscular mycorrhizal fungi associated with drought stress resistance in native plants of arid desert ecosystems: a review. Diversity 15(3):391. https://doi.org/10.3390/d15030391

McGinnies WG (1979) General description of desert areas. In: Goodall DW, Perry RA (eds) Arid land ecosystems: structure, functioning and management, vol 1. Cambridge University Press, Cambridge, pp 5–21

Meteorological Department, Directorate General of Civil Aviation, Kuwait (2023). https://www.met.gov.kw/Climate/climate.php?lang=eng#Annual_Report

Misak RF, Al-Awadhi JM, Omar SA, Shahid SA (2002) Soil degradation in Kabd area, southwestern Kuwait City. Land Degrad Dev 13(5):403–415. https://doi.org/10.1002/ldr.522

Omar SAS (1991) Dynamics of range plants following 10 years of protection in arid rangelands of Kuwait. J Arid Environ 21(1):99–111. https://doi.org/10.1016/S0140-1963(18)30732-8

Omar SA, Misak R, King P, Shahid SA, Abo-Rizq H, Grealish G, Roy W (2001) Mapping the vegetation of Kuwait through reconnaissance soil survey. J Arid Environ 48(3):341–355. https://doi.org/10.1006/jare.2000.0740

Omar S, Misak R, Shahid SA (2002) Sabkhat and halophytes in Kuwait. In: Barth HJ, Boer B (eds) Sabkha ecosystems: volume I: the Arabian peninsula and adjacent countries. Springer Science & Business Media, pp 71–81

Omar SAS, Al-Mutawa Y, Zaman S (2007) Vegetation of Kuwait, 2nd edn. Kuwait Institute for Scientific Research, Kuwait City

Shahid SA, Omar SA (2022) Kuwait soil taxonomy. Springer

Zaman S (1997) Effects of rainfall and grazing on vegetation yield and cover of two arid rangelands in Kuwait. Environ Conserv 24(4):344–350. https://doi.org/10.1017/S0376892997000453

Chapter 11
Native Vegetation and Flora of Kuwait

Arvind Bhatt

Abstract Native vegetation is the key component of any ecosystem, providing multiple direct and indirect benefits. However, the native vegetation is under tremendous pressure throughout the Arabian Peninsula due to anthropogenic and natural factors. Therefore, maintaining native vegetation is a prime concern in order to maintain ecosystem sustainability. An overview of species diversity, vegetation communities, and habitat diversity of Kuwait is presented and discussed. Furthermore, efforts are being made to document the use of native vegetation of Kuwait through a literature survey and prioritizing them based on their use value. Although Kuwait has quite small number of native species, these species offer valuable gene pool. Among the eight dominant vegetation communities, *Stipagrostietum* community has the highest contribution. A total of six habitats (ecosystems) were identified based on the floristic composition, soil characteristics, and landform, and each of these habitats was reported to be supporting unique species. Six species were classified as the high priority species followed by 23 species under moderate priority. About 152 species occurred only in a single habitat, whereas 221 species occupied 2–3 habitats. The inclusive information about Kuwait vegetation could also be helpful for better understanding and appreciating their various use values that could be used by diversified stakeholders as well as policymakers.

Keywords Adaptation · Biodiversity · Conservation · Habitat · Native species · Use value · Prioritization · Salinity · Sustainability · Water resources

A. Bhatt (✉)
Desert Agriculture and Ecosystems Program, Environment and Life Sciences Research Center, Kuwait Institute for Scientific Research, Safat, Kuwait
e-mail: drbhatt79@gmail.com

M. K. Suleiman, S. A. Shahid (eds.), *Terrestrial Environment and Ecosystems of Kuwait*, https://doi.org/10.1007/978-3-031-46262-7_11

11.1 Introduction

Kuwait, covering an area of 17,818 km^2 in the Arabian Peninsula, is one of the smallest countries in the world. Climate of Kuwait is characterized by lengthy, dry, and hot summers (May to October, when temperatures can exceed 50°C) and mild winters (November to February). The precipitation is infrequent and unpredictable (averaging 114 mm annually), concentrated mainly during winter (Omar et al. 2007). Topography of Kuwait is mostly flat with moderately rolling plains, fragmented by small hills, scarps, wadies of ephemeral streams, and shallow but wide inland depressions (Hassan et al. 2021). Al-Az Zor escarpment, Wadi Al-Batin, and Ahmadi Ridge are the major topographical features in Kuwait. Soils of Kuwait are mostly sandy in texture with high quantities of calcareous materials and limited organic matter and low nutrients. Moreover, soil accumulation is shallow and overlain with gravels of various sizes. *Gatch* (subsurface hard setting) layer occurs in several soils of Kuwait, which is identified as a calcic and/or gypsic pan (Omar and Shahid 2013; Shahid and Omar 2022). Soils of Kuwait have been categorized into two orders (i) *Aridisols (Haplocalcids, Petrocalcids, Haplogypsids, Calcigypsids, Petrogypsids*, and *Aquisalids*) and (ii) Entisols (*Torriorthents* and *Torripsamments*) (Omar and Shahid 2013; Shahid and Omar 2022).

Conservation of biological resources especially native plant diversity is vital because they are unique and irreplaceable resources for the future. Therefore, immediate conservation initiatives are required to promote the natural resource management for preserving and sustaining biodiversity in species, ecosystems, human culture, and society. Increasing population pressure, urbanization, overexploitation, and unsustainable utilization of natural resources are severely threatening the sustainability of plant species globally at different levels (Stévart et al. 2019; Wassie 2020). Moreover, other abiotic factors such as drought, high temperature, low and erratic rain, wind, and salinity also play critical roles in limiting plant growth and development in desert environmental conditions, which ultimately leads to low biodiversity and spatial heterogeneity (Almedeij 2014). Moreover, Kuwait is highly vulnerable to the impacts of climate change (FAO 2006), which may further enhance the negative impact on biodiversity conservation. Therefore, identification and prioritization of species and habitats is imperative to optimize the efficacy of conservation action. However, conservation and sustainable utilization of valuable native plants in Kuwait remains as one of the biggest challenges and thus, immediate conservation initiatives are needed.

Prioritizing species for conservation required detailed knowledge of species status (i.e., distribution, population size, and trends) and threats (e.g., habitat loss or fragmentation) (Mace and Lande 1991). Kuwait developed a national biodiversity strategy after signing the Convention on Biological Diversity (CBD) in 1993 (CBD 1993) in order to promote biodiversity conservation. Environment Public Authority of Kuwait (EPA) in co-operation with IUCN Regional Office for West Asia (ROWA) also made efforts to document and analyze the biodiversity of Kuwait and recommend that continuous long-term monitoring should be initiated and implemented immediately for evaluating species status periodically (Amr 2021; Abdullah and Al-Dosari 2022). However, until now, limited efforts have been made for identifying the rare and threatened species at national level in Kuwait. For example, 19 species of vascular plants have been categorized as rare in Kuwait (Alsdirawi and Faraj 2004). Recently, a total of 121 plant species of Kuwait have been categorized under different threat categories based on herbarium specimens preserved at Kuwait University (Nassep 2022). Knowledge of baseline data (distribution, ecology, population biology, values, and threat) is key for determining the species status. Therefore, immediate attention is required to conduct the species survey and long-term monitoring plan to collect more authentic baseline data for prioritizing the species for biodiversity conservation at national as well as regional level.

11.2 Species Diversity and Vegetation Communities

Generally, species diversity depends on the type of ecosystem (Özyavuz et al. 2013) and it is eroding continuously globally due to habitat loss and climate change (Pimm 2008). Therefore, it is imperative to increase the understanding about the species diversity and distribution in order to develop effective conservation planning (Margules and Pressey 2000; Whittaker et al. 2005).

Native plants are indigenous species, evolved naturally under the local environmental conditions (Miller et al. 1996). Approximately, 374 plant species from 55 families have been recorded from Kuwait (Boulos and Al-Dosari 1994). Of 374 species, 256 species are annuals followed by herbaceous perennials (83 species), shrubs (34 species), and trees (1 species) (Boulos and Al-Dosari 1994). However, recently Abdullah (2017) recorded 402 species and Al-Dosari (2021) recorded a total of 452 species from 61 families including the cultivated or naturalized species. Although Kuwait has relatively small number of species, these species offer valuable gene pool for drought and salt-tolerance research (Omar and Bhat 2008). Besides, these species provide various economical (i.e., medicinal, fodder, edible, ornamental) and ecological (soil stabilization, control erosion, improve soil fertility by fixing nitrogen, i.e., legumes, provide shading, etc.) services. Nonetheless, the high anthropogenic pressure together with extreme climatic conditions is severely threatening the Kuwait flora. Therefore, immediate priority should be given to native species conservation. Vegetation composition and covers is typically sparse and discontinuous, but can be remarkably high, depending on locality, especially during winter (Fig. 11.1).

In the past, Kuwait vegetation was classified into four plant communities (i.e., *Haloxyletum, Rhanterietum, Zygophyletum,* and *Cyperetum*) (Dickson 1955). Later on *Panicum turgidum* was included (Kernick 1966; Halwagy and Halwagy 1974). However, Omar et al. (2001, 2007) further updated and classified Kuwait's native vegetation into eight dominant communities (i.e., *Haloxyletum, Rhanterietum, Cyperetum, Stipagrostietum, Zygophyletum, Centropodietum, Halophyletum,* and *Panicetum*) (Fig. 11.2). Among these communities, *Stipagrostietum* had the highest

Fig. 11.1 Overview of flora of Kuwait

contribution and *Zygophyletum* the lowest contribution (Table 11.1). Substantial variations in vegetation communities of Kuwait have been observed since 1974 (Omar and Bhat 2008). However, massive damage to the vegetation was observed during Gulf War (Böer and Sargeant 1998). A recent study on vegetation communities of Kuwait has shown a declining trend in *Haloxyletum* and *Rhanterietum*, while an increasing trend is recorded in *Cyperetum* and *Stipagrostietum* communities (Abdullah 2017).

Fig. 11.2 Dominant native species in Kuwait (**a**) *Haloxylon salicornicum*; (**b**) *Nitraria retusa*; (**c**) *Cyperus conglomerates*; (**d**) *Panicum turgidum*; (**e**) *Rhanterium epapposum*; (**f**) *Centropodia forsskalii*; (**g**) *Stipagrostis plumose*; (**h**) *Zygophyllum qatarense*

Table 11.1 Contribution of dominant vegetation communities

Community	Area (km^2)	Contribution (%)
Haloxyleum	3696	23.94
Rhanterietum	338	2.18
Cyperetum	4381	28.37
Stipagrostietum	6407	41.50
Zygophyletum	41	0.27
Centropodietum	160	1.04
Halophyletum	114	0.74
Panicetum	303	1.96
Total	15,440	100.00

Source: Omar et al. (2007)

Conservation of genetic variability within a species is essential to conserve biodiversity because the presence of high levels of genetic variability in natural plant populations plays an important role in species evolution as well as persistence of future population. Kuwait developed a national biodiversity strategy and committed to conserving biodiversity at genes, species, and ecosystems levels (CBD 1993). However, so far only *Haloxylon salicornicum* and *Rhanterium eppaposum* have been investigated in terms of accessing the genetic diversity and plant succession in Kuwait (Al-Salameen et al. 2018, 2022; Omar and Bhat 2008). Therefore, more efforts are needed to bridge the knowledge gaps regarding genetic diversity conservation.

11.3 Habitat Diversity

Habitats may differ as they contain unique combination of species, where these species interact among themselves and with their environment in distinctive ways. Therefore, habitat conservation is key to maintaining species diversity. Various factors such as expansion of urban, residential, industrial areas, road construction, oil production, overgrazing, quarries, earlier military operations, off-road driving, camping, and land degradation are the main reasons for habitat loss in Kuwait (Al-Awadhi 2001; Al-Dousari et al. 2019; Omar et al. 2007; Omar and Bhat 2008). Therefore, strict regulations are needed to be enforced to regulate camping, grazing, and other damaging human activities. Initially, four desert habitats (ecosystems), namely, sand dunes, swamps and salt depressions, flat desert, and desert plateau were identified in Kuwait (Halwagy and Halwagy 1974). However, later six habitats (ecosystems) were suggested based on the floristic composition, soil characteristics, and landforms (Table 11.2) (Omar et al. 2007). Each of these habitats (ecosystems) supports unique or dominant plant species together with several associated species (Table 11.2).

Various factors, such as natural and/or anthropogenic disturbances, climate change, exotic species invasions, and inconsistent weather patterns are severely affecting plant communities (Simberloff et al. 2013; Komatsu et al. 2019). Adverse changes in plant communities have several negative impacts on nature and people (Quijas et al. 2010). Therefore, emphasis should be given to maintain the native plant communities to obtain sustainable ecosystem services.

11.4 Value of Native Plants

Native plants, including shrubs, ground covers, and trees, have long been recognized as important and valuable resources that provide various ecosystem services (Costanza et al. 1997). Native plants provide various (i) direct economic benefits such as food, medicine, fodder, fuel, fibers, etc. and (ii) indirect benefits such as soil stabilization, phytoremediation, windbreak, shade, erosion control, water quality,

Table 11.2 Species distribution on different desert ecosystems of Kuwait

Habitat (ecosystem)	Dominant species	Existence
Coastal plain and lowland	*Halocnemum strobilaceum, Tamarix aucheriana, Juncus rigidus, Cressa cretica, Cornulaca leucacantha, Salicornia herbacea, Mesembryanthemum nodiflorum,* and *Spergularia diandra*	Coastal area
Alluvial fan	*Haloxylon salicornicum, Citrullus colocynthis, Astragalus spinosus, Schismus barbatus,* and *Cistanche tubulosa*	Northern, western, central part of Kuwait
Escarpments, ridges, and hilly	*Astragalus spinosus, Lycium shawii, Anabasis setifera,* and *Neurada procumbens*	Jal AzZor
Wadis and depressions	*Ochradenus baccatus, Calligonum comosum, Polygonum patulum,* and *Chrozophora*	Wadi Al-Batin and Wadi Umm Al-Rimam
Sand dune, ridge, and terrace	*Zygophyllum qatarense, Seidlitzia Rosmarinus, Atriplex leucoclada, Nitraria retusa, Lycium shawii, Pennisetum divisum, Cistanche tubulosa, Cornulaca monacantha,* and *Calligonum comosum*	Northern and southern coastal strip
Desert plain	*Cyperus community, Rhanterium community, Haloxylon,* and *Panicum community*	Northeast and southwest part

Source: Halwagy and Halwagy (1974), Daoud and Al-Rawi (1985), Omar et al. (2007)

biodiversity, wildlife habitat, and carbon sequestration (Burger 2009; Zhang et al. 2007; Phondani et al. 2017). However, the information about the goods and services provided by native species of Kuwait still remains sketchy. Besides the direct benefits, native species also provide non-use benefits such as (i) existence (the value people receive from simply knowing a resource exists), (ii) altruism (the value derived from having other contemporaries use of a resource), and (iii) bequest (preserving a resource for future generations) (More et al. 1996). However, we have not considered such non-use benefits of native species in this chapter.

11.5 Species and Habitat Prioritization

The native plants of Kuwait have been listed by Omar et al. (2007). The information on the potential uses of these species was collected from published literature (i.e., journals, textbooks, proceedings, websites, periodicals, and databases) and appropriately cited. Prioritization of species was completed based on the combined score of use value for each species. The use value was computed based on the number of uses each species is associated with [i.e., medicinal, edible, fodder, ornamental, fuel, other purposes (i.e., improving soil fertility, rope making, thatching, handicraft making, provision of pollen to bee and refuge), cosmetics, insecticides, detergent, shade, phytoremediation, and windbreak]. Species with the use value range between 4 and 6 were considered as top priority species, whereas those with a total score of 3 and <3 were considered as moderate and low-priority species, respectively.

Species richness, which is the number of species that were present in any given habitat, was determined to identify the rich habitat in terms of the total number of species. The species were further prioritized based on the number of habitats in which they occurred. The habitat information of each native species is extracted from available literature, Google Scholar, and Web of Knowledge as well as by exploring the local flora. The species that occupied one habitat were given high priority. However, those that occupied 2–3 habitats were considered as moderate and those that occurred in >3 were considered as of low priority.

Kuwait plant diversity consists of 374 species, of which forbs are the dominant component (245 species or 65.51%), followed by grasses (79 species or 21.12%), shrubs (46 species or 12.30%), trees (3 species or 0.80%), and single fern species (0.27%). With regard to the life form, 256 species (73.83%) are annual, 2 species (0.53%) are biennial, and 116 species (31.02%) are perennial (Boulos and Al-Dosari 1994; Omar et al. 2007). These species belong to 55 families and *Poaceae* with 74 species is the most dominant family, followed by *Asteraceae* (46 species), *Brassicaceae* (34 species), *Amaranthaceae* (31 species), and *Fabaceae* (28 species). Moreover, six families also have higher number of species (*Caryophyllaceae* – 20 species, *Euphorbiaceae* – 13 species, *Boraginaceous* – 20 species, *Plantaginaceae* – 8 species, *Solanaceae* and *Lamiaceae* – 6 species each). The remaining families contain < 6 species (Fig. 11.3). Of 374 species, we did not find any information on the use value for 2 species.

In terms of uses, 265 species (71.24%) are utilized for fodder followed by medicinal (140 species 37.63%), ornamental (41 species 11.12%), and other purposes (i.e., improving soil fertility, rope making, thatching, handicraft making, provision of honey and refuge) (29 species 7.80%) (Fig. 11.4). The combined score for the species varied from 0 to 6 (Table 11.3). Of 374 species, 6 species namely *Acacia pachyceras*, *Aeluropus lagopoides*, *Alhagi graecorum*, *Anabasis setifera*, *Bassia muricata*, and *Cressa cretica* scored four or more and thus, are classified as the high priority species for their uses. However, 23 and the rest 343 species had a combined score of 3 and less use value and hence, are categorized as moderate and low-priority species, respectively. Conversely, 2 species, namely *Halophila ovalis* and *Orobanche minor*, scored zero due to lack of information on their use value.

In general, the native species occur in different habitats. The highest number of species occurred in sandy soil (239 species 63.90%) followed by gravel (151 species 40.37%), disturbed area (60 species 16.04%), and silt (52 species 13.90%) (Fig. 11.5). In terms of the number of habitats occupied, 152 species occurred only in a single habitat, whereas, 221 occurred in 2–3 habitats, but only one species, i.e., *Senecio vulgaris* L. was found in 4 habitats.

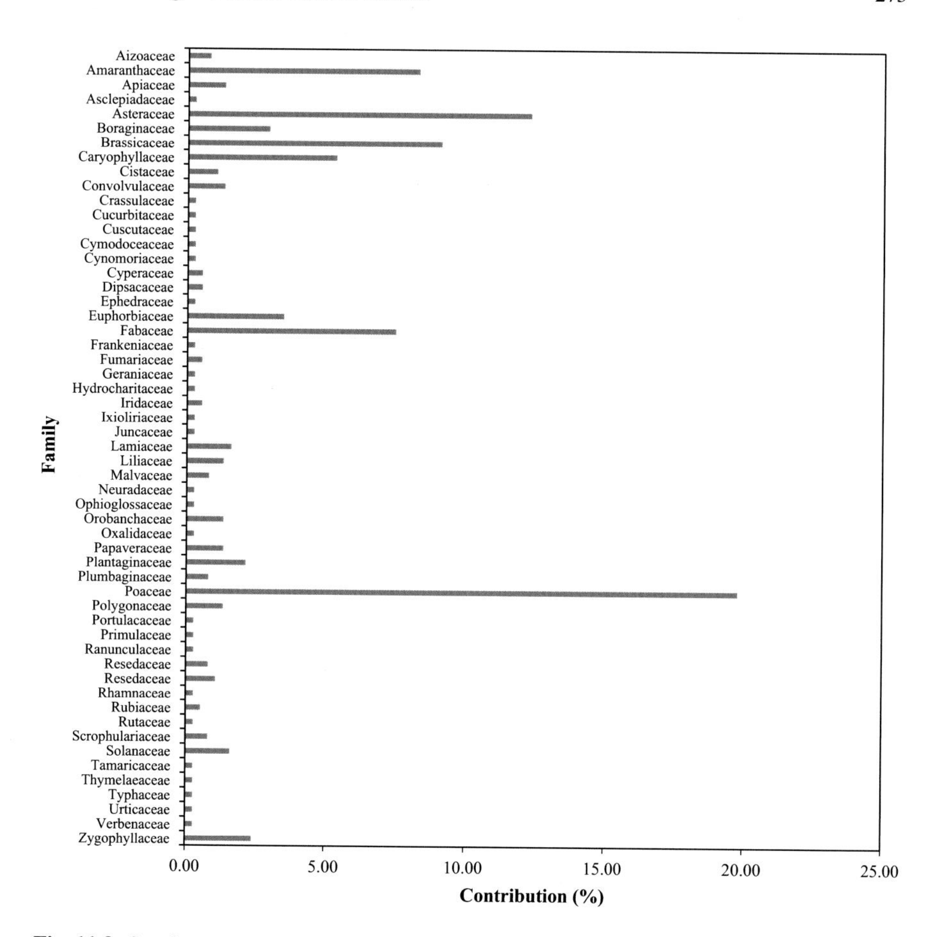

Fig. 11.3 Species contribution percentage of various plant families

11.6 Discussion

The vegetation of Kuwait is characterized by dominance of annual plants (68.45%) followed by perennial (31.02%) and biennial (0.53%). The high diversity of annual species could be linked to the extreme environmental conditions of Kuwait. Evolution of annual life cycle in hot arid desert has been linked to the survival strategy that provides a fitness advantage under such climatic conditions (Friedman and Rubin 2015; Poppenwimer et al. 2022). Moreover, prolonged summer and scarce precipitation events in hot arid climate place severe limitations on plant growth and development from late spring to autumn (Brown 2003). Such circumstances favor annual life form, because annual species exhibit higher seed dormancy and

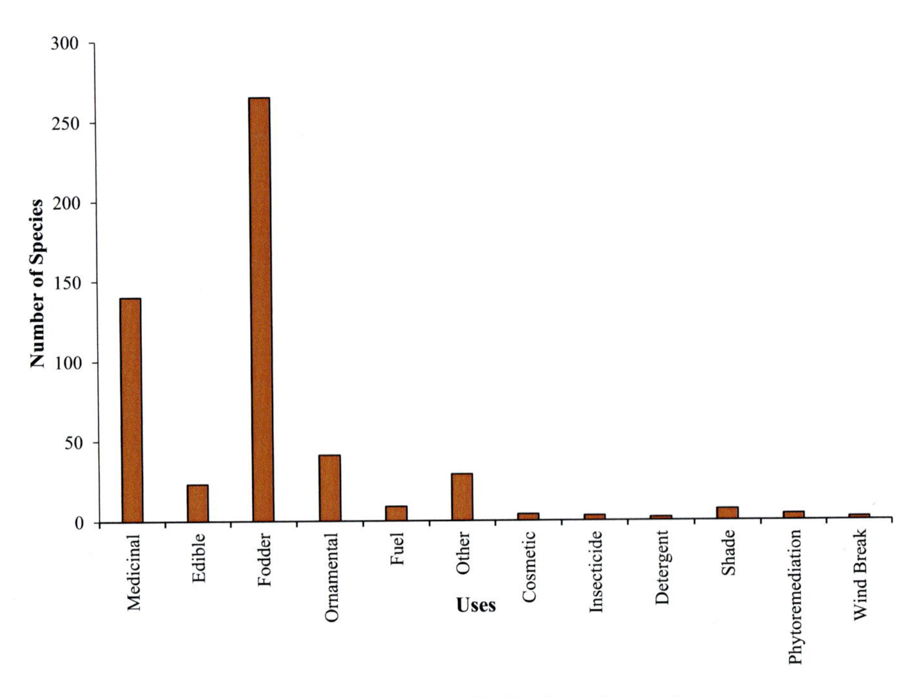

Fig. 11.4 Economic and ecological services provided by the native species

reproductive effort, offering an effective mechanism to survive under such conditions (Aronson et al. 1993; Gutterman 1993; Baskin and Baskin 2014). Therefore, more emphasis should be given to perennial species conservation because they usually provide greater ecosystem services than annual species (Grman et al. 2010; Glover et al. 2010; Poppenwimer et al. 2022).

Forbs (annual or perennial), grasses, and shrubs are the most common growth forms in Kuwait. Forbs accounted for over 65% of total vegetation in Kuwait followed by grasses (21.12%) and shrubs (12.30%). The lower dominance of grasses and shrubs could be linked to the overgrazing that causes severe loss to these growth forms (i.e., grasses and shrubs) (Al-Awadhi et al. 2005). Very low proportion of trees and ferns could be associated with harsh environmental conditions that may affect their recruitment. The chances of tree seedling mortality will be high during extended drought and even if they succeed to establish themselves, they may be subjected to growth suppression from browsing and droughts during the later stages of development (Martin and Moss 1997; Wilson and Witkowski 1998; Midgley and Bond 2001).

Indigenous knowledge on the use value of native plants is under threat due to the unfamiliarity and current modernizing trends among the local people, which is further accelerating their wisdom on the conservation of precious flora. Hence, the existing information on the use value as well as habitat preference of Kuwaiti native plant species is reported. Prioritization is used to determine priorities and implement suitable measures in conservation by regulating the available resources

Table 11.3 Species prioritization and categorization based on habitats of occurrence and use value

No.	Species	Family	Life form	Growth form	Habitat	Number of habitats occupied	Priority	Uses	Score	Priority
1	*Aaronsohnia factorovskyi* Warb. & Eig	Asteraceae	Annual	Forb	Sand, Silt	2	Moderate	Medicinal	1	Low
2	*Acacia pachyceras* O.Schwartz	Fabaceae	Perennial	Tree	Sand	1	High	Fodder, Fuel, Shade Other	4	High
3	*Acantholepis orientalis* Less	Asteraceae	Annual	Forb	Sand, Silt	2	Moderate	Medicinal	1	Low
4	*Achillea fragrantissima* (Forssk.)	Asteraceae	Perennial	Shrub	Sand, Gravel	2	Moderate	Medicinal, Fodder	2	Low
5	*Adonis dentata* Delile.	Ranunculaceae	Annual	Forb	Sand, Wadis	2	Moderate	Ornamental	1	Low
6	*Aegilops bicornis* (Forssk.) Jaub. & Spach	Poaceae	Annual	Grass	Sand, Silt	2	Moderate	Fodder	1	Low
7	*Aegilops kotschyi* Boiss.	Poaceae	Annual	Grass	Disturbed Areas	1	High	Fodder	1	Low
8	*Aegilops triuncialis* L.	Poaceae	Annual	Grass	Sand, Gravel	2	Moderate	Fodder	1	Low
9	*Aeluropus lagopoides* (L.) Trin. ex Thwaites	Poaceae	Perennial	Grass	Sand, Coastal Areas	2	Moderate	Edible, Fodder, Fuel, Other	4	High
10	*Aeluropus littoralis* (Gouan) Parl.,	Poaceae	Perennial	Grass	Sand, Coastal Areas	2	Moderate	Fodder	1	Low

(continued)

Table 11.3 (continued)

No.	Species	Family	Life form	Growth form	Habitat	Number of habitats occupied	Priority	Uses	Score	Priority
11	*Agathophora alopecuroides* (Delile) Fenzl. ex Bunge	Amaranthaceae	Annual	Shrub	Sand, Gravel	2	Moderate	Fodder	1	Low
12	*Aizoon canariense* L.	Aizoaceae	Annual	Forb	Sand, Coastal Areas	2	Moderate	Edible, Fodder, Ornamental	3	Moderate
13	*Aizoon hispanicum* L	Aizoaceae	Annual	Forb	Sand, Gravel, Runnels	3	Moderate	Ornamental	1	Low
14	*Alhagi graecorum* Boiss.	Fabaceae	Perennial	Shrub	Sand, Coastal Areas	2	Moderate	Medicinal, Edible, Fodder, Fuel, Wind Break, Other	6	High
15	*Allium sindjarense* Boiss. & Hausskn. ex Regel	Liliaceae	Perennial	Forb	Sand, Silt	2	Moderate	Medicinal, Ornamental	2	Low
16	*Allium sphaerocephalum* L.	Liliaceae	Perennial	Forb	Sand, Loam	2	Moderate	Medicinal, Edible, Ornamental	3	Moderate
17	*Althaea ludwigii* L.	Malvaceae	Annual	Forb	Sand, Gravel, Loam	3	Moderate	Fodder	1	Low
18	Alyssum homolocarpum (Fisch &Mey.)	Brassicaceae	Annual	Forb	Sand, Silt	2	Moderate	Medicinal, Fodder	2	Low
19	*Alyssum linifolium* Steph. ex Willd.,	Brassicaceae	Annual	Forb	Sand, Gravel	2	Moderate	Medicinal, Fodder	2	Low
20	*Amaranthus graecizans* L.	Amaranthaceae	Annual	Forb	Disturbed Areas	1	High	Medicinal, Ornamental	2	Low

21	*Amaranthus hybridus* L.	Amaranthaceae	Annual	Forb	Disturbed Areas	1	High	Ornamental	1	Low
22	*Amaranthus lividus* L.	Amaranthaceae	Annual	Forb	Disturbed Areas	1	High	Medicinal, Ornamental	2	Low
23	*Ammi majus* L.	Apiaceae	Annual	Forb	Disturbed Areas	1	High	Fodder	1	Low
24	*Ammochloa palaestina* Boiss.	Poaceae	Annual	Grass	Sand	1	High	Fodder	1	Low
25	*Anabasis lachnantha* Aellen & Rech.	Amaranthaceae	Perennial	Shrub	Silt	1	High	Fodder	1	Low
26	*Anabasis setifera* Moq.	Amaranthaceae	Perennial	Shrub	Sand, Coastal Areas	2	Moderate	Medicinal, Fodder, Detergent, Phytoremediation	4	High
27	*Anagallis arvensis* L	Primulaceae	Annual	Forb	Gravel	1	High	Medicinal	1	Low
28	*Anastatica hierochuntica* L	Brassicaceae	Annual	Forb	Sand, Runnels	2	Moderate	Medicinal	1	Low
29	*Anchusa hispida* Forssk.	Boraginaceae	Annual	Forb	Sand, Silt	2	Moderate	Ornamental	1	Low
30	*Andrachne telephioides* L.	Euphorbiaceae	Perennial	Forb	Silt	1	High	Fodder, Insecticide	2	Low
31	*Anisosciadium isosciadium* Bornm.	Apiaceae	Annual	Forb	Sand	1	High	Medicinal, Fodder	2	Low
32	*Anthemis deserti* Boiss.	Asteraceae	Annual	Forb	Silt	1	High	Ornamental	1	Low
33	*Anthemis pseudocotula* Boiss.	Asteraceae	Annual	Forb	Sand, Gravel	2	Moderate	Medicinal, Ornamental	2	Low
34	*Anvillea garcinii* (Burm.f.)	Asteraceae	Perennial	Shrub	Sand	1	High	Ornamental	1	Low
35	*Arnebia decumbens* (Vent.) Coss. & Kralik	Boraginaceae	Annual	Forb	Sand, Coastal Areas	2	Moderate	Medicinal, Cosmetic	2	Low

(continued)

Table 11.3 (continued)

No.	Species	Family	Life form	Growth form	Habitat	Number of habitats occupied	Priority	Uses	Score	Priority
36	*Arnebia linearifolia* A.	Boraginaceae	Annual	Forb	Sand, Wadis	2	Moderate	Medicinal, Fodder, Cosmetic	3	Moderate
37	*Arnebia tinctoria* Forssk	Boraginaceae	Annual	Forb	Sand, Gravel	2	Moderate	Cosmetic	1	Low
38	*Artemisia herba-alba* Asso	Asteraceae	Perennial	Shrub	Sand	1	High	Medicinal	1	Low
39	*Artemisia scoparia* Waldst. & Kit.	Asteraceae	Biennial	Shrub	Sand	1	High	Medicinal	1	Low
40	*Asphodelus tenuifolius* Cav.	Liliaceae	Annual	Forb	Sand, Gravel	2	Moderate	Medicinal, Edible	2	Low
41	*Asphodelus viscidus* L.	Liliaceae	Annual	Forb	Sand, Gravel	2	Moderate	Fodder, Ornamental	2	Low
42	*Aster squamatus* (Spreng.)	Asteraceae	Annual	Shrub	Sand, Gravel	2	Moderate	Fodder	1	Low
43	*Asteriscus hierochunticus* (Michon) Wiklund Syn. *Asteriscus pygmaeus* (DC) Coss.& Durand	Asteraceae	Annual	Forb	Silt	1	High	Fodder	1	Low
44	*Astragalus annularis* Forssk	Fabaceae	Annual	Forb	Sand	1	High	Fodder, Other	2	Low
45	*Astragalus bombycinus* Boiss.	Fabaceae	Annual	Forb	Sand	1	High	Fodder, Other	2	Low
46	*Astragalus corrugatus* Bertol.	Fabaceae	Annual	Forb	Silt	1	High	Fodder, Other	2	Low

47	*Astragalus hauarensis* Boiss.	Fabaceae	Annual	Forb	Sand	1	High	Fodder, Other	2	Low
48	*Astragalus schimperi* Boiss	Fabaceae	Annual	Forb	Sand, Gravel	2	Moderate	Fodder, Other	2	Low
49	*Astragalus sieberi* DC.	Fabaceae	Perennial	Shrub	Sand, Gravel	2	Moderate	Fodder, Other	2	Low
50	*Astragalus spinosus* Muschl.	Fabaceae	Perennial	Shrub	Sand, Gravel	2	Moderate	Medicinal, Ornamental, Other	3	Moderate
51	*Astragalus tribuloides* Delile	Fabaceae	Annual	Forb	Sand, Gravel	2	Moderate	Fodder, Other	2	Low
52	*Atractylis cancellata* L.	Asteraceae	Annual	Forb	Silt	1	High	Medicinal	1	Low
53	*Atractylis carduus* (Forssk.) C. Chr	Asteraceae	Perennial	Forb	Sand, Gravel, Coastal Areas	3	Moderate	Medicinal	1	Low
54	*Atriplex dimorphostegia* Kar. & Kir.	Amaranthaceae	Annual	Forb	Sand, Coastal Areas	2	Moderate	Fodder	1	Low
55	*Atriplex leucoclada* Boiss	Amaranthaceae	Perennial	Shrub	Sand	1	High	Fodder, Phytoremediation	2	Low
56	*Avena barbata* Pott ex Link.	Poaceae	Annual	Grass	Disturbed Areas	1	High	Fodder	1	Low
57	*Avena fatua* L.	Poaceae	Annual	Grass	Disturbed Areas	1	High	Fodder	1	Low
58	*Avena sativa*. L.	Poaceae	Annual	Grass	Sand, Gravel, Silt	3	Moderate	Fodder	1	Low
59	*Avena sterilis* L.	Poaceae	Annual	Grass	Sand, Gravel, Disturbed Areas	3	Moderate	Fodder	1	Low
60	*Bassia eriophora* (Schrad.) Asch.	Amaranthaceae	Annual	Forb	Sand	1	High	Medicinal, Fodder, Ornamental	3	Moderate

(continued)

Table 11.3 (continued)

No.	Species	Family	Life form	Growth form	Habitat	Number of habitats occupied	Priority	Uses	Score	Priority
61	*Bassia muricata* (L.) Asch	Amaranthaceae	Annual	Forb	Sand, Silt	2	Moderate	Medicinal, Fodder, Ornamental, Fuel, Wind Break, Other	5	High
62	*Bassia scoparia* (L.)	Amaranthaceae	Annual	Forb	Sand, Disturbed Areas	2	Moderate	Medicinal, Fodder, Phytoremediation	3	Moderate
63	*Bellevalia saviczii* Woronow	Asparagaceae	Perennial	Grass	Sand, Gravel	2	Moderate	Medicinal, Edible, Fodder	3	Moderate
64	*Beta vulgaris* L.	Amaranthaceae	Annual	Forb	Sand, Coastal Areas	2	Moderate	Medicinal, Edible	2	Low
65	*Bienertia cycloptera* Bge.Ex. Boiss	Amaranthaceae	Annual	Forb	Sand, Coastal Areas	2	Moderate	Fodder, Phytoremediation, Other	3	Moderate
66	*Brachypodium distachyum* (L.) P. Beauv.	Poaceae	Annual	Grass	Sand, Gravel	2	Moderate	Fodder	1	Low
67	*Brassica juncea* L.	Brassicaceae	Annual	Forb	Sand, Silt, Disturbed Areas	3	Moderate	Fodder	1	Low
68	*Brassica tournefortii* Gouan	Brassicaceae	Annual	Forb	Disturbed Areas	1	High	Fodder	1	Low
69	*Bromus catharticus* Vahl.	Poaceae	Annual	Grass	Disturbed Areas	1	High	Fodder	1	Low

70	*Bromus danthoniae* Trin.	Poaceae	Annual	Grass	Sand, Silt, Disturbed Areas	3	Moderate	Fodder	1	Low
71	*Bromus madritensis* L.	Poaceae	Annual	Grass	Sand, Silt	2	Moderate	Fodder	1	Low
72	*Bromus sericeus* Drobov	Poaceae	Annual	Grass	Sand	1	High	Fodder	1	Low
73	*Bromus tectorum* L.	Poaceae	Annual	Grass	Sand, Silt, Disturbed Areas	3	Moderate	Fodder	1	Low
74	*Bupleurum semicompositum* L.	Apiaceae	Annual	Forb	Sand, Gravel, Disturbed Areas	3	Moderate	Fodder	1	Low
75	*Cakile arabica* Velen	Brassicaceae	Annual	Forb	Sand, Gravel	2	Moderate	Fodder, Ornamental	2	Low
76	*Calendula arvensis* (Vaill.) L.	Asteraceae	Annual	Forb	Silt	1	High	Fodder, Ornamental	2	Low
77	*Calendula tripterocarpa* Rupr	Asteraceae	Annual	Forb	Plantation, Roadsides	2	Moderate	Fodder, Ornamental	2	Low
78	*Calligonum polygonoides* L.	Polygonaceae	Perennial	Shrub	Sand, Roadsides	2	Moderate	Ornamental, Shade	2	Low
79	*Calotropis procera* (Aiton) W.T. Aiton	Asclepiadaceae	Perennial	Shrub	Sand, Gravel	2	Moderate	Medicinal	1	Low
80	*Cardaria draba* (L.) Desv.	Brassicaceae	Perennial	Forb	Sand, Gravel, Disturbed Areas	3	Moderate	Medicinal, Edible	2	Low
81	*Carduus pycnocephalus* L.	Asteraceae	Annual	Forb	Silt	1	High	Fodder, Ornamental	2	Low
82	*Carrichtera annua* (L.) DC	Brassicaceae	Annual	Forb	Disturbed Areas	1	High	Fodder, Ornamental	2	Low
83	*Carthamus oxyacantha* M.Bieb.	Asteraceae	Annual	Forb	Sand, Gravel	2	Moderate	Fodder, Other	2	Low

(continued)

Table 11.3 (continued)

No.	Species	Family	Life form	Growth form	Habitat	Number of habitats occupied	Priority	Uses	Score	Priority
84	*Caylusea hexagyna* (Forssk.) M.L.Green	Resedaceae	Biennial	Forb	Sand, Gravel	2	Moderate	Medicinal	2	Low
85	*Cenchrus ciliaris* L	Poaceae	Perennial	Grass	Sand, Gravel	2	Moderate	Fodder, Ornamental	2	Low
86	*Cenchrus setigerus* Vahl	Poaceae	Perennial	Grass	Sand, Gravel	2	Moderate	Fodder, Ornamental	2	Low
87	*Centaurea bruguierana* (DC)	Asteraceae	Annual	Forb	Sand, Gravel, Disturbed Areas	3	Moderate	Fodder, Ornamental	2	Low
88	*Centaurea mesopotamica* Bornm.	Asteraceae	Annual	Forb	Sand, Gravel	2	Moderate	Fodder, Ornamental	2	Low
89	*Centaurea psedosiniaca* Czerep	Asteraceae	Annual	Grass	Sand	1	High	Fodder, Ornamental	2	Low
90	*Centropodia forsskalii* (Vahl) Cope	Poaceae	Perennial	Grass	Sand	1	High	Fodder	1	Low
91	*Chenopodium album* L.	Amaranthaceae	Annual	Forb	Sand	1	High	Edible, Fodder	2	Low
92	*Chenopodium glaucum* L	Amaranthaceae	Annual	Forb	Gravel, Disturbed Areas	2	Moderate	Fodder	1	Low
93	*Chenopodium murale* (L.) S. Fuentes, Uotila & Borsch	Amaranthaceae	Annual	Forb	Disturbed Areas	1	High	Fodder	1	Low

94	*Chenopodium opulifolium* Schrad.	Amaranthaceae	Annual	Shrub	Disturbed Areas	1	High	Fodder	1	Low
95	*Chrozophora obliqua* (Vahl)	Euphorbiaceae	Perennial	Forb	Sand	1	High	Fodder, Ornamental	2	Low
96	*Chrozophora tinctoria* (L.) Raf	Euphorbiaceae	Annual	Forb	Sand, Gravel	2	Moderate	Fodder, Ornamental	2	Low
97	*Chrozophora verbascifolia* (Willd.)	Euphorbiaceae	Annual	Forb	Rocky slopes	1	High	Fodder, Ornamental	2	Low
98	*Chrysanthemum coronarium* L.	Asteraceae	Annual	Forb	Plantation, Roadsides	2	Moderate	Medicinal, Edible	2	Low
99	*Cistanche tubulosa* (Schrenk) Wight	Orobanchaceae	Perennial	Forb	Sand, Coastal Areas	2	Moderate	Medicinal	1	Low
100	*Citrullus colocynthis* (L.) Schrad.	Cucurbitaceae	Perennial	Forb	Sand, Gravel	2	Moderate	Medicinal	1	Low
101	*Convolvulus arvensis* L	Convolvulaceae	Perennial	Forb	Wadis, Roadsides	2	Moderate	Medicinal, Fodder, Fuel	3	Moderate
102	*Convolvulus cephalopodus* Boiss.	Convolvulaceae	Perennial	Forb	Sand	1	High	Fodder, Ornamental	2	Low
103	*Convolvulus oxyphyllus* Boiss.	Convolvulaceae	Perennial	Shrub	Sand, Gravel	2	Moderate	Medicinal, Ornamental	2	Low
104	*Convolvulus pilosellifolius* Desr	Convolvulaceae	Perennial	Forb	Sand, Gravel	2	Moderate	Medicinal, Fodder, Ornamental	3	Moderate
105	*Conyza bonariensis* (L.)	Asteraceae	Annual	Forb	Wadis	1	High	Medicinal	1	Low
106	*Cornulaca aucheri* Moq	Amaranthaceae	Annual	Forb	Sand, Coastal Areas	2	Moderate	Fodder	1	Low

(continued)

Table 11.3 (continued)

No.	Species	Family	Life form	Growth form	Habitat	Number of habitats occupied	Priority	Uses	Score	Priority
107	*Cornulaca monacantha* Delile	Amaranthaceae	Perennial	Shrub	Sand, Coastal Areas	2	Moderate	Medicinal, Fodder	2	Low
108	*Coronilla scorpioides* (L.) W.D.J. Koch	Fabaceae	Annual	Forb	Sand, Gravel	2	Moderate	Medicinal	1	Low
109	*Coronopus didymus* (L.) J. E. Smith	Brassicaceae	Annual	Forb	Disturbed Areas	1	High	Fodder	1	Low
110	*Crassula alata* (Viv.) A. Berger	Crassulaceae	Annual	Forb	Gravel	1	High	Fodder	1	Low
111	*Cressa cretica* L.	Convolvulaceae	Perennial	Forb	Sand, Coastal Areas	2	Moderate	Medicinal, Fodder, Ornamental, Fuel	4	High
112	*Crucianella membranacea* Boiss.	Rubiaceae	Annual	Forb	Sand, Silt	2	Moderate	Fodder	1	Low
113	*Cuscuta planiflora* Ten	Cuscutaceae	Annual	Forb	Gravel	1	High	Fodder	1	Low
114	*Cutandia dichotoma* (Forssk.)	Poaceae	Annual	Grass	Sand	1	High	Fodder	1	Low
115	*Cutandia memphitica* (Spreng.) Benth.	Poaceae	Annual	Grass	Sand	1	High	Fodder	1	Low
116	*Cymbopogon commutatus* (Steud.) Stapf	Poaceae	Perennial	Grass	Silt	1	High	Insecticide	1	Low
117	*Cynodon dactylon* (L.) Pers.	Poaceae	Perennial	Grass	Sand	1	High	Medicinal, Fodder	2	Low

118	*Cynomorium coccineum* L	Cynomoriaceae	Perennial	Forb	Coastal Areas, Salt Marsh	2	Moderate	Fodder	1	Low
119	*Cyperus conglomeratus* Rottb.,	Cyperaceae	Perennial	Grass	Sand	1	High	Medicinal, Fodder, Other	3	Moderate
120	*Cyperus rotundus* L.	Cyperaceae	Perennial	Grass	Sand	1	High	Medicinal, Fodder	2	Low
121	*Dactyloctenium aegyptium* (L.) Willd.	Poaceae	Annual	Grass	Sand, Gravel	2	Moderate	Medicinal, Fodder	2	Low
122	*Datura innoxia* Mill.	Solanaceae	Annual	Forb	Sand, Gravel	2	Moderate	Medicinal	1	Low
123	*Daverra triradiata* Hochst. ex Boiss.	Apiaceae	Perennial	Forb	Sand, Gravel	2	Moderate	Fodder	1	Low
124	*Dichanthium annulatum* (Forssk.) Stapf	Poaceae	Perennial	Grass	Sand	1	High	Fodder	1	Low
125	*Dichanthium foveolatum* (Delile) Roberty	Poaceae	Perennial	Grass	Sand	1	High	Fodder	1	Low
126	*Digitaria ciliaris* (Retz.) Koeler	Poaceae	Annual	Grass	Sand, Loam	2	Moderate	Fodder	1	Low
127	*Digitaria sanguinalis* (L.) Scop.	Poaceae	Annual	Grass	Sand	1	High	Fodder	1	Low
128	*Dinebra retroflexa* (Vahl) Panzer	Poaceae	Annual	Grass	Disturbed Areas	1	High	Fodder	1	Low
129	*Dipcadi erythraeum* Webb & Berthel	Liliaceae	Annual	Grass	Sand	1	High	Medicinal, Ornamental	2	Low
130	*Diplotaxis acris* (Forssk.) Boiss.	Brassicaceae	Annual	Forb	Sand	1	High	Fodder, Ornamental	2	Low

(continued)

Table 11.3 (continued)

No.	Species	Family	Life form	Growth form	Habitat	Number of habitats occupied	Priority	Uses	Score	Priority
131	*Diplotaxis harra* (Forssk.) Boiss	Brassicaceae	Annual	Forb	Gravel	1	High	Fodder, Ornamental	2	Low
132	*Ducrosia anethifolia* (DC.) Boiss.	Apiaceae	Perennial	Forb	Sand, Gravel	2	Moderate	Medicinal, Ornamental	2	Low
133	*Echinochloa colona* (L.) Link	Poaceae	Annual	Grass	Disturbed Areas	1	High	Medicinal, Fodder	2	Low
134	*Echinops blancheanis* Boiss	Asteraceae	Perennial	Forb	Silt	1	High	Medicinal	1	Low
135	*Echium rawolfii* Delile	Boraginaceae	Annual	Forb	Sand	1	High	Fodder, Ornamental	2	Low
136	*Emex spinose* (L.)	Polygonaceae	Annual	Forb	Sand, Silt, Disturbed Areas	3	Moderate	Medicinal, Edible, Fodder	3	Low
137	*Ephedra alata* Decne	Ephedraceae	Perennial	Forb	Sand, Gravel, Rocky slopes	3	Moderate	Medicinal	1	Low
138	*Eragrostis barrelieri* Daveau	Poaceae	Annual	Grass	Sand, Disturbed Areas	2	Moderate	Fodder	1	Low
139	*Eragrostis minor* Host,	Poaceae	Annual	Grass	Sand, Gravel	2	Moderate	Fodder	1	Low
140	*Eremobium aegyptiacum* (Spreng.) Asch.	Brassicaceae	Annual	Forb	Sand	1	High	Fodder	1	Low
141	*Eremopoa persica* (Trin.) Rozhev.	Poaceae	Annual	Grass	Sand, Gravel	2	Moderate	Fodder	1	Low

142	*Eremopyrum bonaepartis* (Spreng.) Nevski	Poaceae	Annual	Grass	Gravel	1	High	Fodder	1	Low
143	*Eremopyrum distans* (C. Koch) Nevski	Poaceae	Annual	Grass	Silt	1	High	Fodder	1	Low
144	*Erodium byryonilifolium*	Poaceae	Annual	Forb	Sand, Gravel	2	Moderate	Fodder	1	Low
145	*Erodium ciconium* (L.) L'Hér.	Poaceae	Annual	Forb	Sand, Gravel	2	Moderate	Fodder	1	Low
146	*Erodium cicutarium* (L.) L'Hér.	Poaceae	Annual	Forb	Disturbed Areas	1	High	Fodder	1	Low
147	*Erodium glaucophyllum* (L.) L'Hér.	Poaceae	Annual	Forb	Rocky slopes	1	High	Fodder	1	Low
148	*Erodium laciniatum* (Cav.) Willd	Poaceae	Annual	Forb	Gravel	1	High	Fodder	1	Low
149	*Eruca sativa* Mill.	Brassicaceae	Annual	Forb	Sand, Coastal Areas	2	Moderate	Medicinal, Edible, Ornamental	3	Low
150	*Euphorbia densa* Schrenk	Euphorbiaceae	Annual	Forb	Sand, Gravel, Silt	3	Moderate	Medicinal	1	Low
151	*Euphorbia granulata* Forssk., Fl.	Euphorbiaceae	Annual	Forb	Gravel	1	High	Medicinal	1	Low
152	*Euphorbia grossheimii* (Prokh.)	Euphorbiaceae	Annual	Forb	Gravel, Wadis	2	Moderate	Medicinal	1	Low
153	*Euphorbia helioscopia* L.	Euphorbiaceae	Annual	Forb	Silt	1	High	Medicinal	1	Low

(continued)

Table 11.3 (continued)

No.	Species	Family	Life form	Growth form	Habitat	Number of habitats occupied	Priority	Uses	Score	Priority
154	*Euphorbia hirta* L.	Euphorbiaceae	Annual	Forb	Sand, Silt, Disturbed Areas	3	Moderate	Medicinal	1	Low
155	*Euphorbia indica* Lam.,	Euphorbiaceae	Annual	Forb	Sand, Silt, Disturbed Areas	3	Moderate	Medicinal	1	Low
156	*Euphorbia peplus* L.	Euphorbiaceae	Annual	Forb	Disturbed Areas, Plantations	2	Moderate	Medicinal	1	Low
157	*Euphorbia serpens* Kunth	Euphorbiaceae	Annual	Forb	Sand, Disturbed Areas	2	Moderate	Medicinal	1	Low
158	*Euphorbia supina* Raf.	Euphorbiaceae	Perennial	Forb	Gravel	1	High	Medicinal	1	Low
159	*Fagonia bruguieri* DC.	Zygophyllaceae	Perennial	Shrub	Sand, Gravel	2	Moderate	Medicinal	1	Low
160	*Fagonia glutinosa* Delile	Zygophyllaceae	Perennial	Forb	Sand, Gravel	2	Moderate	Medicinal	1	Low
161	*Fagonia indica* Burm. f	Zygophyllaceae	Perennial	Forb	Sand, Gravel	2	Moderate	Medicinal	1	Low
162	*Farsetia aegyptia* Turra	Brassicaceae	Perennial	Shrub	Sand, Silt	2	Moderate	Fodder	1	Low
163	*Farsetia burtonae* Oliv.	Brassicaceae	Perennial	Shrub	Sand, Gravel	2	Moderate	Fodder	1	Low
164	*Filago pyramidata* L.	Asteraceae	Annual	Forb	Sand, Disturbed Areas	2	Moderate	Fodder	1	Low
165	*Flaveria trinervia* (Sprengel) C. Mohr	Asteraceae	Annual	Forb	Disturbed Areas, Coastal Areas	2	Moderate	Fodder	1	Low

166	*Frankenia pulverulenta* L.	Frankeniaceae	Annual	Forb	Sand, Gravel	2	Moderate	Fodder	1	Low
167	*Fumaria parviflora* Lam.	Fumariaceae	Annual	Forb	Sand, Silt	2	Moderate	Medicinal	1	Low
168	*Gagea reticulata* (Pall.) Schult. & Schult.f.	Liliaceae	Perennial	Grass	Sand, Gravel	2	Moderate	Medicinal	1	Low
169	*Galium tricornutum* Dandy	Rubiaceae	Annual	Forb	Sand, Silt	2	Moderate	Medicinal	1	Low
170	*Gladiolus italicus* Mill.	Iridaceae	Perennial	Grass	Sand, Silt	2	Moderate	Medicinal	1	Low
171	*Glaucium corniculatum* (L.) Rudolph.	Papaveraceae	Annual	Forb	Disturbed Areas	1	High	Fodder	1	Low
172	*Gymnarrhena micrantha* Desf	Asteraceae	Annual	Forb	Silt	1	High	Fodder	1	Low
173	*Gynandriris sisyrinchium* (L.)	Iridaceae	Perennial	Grass	Sand, Gravel	2	Moderate	Medicinal, Fodder	2	Low
174	*Gypsophila capillaris* (Forssk.)	Caryophyllaceae	Annual	Forb	Coastal Areas	1	High	Fodder	1	Low
175	*Halocnemum strobilaceum* (Pall.) M. Bieb.	Amaranthaceae	Perennial	Shrub	Coastal Areas	1	High	Fodder	1	Low
176	*Halodule uninervis* (Forssk.) Asch.	Cymodoceaceae	Perennial	Forb	Aquatic	1	High	Medicinal, Other	2	Low
177	*Halophila ovalis* (R. Br.) Hook. f	Hydrocharitaceae	Perennial	Forb	Aquatic	1	High	-	0	

(continued)

Table 11.3 (continued)

No.	Species	Family	Life form	Growth form	Habitat	Number of habitats occupied	Priority	Uses	Score	Priority
178	*Halothamnus iraqensis* Botsch.	Amaranthaceae	Perennial	Forb	Sand	1	High	Fodder	1	Low
179	*Haloxylon salicornicum* (Moq.) Bunge ex Boiss Syn. *Hammada salicornica* (Moq.) Iljin	Amaranthaceae	Perennial	Shrub	Sand, Gravel	2	Moderate	Medicinal, Fodder, Other	3	Moderate
180	*Haplophyllum tuberculatum* (Forssk.) Juss.	Rutaceae	Perennial	Shrub	Gravel	1	high	Medicinal, Fodder, Insecticide	3	Moderate
181	*Helianthemum kahiricum* Delile	Cistaceae	Perennial	Shrub	Silt	1	high	Fodder	1	Low
182	*Helianthemum ledifolium* (L.)	Cistaceae	Annual	Forb	Sand, Gravel, Silt	3	Moderate	Fodder	1	Low
183	*Helianthemum lippii* (L.) Dum.-Cours	Cistaceae	Perennial	Forb	Sand, Gravel, Silt	3	Moderate	Medicinal	1	Low
184	*Helianthemum salicifolium* (L.)	Cistaceae	Annual	Forb	Silt	1	High	Fodder	1	Low
185	*Heliotropium bacciferum* Forssk.	Boraginaceae	Perennial	Shrub	Sand, Gravel	2	Moderate	Medicinal, Edible, Fuel	3	Moderate
186	*Heliotropium kotschyi* Gürke	Boraginaceae	Perennial	Shrub	Sand, Gravel	2	Moderate	Medicinal	1	Low
187	*Heliotropium lasiocarpum* Fisch. & C.A. Mey	Boraginaceae	Annual	Shrub	Sand	1	High	Medicinal, Fuel	2	Low

188	*Herniaria hemistemon* J. Gay	Caryophyllaceae	Perennial	Forb	Sand	1	High	Medicinal	1	Low
189	*Herniaria hirsuta* L	Caryophyllaceae	Annual	Forb	Sand, Gravel	2	Moderate	Medicinal	1	Low
190	*Hippocrepis areolata* Desv.	Fabaceae	Annual	Forb	Sand, Gravel	2	Moderate	Fodder	1	Low
191	*Hippocrepis unisiliquosa* L.	Fabaceae	Annual	Forb	Gravel	1	High	Fodder	1	Low
192	*Hordeum marinum* Huds.	Poaceae	Annual	Grass	Sand, Disturbed Areas	2	Moderate	Fodder	1	Low
193	*Horwoodia dicksoniae* Turrill	Brassicaceae	Annual	Forb	Sand	1	High	Fodder	1	Low
194	*Hyoscyamus muticus* L.	Solanaceae	Annual	Forb	Gravel	1	High	Medicinal	1	Low
195	*Hyoscyamus pusillus* L.	Solanaceae	Annual	Forb	Sand, Wadis	2	Moderate	Medicinal	1	Low
196	*Hypecoum littorale* Wulfen in Jacq.,	Papaveraceae	Annual	Forb	Sand, Gravel	2	Moderate	Fodder	1	Low
197	*Hypecoum pendulum* L.	Papaveraceae	Annual	Forb	Sand, Gravel	2	Moderate	Fodder	1	Low
198	*Ifloga spicata* (Forssk.) Sch. Bip.	Asteraceae	Annual	Forb	Sand, Gravel	2	Moderate	Fodder	1	Low
199	*Imperata cylindrica* (L.) Raeusch.	Poaceae	Perennial	Grass	Sand, Disturbed Areas	2	Moderate	Fodder	1	Low
200	*Ixiolirion tataricum* (Pall.) Herb.	Ixioliriaceae	Perennial	Forb	Sand, Gravel, Silt	3	Moderate	Medicinal	1	Low
201	*Juncus rigidus* Desf.	Juncaceae	Perennial	Grass	Coastal Areas	1	High	Fodder, Other	2	Low
202	*Koelpinia linearis* Pall.	Asteraceae	Annual	Forb	Sand, Gravel	2	Moderate	Medicinal	1	Low
203	*Lactuca serriola* L	Asteraceae	Annual	Forb	Gravel	1	High	Medicinal	1	Low

(continued)

Table 11.3 (continued)

No.	Species	Family	Life form	Growth form	Habitat	Number of habitats occupied	Priority	Uses	Score	Priority
204	*Lallemantia royleana* (Benth.)	Lamiaceae	Annual	Forb	Gravel	1	High	Medicinal	1	Low
205	*Lappula spinocarpos* (Forssk.) Asch. ex Kuntze	Boraginaceae	Annual	Forb	Sand, Gravel	2	Moderate	Fodder	1	Low
206	*Lasiurus scindicus* Henrard	Poaceae	Perennial	Grass	Sand	1	High	Fodder	1	Low
207	*Launaea angustifolia* (Desf.)	Asteraceae	Annual	Forb	Sand, Gravel	2	Moderate	Fodder	1	Low
208	*Launaea capitata* (Spreng.) Dandy	Asteraceae	Annual	Forb	Sand	1	High	Fodder	1	Low
209	*Launaea mucronata* (Forssk.) Muschl.	Asteraceae	Perennial	Forb	Sand, Disturbed Areas	2	Moderate	Fodder	1	Low
210	*Launaea nudicaulis* (L.) Hook.f.	Asteraceae	Perennial	Forb	Sand, Disturbed Areas	2	Moderate	Fodder	1	Low
211	*Leontodon laciniatus* (Bertol.) Widder	Asteraceae	Annual	Forb	Sand, Gravel	2	Moderate	Fodder	1	Low
212	*Lepidium aucheri* Boiss.	Brassicaceae	Annual	Forb	Wadis	1	High	Medicinal	1	Low
213	*Lepidium sativum* L.	Brassicaceae	Annual	Forb	Sand, Gravel	2	Moderate	Medicinal, Edible	2	Low
214	*Leptaleum filifolium* (Willd.) DC	Brassicaceae	Annual	Forb	Sand, Gravel	2	Moderate	Medicinal	1	Low
215	*Leptochloa fusca* (L.) Kunth	Poaceae	Perennial	Grass	Sand, Coastal Areas	2	Moderate	Fodder, Other	2	Low
216	*Limonium carnosum* Kuntze	Plumbaginaceae	Perennial	Shrub	Silt, Coastal Areas	2	Moderate	Medicinal	1	Low

217	*Limonium thouini* (Viv.) O. Kuntze	Plumbaginaceae	Annual	Forb	Gravel	1	High	Medicinal	1	Low
218	*Linaria albifrons* Spreng.	Scrophulariaceae	Annual	Forb	Sand, Gravel	2	Moderate	Fodder	1	Low
219	*Linaria simplex* (Willd.) DC.	Scrophulariaceae	Annual	Forb	Gravel	1	High	Fodder	1	Low
220	*Loeflingia hispanica* L	Caryophyllaceae	Annual	Forb	Sand, Gravel	2	Moderate	Fodder	1	Low
221	*Lolium multiflorum* Lam.	Poaceae	Annual	Grass	Sand Gravel	2	Moderate	Fodder	1	Low
222	*Lolium rigidum* Gaudin	Poaceae	Annual	Grass	Disturbed Areas, Plantations, Roadsides	3	Moderate	Fodder	1	Low
223	*Lolium temulentum* L.	Poaceae	Annual	Grass	Sand, Disturbed Areas	2	Moderate	Fodder	1	Low
224	*Lotus halophilus* Boiss. & Spruner	Fabaceae	Annual	Forb	Sand, Gravel	2	Moderate	Medicinal, Other	2	Low
225	*Lycium shawii* Roem. & Schult.	Solanaceae	Perennial	Shrub	Gravel, Wadis	2	Moderate	Fodder, Shade, Other	3	Moderate
226	*Malcolmia africana* (L.)	Brassicaceae	Annual	Forb	Disturbed Areas	1	High	Fodder	1	Low
227	*Malcolmia grandiflora* (Bunge) Kuntze,	Brassicaceae	Annual	Forb	Disturbed Areas	1	High	Fodder	1	Low
228	*Malcolmia pygmaea* (DC.)	Brassicaceae	Annual	Forb	Sand Gravel	2	Moderate	Fodder	1	Low
229	*Malva nicaeensis* All.	Malvaceae	Annual	Forb	Gravel, Rocky Slopes	2	Moderate	Fodder	1	Low

(continued)

Table 11.3 (continued)

No.	Species	Family	Life form	Growth form	Habitat	Number of habitats occupied	Priority	Uses	Score	Priority
230	*Malva parviflora* L.	Malvaceae	Annual	Forb	Sand, Gravel, Disturbed Areas	3	Moderate	Medicinal, Edible, Fodder	3	Moderate
231	*Maresia pygmaea* (Delile)	Brassicaceae	Annual	Forb	Sand, Gravel	2	Moderate	Fodder	1	Low
232	*Matricaria aurea* (Loefl.) Sch. Bip.	Asteraceae	Annual	Forb	Gravel	1	High	Medicinal, Fodder	2	Low
233	*Matthiola longipetala* (Vent.) DC.	Brassicaceae	Annual	Forb	Sand, Silt	2	Moderate	Fodder	1	Low
234	*Medicago laciniata* (L.) Mill.	Fabaceae	Annual	Forb	Gravel, Plantations, Roadsides	3	Moderate	Fodder, Other	2	Low
235	*Medicago polymorpha* L.	Fabaceae	Annual	Forb	Sand, Gravel	2	Moderate	Edible, Fodder, Other	3	Moderate
236	*Melilotus indicus* (L.) All.	Fabaceae	Annual	Forb	Plantation Roadsides	2	Moderate	Medicinal, Other	2	Low
237	*Mesembryanthemum nodiflorum* L.	Aizoaceae	Annual	Forb	Sand, Coastal Areas	2	Moderate	Fodder	1	Low
238	*Moltkiopsis ciliata* (Forssk.) I.M. Johnst.	Boraginaceae	Perennial	Forb	Sand	1	High	Fodder	1	Low
239	*Monsonia nivea* (Decne.) Webb	Geraniaceae	Perennial	Forb	Sand	1	High	Medicinal, Fodder	2	Low
240	*Neotorularia torulosa* (Desf.)	Brassicaceae	Annual	Forb	Sand, Gravel, Wadis	3	Moderate	Fodder	1	Low

241	*Neurada procumbens* L.	Neuradaceae	Annual	Forb	Sand	1	High	Edible, Fodder	2	Low
242	*Nitraria retusa* (Forssk.) Asch.	Zygophyllaceae	Perennial	Shrub	Sand, Coastal Areas	2	Moderate	Fodder, Shade	2	Low
243	*Notoceras bicorne* (Aiton)	Brassicaceae	Annual	Forb	Gravel, Wadis	2	Moderate	Fodder	1	Low
244	*Ochradenus baccatus* Delile	Resedaceae	Perennial	Shrub	Sand, Gravel	2	Moderate	Medicinal	1	Low
245	*Ogastemma pusillum* (Coss. & Durand ex Bonnet & Baratte) Brummitt	Boraginaceae	Annual	Forb	Sand, Gravel	2	Moderate	Fodder	1	Low
246	*Oligomeris linifolia* (Vahl) J.F. Macbr	Resedaceae	Annual	Forb	Gravel	1	High	Fodder	1	Low
247	*Oligomeris subulata* (Webb & Berth.)	Resedaceae	Annual	Forb	Sand, Gravel	2	Moderate	Fodder, Other	2	Low
248	*Onobrychis ptolemaica* DC.	Fabaceae	Perennial	Forb	Sand, Gravel	2	Moderate	Fodder, Other	2	Low
249	*Ononis reclinata* L.	Fabaceae	Annual	Forb	Sand, Gravel	2	Moderate	Fodder, Other	2	Low
250	*Ononis serrata* Forssk.	Fabaceae	Annual	Forb	Sand	1	High	Fodder, Other	2	Low
251	*Ophioglossum aitchisonii*	Ophioglossaceae	Annual	Fern	Sand	1	High	Edible	1	Low
252	*Orobanche aegyptiaca* Pers.	Orobanchaceae	Annual	Forb	Rocky slopes	1	High	Medicinal	1	Low

(continued)

Table 11.3 (continued)

No.	Species	Family	Life form	Growth form	Habitat	Number of habitats occupied	Priority	Uses	Score	Priority
253	*Orobanche cernua* Loefl.	Orobanchaceae	Annual	Forb	Gravel, Wadis	2	Moderate	Medicinal	1	Low
254	*Orobanche minor* Sm.	Orobanchaceae	Annual	Forb	Disturbed Areas, Plantations	2	Moderate	-	0	
255	*Orobanche ramosa* L.	Orobanchaceae	Annual	Forb	Disturbed Areas, Plantations	2	Moderate	Medicinal	1	Low
256	*Oxalis corniculata* L.	Oxalidaceae	Perennial	Forb	Plantations	1	High	Fodder	1	Low
257	*Panicum antidotale* Retz.	Poaceae	Perennial	Grass	Sand	1	High	Fodder	1	Low
258	*Panicum turgidum* Forssk.	Poaceae	Perennial	Grass	Sand, Gravel	2	Moderate	Fodder	1	Low
259	*Papaver rhoeas* L	Papaveraceae	Annual	Forb	Disturbed Areas, Roadsides	2	Moderate	Medicinal	1	Low
260	*Parapholis incurva* (L.) C.E. Hubb.	Poaceae	Annual	Grass	Disturbed Areas	1	High	Fodder	1	Low
261	*Paronychia arabica* (L.) DC	Caryophyllaceae	Annual	Forb	Sand, Silt, Wadis	3	Moderate	Medicinal, Fodder	2	Low
262	*Peganum harmala* L.	Zygophyllaceae	Perennial	Shrub	Sand, Gravel, Wadis	3	Moderate	Medicinal, Other	2	Low
263	*Pennisetum divisum* (J.F. Gmel.) Henrard	Poaceae	Perennial	Grass	Sand	1	High	Fodder	1	Low
264	*Phalaris minor* Retz.	Poaceae	Annual	Grass	Plantations, Roadsides	2	Moderate	Fodder	1	Low
265	*Phalaris paradoxa* L	Poaceae	Annual	Grass	Disturbed Areas	1	High	Fodder	1	Low

266	*Phragmites australis* (Cav.) Trin. ex Steud.	Poaceae	Perennial	Grass	Wadis	1	High	Medicinal, Fodder	2	Low
267	*Phyla nodiflora* (L.) Greene	Verbenaceae	Perennial	Forb	Wadis, Plantations	2	Moderate	Medicinal	1	Low
268	*Picris babylonica* Hand.-Mazz	Asteraceae	Annual	Forb	Sand, Gravel, Wadis	3	Moderate	Fodder	1	Low
269	*Plantago amplexicaulis* Cav	Plantaginaceae	Annual	Forb	Sand	1	High	Fodder	1	Low
270	*Plantago boissieri* Hausskn. & Bornm	Plantaginaceae	Annual	Forb	Sand	1	High	Fodder	1	Low
271	*Plantago ciliata* Desf.	Plantaginaceae	Annual	Forb	Sand, Gravel	2	Moderate	Fodder	1	Low
272	*Plantago coronopus* L	Plantaginaceae	Annual	Forb	Sand, Gravel	2	Moderate	Fodder	1	Low
273	*Plantago lanceolata* L.	Plantaginaceae	Perennial	Forb	Sand, Gravel	2	Moderate	Fodder	1	Low
274	*Plantago notata* Lag.	Plantaginaceae	Annual	Forb	Gravel	1	High	Fodder	1	Low
275	*Plantago ovata* Forssk	Plantaginaceae	Annual	Forb	Sand	1	High	Medicinal	1	Low
276	*Plantago psammophila* Agnew & Chal.-Kabi	Plantaginaceae	Annual	Forb	Sand, Gravel	2	Moderate	Fodder	1	Low
277	*Poa annua* L.	Poaceae	Annual	Grass	Plantations	1	High	Fodder	1	Low
278	*Poa infirma* Kunth.	Poaceae	Annual	Grass	Sand, Silt	2	Moderate	Fodder	1	Low
279	*Poa sinaica* Steud.,	Poaceae	Perennial	Grass	Sand, Gravel	2	Moderate	Fodder	1	Low

(continued)

Table 11.3 (continued)

No.	Species	Family	Life form	Growth form	Habitat	Number of habitats occupied	Priority	Uses	Score	Priority
280	*Polycarpaea repens* (Forssk.) Asch. & Schweinf	Caryophyllaceae	Perennial	Forb	Sand, Gravel	2	Moderate	Medicinal, Fodder	2	Low
281	*Polycarpaea robbairea* (Kuntze) Greuter & Burdet	Caryophyllaceae	Annual	Forb	Sand, Gravel, Silt	3	Moderate	Fodder	1	Low
282	*Polycarpon tetraphyllum* (L.) L	Caryophyllaceae	Annual	Forb	Gravel	1	High	Fodder	1	Low
283	*Polygonum patulum* M. Bieb.	Polygonaceae	Annual	Forb	Disturbed Areas	1	High	Fodder	1	Low
284	*Polypogon monspeliensis* (L.) Desf.	Poaceae	Annual	Grass	Plantations	1	High	Fodder	1	Low
285	*Portulaca oleracea* L.	Portulacaceae	Annual	Forb	Disturbed Areas, Plantations	2	Moderate	Fodder	1	Low
286	*Prosopis farcta* (Banks & Sol.)	Fabaceae	Perennial	Tree	Sand, Gravel	2	Moderate	Medicinal, Fuel, Shade	3	Moderate
287	*Psylliostachys spicata* (Willd.) Nevski	Plumbaginaceae	Annual	Forb	Sand, Gravel	2	Moderate	Fodder	1	Low
288	*Pteranthus dichotomus* Forssk	Caryophyllaceae	Annual	Forb	Sand, Wadis	2	Moderate	Fodder	1	Low
289	*Pulicaria undulata* (L.) C.A. Mey.	Asteraceae	Perennial	Shrub	Sand, Silt, Wadis	3	Moderate	Medicinal	1	Low
290	*Reichardia tingitana* (L.) Roth	Asteraceae	Annual	Forb	Wadis	1	High	Medicinal, Fodder	2	Low

291	*Reseda arabica* Boiss.	Resedaceae	Annual	Forb	Gravel	1	High	Fodder	1	Low
292	*Reseda decursiva* Forssk.	Resedaceae	Annual	Forb	Sand, Gravel	2	Moderate	Fodder	1	Low
293	*Reseda muricata* C. Presl.	Resedaceae	Annual	Forb	Sand, Gravel	2	Moderate	Fodder	1	Low
294	*Rhanterium epapposum* Oliv.	Asteraceae	Perennial	Shrub	Sand, Gravel	2	Moderate	Fodder, Fuel	2	Low
295	*Rhynchelytrum repens* (Willd.) C.E. Hubb	Poaceae	Annual	Grass	Disturbed Areas, Plantations	2	Moderate	Medicinal	1	Low
296	*Roemeria hybrida* (L.) DC.	Papaveraceae	Annual	Forb	Sand	1	High	Medicinal	1	Low
297	*Rostraria cristata* (L.) Tzvelev.	Poaceae	Annual	Grass	Disturbed Areas, Plantations	2	Moderate	Fodder	1	Low
298	*Rostraria pumila* (Desf.) Tzvelev	Poaceae	Annual	Grass	Sand, Silt	2	Moderate	Fodder	1	Low
299	*Rumex pictus* Forssk.	Polygonaceae	Annual	Forb	Sand	1	High	Fodder	1	Low
300	*Rumex vesicarius* L	Polygonaceae	Annual	Forb	Gravel, Wadis	2	Moderate	Medicinal, Edible, Fodder	3	Moderate
301	*Salicornia europaea* L	Amaranthaceae	Annual	Forb	Coastal Areas, Salt Marsh	2	Moderate	Fodder	1	Low
302	*Salsola cyclophylla* Baker	Amaranthaceae	Perennial	Shrub	Sand, Coastal Areas	2	Moderate	Fodder	1	Low
303	*Salsola imbricata* Forssk.	Amaranthaceae	Perennial	Shrub	Sand, Coastal Areas	2	Moderate	Medicinal	1	Low

(continued)

Table 11.3 (continued)

No.	Species	Family	Life form	Growth form	Habitat	Number of habitats occupied	Priority	Uses	Score	Priority
304	*Salsola jordanicola* Eig.	Amaranthaceae	Annual	Forb	Sand, Coastal Areas	2	Moderate	Fodder	1	Low
305	*Salvia aegyptiaca* L.	Lamiaceae	Perennial	Forb	Gravel, Wadis	2	Moderate	Medicinal	1	Low
306	*Salvia lanigera* Poir.	Lamiaceae	Perennial	Forb	Sand, Gravel	2	Moderate	Medicinal	1	Low
307	*Salvia spinosa* L.	Lamiaceae	Perennial	Forb	Gravel, Wadis	2	Moderate	Medicinal	1	Low
308	*Savignya parviflora* (Delile) Webb	Brassicaceae	Annual	Forb	Sand, Gravel	2	Moderate	Fodder	1	Low
309	*Scabiosa olivieri* Coult.,	Dipsacaceae	Annual	Forb	Silt	1	High	Fodder	1	Low
310	*Scabiosa palaestina* L.	Dipsacaceae	Annual	Forb	Disturbed Areas	1	High	Fodder	1	Low
311	*Schimpera arabica* Hochst. & Steud. ex Steud.	Brassicaceae	Annual	Forb	Roadsides	1	High	Fodder	1	Low
312	*Schismus arabicus* Nees	Poaceae	Annual	Grass	Sand, Silt	2	Moderate	Fodder	1	Low
313	*Schismus barbatus* (L.) Thell.	Poaceae	Annual	Grass	Sand, Silt	2	Moderate	Fodder	1	Low
314	*Sclerocephalus arabicus* Boiss	Caryophyllaceae	Annual	Forb	Sand	1	High	Fodder	1	Low
315	*Scorpiurus muricatus* L.	Fabaceae	Annual	Forb	Sand, Gravel, Wadis	3	Moderate	Fodder	1	Low
316	*Scorzonera papposa* DC.	Asteraceae	Perennial	Forb	Rocky slopes	1	High	Edible, Fodder	2	Low
317	*Scorzonera tortuosissima* Boiss.,	Asteraceae	Perennial	Forb	Rocky slopes	1	High	Fodder	1	Low

318	*Scrophularia deserti* Del.	Scrophulariaceae	Perennial	Shrub	Gravel, Wadis	2	Moderate	Medicinal	1	Low
319	*Seetzenia orientalis* Pecne.	Zygophyllaceae	Perennial	Forb	Sand, Gravel	2	Moderate	Medicinal	1	Low
320	*Seidlitzia rosmarinus* Bunge ex Boiss.	Amaranthaceae	Perennial	Shrub	Sand, Coastal Areas	2	Moderate	Medicinal, Detergent	1	Low
321	*Senecio glaucus* L.	Asteraceae	Annual	Forb	Sand	1	High	Medicinal, Fodder	2	Low
322	*Senecio vulgaris* L.	Asteraceae	Annual	Forb	Sand, Gravel, Disturbed Areas, Plantations	4	Low	Medicinal, Fodder	2	Low
323	*Setaria verticillata* (L.) P. Beauv	Poaceae	Annual	Grass	Plantations	1	High	Fodder	1	Low
324	*Setaria viridis* (L.) P. Beauv	Poaceae	Annual	Grass	Plantations	1	High	Medicinal, Fodder	2	Low
325	*Silene arabica* Boiss	Caryophyllaceae	Annual	Forb	Sand, Gravel	2	Moderate	Fodder	1	Low
326	*Silene arenosa* C. Koch	Caryophyllaceae	Annual	Forb	Sand	1	High	Fodder	1	Low
327	*Silene conoidea* L	Caryophyllaceae	Annual	Forb	Sand, Gravel	2	Moderate	Fodder	1	Low
328	*Silene villosa* Forssk	Caryophyllaceae	Annual	Forb	Sand	1	High	Fodder	1	Low
329	*Sinapis arvensis* L	Brassicaceae	Annual	Forb	Roadsides	1	High	Fodder	1	Low
330	*Sisymbrium erysimoides* Desf.	Brassicaceae	Annual	Forb	Plantations	1	High	Fodder	1	Low
331	*Sisymbrium irio* L.	Brassicaceae	Annual	Forb	Plantations	1	High	Medicinal, Fodder	2	Low
332	*Sisymbrium orientale* L.	Brassicaceae	Annual	Forb	Disturbed Areas, Plantations	2	Moderate	Fodder	1	Low

(continued)

Table 11.3 (continued)

No.	Species	Family	Life form	Growth form	Habitat	Number of habitats occupied	Priority	Uses	Score	Priority
333	*Sisymbrium septulatum* DC.	Brassicaceae	Annual	Forb	Sand, Gravel	2	Moderate	Fodder	1	Low
334	*Solanum nigrum* L.	Solanaceae	Annual	Forb	Plantations	1	High	Medicinal, Cosmetic	2	Low
335	*Sonchus oleraceus* L	Asteraceae	Annual	Forb	Plantations	1	High	Medicinal, Edible	2	Low
336	*Sonchus tenerrimus* L.	Asteraceae	Annual	Forb	Disturbed Areas	1	High	Ornamental	1	Low
337	*Sorghum halepense* (L.) Pers.	Poaceae	Perennial	Grass	Plantations	1	High	Fodder	1	Low
338	*Spergula fallax* (Lowe) E.H.L. Krause	Caryophyllaceae	Annual	Forb	Silt	1	High	Fodder	1	Low
339	*Spergularia diandra* (Guss.) Boiss	Caryophyllaceae	Annual	Forb	Disturbed Areas, Plantations	2	Moderate	Fodder	1	Low
340	*Spergularia marina* (L.) Griseb	Caryophyllaceae	Annual	Forb	Plantations	1	High	Fodder	1	Low
341	*Sphenopus divaeciatus* (Gouan) Reichb	Poaceae	Annual	Grass	Sand, Silt	2	Moderate	Fodder	1	Low
342	*Sporobolus arabicus* Boiss.,	Poaceae	Perennial	Grass	Sand, Silt	2	Moderate	Fodder	1	Low
343	*Stellaria media* (L.) Vill	Caryophyllaceae	Annual	Forb	Plantations	1	High	Medicinal, Fodder	2	Low
344	*Stipa capensis* Thunb.	Poaceae	Annual	Grass	Sand, Gravel	2	Moderate	Fodder	1	Low

345	*Stipagrostis ciliata* (Desf.) De Winter	Poaceae	Perennial	Grass	Sand, Gravel	2	Moderate	Fodder	1	Low
346	*Stipagrostis drarii* (Täckh.) De Winter.	Poaceae	Perennial	Grass	Sand	1	High	Fodder	1	Low
347	*Stipagrostis obtusa* (Delile) Nees	Poaceae	Perennial	Grass	Sand, Gravel	2	Moderate	Fodder	1	Low
348	*Stipagrostis plumosa* (L.) Munro ex T. Anderson	Poaceae	Perennial	Grass	Sand, Gravel	2	Moderate	Fodder	1	Low
349	Suaeda aegyptiaca (Hasselq.) Zohary	Amaranthaceae	Annual	Forb	Sand, Coastal Areas	2	Moderate	Medicinal, Edible, Fodder	3	Moderate
350	*Suaeda vermiculata* Forssk. ex J.F. Gmel	Amaranthaceae	Perennial	Shrub	Coastal Areas	1	High	Medicinal, Fodder	2	Low
351	*Tamarix aucheriana* (Decne.) Baum	Tamaricaceae	Perennial	Shrub	Coastal Areas	1	High	Medicinal, Shade	2	Low
352	*Telephium sphaerospermum* Boiss	Caryophyllaceae	Annual	Forb	Gravel	1	High	Fodder	1	Low
353	*Teucrium oliverianum* Ging. ex Benth.	Lamiaceae	Perennial	Shrub	Sand, Wadis	2	Moderate	Fodder	1	Low
354	*Teucrium polium* L.	Lamiaceae	Perennial	Shrub	Gravel	1	High	Medicinal	1	Low
355	Thymelaea mesopotamica (C. Jeffrey) B. Peterson	Thymelaeaceae	Annual	Forb	Sand, Gravel	2	Moderate	Medicinal	1	Low
356	*Traganum nudatum* Delile	Amaranthaceae	Perennial	Shrub	Sand, Rocky Slopes	2	Moderate	Fodder	1	Low

(continued)

Table 11.3 (continued)

No.	Species	Family	Life form	Growth form	Habitat	Number of habitats occupied	Priority	Uses	Score	Priority
357	*Tribulus macropterus* Boiss.,	Zygophyllaceae	Annual	Forb	Sand, Gravel	2	Moderate	Medicinal	1	Low
358	*Tribulus terrestris* L.	Zygophyllaceae	Annual	Forb	Sand, Silt	2	Moderate	Medicinal, Fodder	2	Low
359	*Trifolium lappaceum* L.	Fabaceae	Annual	Forb	Roadsides	1	High	Fodder	1	Low
360	*Trifolium resupinatum* L.	Fabaceae	Annual	Forb	Gravel, Wadis	2	Moderate	Fodder	1	Low
361	*Trigonella anguina* Delile	Fabaceae	Annual	Forb	Sand, Gravel	2	Moderate	Fodder	1	Low
362	*Trigonella hamosa* L.	Fabaceae	Annual	Forb	Sand	1	High	Fodder	1	Low
363	*Trigonella stellata* Forssk.	Fabaceae	Annual	Forb	Gravel	1	High	Fodder	1	Low
364	*Trisetaria linearis* Forssk.	Poaceae	Annual	Grass	Sand	1	High	Fodder	1	Low
365	*Typha domingensis* (Pers.) Poir. ex Steud.	Typhaceae	Perennial	Grass	Wadis	1	High	Medicinal	1	Low
366	*Urospermum picroides* (L.) F.W. Schmidt	Asteraceae	Annual	Forb	Wadis, Plantations	2	Moderate	Fodder	1	Low
367	*Urtica urens* L.	Urticaceae	Annual	Forb	Disturbed Areas, Plantations	2	Moderate	Medicinal	1	Low
368	*Vaccaria hispanica* (Mill.) Rauschert	Caryophyllaceae	Annual	Forb	Disturbed Areas, Plantations	2	Moderate	Medicinal	1	Low

369	*Vicia sativa* L.	Fabaceae	Annual	Forb	Plantations	1	High	Medicinal, Fodder	2	Low
370	*Withania somnifera* (L.)	Solanaceae	Perennial	Shrub	Plantations	1	High	Medicinal	1	Low
371	*Xanthium strumarium* L.	Asteraceae	Annual	Forb	Disturbed Areas, Plantations	2	Moderate	Fodder	1	Low
372	*Zilla spinosa* (L.) Prantl	Brassicaceae	Perennial	Shrub	Sand, Gravel	2	Moderate	Fodder, Shade	2	Low
373	*Zizyphus spina-christi* (L.) Willd.	Rhamnaceae	Perennial	Tree	Sand, Silt, Coastal Areas	3	Moderate	Medicinal, Edible, Fodder	3	Moderate
374	*Zygophyllum qatarense* Hadidi.	Zygophyllaceae	Perennial	Shrub	Sand, Coastal Areas	2	Moderate	Medicinal, Fodder	2	Low

Source: Omar et al. (2007)

allocation to the diverse components of biodiversity (Le Berre et al. 2019). Prioritization can be completed based on species use value, habitats, etc., depending on the targets and goals. In general, conservation of biodiversity is important for human well-being because it provides various direct and indirect benefits to human being and also plays an important role in ecosystem functions. Therefore, engagement and participation of different stakeholders such as government organizations, non-governmental organizations (NGOs), scientists, and local users of natural resources is essential. The information in this chapter will be helpful in bridging the research gap by developing and applying conservation strategies and to improve the effectiveness of conservation actions in a scientific and informed way.

The native plants are widely utilized in Kuwait for various economic and ecological purposes. Among the use categories, fodder, medicinal, and ornamental were the main use categories, indicating their economic and environmental potential. Most plant species used in Kuwait are used for only one or two categories, whereas a few plant species such as *Acacia pachyceras*, *Aeluropus lagopoides*, *Alhagi graecorum*, *Anabasis setifera*, *Bassia muricata,* and *Cressa cretica* have multiple usages (>3), indicating that these species have higher potential for utilization and development and thus they should be given extra attention. Prioritization of species can be very useful for the ranking of species based on the use value and additionally it can be used for developing future management decisions. However, most young people are unaware about the importance of native plants. Therefore, this study could be helpful in fulfilling the educational and awareness gaps and promote them to understand the importance of conservation and sustainable utilization of native plants. Moreover, initiation of cultivation of prioritized species may not only reduce the pressure on their wild population but will also provide additional income.

Thirteen different habitats have been identified namely sand, gravel, silt, wadis, disturbed areas, coastal area, runnels, loam, plantations, rocky slopes, salt marsh, roadsides, and aquatic (Fig. 11.5). Among these habitats, sandy soil had the highest number of species followed by gravel, disturbed area, and silty soil. These findings indicate that difference in soil substrate structure among the habitats is responsible for variation in the species richness. Previous studies also suggested that species richness is influenced by soil at the local scale by regulating the water and nutrient supplies (Kammer et al. 2013; Dingaan et al. 2017). Variation in particle size among the habitat may have affected the soil permeability, aeration, soil reaction (pH), nutrients availability, cation exchange capacity, soil fertility, and presence and richness of mycorrhizae (Bednarek et al. 2004; Carrenho et al. 2007; Alday et al. 2012). Overall, the maximum number of species preferred sandy soil as compared to other habitats. Usually, sandy soils are typically low in nutrients due to relatively coarse particle size and larger pore spaces wash out any available nutrients. However, most of the desert plant species have ability to cope with infertile sandy soils (Chen et al.

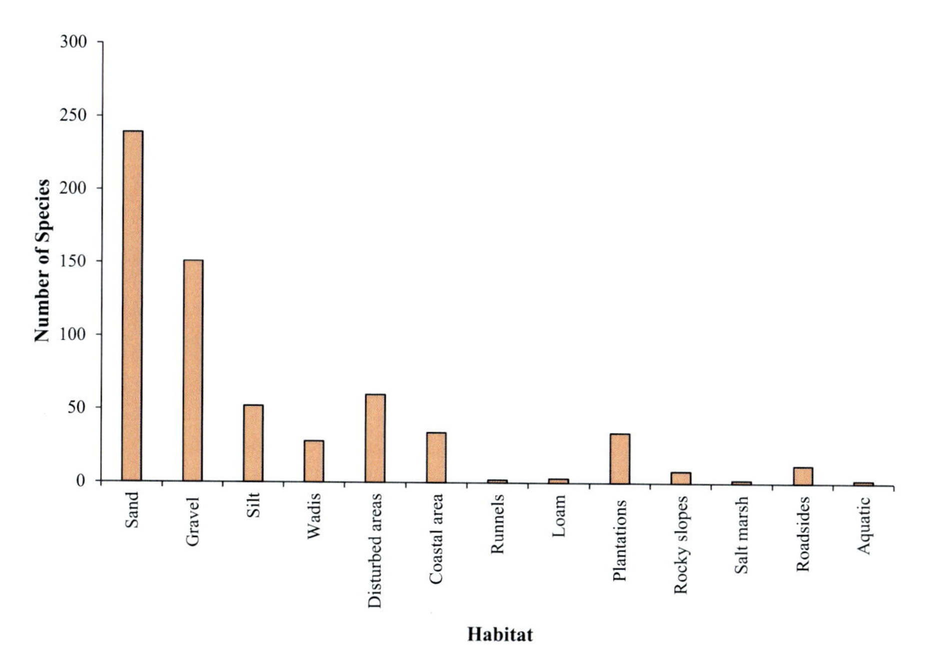

Fig. 11.5 Distribution of native species in major habitat type

2014). Moreover, the fluctuation in species diversity among the habitats could be linked to the differences in physical and edaphic factors among the habitats (He et al. 2022).

Usually, information of species habitats is necessary for effective conservation planning for protection and species recovery plans. Presently, the information about the habitats of most of the native species of Kuwait has remained incomplete. In the present study, 152 species are occupied only in single habitat indicating their habitat specificity. The necessity of special habitat requirement for growth and proliferation can make them prone to extinction, if their habitat is lost due to anthropogenic or natural factors (Işik 2011). On the other hand, species that occurred in different habitats are more likely to have higher chances of maintaining their population in the near future. Moreover, such species will have lower chances of extinction because of their ability to grow in specific habitats. Generally, habitat loss and destruction are recognized as the main direct threats to biodiversity. Therefore, special attention should be given for the conservation for those species which are restricted to specific habitat condition. However, further investigations about the species-specific habitat requirement are essential to predict the likely effects of future climate change impact on these species.

11.7 Conclusions and Recommendations

Overall, Kuwait-native plants play a significant role in safeguarding the stability of the natural environment. The diverse flora also provides various essential economic and ecological services to humans. However, most of the areas in Kuwait are prone to grazing and human impacts and ultimately, posing serious threats to floral diversity. Therefore, it is necessary to set up a comprehensive conservation program to avoid the extinction of native flora and to protect their habitats. Moreover, species survey and long-term monitoring plan need to be given priority in order to collect more authentic baseline data for prioritizing the species for biodiversity conservation at both national as well as regional levels.

References

Abdullah MT (2017) Conserving the biodiversity of Kuwait through DNA barcoding the flora. PhD Dissertation University of Edinburgh UK

Abdullah MT, Al-Dosari ME (2022) Vegetation of the State of Kuwait. IUCN and Kuwait, State of Kuwait: Environment Public Authority, Gland

Al Salameen F, Habibi N, Kumar V, Al Amad S, Dashti J, Talebi L, Al Doaij B (2018) Genetic diversity and population structure of *Haloxylon salicornicum* moq. in Kuwait by ISSR markers. PloS One 13(11):e0207369

Al Salameen F, Habibi N, Al Amad S, Al Doaij B (2022) Genetic Diversity of *Rhanterium epapposum* Oliv. Populations in Kuwait as revealed by GBS. Plants 11(11):1435

Al-Awadhi JM (2001) Impact of gravel quarrying on the desert environment of Kuwait. Environmental Geology 41(3):365–371

Al-Awadhi JM, Omar SA, Misak RF (2005) Land degradation indicators in Kuwait. Land Degrad Dev 16(2):163–176

Alday JG, Marrs RH, Martínez-Ruiz C (2012) Soil and vegetation development during early succession on restored coal wastes: a 6-year permanent plot study. Plant Soil 353(1):305–320

Al-Dosari M (2021) Revision of the Flora of Kuwait, Nomenclature and Revisiting the Taxonomy of *Arnebia* (Boraginaceae). PhD Dissertation. Arabian Gulf University Bahrian

Al-Dousari AM, Alsaleh A, Ahmed M, Misak R, Al-Dousari N, Al-Shatti F, William T (2019) Off-road vehicle tracks and grazing points in relation to soil compaction and land degradation. Earth Syst Environ 3(3):471–482

Almedeij J (2014) Drought analysis for Kuwait using standardized precipitation index. Sci World J 1(1):1–9

Alsdirawi F, Faraj M (2004) Establishing a transboundary peace park in the demilitarised zone (DMZ) on the Kuwaiti/Iraqi borders. Parks 14(1):48–55

Amr ZS (2021) The state of biodiversity in Kuwait. IUCN and Kuwait, State of Kuwait: Environment Public Authority, Gland

Aronson J, Kigel J, Shmida A (1993) Reproductive allocation strategies in desert and Mediterranean populations of annual plants grown with and without water stress. Oecologia 93(3):336–342

Baskin CC, Baskin JM (2014) Seeds: ecology, biogeography, and evolution of dormancy and germination, 2nd edn. Academic Press, San Diego

Bednarek R, Dziadowiec H, Pokojska U, Prusinkiewicz Z (2004) Ecopedological studies. PWN, Polish Scientific Publisher, Warsaw, p 344

Böer B, Sargeant D (1998) Desert perennials as plant and soil indicators in Eastern Arabia. Plant and Soil 199(2):261–266

Boulos L, Al-Dosari M (1994) Checklist of the flora of Kuwait. Kuwait J Sci 21(2):203–218
Brown G (2003) Species richness, diversity and biomass production of desert annuals in an ungrazed *Rhanterium epapposum* community over three growth seasons in Kuwait. Plant Ecol 165(1):53–68
Burger JA (2009) Management effects on growth, production and sustainability of managed forest ecosystems: past trends and future directions. Forest Ecol Manag 258(10):2335–2346
Carrenho R, Trufem SFB, Vlr B, Silva ES (2007) The effect of different soil properties on arbuscular mycorrhizal colonization of peanuts, sorghum and maize. Acta Botanica Brasilica 21(3):723–730
CBD-Convention on Biological Diversity (1993) United Nations Treaty Series 1760: 1–30619
Chen G, Zhao J, Zhao X, Zhao P, Duan R, Nevo E, Ma X (2014) A psammophyte *Agriophyllum squarrosum* (L.) Moq.: a potential food crop. Genet Resour Crop Evol 61(3):669–676
Costanza R, dArge R, de Groot R, Farber S, Grasso M, Hannon B, Limburg K, Naeem S, Oneill RV, Paruelo J, Raskin RG, Sutton P, Vanden BM (1997) The value of the world's ecosystem services and natural capital. Nature 387(6630):253–260
Daoud HS, Al-Rawi A (1985) Flora of Kuwait. v. 1: Dicotyledoneae
Dickson V (1955) Wild flowers of Kuwait and Bahrain. Allen and Unwin
Dingaan MNV, Tsubo M, Walker S, Newby T (2017) Soil chemical properties and plant species diversity along a rainfall gradient in semi-arid grassland of South Africa. Plant Ecol Evol 150(1):35–44
FAO (2006) Forests and climate change in the Near East Region. Rome, Research and Extension, Food and Agriculture Organization of the United Nations
Friedman J, Rubin MJ (2015) All in good time: understanding annual and perennial strategies in plants. Am J Bot 102(32):497–499
Glover JD, Reganold JP, Bell LW, Borevitz J, Brummer EC, Buckler ES, Xu Y (2010) Increased food and ecosystem security via perennial grains. Science 328(5986):1638–1639
Grman E, Lau JA, Schoolmaster DR Jr, Gross KL (2010) Mechanisms contributing to stability in ecosystem function depend on the environmental context. Ecol Lett 13(11):1400–1410
Gutterman Y (1993) Seed germination in desert plants. Adaptations of desert organisms. Springer, Berlin, p 253
Halwagy R, Halwagy M (1974) Ecological studies on the desert of Kuwait. II. The vegetation. Kuwait J Sci 1(1):87–95
Hassan A, Alfaraj M, Fayad M, Allen CD (2021) Optimizing site selection of new cities in the desert using environmental geomorphology and GIS: a case study of Kuwait. Appl Geomat 13(4):953–968
He C, Fang L, Xiong X, Fan F, Li Y, He L, Zhu J (2022) Environmental heterogeneity regulates species-area relationships through the spatial distribution of species. For Ecosyst 9:100033
Işik K (2011) Rare and endemic species: why are they prone to extinction? Turk J Bot 35(4):411–417
Kammer PM, Schöb C, Eberhard G, Gallina R, Meyer R, Tschanz C (2013) The relationship between soil water storage capacity and plant species diversity in high alpine vegetation. Plant Ecol Divers 6(3-4):457–466
Kernick MD (1966) Plant resources, range ecology and fodder plant introduction. Report to the Government of Kuwait. FAO, TA, 181:14
Komatsu KJ, Avolio ML, Lemoine NP, Isbell F, Grman E, Houseman GR, Zhang Y (2019) Global change effects on plant communities are magnified by time and the number of global change factors imposed. Proc Natl Acad Sci U S A 116(36):17867–17873
Le Berre M, Noble V, Pires M, Médail F, Diadema K (2019) How to hierarchise species to determine priorities for conservation action? A critical analysis. Biodivers Conserv 28(12):3051–3071
Mace GM, Lande R (1991) Assessing extinction threats: toward a reevaluation of IUCN threatened species categories. Conserv Biol 5(2):148–157
Margules CR, Pressey RL (2000) Systematic conservation planning. Nature 405(6783):243–253
Martin DM, Moss JMS (1997) Age determination of *Acacia tortilis* (Forsk.) Hayne from northern Kenya. Afr J Ecol 35(3):266–277

Midgley JJ, Bond WJ (2001) A synthesis of the demography of African acacias. J Trop Ecol 17(6):871–886
Miller AG, Cope TA, Nyberg JA (1996) Flora of the Arabian Peninsula and Scotra. Edinburg University Press, Edinburgh
More TA, Averill JR, Stevens TH (1996) Values and economics in environmental management: a perspective and critique. J Environ Manag 48(4):36–44
Nassep M (2022) Conservation of native plants as a means of resource utilization and desertification mitigation in Kuwait. Dissertation, University of Birmingham, UK
Omar SAS, Bhat NR (2008) Alteration of the *Rhanterium epapposum* plant community in Kuwait and restoration measures. Int J Environ Stud 65(1):139–155
Omar SAS, Shahid SA (2013) Reconnaissance soil survey for the State of Kuwait. In: Shahid SA, Taha FK, Abdelfattah MA (eds) Developments in soil classification, land use planning and policy implications: innovative thinking of soil inventory for land use planning and management of land resources. Springer Science and Business Media, Dordrecht, pp 85–107
Omar SAS, Misak R, King P, Shahid SA, Abo-Rezq H, Grealish G, Roy W (2001) Mapping the vegetation of Kuwait through reconnaissance soil survey. J Arid Environ 48(3):341–355
Omar SAS, Al-Mutawa YAA, Zaman S (2007) Vegetation of Kuwait: A comprehensive illustrative guide to the flora and ecology of the desert of Kuwait. Aridland Agriculture Department, Food Resources Division, Kuwait Institute for Scientific Research
Özyavuz M, Korkut AB, Özyavuz A (2013) Native vegetation. In: Advances in landscape architecture. IntechOpen
Phondani PC, Bhatt A, Elsarrag E, Horr YA (2017) Ethnobotanical magnitude towards sustainable utilization of wild foliage in Arabian Desert. J Tradit Complement Med 6(3):209–218
Pimm SL (2008) Biodiversity: climate change or habitat loss—which will kill more species? Curr Biol 18(3):R117–R119
Poppenwimer T, Mayrose I, DeMalach N (2022) Revising the global biogeography of plant life cycles. BioRxiv
Quijas S, Schmid B, Balvanera P (2010) Plant diversity enhances provision of ecosystem services: a new synthesis. Basic Appl Ecol 1(7):582–593
Shahid SA, Omar SAS (2022) Kuwait soil taxonomy. Springer, p 149
Simberloff D, Martin JL, Genovesi P, Maris V, Wardle DA, Aronson J, Vilà M (2013) Impacts of biological invasions: what's what and the way forward. Trends Ecol Evol 28(1):58–66
Stévart T, Dauby G, Lowry PP, Blach-Overgaard A, Droissart V, Harris DJ, Couvreur TL (2019) A third of the tropical African flora is potentially threatened with extinction. Sci Adv 5(11):eaax9444
Wassie SB (2020) Natural resource degradation tendencies in Ethiopia: a review. Environ Syst Res 9(1):1–29
Whittaker RJ, Araújo MB, Paul J, Ladle RJ, Watson JEM, Willis KJ (2005) Conservation biogeography: assessment and prospect. Divers Distrib 11(1):3–23
Wilson TB, Witkowski ETF (1998) Water requirements for germination and early seedling establishment in four African savanna woody plant species. J Arid Environ 38(4):541–550
Zhang J, Ge Y, Chang J, Jiang B, Jiang H, Peng C, Zhu J, Yuan W, Qi L, Yu S (2007) Carbon storage by ecological service forests in Zhejiang Province, subtropical China. Forest Ecol Manag 245(1–3):64–75

Chapter 12
Wildlife of the Terrestrial Ecosystems of Kuwait

Matrah Abdulrazag Al-Mutairi

Abstract Arid land environment exerts tremendous pressure on wild animals living within it. The dry and hot conditions, in addition to limited and restricted resources, make the survival of the terrestrial animals very challenging. Furthermore, human impact and the expansion of urban activities negatively impact the survival and distribution of native fauna in the State of Kuwait. The dry areas in the Arabian Peninsula had witnessed the extinction of several faunal species in the last century due to the previously mentioned reasons. Many large ungulates and predators were lost or their current distribution is restricted. In attempts to restore loss of wildlife species, many countries in the region developed conservation and reintroduction programs to protect the wildlife of the region. The situation in Kuwait is similar but more focus and efforts should be directed to the reintroduction and rehabilitation programs, especially that Kuwait had joined many international treaties and conventions to protect the natural resources including wildlife. Kuwait should maintain its active role in these conventions to network with international scientific communities for mutual benefits and to further the objectives of the conventions to accomplish the set goals.

Keywords Fauna · Arid land · Urban · Survival · Reintroduction · Conservation · Conventions

M. A. Al-Mutairi (✉)
Desert Agriculture and Ecosystems Program, Environment and Life Sciences Research Center, Kuwait Institute for Scientific Research, Safat, Kuwait
e-mail: mmutairy@kisr.edu.kw

M. K. Suleiman, S. A. Shahid (eds.), *Terrestrial Environment and Ecosystems of Kuwait*, https://doi.org/10.1007/978-3-031-46262-7_12

12.1 Introduction

The State of Kuwait is located at the northwestern corner of the Arabian Gulf and covers an area of 17,818 km^2. During the summer, the terrestrial parts are extremely harsh and dry with temperature that can reach 48 °C and prevailing northwestern winds accompanied sometimes with sand storms. Temperatures drop to be pleasant in winter seasons and may drop to 8 °C (Omar 2007). Kuwait is a relatively flat country with small escarpments. The ecosystem is classified into six categories (Omar 2007), (i) *Coastal plain and lowland ecosystem* (characterized by salt marshes, saline depressions, and covers the areas of coastline and islands), (ii) *Desert plains ecosystem* (extending from the coastal line and toward the southwestern borders and northwestern part of the State of Kuwait), (iii) *Alluvial fan ecosystem* (this occurs at the northern, central, and western parts of the country and is mostly known as gatch with high concentration of gypsum and minimum growth of plants only in areas with accumulated sands), (iv) *Escarpments and ridges ecosystem* (distinguished by relatively elevated features such as Jal Az-Zor, the Ahmadi ridge, and Albahra. The highest is Jal Az-Zor which rises to 135 m), (v) *Wadis and depression ecosystem* (these are scattered within the drainage system of rain water and usually very rich in vegetation due to the high and prolonged moisture compared to other ecosystems), and (vi) *Sand dune ecosystem* (originating from the northwestern parts of Kuwait, the sand dunes extend to the southwest parts in accordance with the prevailing northwestern winds).

Due to the flat topography of Kuwait and the small country size, there are less significant differences in the distribution of wildlife within different ecosystems. Marine birds and most of the migratory birds prefer to utilize the coastal plain ecosystems due to their feeding habits and migratory routes. On the other hand, small mammals avoid coastal ecosystems due to habitat isolation and human interference and prefer the desert plains and sand dunes. Larger mammals are mostly restricted to protected areas and prefer dense vegetation cover associated with escarpments and wadis. Lizards and reptiles are distributed around different ecosystems but with higher numbers in the areas which are protected and access is restricted. Snakes in particular prefer sand dunes due to the abundance of mice colonies in these areas (Al-Mutairi et al. 2014, 2017, 2018). It is essential to emphasize on the importance of protected areas to conserve biodiversity of living organisms in general and wildlife in particular at the State of Kuwait. The biodiversity and density of wildlife species are greater in protected areas compared to non-protected and open areas (Al-Mutairi 2012).

12.2 Fauna of Kuwait

In Kuwait, the fauna adapted to arid zones suffer from the human impact and urban development, that has deteriorated their natural habitats and consequently reduced their number and distribution to critical levels. The habitat degradation and fragmentation caused the reduction and decline in the recorded species, their population size, and distribution. This is mainly due to urbanization and habitat fragmentation. Most desert animals are localized to protected areas, such as natural reserves, research station, protected islands, and open areas of industrial and military regions where access is limited (Omar et al. 2001; Misak et al. 2002; Al-Dousari 2005). Plant and animal communities are disappearing rapidly from open access areas. Habitat loss and land deterioration are not the only factors contributing to the diminished wildlife populations in Kuwait. Other factors include poaching and intended destruction of animal natural habitats. Mammals are suffered the most from these natural and anthropogenic factors. They lack the ability to migrate to other regions with favorable living conditions and they are too big to hide compared to arthropods and lizards. For the past 100 years, the desert of Kuwait has witnessed local extinctions of several large mammals such as Arabian wolf (*Canis lupus arabs*), Arabian oryx (*Oryx leucoryx*), striped hyaena (*Hyaena hyaena*), jackal (*Canis aureus*), honey badger (*Mellivora capensis*), *Gazelle subgutturosa* and *G. gazelle*, sand cat (*Felis margarita*), Ruppell's fox (*Vulpes rueppellii*), and Cape hare (*Lepus capensis*). Species of bird and reptile are also recorded to be threatened in Kuwait (Alsdirawi 1989).

Although Kuwait lies within harsh and dry region, fauna shows remarkable biodiversity within the terrestrial component of the country (Fig. 12.1). Records show the presence of more than 420 species of birds recorded in Kuwait, both residents and migratory. The number of mammals recorded in Kuwait is 32 species, including extinct and rare species. The recorded number of reptile species is 42 while recorded species of arthropods can reach more than 800 (Al-Mutairi et al. 2014). This relatively high biodiversity status compared to poor habitats is fundamental and needs to be maintained.

As biodiversity and habitat loss increase, public awareness and governmental measures are being enforced. To reverse this situation, there is a need to preserve and restore what has been lost due to human interference and competition on limited natural resources. Most conservation ecological studies have been focused on finding and validating mechanisms capable of restoring damaged ecosystem components, rather

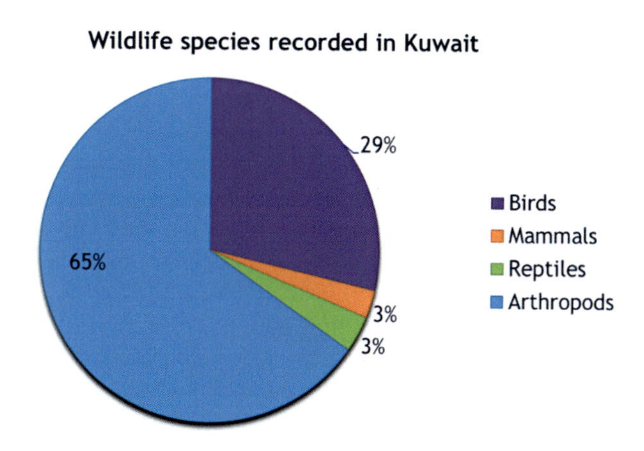

Fig. 12.1 Distribution of the recorded fauna species in the State of Kuwait

than reintroducing animals. This is justified, since these components are important for the animal's habitat and restoring this habitat will eventually benefit wildlife.

Most restoration ecology projects in the Middle East depend on conserving and protecting the natural ecosystems, rather than managing. This process does not allow practical implementation of scientific information. In addition, it does not prevent the transfer of degradation factors from the surrounding damaged ecosystems. Protecting natural ecosystems without management may even enhance the deterioration of the system. A perfect example is the domination of floral and faunal invasive species and competition with local species on the limited resources of the protected areas (Puddu et al. 2009). A common mistake occurring when choosing and protecting specific areas is the lack of natural corridors between these parts. This will isolate the animals inhabiting each area and prevent any communication and interaction between animals. Eventually, genetic diversity will be lost and this will negatively affect the biodiversity. In the Arabian Peninsula, including Kuwait, wild mammals are of special historical importance since they existed in healthier ecosystems in the past. Most of these animals are now extinct or found only in zoos and private collections. The restoration of these large mammal communities was a major point of interest in the past few decades, therefore, some mammals were reintroduced into protected areas where they reproduced and that was a huge success. Such mammals were the Arabian oryx (*Oryx leucoryx*) that was reintroduced in Bahrain, Qatar, United Arab Emirates, Kingdom of Saudi Arabia, and Oman. The success was highly dependent on human support. The reintroduced large mammals were regularly supported with food, water supply, medical care, and even in vitro fertilization programs. Furthermore, these reintroduction programs are of high cost. Franco et al. (2007) highlighted the importance of evaluating the techniques used for restoration purposes and insisted on choosing the most effective and most cost-effective ones. This will allow restoration programs to last for long times.

12.3 The Importance of Arid Land Ecosystems

Arid land ecosystems are of special importance because they host a unique blend of living organisms highly adapted and tolerant to drought conditions and harsh environments. The native vegetation of Kuwait is of special importance because it represents a transition between semi-arid desert and arid desert vegetation (Omar et al. 2001). It also offers a valuable gene pool and plant materials for drought, heat, and salt tolerance research (Omar and Bhat 2008). There are several factors that contribute toward making arid lands a harsh environment for animals to survive. These factors include large fluctuation in temperature between day and night as well as summer and winter, long drought periods followed by comparatively heavy rain, food scarcity and high competition between animals to acquire food resources, open areas, and the danger of being exposed to predators and lacking solid substrates for shelter and cover. Desert animals have evolved survival techniques and the unique adaptations enabled them to live in harsh environments. Their survival mechanisms may help, and inspire, to overcome some of the world's most frightening problems such as food scarcity, harsh environmental changes, and drought. They also provide a fundamental block in the diversity of our planet, and an important gene pool for any future studies. Moreover, arid ecosystems in Kuwait and other parts of the world cover almost 30% of the world's land area and are inhabited by almost two billion people (UN 2012). Any loss or degradation of these ecosystems will eventually impact not only on the people inhabiting them, but the whole world. In addition, diminished plant cover will reduce the absorption of atmospheric carbon dioxide (White et al. 1999).

12.4 Interaction of Fauna with Surrounding Environmental Components

The identification of important habitat types for wild animals is a key toward planning sustainable land management strategies. A habitat is defined as an area of the landscape that satisfies an animal's living requirements for food, shelter, and reproduction (Anderson 1991). All animals live under physiological and environmental constraints. Habitat selection reveals the delicate balance that an animal practices in using its surrounding habitat components. This balance should include consideration of costs related to both predation risk and safety on one hand, and benefits related to foraging, shelter, and reproduction on the other (Stephens and Krebs 1986). For example, many morphological defenses from predation, such as crypsis (the ability to conceal itself especially from a predator), require animals to restrict themselves to suitable habitats (Grand and Dill 1999; Leaver and Daly 2003; Caro 2005). Granivorous rodents (eating grain and seeds) inhabiting arid environments tend to show stable foraging behavior and need to cover large areas while searching for food (Tchabovsky et al. 2004).

Desert fauna can play a role in the restoration of degraded ecosystems. Some animals work as environmental engineers by dispersing natural seeds and relocating them in favorable places. Figure 12.2 provides some insights on how the seed consumers may help the seed dispersal process. In addition, some rodents and ant species redesign their habitat by creating small seed banks within their burrows or next to their areas of activities (Fig. 12.3). The created burrows catch and accumulate seeds transferred by natural elements. This seed pool will provide stability for seeds to revegetate (Fig. 12.2) (Al-Mutairi 2012).

Habitat utilization is of fundamental importance to understand a species' ecology. Species that rank their habitats utilization with respect to resource productivity and predation can promote their coexistence and survival over time (Grand 2002). Animals balance foraging benefits with predation risk in order to survive. Avoiding habitats with high predation risk is a natural behavior when favored habitats with high food quality and quantity are common (Tew et al. 1992), but when favored habitats are scarce, habitats which provide foraging resources and expose animals

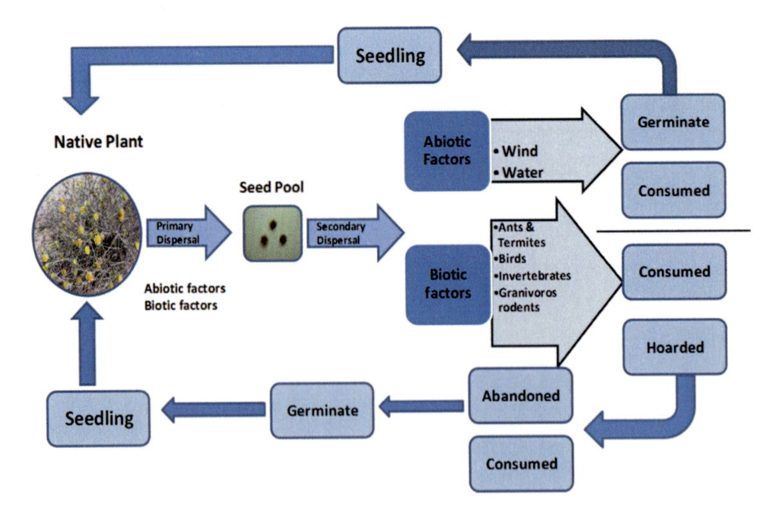

Fig. 12.2 The possible fates of native seeds in arid ecosystems (Al-Mutairi 2012)

Fig. 12.3 Natural catchment created by desert rodents in the protected desert area (Source: Al-Mutairi 2012)

to predation risk may need to be used. Other factors influencing habitat selection by animals are the provision of shelter, mating opportunities, and camouflage. Preferred habitats are not always available, thus, habitat fragmentation, isolation, and habitat destruction may force animals to adapt to available habitats or to choose sub-optimal habitats (Prugh et al. 2008).

12.5 Impact of Habitat Loss in Kuwait on Wildlife

Some studies showed that desert animals were more abundant in the intact natural habitats compared to degraded habitats. Population of wild animals' density was higher in the intact areas compared to degraded areas. Habitat selection was significantly affected by degradation and human activities; both were consistently avoided as proved by studies conducted in Kuwait (Misak et al. 2002; Al-Mutairi et al. 2018). A study (Al-Mutairi 2012) examined the population density of the desert rodent, and lesser jerboa (*Jaculuc jaculus*) at the arid land of Kuwait. Three degrees of degradation were tested to reflect the correlation between habitat loss and population density (Fig. 12.4). The population of lesser jerboa was higher in intact and protected areas compared to semi-degraded or completely degraded lands. The study also showed clear evidence of the wild animal avoiding human settlements and activities (Al-Mutairi 2012).

Predation risk is thought to be a key factor determining when and where animals forage in arid lands (Sivy et al. 2011). However, predators and prays usually follow similar habitat selection, both avoid sites of human activities and settlements. Areas covered with moving, unstable sand dunes or lie within wind corridors were avoided by most animal species within the arid lands. Those areas usually have low density of vegetation cover and, therefore, less food resources for animals in all trophic (process of getting and eating food) levels. Different wildlife species may coexist with other competing species with similar feeding habits, suggesting that there is a balance between those coexisting species and their food resources (Fig. 12.5). Traba et al. (2010) suggested that coexisting rodent species in desert environments follow one of two strategies. Both coexisting species are specialists and therefore each feeds on a specific type of food and their diets do not overlap. Alternatively, generalist species coexist with specialists, when generalists feed on food items that are less preferred by specialists. Limited dietary resources in arid environments force

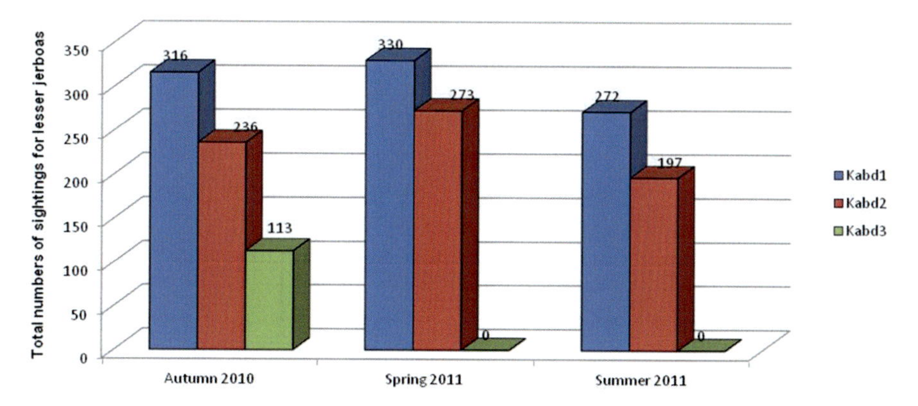

Fig. 12.4 The total number of lesser jerboa sightings on nocturnal transects in Kabd1 (intact), Kabd2 (semi-intact), and Kabd3 (degraded) in different seasons and years. (Source: Al-Mutairi 2012)

Long-eared Owl (*Asio otus*) at Kabd area, west of Kuwait city

Black giant ant (*Camponotus xerxes*)

Desert Wheatear (*Oenanthe deserti)*

Lesser spotted Eagle (*Aquila pomarina*)

Red fox (*Vulpes vulpes*)

Fig. 12.5 Wildlife of Kuwait

animals to exhibit more diverse physiological and ecological behaviors in order to survive. Recent studies have shown that some wild animals living in partially degraded lands were smaller in size and had shorter limbs compared to the same species living in intact adjacent areas (Al-Mutairi et al. 2012).

Land degradation due to increased stocking rate of grazing animals is a threat to many landscapes and habitats (Desbiez et al. 2009). Cumming and Cumming (2003) studied the relationships between body mass, shoulder height, hoof area, stride length, and daily ranging distance in African ungulates, and found that the larger grazing mammals destroyed a greater area. The same case is obvious in Kuwait. The unprotected areas inhabit large grazing animals such as camels and medium-sized grazing animals such as sheep and goat but in large numbers. All these animals form a destructive effect over open arid lands.

12.6 Human-Induced Factors Affecting the Terrestrial Wildlife in Kuwait

Globally, there is a noticed decline at all aspects of biodiversity levels. Several factors induced this decline either by number of species or their distribution. These factors are generally attributed to human factors in addition to other minor factors such as global warming and shifting of seasons. In Kuwait, biodiversity loss of arid lands is mainly caused by human interaction with its natural surroundings. The loss of natural habitats to urbanization is the main factor contributing to habitat loss or fragmentation. Most of the wildlife species in Kuwait are now restricted to remote protected areas such as natural reserves or fenced border that surround the borders of Kuwait with adjacent countries. Human activities such as the expansion of new cities and settlements into the natural habitats for fauna caused the extinction of some species and decline in the number and distribution of other species. Oil drilling and oil industry also caused severe damage to the welfare of wild animals. In addition to driving the wildlife away from areas with active drilling and excavation, the pollution caused by oil industry significantly reduced the biodiversity in certain areas. On the other hand, areas under oil companies provide indirect protection for wildlife species, provided that the area has no activities related to oil industry (Al-Mutairi et al. 2014). Oil companies in the State of Kuwait are known for fencing and protecting large areas for potential use in the future. Quarries, on the other hand, were a very common practice in Kuwait in the past. Although this was banned and stopped in the past few years, the quarry areas still suffer from the loss of topsoil and consequently the loss of both flora and fauna. Several attempts were made to restore the quarry areas but they were not very successful due to the loss of original topsoil. The state of Kuwait has announced the increase of protected areas throughout different regions of the country in an attempt to restore vegetation and biodiversity. The areas cover large portion of the country and will definitely elevate the biodiversity of wildlife. However, most of these areas are not connected and have no natural corridors which will result in gene isolation of the animal's populations within each area. This situation will induce population decline within the species living in the areas due to gene pooling and accumulation of genetic disorders.

Winter camping is a common practice in Kuwait and is part of the culture and heritage of Kuwaiti people. This practice was simple and environmentally friendly in the past but recently, it turned to be very damaging to the surrounding natural environment. This practice involves off-road driving and establishing temporary settlement in remote natural habitats which causes the loss of natural habitat for fauna especially destroying their shelters and damaging their food resources such as vegetation and other faunal species. Roads exert a range of harmful effects on biological communities: they may cause the loss and fragmentation of natural habitats, and introduce mortality due to collisions with vehicles (Forman and Alexander 1998; Slater 2002; Riley et al. 2006). The impact of individual roads increases with increasing magnitude as a consequence of increased width, traffic volume, and vehicle speed (Dowding et al. 2010). Overgrazing is another major factor for habitat loss to wildlife. Large herds of sheep and camels are kept in the open areas and fed mainly with supplements but released to feed on natural vegetation and cause the loss of habitats and food sources that support wildlife. Another major problem is the illegal hunting which causes the decline in the populations of both residents and migratory birds. Although laws in the State of Kuwait ban hunting and harming birds or other faunal species, it is still very common practice especially during fall and spring season when birds migrate through the country.

12.7 Future Plans for Restoration of Fauna in Kuwait

Restoration of fauna in the State of Kuwait is always combined within the efforts to restore damaged arid lands. This approach is logical since wild animals can only flourish and increase in species diversity and population levels within intact and supportive habitats. Revegetation and restoration of damaged lands is planned in large scales covering vast areas around the State of Kuwait. Future plans include the systemic revegetation of remote open areas, damaged areas within the oil-producing sector, military areas, and natural reserves with partial degradation in collaboration with the United Nations (Al-Enezi et al. 2022; Madouh et al. 2022). Most of these areas have lost their vegetation cover due to human activities, military impact, oil contamination, and global warming. Several governmental and non-governmental bodies within the country collaborate to achieve this goal of restoring damaged ecosystems. Restoring fauna to its original status requires gathering sufficient information and data derived from indigenous knowledge and literature based on the heritage of the country. Several extinct species were mentioned within the Kuwaiti heritage and others were documented by scientific records. It is essential to start with compiling bibliographic record of all wildlife species related to Kuwait. The following steps are to be adopted to determine invasive species that have to be maintained under control or eliminated if they negatively impact the existence of native fauna.

12.7.1 Monitoring

Monitoring is the first step in any restoration process. Time-based *monitoring* of the project progress helps to identify potential issues that can be resolved to further the project, track progress, and measure outcomes to assure the project is leading to meet the set goals and the objectives. This should start prior to restoration process and continue throughout the whole process and even after the completion of restoration efforts. The collected data will reflect the success or failure of any restoration process. The data should be collected on the species identification level and the density of each species. Distribution and seasonal changes are also essential to determine the success or failure. Monitoring includes but not limited to detailed information for surveyed species (regulations, laws, etc.) defining the study area, obtaining proper licenses, stratifying the study sites, site preparation, habitat assessment, and previous records of fauna.

The monitoring process varies according to the surveyed species. Different species require different methods that can provide data about the existence and distribution of each species. The survey methods are decided according to the targeted group of native animals such as migratory birds, resident birds, game animals, carnivores, prey species, rodents, insects, and reptiles.

12.7.2 Development and Use of National Databases

To serve the data needs of the conservation projects, it is essential to create database management system (DMS). The database should be designed to accommodate all the required themes in any restoration study, which have facilitated the designing of the restoration plan. The objective of establishing database is to collect and manage the ecological data during the monitoring process. The system should be capable of managing and tracking information related to the project data generated from the environmental assessment and surveys conducted during the data collection period. This information includes aerial images, climate data, soil, vegetation, sand movement, contamination, flora, and fauna. The database can efficiently serve the needs of any project by integrating and analyzing data utilizing geographical information system (GIS) tool, remote sensing, photogrammetry, and Microsoft Office applications. Data management plays an essential role in the success of restoration projects, especially the data collection and analysis tasks. Initially, all pre-documented data and sources of information should be integrated into a centralized database.

12.7.3 Development of Conservation and Reintroduction Programs

It is vital to determine the feasibility of re-introduced species that used to live freely in Kuwaiti land and disappeared from the country due to overhunting and habitat loss. It is also important to start establishing a captive breeding facility that supports the welfare and reproduction of the acquired species for reintroduction purposes. There are two options for the reintroduction process: the first one is to acquire animal species that are adapted to live in open areas and are treated as wild animals from the origin, and the second option is to buy the acquired species, breed them, and release the second or third generation to the wild. Kuwait has large and diverse areas; some of them represent suitable releasing areas for extinct species to support conservation and reintroduction objectives. Further studies are required before deciding the optimal area for release. At the current stage, areas within the borders or controlled by the oil sector represent potential areas for breeding programs.

Some species are considered possible candidates for captive-breeding and release, based on ethical and cultural values since they were part of the cultural heritage of the area and some of them were successfully re-introduced in neighboring countries with similar geographical characteristics. The most likely species to be re-introduced are listed in Table 12.1.

The captive-breeding and future release of animals bred at the proposed Wildlife Breeding Center is suggested to be divided into two stages. Stage 1 is to concentrate on medium-sized mammalian carnivores and avian species for captive-breeding and possible release in the designated area, which is suggested to be done in the first 3 years. The second stage is to breed large native wildlife, such as large ungulates, which is suggested after 3 years from the start of the reintroduction projects. The selected population of reintroduced animals should be genetically diverse of wild

Table 12.1 Most successful wildlife species suggested for re-introduction programs

ID	Species	Family	Common name	Conservation status in Kuwait
1	*Oryx leucoryx*	Bovidae	Arabian oryx	Extinct
2	*Gazella subgutturosa marica*	Bovidae	Sand gazelle	Extinct
3	*Gazella dorcas saudiya*	Bovidae	Arabian gazelle	Extinct
4	*Felis silvestris iraki*	Felidae	Wild cat	Extinct
5	*Caracal caracal schmitzi*	Felidae	Caracal	Extinct
6	*Herpestes edwardsi*	Viveridae	Indian grey mongoose	Extinct
7	*Chlamydotis undulata macqueeni*	Otididae	Houbara Bustard	Rare
8	*Struthio camelus syriacus*	Struthiodidae	Ostrich	Extinct

animals. This could be done by having different sources of the original reintroduced populations. GCC countries have different public and private populations that could be used. In addition, some zoos in the United States of America (USA) have an original collection, especially the Arabian Oryx that was captured from the Arabian Peninsula and bred for decades. International, regional, and local expertise is required for such major projects; in addition, continuous training on reintroduction guidelines and protocols is a major component of any sustainable reintroduction process (Al-Mutairi et al. 2014).

12.7.4 Enforcement of Environmental Laws

This section summarizes the Kuwait national and international legal authority's policies that protect the native habitats and ecosystems. The main entity in Kuwait for enforcing the environmental laws is the Environment Public Authority (EPA). The mandates are utilized in the development of the management plan to identify and define basic management responsibilities for arid lands in Kuwait. All natural resources within the terrestrial lands fall within the State of Kuwait Environmental Protection Law (EPA) No. (42); entitled "Natural Resources". The following are the key elements in selected Articles of the law which stipulate conservation and protection themes of Kuwait's natural resources:

Article 1: Clauses in Article 1 refer to the conservation of the environment, its natural resources, and biodiversity, the management of indicator species to measure the health of natural habitats or ecosystems, and rehabilitation of degraded areas. Additional excerpts cover integrated environmental management of coastal areas, requiring coordination among concerned governmental authorities, which guarantees the conservation of the environment in coastal areas.

Article 3: Calls for the "conservation of biodiversity in the full territory of the state".

Article 15: Specifies government funding for projects aiming at the protection of the environment and the rehabilitation of damaged sites.

Article 113: "Environmental Management" addresses the Environment Police as a martialed special unit established under the Ministry of Interior to be responsible for "the enforcement of environmental laws and regulations".

Article 125: Additionally, defines responsibilities to establish a natural history museum as follows: "The government shall establish a museum for the natural history of the State of Kuwait for enhancing environmental citizenship, and preserving the natural history of the State of Kuwait in all forms, whether geological, geomorphological, biological or marine, and whether currently existing, lost or extinct". This museum shall be established within five years as of the date of issuing this law as a maximum. The Supreme Council shall determine the governmental entity responsible for the establishment of this museum, and the governmental authority to be concerned with its management".

State of Kuwait Environmental Protection Law (EPA) No 42: Cultural Heritage. The following is a selected Article of the law that elaborates on the conservation and protection of Kuwait's cultural and historic resources (under the authority of the NCCLA/Antiquities Department).

Article 124: Says that "The damage or prejudice or trafficking of cultural heritage, whether movable or immovable, is strictly prohibited". The same applies to the establishment of military or civilian facilities in archaeological sites of valuable historical, touristic, and religious nature".

In summary, the legislations to conserve wildlife and their habitats in the State of Kuwait are strongly and clearly formulated. However, the enforcement element requires more efforts to be effective. In the past few years, environmental rangers were trained and authorized to enforce the environmental laws at different sectors of the country. In addition, a recent section known as the Environmental Police was established to help reducing trespassing on natural environmental components including wildlife.

12.8 International Commitments

In addition to the aforementioned state legislative and policy mandates, The state of Kuwait has signed international conservation and biodiversity agreements, thereby committing to the protection of policies and agreements, which have application in principle to natural ecosystems, the most well-known convention is:

Convention on Biodiversity Conservation (CBD) – Kuwait signed the CBD in 1992 and formally ratified it on August 2, 2002, obligating the country to conserve important habitats and to establish a database on natural resources.

As a part of the international community, the State of Kuwait is obliged to participate and take action in major international conventions related to wildlife protection and promotion. Since it was established, the Environment Public Authority (EPA) of Kuwait was assigned the task of studying and signing major conventions, treaties, and protocols related to the environment, including fauna and flora, which are key environmental components. Therefore, all the international treaties were considered and referred to in this report including those which were only once signed and approved by EPA. Table 12.2 lists the international treaties related to wildlife and officially signed by the State of Kuwait.

Kuwait has been an integral part of its surrounding geographical region since it was established by sharing similar environmental stress and habitat loss. Therefore, it signed several environmental treaties related to wildlife that focused on issues of regional interests such as of the Gulf countries, Middle East, or the continent of Asia. Table 12.3 summarizes the regional conventions or protocols related to wildlife that Kuwait is committed to.

Table 12.2 International environmental treaties, conventions, protocols, and agreements related to wildlife signed and ratified by the State of Kuwait

Convention and treatise	Date of signature	Date of ratification
United Nations Convention to Combat Desertification (UNCCD) (1994)	22 September 1995	27 June 1997
United Nations Convention on Biological Diversity (UNCBD), Rio de Janeiro (1992)	9 June 1992	2 August 2002
Convention on the International Trade in Endangered Species of wild fauna and flora (CITES), Washington (1973)	9 April 1973	12 August 2002
Kyoto Protocol to the United Nations, framework convention on climate change, Japan (1997)	–	11 March 2005

Table 12.3 Regional environmental treaties, conventions, protocols and agreements related to wildlife signed and ratified by the State of Kuwait

Conventions and protocols	Date of signature	Date of ratification
Kuwait Regional Convention for Cooperation on the Protection of the Marine Environment from pollution (ROPME 1978)	24 April 1987	7 November 1978
ROPME protocol for the protection of the marine environment against pollution from land-based sources (1990)	21 February 1990	13 May 1992
Convention on the protection and conservation of wildlife and natural habitats in GCC countries	–	31 December 2002

12.9 Conclusions and Recommendations

The critical situation of wild animals living in the terrestrial habitats of Kuwait requires urgent measures to slow the deterioration of biodiversity. Loosing large predators and ungulates in the last century, decline in the number and distribution of species to limited habitats has been documented and thoroughly studied by different bodies and organizations in the region. The international, regional, and national collaboration is needed at this stage to implement conservation and reintroduction programs at the national level. The captive-breeding and future release of animals bred at the Wildlife Breeding Center is suggested to be divided into two stages. Stage 1 is to concentrate on medium-sized mammalian carnivores and avian species for captive-breeding and possible release in the designated area. The second stage is to breed large native wildlife, such as large ungulates. The selected population of reintroduced animals should be genetically diverse of wild animals. This could be done by having different sources of the original reintroduced populations. Gulf Cooperation Council (GCC) countries have different public and private populations that could be used. In addition, some zoos in the United States of America (USA) have wildlife species that are native to the Arabian Peninsula and were transformed to the USA before decades, especially the Arabian Oryx that was captured from the

Arabian Peninsula and bred for decades. International, regional, and local expertise is required for such major projects; in addition, continuous training on the reintroduction guidelines and protocols is a major component of any sustainable reintroduction process.

References

Al-Dousari A (2005) Causes and indicators of land degradation in the north-western parts of Kuwait. Arab Gulf J Sci Res 23:69–79

Al-Enezi A, Quoreshi AM, Bhat NR, Shahid SA, Al-Mutairi M, Petrov P, Madouh T, Suleiman MK, Al-Dousari A, Baroon HJ, Othman A, Dashti A, Al-Jeri M, Al-Dousari N, Grina R, Al-Mansour H, Manuvel AJ, Sivadasan MT, Jacob S, Thomas R, Al-Sayegh M, Cruz GD (2022) Ecological monitoring and evaluation of success of active revegetation of degraded terrestrial ecosystems under KERP (SP003EC). Kuwait Institute for Scientific Research Kuwait, progress report 2. KISR # 17538

Al-Mutairi M (2012) The impact of habitat degradation on the ecology of Lesser Jerboa *(Jaculus jaculus)* in the arid environments of the State of Kuwait. Ph.D. dissertation, University of Bristol, UK

Al-Mutairi M, Mata F, Bhuller R (2012) The effect of habitat degradation, season and gender on morphological parameters of lesser jerboas (*Jaculus jaculus L.*) Kuwait. Anim Biodivers Conserv 35:119–124

Al-Mutairi M, Khalil F, Almulla L, Al-Ballam Z, Al-Dossery S, Tawfik H, Ahmad I (2014) Develop a program for establishing an effective wildlife habitat management for the Kuwait Oil Company (KOC). Kuwait Institute for Scientific Research Kuwait, final report, KISR # 12204

Al-Mutairi M, Abu-Rezq H, Khalil F, Dcruz D, Ragheb G, Tawfik H, Al-Hebini, K (2017) Habitat assessment for the agricultural research station at Kabd area. Kuwait Institute for Scientific Research Kuwait, final report, KISR # 14175

Al-Mutairi M, Grina R, Al-Osiri Y, Awad M, Al-Mulla L, Kumar N, Subrahmanyam M, Al-Awadhi M, Al-Ragam O, Tawfiq H, Bhat N (2018) Ecological assessment and restoration plan of terrestrial biodiversity in Umm Al-Namil Iiland, Kuwait. Kuwait Institute for Scientific Research Kuwait, final report, KISR # 14623

Alsdirawi F (1989) Wildlife resources of Kuwait: historic trends and conservation potentials. PhD dissertation, University of Arizona, USA

Anderson S (1991) Managing our wildlife resources, 2nd edn. Prentice Hall, Upper Saddle River

Caro T (2005) Antipredator defenses in birds and mammals. University of Chicago Press, Chicago

Cumming D, Cumming G (2003) Ungulate community structure and ecological processes: body size, hoof area and trampling in African Savannas. Oecologia 134:560–568

Desbiez A, Bodmer R, Santos S (2009) Wildlife habitat selection and sustainable resources management in a Neotropical wetland. Int J Biodiv Conserv 1:11–20

Dowding C, Harris S, Poulton S, Baker P (2010) Nocturnal ranging behaviour of urban hedgehogs, *Erinaceus europaeus*, in relation to risk and reward. Anim Behav 80:13–21

Forman R, Alexander L (1998) Roads and their major ecological effects. Annu Rev Ecol Syst 29:207–231

Franco A, Palmeirim J, Sutherland W (2007) A method for comparing effectiveness of research techniques in conservation and applied ecology. Biol Conserv 1:96–105

Grand T (2002) Foraging-predation risk trade-offs, habitat selection, and the coexistence of competitors. Am Nat 159:106–112

Grand T, Dill L (1999) The effect of group size on the foraging behaviour of *juvenile coho* salmon: reduction of predation risk or increased competition. Anim Behav 58:443–451
Leaver L, Daly D (2003) Effect of predation risk on selectivity in heteromyid rodents. Behav Process 64:71–75
Madouh T, Suleiman MK, Quoreshi A, Bhat NR, Shahid SA, Al-Mutairi M, Al-Enezi A, Grina R, Al-Jeri M, Al-Dousari N, Othman A, Al-Mansour H, Baroon H, Al-Sayegh M, Mazyad, Sivadasan MT, Manuvel AJ, Jacob S, Thomas R, Khalifah N, Cruz GD (2022) Ecological monitoring and evaluation of restoration and revegetation success of areas damaged by military fortifications (element 1 of claim 5000450) Sheikh Sabah Al Ahmad Nature Reserve under KERP (SP004EC). Kuwait Institute for Scientific Research, Kuwait, progress report 2. KISR # 17539
Misak R, Al-Awadhi J, Omar S, Shahid SA (2002) Soil degradation in Kabd area, southwestern Kuwait City. Land Degrad Dev 13:403–415
Omar S (2007) Vegetation of Kuwait: a comprehensive illustration guide to the flora and ecology of the desert of Kuwait. Kuwait Institute for Scientific Research, Kuwait
Omar S, Bhat N (2008) Alteration of the *Rhanterium epapposum* plant community in Kuwait and restoration measures. Int J Environ Stud 65:139–155
Omar S, Misak R, King P, Shahid SA, Abo-Rizq H, Grealish G, Roy W (2001) Mapping the vegetation of Kuwait through reconnaissance soil survey. J Arid Environ 48:341–355
Prugh LR, Hodges KE, Sinclair ARE, Brashares JS (2008) Effect of habitat area and isolation on fragmented animal populations. PNAS 105(52):20770–20775. https://doi.org/10.1073/pnas.08060801
Puddu G, Maiorano L, Falcucci A, Corsi F (2009) Spatial-explicit assessment of current and future conservation options for the endangered Corsican Red Deer (*Cervus elaphus corsicanus*) in Sardinia. Biodivers Conserv 18:2001–2016
Riley S, Pollinger J, Sauvajot R, York E, Bromley C, Fuller T, Wayne R (2006) A southern California freeway is a physical and social barrier to gene flow in carnivores. Mol Ecol 15:1733–1741
Sivy K, Ostojab S, Schuppa E, Durham S (2011) Effects of rodent species, seed species, and predator cues on seed fate. Acta Oecol 37:321–328
Slater F (2002) An assessment of wildlife road casualties–the potential discrepancy between numbers counted and numbers killed. Web Ecol 3:33–42
Stephens D, Krebs J (1986) Foraging theory. Princeton University Press, Princeton
Tchabovsky A, Merritt J, Alleksandrov D (2004) Ranging patterns of two *syntopic gerbillid* rodents: a radiotelemetry and live-trapping study in semi-desert habitat of Kalmykia, Russia. Acta Theriol 49:17–31
Tew T, Macdonald D, Rands M (1992) Herbicide application affects microhabitat use by arable wood mice (*Apodemus sylvaticus*). J Appl Ecol 29:532–539
Traba J, Acebes P, Campos V, Giannoni S (2010) Habitat selection by two sympatric rodent species in the Monte desert, Argentina. First data for *Eligmodontia moreni* and *Octomys mimax*. J Arid Environ 74:79–185
UN (2012) From Rio to Rio+20: progress and challenges since the 1992 Earth summit. Rio+20, United Nations Conference on Sustainable Development. http://www.un.org/en/sustainablefuture/pdf/Rio+20_FS_RiotoRio.pdf
White A, Cannel M, Friend F (1999) Climate change impacts on ecosystems and the terrestrial carbon sink: a new assessment. Glob Environ Chang 9:S21–S30

Chapter 13
Major Threats to the Terrestrial Ecosystems of Kuwait and Proposed Conservation Practices

M. Anisul Islam and Sheena Jacob

Abstract Ecological restoration and conservation of terrestrial ecosystems in Kuwait, which is comprised of 374 plant species, are critically important to preserve the ecological balance of the country as both natural factors and anthropogenic disturbances have contributed to declining plant cover over the past decades. It is well documented that the recovery process of arid ecosystems from disturbances is slower than other ecosystems and can take a long time, and most ecosystems rely on permanent seed banks of native species. However, most late-successional species in the arid environment cannot create a permanent and viable seed bank. Therefore, restoration and conservation activities have to rely on sowing seeds or transplanting nursery-grown seedlings. Currently, aside from anthropogenic threats gradual climate change appears to be a major threat to terrestrial ecosystems in Kuwait, which in future will likely cause a decline in net primary productivity, reduction of ecosystem functions and services, and loss of wildlife habitat. Although numerous protected areas and seed banks have been established, further efforts should be focused on functional vegetation distribution database development, increasing the number of protected areas, identifying endangered species and their propagation and conservation, and a public campaign to increase awareness toward conservation of the Kuwait's terrestrial ecosystems.

Keywords Anthropogenic · Climate change · Habitat loss · Native plant species · Protected areas · Soil degradation · Ecological restoration

---

M. A. Islam (✉) · S. Jacob
Desert Agriculture and Ecosystems Program, Environment and Life Sciences Research Center, Kuwait Institute for Scientific Research, Safat, Kuwait
e-mail: aislam@kisr.edu.kw; sjacob@kisr.edu.kw

M. K. Suleiman, S. A. Shahid (eds.), *Terrestrial Environment and Ecosystems of Kuwait*, https://doi.org/10.1007/978-3-031-46262-7_13

13.1 Introduction

Arid lands, which cover about 26–30% of the world's land surface area, are characterized by prolonged drought and overexploitation of resources and are considered the most fragile lands on the earth (Gaur and Squires 2018). The aridity index, a numerical indicator of the degree of dryness of the climate of a given geographic location, is usually used to classify different categories of dry lands, and Kuwait is considered a hyper-arid area because of the lower aridity index (≤0.3). Hyper-arid areas cover approximately 8% of the earth's surface areas (UNESCO 1977; Maghchiche 2022) and are characterized by very low and irregular rainfall where shrubs are the only perennial vegetation observed in limited areas.

Most ecosystems around the globe are facing increased levels of threats in recent years due to climate change, habitat loss and degradation, pollution, overexploitation of natural resources, urban and agricultural expansion, and other anthropogenic disturbances. Despite Kuwait being a small country by land size at 17,818 km^2, Halwagy and Halwagy (1974) classified four ecosystems in Kuwait based on the variations in the habitat and floristic composition. However, Omar (2007) suggested that Kuwait has the following ecosystems: (1) the coastal plain and lowland ecosystem; (2) the desert plain and lowland ecosystem; (3) the alluvial fan ecosystem; (4) the escarpment, ridges, and hilly ecosystems; (5) the wadi and depressions ecosystem; and (6) the barchan sand dune ecosystem. Although, as a whole, the unique desert ecosystems of Kuwait are comprised of 374 plant species, most of which are annuals, and only 34 are shrub species. The primary plant communities in Kuwait include eight different vegetation groups such as *Haloxyletum, Rhanterietum, Cyperetum, Stipagrostietum, Zygophylletum, Centropodietum, Panicetum,* and *Halophyletum* (Omar 2007). Therefore, it is obvious that the ecosystems in Kuwait are relatively diverse and represent a complex structure through which they function and provide services. Each of the above-mentioned plant communities is host to specific fauna and flora. Each shows signs of land degradation due to overuse of resources, lack of significant natural regeneration of native species, conversion of land for urban expansions, invasion-related issues, and most importantly, gradual rise in atmospheric temperature because of global climate change effects.

13.2 Ecosystem Functions and Services

Ecosystem function is the capacity of natural components and their interactive processes to provide goods and services to sustain animal and human wellbeing, either directly or indirectly (de Groot et al. 2002). In contrast, ecosystem services are the benefits provided to humans by transforming resources (or environmental assets, including land, water, vegetation, and atmosphere) into a flow of essential goods and services such as clean air, water, and food (Costanza et al. 1997). The Millennium Ecosystem Assessment (MEA 2005), a major UN-sponsored effort to analyze the impact of human actions on ecosystems and human wellbeing, identified four major categories of ecosystem services: (i) provisioning, (ii) regulating, (ii) cultural, and (iv) supporting.

Like many other ecosystems, such as a forest ecosystem, mountain ecosystem, or agroecosystem, desert ecosystem around the world and particularly in Kuwait offers valuable ecosystem services in the form of food and medicinal ingredients for humans, forage for wildlife and farm animals, camouflage and shelter for wildlife, and cultural heritage such as falconry. However, due to biotic and abiotic threats and disturbances, the services provided by the desert ecosystems are dwindling gradually. In the following section, the major threats and disturbances to the desert ecosystems of Kuwait are discussed in detail.

13.2.1 Trends of Threats and Disturbances to Ecosystems, Including Invasion-Associated Damages

People have interacted with different ecosystems for centuries and relied on them for food, shelter, medicine, and other needs. Many cultures have developed around nearby ecosystems. However, as human populations have steadily increased over time, most ecosystems have been overexploited and in certain cases, have been converted to farm or agricultural lands. In recent years, urban developments and other needed infrastructure developments have also contributed toward threatening the functioning and overall existence of the ecosystems in Kuwait. However, a major ecosystem disturbance event occurred in 1990 during Iraq's invasion and occupation of Kuwait, resulting in significant and widespread environmental damages, including soil contamination and degradation, habitat loss, and major disruption in ecosystem functioning. One of the serious sources of damage was when about 700 oil wells were set on fire, which resulted in the release of oil and very toxic and carcinogenic compounds into the land and atmosphere. Kuwait's desert environment also suffered due to military and heavy machinery movements. Moreover, seepage of toxic chemicals and oil also contaminated the groundwater deposits. In summary, the following environmental damages were recognized by the United Nations Compensation Commission (UNCC 2023):

(a) Oil and chemical contamination of freshwater aquifers in North Kuwait
(b) Extensive oil contamination of coastal areas and potential damage to the marine environment
(c) Extensive damage to the desert ecosystem through oil and tarcrete deposition covering an area of 114 km^2
(d) 163 wellhead pits resulting from blown-out oil wells on land
(e) In addition to the above, further damage resulted from the disposal of ordnance and unexploded ordnance (UXO) in the desert, coastal, and marine ecosystems.

13.2.1.1 Natural and Anthropogenic Threats to the Ecosystems

There are a number of anthropogenic and natural threats to the ecosystems. In addition, these threats can be both biotic and abiotic. The potential threats to the ecosystems in general and the ecosystems of Kuwait in particular are described in the following sections.

Climate, Climate Change, and Land Degradation

Kuwait has a hyper-arid desert climate characterized by extremely hot temperatures during summer months with no rainfall. Summer temperature can rise up to 50 °C during mid-summer. December and January are the colder winter months when the mean low temperature can be as low as 14 °C. The average annual rainfall is 113 mm between November and April, and the average wind speed is 12 km/h (Almedeij 2012). In coming years, Kuwait's temperature is expected to rise by 1.8 °C to 2.57 °C compared to a 2000–2009 baseline (Alahmad et al. 2022). Additionally, drastic changes in rainfall amount and pattern could be linked to the effects of global climate change. Moreover, the frequency and duration of dust storms have increased in recent years (Asem and Roy 2010), indicating the potential role of climate change in Kuwait and surrounding countries.

Desertification, land degradation, and annual temperature rise, along with reduction and erratic rainfall patterns, have negatively affected the terrestrial plant population distribution in Kuwait. For example, in 1953, *Rhanterium epapposum* Oliv. community (the dominant native species of Kuwait) accounted for 40% of the total vegetated area in Kuwait, whereas it reduced to a mere 2.1% in 2008. Currently, this species can be found in protected areas only (Omar and Bhat 2008). A similar decline in coverage has also been observed with *Farsetia aegyptia* Turra. and *Haloxylon salicornicum* (Moq.) Bunge ex Boiss. The decline in vegetation cover in Kuwait could be attributed to soil degradation exacerbated by persistent high temperatures, high evaporation, removal of the top soil layer, and scanty soil organic matter. Furthermore, as the climate became harsher along with less and infrequent precipitation, and wind induced soil erosion that degrades the soil further, the production of healthy and viable seeds also declined, ultimately resulting in extremely poor natural recruitment of perennials, and eventual loss of vegetation cover.

Loss of Species and Net Primary Productivity

Net primary productivity (NPP), as the base for the research of matter recycling and energy flow in terrestrial ecosystems, is sensitive to the environment and climate changes in arid regions, and is an important indicator of eco-environmental characteristics. Compared to the NPP of 871–1098 g C ha^{-1} $year^{-1}$ in tropical forest ecosystems, desert ecosystems have an NPP of 28–151 g C ha^{-1} $year^{-1}$ (Gough 2011).

Scarce and unpredictable precipitation, high soil surface temperature, high evaporation, and high solar irradiance coupled with steady to strong gusty wind typify a warm desert climate. These conditions force perennial plants to survive in soils with limited available moisture which limits their photosynthesis and net primary productivity.

Loss of Wildlife Habitat

Climate change and deforestation could destroy 23% of all habitats worldwide by the end of this century (Di Marco et al. 2018). The drastic habitat loss would translate to a rapid extinction rate of the already vulnerable species. Currently, the habitat shrinkage for mammals, amphibians, and birds accounts for an 18% loss of previous natural ranges. At present, no data is available on the loss of wildlife habitat in Kuwait. However, it is presumed from many field observations that due to a lack of vegetation cover, the abundance or sighting of certain wildlife species has also declined in Kuwait.

Reduction of Ecosystem Functionality/Services

Kuwait's desert terrestrial ecosystem consists of many native plant species that provide shelter for wildlife and medicinal ingredients for human health. For example, *Farsetia aegyptia*, a key species of Kuwait's desert ecosystems, is used for its anti-diabetic and antispasmodic properties (Ismail et al. 2013). Moreover, plant extract of *F. aegyptia* is also taken to relieve rheumatic pains. *Haloxylon salicornicum* is another important native species in Kuwait used as forage for livestock and a food source for wildlife. In addition, *H. salicornicum* can help stabilize the soil surface and improve microclimate, providing camouflage and shelter for various animals (Böer and Sargeant 1998). Another dominant species in Kuwait is *Rhanterium epapposum* Oliv., a bushy perennial shrub commonly grazed by camels and sheep (Demirci et al. 2017). Its abundance has declined significantly during the past decades. However, it has medicinal values to treat skin infections, antioxidant activity due to polyphenolic content (Shahat et al. 2014), and the presence of essential oils (Al-Yahya et al. 1990).

With continuous overgrazing and due to land degradation, the abundance of *F. aegyptia, H. salicornicum,* and *R. epapposum* has declined, thereby, reducing the provision of ecosystem services such as the medicinal use of different plant parts of these species. In addition, the reduction of vegetation cover due to overgrazing or land degradation also resulted in habitat loss or change of habitat condition and quality for certain wildlife species in the desert ecosystems altering their functionality and services.

Camping and Off-Road Activities

Camping activity is a common practice in Kuwait, especially during winter. Kuwait municipality has designated 18 camping sites in the North, West, and South of the country, covering an area of 616 km^2 which is 3.45% of the total land area of Kuwait. According to available information, it was reported that 1217 camping permits were issued in 2017. Normally during winter, precipitation occurs in Kuwait, facilitating the seed germination of native annual and perennial plant species. However, camping activities, which coincide with the winter months, can cause severe damage to the desert ecosystems by affecting soil infiltration capacity and consequently disturbing the seed germination and seedlings growth of native species. This, in turn, causes loss of wildlife habitat that drives wildlife populations further away from the areas of campsites. However, human movement and activities often go beyond these areas, further exacerbating the impacts on the ecosystems and surrounding environments through land compaction and the resultant decline in water infiltration. Off-road activities in Kuwait include vehicle movement to supervise camel herds throughout the year, all-terrain vehicle (ATV) driving for leisure and transportation, and setting up temporary camps during winter months for recreation. The negative impacts of off-road activities on the ecosystem, wildlife habitat, and population are almost identical to that of camping activities.

Urban Development and Land Uses

With the increasing local population, the demand for housing, recreational areas, parks, and business establishments has increased over the past decade. In this process, natural desert ecosystems are lost because of urban expansion, road construction, and other related activities. Accurate information on the extent of urban development and the associated loss in biodiversity is not well documented yet.

Overexploitation and Uncontrolled Grazing

It is well documented that overgrazing by camels has been linked to loss of vegetation cover and associated loss of biodiversity. Overgrazing is one factor that can be attributed to the huge decline of *R. eppaposum* and to some extent, *F. aegyptia* cover in Kuwait. It is also presumed that the loss of vegetation cover may also have contributed to the decline of the wildlife habitat and loss of biodiversity in Kuwait.

Landfilling

Kuwait has a total of 19 landfill sites, of which six are currently in operation, three are rehabilitated, and the rest are closed. One of the landfill sites in Mina Abdullah has an area of about two million square meters and receives municipal waste (solid, household, agricultural, commercial, and non-hazardous industrial waste). In order to prepare land for landfills, soils need to be excavated, causing disturbance to the soil and surrounding environment. In addition, continuous heavy-duty vehicle movement also causes noise pollution and surface vibration, which could disturb the wildlife population living close to these landfill areas. Moreover, these landfill areas generate toxic gases deemed harmful to the environment and for animal and human health.

13.2.2 Current Conservation Practices

In order to conserve and enhance biodiversity of the terrestrial ecosystems in Kuwait, the following measures are being taken:

13.2.2.1 Native Seed Bank

Public Authority for Agriculture Affairs and Fish Resources (PAAFR) and Kuwait Institute for Scientific Research (KISR) actively maintain native seed banks. Currently, seeds of 17 key native species collected over time are stored in the KISR seed bank. Moreover, some of the large-scale private nurseries also collect native seeds locally every year and store them for future propagation purposes.

13.2.2.2 Revegetation with Native Species

Natural recovery from desert disturbance is typically very slow. For example, Abella (2010) reported that the average time for the re-establishment of perennial plant cover following a variety of disturbances in North America's Mojave and Sonoran Deserts was 76 years, and even partial recovery of species composition required over two centuries. Therefore, revegetation of desert areas is imperative to prevent the progression of desertification and to restore plant cover. Revegetation of desert areas can be possible either by sowing seeds or by outplanting nursery-grown seedlings. KISR routinely mass propagates key native species for revegetation of desert areas. Revegetation is currently a high priority in Kuwait as it has embarked on revegetating four protected areas (Al Huwaimliya, Wadi Al-Batin, Um Ghudair, and Al-Khuwaisat) with native plant species under the Kuwait Environmental Remediation Program (KERP) covering an area of 1156 km^2 (Alenezi et al. 2021).

13.2.2.3 Controlled Grazing

As stated earlier, overgrazing and human activities cause disturbances in desert ecosystems and the mortality of plants. One effective way to prevent overgrazing is fencing designated areas for the protection of native species and to ensure adequate natural recruitment of native species on a regular basis. Currently, many protected areas as well as other important areas in Kuwait, are fenced to minimize grazing by camel and other livestock.

13.2.2.4 Protected Areas

Increased anthropogenic disturbances and urban development coupled with climate change effects have made the desert ecosystem very vulnerable. Therefore, to preserve and enhance local biodiversity, allow future generations to enjoy nature and cherish their rich heritage, and safeguard the overall environment, the government has designated certain areas as protected. In situ conservation of native species is ensured in Kuwait in nature reserves, parks, and protected areas. Kuwait currently has 18 terrestrial protected areas and natural reserves (EPA 2023) covering an area of 2979 km^2 (UNEP-WCMC 2023).

13.3 Proposed Conservation Practices

From the conservation of biodiversity and land revegetation perspective, Kuwait is heading in the right direction, backed up by the government's commitment and policy formulation. However, more can be done, considering that Kuwait's terrestrial ecosystems are very fragile and vulnerable to the overall increase in global temperature due to climate change effects. The following steps should be considered and undertaken on an urgent basis to prevent further degradation of Kuwait's ecosystems and to ensure all-around environmental wellness:

13.3.1 Review and Gap Analyses of Current Conservation Practices to Set Up Conservation Strategy

A national conservation strategy should be developed for at least a 10-year period or more to ensure that all stakeholders are informed and coordinated for the overall conservation of Kuwait's terrestrial ecosystems. This conservation strategy should align with global climate changes and should have the flexibility to adjust as needed for Kuwait's ecosystem status and state of functioning.

13.3.2 Increasing Protected Areas to Conserve Native Vegetation

As mentioned earlier, currently an area of 2979 km^2 is designated as terrestrial protected area (UNEP-WCMC 2023) out of 17,818 km^2 of total land area in Kuwait, which is 16.72% of the total land area. The terrestrial ecosystems, through their services such as carbon storage, microclimate amelioration, improved air quality, and habitat for wildlife, ensure maintaining the sustainability of a healthy environment. Vegetation cover can be increased by increasing the total expanse of protected areas in Kuwait up to 25% of the total land area. This would allow the conservation of especially endangered species and prevent further biodiversity loss.

13.3.3 Functional Vegetation Distribution Database Development

In recent years, efforts were made by the Kuwait Institute for Scientific Research (KISR) to develop a comprehensive and functional vegetation distribution database. This database would help scientists and policymakers to continuously monitor the distribution of key native plant species as well as the rare species with changing climate and other associated factors and aid formulating policies that are aligned with the protection of natural resources and biodiversity of Kuwait. It is essential to ensure that the vegetation database is to be updated regularly.

13.3.4 Identification of Flagship and Extinct Species for Their Conservation

While flagship or keystone species have been identified in Kuwait for revegetation purposes, no studies have been conducted to identify Kuwait's rare and endangered plant species for their conservation and appropriate propagation. It is considered that the only tree species *Vachellia pachyceras* (Talha) of Kuwait is on the verge of becoming extinct (Suleiman et al. 2018) as it is currently represented by only one naturally occurring mature tree in Sabah Al Ahmed Nature Reserve (SAANR). Currently, efforts are being made to propagate this species from seeds at KISR. It is worth mentioning here that, as of now, Kuwait does not have a red list of plants that are in danger of extinction, and it is about time that efforts should be made to prepare such a list based on field observations throughout Kuwait.

13.3.5 Sustaining Rare and Endangered Species Through Propagation for Desert Rehabilitation

Many native plant species are facing the threats of extinction due to soil degradation, anthropogenic disturbances, rising temperature, and atmospheric carbon dioxide level because of global warming. For example, PAAFR staff through field observations have suggested that the presence of *Calligonum comosum* (arta) and *Ephedra alata* (alanda) has declined alarmingly over the past years in Kuwait. It has been reported that *Ephedra alata* (alanda) propagation is very challenging, and

PAAFR is currently working on its reproduction. As mentioned earlier, preparing a list for rare and endangered species in Kuwait is imperative, and efforts should be made to find appropriate propagation techniques for these rare and endangered species. Eventually, these should be introduced back to the desert ecosystem through seed sowing or by outplanting nursery-grown seedlings for their effective conservation and rehabilitation of the desert ecosystem.

13.3.6 Awareness Development Toward Conservation of Ecosystems

Along with policy formulation and execution to protect and conserve biodiversity, creating mass awareness for biodiversity conservation is an effective way for the conservation of biodiversity and desert ecosystem rehabilitation. This can be achieved by highlighting the values and services provided by healthy ecosystems to animals as well as humankind. This information can be part of the elementary school curriculum also. Mass awareness can also be achieved through publishing and disseminating conservation-related messages through social media such as Instagram, Facebook, and local radio and television channels.

The laws for protecting and preserving biodiversity in the terrestrial ecosystems of Kuwait are to be implemented strictly. Kuwait passed legislation in 2014, law no. 42/2014 on environmental protection. This law sets sanctions and large fines for those who destroy, uproot, or cut plants in desert ecosystems. However, unfortunately, grazing and off-road activities still take place in desert areas, often in protected areas. Therefore, stricter enforcement of the law is imperative to protect fragile desert ecosystems and conserve and sustain biodiversity.

13.4 Conclusions and Recommendations

Kuwait is a small country enriched by 374 native plant species. However, like many other countries in the world, the fragile desert ecosystems of Kuwait are vulnerable to the effects of global climate change with increasing temperature and ambient carbon dioxide levels, along with prevailing anthropogenic threats such as uncontrolled grazing, off-road activities, camping, landfilling, and infrastructure development. These threats will negatively affect ecosystem functioning and the services that they offer to humans and animals. To prevent further deterioration of ecosystem functions and services and to preserve and boost local biodiversity, Kuwait needs to formulate and execute conservation strategies aligned with global climate changes and local environmental threats, develop a vegetation database indicating the status of rare and endangered species, increase propagation of rare or threatened native species, and increase protected areas for the conservation of native species and biodiversity along with the stricter implementation of environmental law. In addition, establishment of a genebank to conserve native plant seeds is highly recommended. Since Kuwait joined Convention on the Protection and Conservation of Wildlife and Natural Habitats in Gulf Cooperation Council countries, it would be very important to develop international and regional linkages to promote the exchange of native plant seeds.

References

Abella SR (2010) Disturbance and plant succession in the Mojave and Sonoran Deserts of the American Southwest. Int J Environ Res Public Health 7:1248–1284

Al-Yahya MA, Al-Meshal IA, Mossa JS, Al-Badr AA, Tariq M (1990) Saudi plants: a phytochemical and biological approach, General Directorate of Research Grants Programs. KACST, Riyadh, pp 75–80

Alahmad B, Vicedo-Cabrera AM, Chen K, Garshick E, Bernstein AS, Schwartz J, Koutrakis P (2022) Climate change and health in Kuwait: temperature and mortality projections under different climatic scenarios. Environ Res Lett 17:074001. https://doi.org/10.1088/1748-9326/ac7601

Alenezi A, Quoreshi A, Bhat N, Shahid SA, Omar S (2021) Ecological monitoring and evaluation of active revegetation of degraded terrestrial ecosystems, Publication No. 17406. Kuwait Institute for Scientific Research, Kuwait

Almedeij J (2012) Modeling rainfall variability over urban areas: a case study for Kuwait. Sci World J. Article ID 980738:8. https://doi.org/10.1100/2012/980738

Asem S, Roy W (2010) Biodiversity and climate change in Kuwait. Int J Clim Change Strategies Manage 2:68–83. https://doi.org/10.1108/17568691011020265

Böer B, Sargeant D (1998) Desert perennials as plant and soil indicators in eastern Arabia. Plant Soil 99(2):261–266

Costanza R, d'Arge R, de Groot R, Farber S, Grasso M, Hannon B, Limburg K, Naeem S, O'Neill RV, Paruelo J, Raskin RG, Sutton P, van den Belt M (1997) The value of the world's ecosystem services and natural capital. Nature 387:253–260. https://doi.org/10.1038/387253a0

de Groot RS, Wilson MA, Boumans RMJ (2002) A typology for the classification, description and valuation of ecosystem functions, goods and services. Ecol Econ 41:393–408

Demirci B, Yusufoglu HS, Tabanca N, Temel HE, Bernier UR, Agramonte NM, Alqasoumi SI, Al-Rehaily AJ, Başer KHC, Demirci F (2017) *Rhanterium epapposum* Oliv. essential oil: chemical composition and antimicrobial, insect-repellent and anticholinesterase activities. Saudi Pharm J 25(5):703–708

Di Marco M, Venter O, Possingham HP, Watson JEM (2018) Changes in human footprint drive changes in species extinction risk. Nat Commun 9:4621. https://doi.org/10.1038/s41467-018-07049-5

EPA (2023) Maps and reserves. Environment Public Authority, Kuwait. https://epa.gov.kw/en-us/Maps. Accessed on 10 Apr 2023

Gaur M, Squires V (2018) Geographic extent and characterization of the world's arid zones and their peoples. https://doi.org/10.1007/978-3-319-56681-8_1

Gough CM (2011) Terrestrial primary production: fuel for life. Nat Educ Knowl 3(10):28

Halwagy R, Halwagy H (1974) Ecological studies on the desert of Kuwait. II. The vegetation. J Univ Kuwait (Sci) 1:87–95

Ismail WM, Motaal AA, Sokkar NM, El-Fishawy AM (2013) Botanical and genetic characteristics of *Farsetia aegyptia* Turra growing in Egypt. Bull Fac Pharm Cairo Univ. https://doi.org/10.1016/j.bfopcu.2013.04.001

Maghchiche A (2022) Natural resources conservation and advances for sustainability. https://doi.org/10.1016/C2019-0-03763-6

MEA (2005) Millennium ecosystem assessment. https://www.millenniumassessment.org/en/index.html. Accessed on 10 Apr 2023

Omar SAS (2007) Vegetation of Kuwait, 2nd edn. Kuwait Institute for Scientific Research, Kuwait. isbn: 978-99906-41-76-9

Omar SAS, Bhat NR (2008) Alteration of the *Rhanterium epapposum* plant community in Kuwait and restoration measures. Int J Environ Stud 65(1):139–155

Shahat AA, Ibrahim AY, Elsaid MS (2014) Polyphenolic content and antioxidant activity of some wild Saudi Arabian Asteraceae plants. Asia Pac J Trop Med 7:545–551

Suleiman MK, Dixon K, Commander L, Nevill P, Bhat NR, Islam MA, Jacob S, Thomas RR (2018) Seed germinability and longevity influences regeneration of *Acacia gerrardii*. Plant Ecol 219:591–609. https://doi.org/10.1007/s11258-018-0820-8

UNCC (2023) United Nations Compensation Commission. Follow up program for environmental awards. State of Kuwait. https://uncc.ch/state-kuwait. Accessed on 10 Apr 2023

UNEP-WCMC (2023) Protected area profile for Kuwait from the world database on protected areas, August 2023. Available at: www.protectedplanet.net. Accessed on 10 Apr 2023

UNESCO (1977) Map of the world distribution of arid regions. MAB technical notes 7. isbn: 92-3-101484-6. https://unesdoc.unesco.org/ark:/48223/pf0000032661

Chapter 14
Kuwait Deserts and Ecosystems in the Context of Changing Climate

Ali M. Quoreshi and Tareq A. Madouh

Abstract Desert ecosystems occupy one-third of the world's land surface and are highly vulnerable to natural and human effects and are unpredictable. The atmospheric concentration of most dominant *greenhouse gas* carbon dioxide (CO_2) has increased considerably as a result of anthropogenic activities. Modeling studies suggest that the Arab region will confront a rise of 2 °C to 5.5 °C in the surface temperature by the end of this century. Small alterations in temperature and precipitation because of climate change could extensively influence biodiversity living in the desert and disrupt the ecology of the already fragile desert ecosystems. However, it is difficult to visualize that global climate change would have a further effect on inherent unstable desert surfaces. Soil microbiomes accomplish an essential function in biogeochemical cycles and predict that global climate change may affect microbial biota that operate numerous functions on the Earth. Scientists started to recognize that rhizosphere processes are integral to understanding climate change impacts such as raised *greenhouse gas emissions*, heating, and drought on soil carbon sequestration. However, responses of microbiome to climate change are poorly understood. This publication aims to provide a review and discussion pertaining to the effect of climate change threats on the ecosystems in general and the Kuwait desert ecosystems in specific.

Keywords Arbuscular mycorrhiza · Biodiversity · Elevated CO_2 · Greenhouse gas · Increased temperature · Microbiomes · Sea level rise

---

A. M. Quoreshi (✉) · T. A. Madouh
Desert Agriculture and Ecosystems Program, Environment and Life Sciences Research Center, Kuwait Institute for Scientific Research, Safat, Kuwait
e-mail: aquoreshi@kisr.edu.kw; tmadouh@kisr.edu.kw

M. K. Suleiman, S. A. Shahid (eds.), *Terrestrial Environment and Ecosystems of Kuwait*, https://doi.org/10.1007/978-3-031-46262-7_14

14.1 Introduction

Kuwait is a small country with a total area of 17,818 km^2, located in the north-eastern part of the Arabian Peninsula with inhabitants of 4.1 million people. The country is surrounded in the north and northeast by Iraq, in the west and south side by Saudi Arabia, and it faces the Arabian Gulf to the east side. The climate of Kuwait can be characterized as harsh arid to hyper-arid environment with extended dry and extremely warm summer, with a mean temperature of 46.2 °C. The average minimum daily temperature can drop to 8.7 °C during the winter months (Dec–Feb 1999–2020) (https://climateknowledgeportal.worldbank.org/country/kuwait/climate-data-historical). Kuwait has scanty and erratic rainfall with an average of 110 mm/annum, low humidity, and frequent sandstorms (Suleiman et al. 2011). The climate is of BWh (hot desert climate) type according to Köppen climate classification (Köppen 1936). In recent periods, there has been considerable variation in precipitation and temperature from 1 year to another. Recently, Kuwait experienced over 50 °C on several occasions and a temperature of 54 °C was recorded in 2016 summer (Merlone et al. 2019; Alahamad et al. 2022). The desert landscape is characterized by patchy perennial vegetation in which dwarf shrubs and grasses predominate (Halwagy and Halwagy 1974). The plant communities of Kuwait comprise scarce perennials and annuals (Sudhersan et al. 2003).

Climate change is believed one of the utmost risks for ecosystems globally. Combine arid, semi-arid, and dry sub-humid lands presently include about 46.2% (±0.8%) of the worldwide terrestrial area with 3 billion inhabitants (Mirzabaev et al. 2019). Desert ecosystems are the most widespread ecosystems worldwide (Cherlet et al. 2018) and are believed highly vulnerable, most responsive, and unpredictable ecosystems to global climate alteration (Mellillo et al. 1993; Loarie et al. 2009). Anthropogenic activity and utilization of available resources by humans and organisms are utmost and have a significant impact on climate change that influences plant and belowground microbial diversity on our globe (Shree et al. 2022). Arid ecosystems are forecasted to be single most delicate systems to climate change. Increase in temperature, lessening precipitation, and enhancing atmospheric greenhouse gases (GHGs) concentrations have already caused considerable shifts in global climate patterns (Orlowsky and Seneviratne 2012) and are predictable to profoundly affect the composition of desert ecosystems (Smith et al. 2000; Bombi et al. 2021). As per most global climate change studies, a sustained build-up of carbon dioxide (CO_2) and other GHGs [(water vapor (H_2O), methane (CH_4), nitrous oxide (N_2O), ozone (O_3), and fluorinated gases] will advance substantial alterations in temperature and precipitation patterns over the globe (IPCC 2021). CO_2 is the major GHG from human activities and initiates global warming and climate change situation (IPCC 2007).

There is no dispute among scientists, whether the climate is altering, growing the atmospheric CO_2, or climate is becoming further warmer (IPCC 2007). Although most of the desert lands are already warmer, substantial warming has occurred through drylands globally (IPCC 2021). However, deserts are often ignored when are considered to climate change issues, but as temperatures increase, they are among the most susceptible places on the Earth. The modeling studies suggest that the Arab region will confront a rise of 2.0 °C to 5.5 °C in the surface temperature by the end of the twenty-first century (AFED 2009). Consequently, Kuwait is not immune from the global changes in climatic patterns, and according to the IPCC (2007), is likely to be moderately affected, possibly by slightly higher mean temperatures and lower rainfall. However, the latest IPCC Special Report on Climate Change and Land under Desertification Chapter (Mirzabaev et al. 2022) has confirmed that deserts and semiarid lands have already shown clear impacts of climate change effect with an escalation of aridity in some areas. The report mentioned with high confidence that drylands are shaped by climate change through increased temperature and further erratic rainfalls. The Climate Change Synthesis Report (2023) of the IPCC sixth assessment report (IPCC 2021) recognizes that sustained GHG emissions will advance to intensifying global warming, with the fairest estimation of reaching 1.5 °C in the near future based on various circumstances and modeled pathways considered and projected increase up to 4.4 °C for a very high GHG emissions scenario.

Global warming issues such as increasing atmospheric CO_2 and warming will be expected to amend ecosystem properties and processes. Continuous increases in temperature and drought will result in decreased vegetation yield and changes in vegetation survival and growth rates (Fernández-Martínez et al. 2019; Maurer et al. 2020). The trend may start to be reversing in some areas with effective mitigation approaches (Wang et al. 2020). Global climate change will not only influence vegetation communities and fauna but will also affect microbiome populations that perform numerous functions which are important to living on the Earth. However, the function of the microbiomes that control ecosystem processes in relation to global warming is less predictable. This publication reviews and addresses the climate change effect on the desert ecosystems and highlights the challenges to desert ecosystems with respect to global climate change for researchers, ecologists, and policymakers.

14.2 Understanding Climate Change Effects and Challenges to Desert Ecosystems

Climate change is often interchangeable to global warming to the public, the media, and the policymakers. Scientists nevertheless tried to use the term in the broader sense to also include natural variations in climate over time, which refer to climate change as a statistically significant variation, persisting for an extended period (Houghton et al. 2001) usually a minimum of 30-year period. Therefore, global warming is not only an alteration in temperature patterns but also a change in other variables in the climate system such as precipitation and prolonged drought. Desert lands are often disregarded when these are considered for climate change issues, but as temperature increase in an already warmer environment, deserts are the most vulnerable places on the Earth. Current studies reported that the Arab region has suffered an irregular increase in surface air temperature and rainfall patterns during the last few decades (IPCC 2007; AFED 2009). Arid ecosystems are projected to be some of the most sensitive systems to global climate change (AFED 2009). If temperatures in the region get increase, or rainfall gets lower, pressure on biological and physical systems would be deepened. Recently, global climate change has become a center of media and political attention around the world, and governments are now beginning to take the problem more seriously. Climate change not only modifies normal atmosphere but also other weather extremes (Fischer and Knutti 2015). Human-induced climate change is currently changing many weather and climate extremities in all areas including desert and semi-desert areas across the globe (IPCC 2023). These climate extreme situations have led to irreversible impacts that are reflected in associated losses and damages to nature and humans (IPCC 2023). The report exhibits a comprehensive assessment of up-to-date information on the perceived influences and predictable hazards of global climate change effect and the possible mitigation and adjustment options.

Anthropogenic activities, particularly discharges of CO_2, have unambiguously triggered warming, consequently, global surface temperature reaching 1.1 °C above 1850–1900 in 2011–2020 (IPCC 2021, 2023). Substantial warming has occurred across arid lands globally (IPCC 2021-AR6). Based on the scientific understandings, severe temperatures, inconsistent rainfall, and increasing evapotranspiration need to be linked with high intensity of anthropogenic activities, and are likely to outstrip the resilience limits of many ecosystems, and trigger irreversible landscape alteration (IPCC 2019). Global warming rates are believed to be double in drylands in contrast to humid lands, because scanty vegetation communities and diminished soil moisture content of dryland ecosystems intensify further temperature and aridity enhances (Huang et al. 2016). This projected increase warming trend is predicted to prolong in the coming decades. According to the latest IPCC Special Report on Climate Change and Land under Desertification Chapter 3 (Mirzabaev et al. 2022), surface warming over drylands is predicted to range ~6.5 °C under a high emission situation by the end of this century. Climate has notably altered in the Arabian Gulf countries, where the temperature has increased during the last 50 years, exhibiting decreased cold days and cooler nights, and the area encountered warmer nights (Al-Maamary et al. 2017).

There is scanty information in relation to climate change and Kuwait desert ecosystems. However, the global warming issue is one of the most demanding issues that Kuwait currently confronts due to the nature of the country's climate characteristics, high greenhouse gas emission rates, and high dependence on unsustainable and non-renewable energy sources. This publication is mainly focused on dryland ecosystems in general and in relation to Kuwait ecosystems. Kuwait is already experiencing a varied range of climate change issues including increased desertification, loss of biodiversity, sea level rise (SLR), and water scarcity (AFED 2009). Kuwait has experienced a significantly higher increase in temperature (1.5 °C to 2 °C) since 1975 compared to the global average (Zafar 2023). Recently, the rainfall pattern in Kuwait has changed which may be attributed to overall climate change effects. One of the most visible effects is the SLR on the coastal areas of Arabian Gulf states. Kuwait is also susceptible to the influences of SLR which may lead to submerging of low-level city areas, affecting coastal ecosystems and groundwater quality (AFED 2009). The SLR is predicted to affect Kuwait's islands, such as Boubyan which may be fully inundated within the next few decades. Similarly, one-third of Failaka Island is predicted to be submerged. The extreme weather episodes observed in drylands during the past decades caused many ecosystem alterations and show impacts on the economy and social influences (Al-Blooshi et al. 2020). Scientists predict that super and ultra-extreme heat episodes with temperatures above 56 °C will become recurrent in the Middle Eastern region and North African desert in the second half of this century (IPCC 2023). A few decades ago, it was considered that climate may not have considerable effect on the ecology of hyper-arid deserts as the climate is already extreme and nothing much to be altered (Warren et al. 1996). Currently, this belief has changed considerably. The ambiguity concerning the forthcoming effects of climate change on terrestrial biodiversity in the Arabian Peninsula is considered very high and makes prediction very difficult despite quality data (AGEDI 2016).

Different anthropogenic activities are converting more areas into deserts, a process called desertification. Global warming is decreasing water resources and is accelerating the incidence of drought, which dries up soils that are already dry causing damage to desert habitats and accelerating desertification. Desertification has been considered as the extreme environmental challenge of present time, and climate warming is further making it more severe. Reinstate arid ecosystems are the greatest challenge because of their inherent harsh environments, dry soils, sandstorms, and soil seed banks consumed by predation, herbivory, and reduced plant colonization, growth, and establishment (Abella 2012). This situation is further challenged by the climate change effects. However, regardless of these complications, restoration of desert ecosystems to some extent is possible using proper site-adapted techniques (Abella and Newton 2009). Further research in desert ecosystem restoration may assist to enhance restoration techniques and improve our understanding in this area.

14.3 Desert Plant Species in Response to Climate Change

Deserts comprise arid and semi-arid lands where evaporation and transpiration (water lost from plants to the air) surpass precipitation (Abella 2017). Desert ecosystem is home to unexpectedly high diversity vegetation communities in regions with low precipitation and elevated evapotranspiration due to extreme temperatures (Ward 2009). Deserts are found all over the globe on each continent, they vary in size, landscape, environment, and species diversity. Despite their characteristic aspect, deserts are home to many vital species, rich in biodiversity (Maestre et al. 2015), including some species not recorded elsewhere on the Earth. It is scientifically accepted that changes in climatic patterns are occurring as a direct result of anthropogenic activities and land surface changes (IPCC 2007). These phenomena are likely to have far-reaching consequences not only for many aspects of human life but also for vegetation cover and biodiversity in general, water, and soil microbial community (IPCC 2007). Despite characteristics of desert climatic conditions, climate change has considerably impacted desert ecosystems, causing further droughts and other extreme weather consequences that make deserts additional inhospitable and reduce productivity and soil fertility. It is anticipated that climate change is also driving the expansion of deserts, which can lead to a decline in biodiversity and amplified GHG emissions. Climate change is extensively recognized as a vital warning to biodiversity, which probably enhance the rate of its damage over time (Bellard et al. 2014). In general, arid ecosystems are one of the toughest territories on this planet and exhibit enormous challenges in keeping vegetation survival, growth, and establishment (Suleiman et al. 2019). In a recent report, drylands have exhibited varied trends of declines and enhancement in vegetation and biodiversity, based on the time, regions, and vegetation types (Mirzabaev et al. 2022). Kuwait is bestowed with abundant biodiversity of terrestrial flora and fauna; nevertheless, the possible decline of terrestrial and marine biodiversity because of climate change is a great concern in Kuwait.

The presence of perennials coverage in Kuwait is scarce and is typically fewer than 10% (Brown 2001). The indigenous plant community of Kuwait comprises scanty perennials and some annuals in spring (Sudhersan et al. 2003). As compared to other arid regions, annuals are the prominent native plant species (256 species) then herbaceous perennials (83 species), bushes and undershrub (34 species), and trees (one) species existing in Kuwait (Omar and Bhat 2008). The indigenous plants of Kuwait are of immense scientific importance as they signify a transition between semi-desert and desert vegetation. Furthermore, the vegetation community is well tolerant to severe climatic conditions. Some desert vegetation plants are considered vulnerable and are challenging risk of disappearance due to anthropogenic disturbances (Sudhersan et al. 2003). Desert vegetation is considered under enormous stress and is subject to huge variations over the period because of a greatly instable environment in relation to water accessibility, fairly brief phase, and intense aridity (Omar and Bhat 2008).

Kuwait has experienced land degradation and loss of vegetation cover primarily because of the remarkably erratic atmosphere, water insufficiency, saline soils, and recent anthropogenic activities, such as overgrazing, recreational camping, and Gulf war events associated with the Iraqi-Kuwait war (Brown 2003; Omar and Bhat 2008; Al-Shehabi and Murphy 2017). Evidence suggests aforesaid land degradation has a straight consequence to the biodiversity loss, allowing increased soil erosion, damage of essential soil properties comprising soil microbial population, and devastation of soil biogeochemical performance. Loss of biodiversity is considered one of the most significant issues of current environmental burdens. Desert-surviving species could be specifically exposed to climate change since these species are inhabiting an environment with their physiological limits (Vale and Brito 2015). Extreme unpredictable environment, low precipitation, saline soils, and acute drought are foremost challenges encountered in Kuwait. Native vegetation communities are very much acclimatized to the local environments, where flora, fauna, microbes, and human beings have exploited arid environments regardless of their inherently indefinite accessibility of resources (Stahlschmitd et al. 2011). As Kuwait's native plants are susceptible to their ecological habitation and have initiated threats at a shocking rate, it is imperative to investigate the reasons and mitigation measures as reestablishment and effective plantation plans are indispensable to setback the adverse tendencies of ecosystem degradation and to preserve these valuable ecosystems.

In general, climate variables and plant systems interact strongly. In a recent study, the role of temporal climate variations in the dynamics of winter annual species in a Chihuahuan Desert ecosystem was investigated (Ignace et al. 2018). In arid ecosystems, precipitation intensely influences biological processes and is considered a key predictive variable for desert vegetation dynamics (Ignace et al. 2018). The research concluded that the total abundance of a desert annual plant community was intensely related to precipitation, particularly winter precipitation with minimal correlation to other climate variables. However, the species quantity was greater convincingly predicted by average temperature during the fall and winter seasons showing cooler months favoring more species (Ignace et al. 2018). A similar trend may be pertinent to the winter annuals of the Kuwait desert ecosystem. Despite forecasts of greater temperature and better flexibility in rainfalls magnitude and frequency within the arid ecosystems (Bernstein et al. 2007), not much efforts are undertaken so far in Kuwait to investigate the effect of seasonal shifts in local climatic fluctuations and land degradation on biodiversity, particularly assessing soil biological community responses to recent environmental changes. A long-term investigation based on time series research data using similar and simple procedures is required in the Kuwait desert for further confirmation. Vegetation in terrestrial ecosystems is generally regarded as an indicator to assess climate change and environmental restoration attempts (Sun et al. 2021). In a recent experiment in China, the consequences of climate change on vegetation dynamics were investigated for a 20-year study period (Sun et al. 2021). In this investigation, the connection linking vegetation community and climate change was investigated by using normalized difference vegetation index (NDVI) and detected the basis of vegetation alteration patterns by using multiple regression analysis during the 20-year study period. Results of this long-term research revealed that a substantial increase ($p < 0.0001$) in the mean NDVI happened in the Mu Us desert of north central China. Furthermore, correlation analysis observed between NDVI and climatic factors indicated that environmental attributes (rainfall, temperature) and relative humidity had a considerable positive correlation (Sun et al. 2021). These observations may be useful to assist scientific corroboration for decision-making regarding environmental restoration management and policy making.

In Kuwait, precipitation has varied considerably in recent years, and the desert annual vegetation indicates such variations with a corresponding increase or decrease in density and biomass production (Zaman 1997). Omar (1991) noted the effects of extended drought on vegetation growth, species structure, and other environmental factors in Kuwait. In contrast, long-term shifts in vegetation composition as a result of gradual changes in the prevailing climatic conditions could entail a change in species composition to favor certain groups of plants, including dominant perennials, at the expense of others. The monitoring of vegetation is an important tool in environmental studies, including the investigation of vegetation succession. Furthermore, vegetation monitoring has recently been recognized as a vital aid for studying the effects of climate change (Grime 1997; Dunnett et al. 1998), but there is a general lack of adequate data, impairing the capacity to determine the effects of climate change on vegetation (Smith et al. 1997). Recently, a study was undertaken to evaluate and describe the native vegetation communities, including species richness and diversity, vegetation cover, and biomass production in the four selected sites across Kuwait's desert in relation to protected and unprotected (disturbed) areas (Quoreshi et al. 2018). The results demonstrated that despite the severe climate prevailing in Kuwait, the dominant species harboring in the protected sites were completely absent in the disturbed sites by anthropogenic pressures (Quoreshi et al. 2018, 2019). Current anthropogenic disturbance is now influencing biodiversity at different degrees around the world, for example by adding extinction rates and leading to shifts in distribution patterns (Thuiller et al. 2005). In fact, IPCC (2007) has clearly stated that regions suffering from environmental problems such as desertification are probably the most sensitive to the effects of climate change. Changes in vegetation composition will not only affect biodiversity but will also have a profound influence on key environmental processes, such as wind erosion of soils (Goudie 2003) and the hydrologic cycle (Ludwig et al. 2005; Scanlon et al. 2005), soil microbial community and processes (Bell et al. 2009), with far-reaching implications for the natural environment, including regional weather patterns (Pielke et al. 1998).

14.4 Impact of Climate Change on Soil Microbiota and Plant-Soil-Microbes' Relationships and Their Consequences Relevant to Ecosystem Functions

Although considerable investigation has been dedicated to the potential effect of global climate changes on vegetation dynamics, little consideration has been devoted to possible alterations in the soil microbial populations and their activities in the ecosystem (Quoreshi et al. 2018). Soil microbiomes perform essential functions of terrestrial ecosystems; nevertheless, current knowledge of their responses and interaction to climate change lags considerably behind compared to other organisms (Maestre et al. 2015). Despite global climate change projections of greater soil temperature and intensified inconstancy in precipitations level and frequency within the arid ecosystems (Bernstein et al. 2007), not much efforts are undertaken in Kuwait to evaluate the consequences of climate change on biodiversity, particularly evaluating soil biological community responses to ecological changes. In any ecosystem, sustainability is dependent on effective biological processes in the soil system, which is mainly regulated by the activity of functional microbial communities. Soil microbial communities conserve soil structure and facilitate the degradation of organic matter, nutrients, and carbon cycling in all ecosystems (Bell et al. 2009). The soil microbes and mycorrhizas can act a crucial activity in plant establishment, growth, and soil stabilization by increasing a plant's capacity to acquire resources from its environment, nitrogen sequestration, drought, contamination, and buffer in protecting from stress conditions (Davies et al. 1992; Smith and Read 1997; Quoreshi 2008). Therefore, soil ecosystem and microbial diversity are essential components of terrestrial ecosystem functions (Coleman and Crossley 1996). A recent investigation attempted to obtain baseline data for microbial community composition and diversity in soils of four selected sites in Kuwait through classical microbiology, *phospholipid* fatty acid (*PLFA*) analysis, and advanced next-generation sequencing (NGS) and bioinformatics analyses (Quoreshi et al. 2018, 2019). The results revealed highly variable diversity indices for microbial communities among Kuwait's desert study sites with differences between winter and summer months.

Soil microbiomes perform an indispensable role in biogeochemical cycles and enhancement of soil fertility, consequently, support for plant growth and establishment (Quoreshi et al. 2019; Suleiman et al. 2019; Naylor et al. 2020). The role of soil microorganisms in GHG emissions and arbitrating soil organic carbon (SOC) is of specific attention in relation to global climate change (Singh et al. 2010). Climate change and alteration in land management practices can negatively impact soil fertility and SOC (Lal 2004), consequently, impacting the soil microbiome and their impact on SOC sequestration. The limited information regarding the potential of soil microbial activity and soil-microbe-plant interactions makes it challenging to precisely justify the changes in microbial activities and interactions that happen in relation to climate change. Microbial activities that have a vital role in the global fluxes of major GHG emissions are expected to respond to climate change (Singh et al. 2010). The results suggest that climate change-mediated disturbances can profoundly alter microbial population structure and functional properties (Jansson and Hofmockel 2020). Interestingly, when carbon and/or nitrogen cycling are influenced, this situation can in turn influence climate change either through positive response to the atmosphere (GHG emissions) or negative feedback, such as carbon immobilization into soil or plant biomass (Sulman et al. 2014). The global climate change model is predicted to alter various climate variables significantly over the next century. The atmospheric CO_2 concentration is predicted to increase continuously along with an increase in surface temperature (IPCC 2022). Compant et al. (2010) evaluated the outcomes of 135 research papers examining the influences of climate change on valuable microbes and their relations with host plants. This review reported that the majority of investigations revealed elevated CO_2 concentration had shown a favorable impact on the richness of arbuscular mycorrhizal (AM) and ectomycorrhizal (ECM) fungi. In contrast, the effect on plant growth-promoting bacteria and other endophytes was found quite changeable. In general, plant-accompanying microorganisms had shown beneficial effects on plants when under enhanced CO_2 concentration but in the case of increased temperature, the effect had shown more variable, positive, neutral, and negative depending on the study method and temperature span considered (Compant et al. 2010). This article portrayed that plant microbiomes are an essential feature affecting the response of plant communities to climate change.

Fundamental ecosystem processes depend on different soil biodiversity including bacteria and fungi (Zhou et al. 2011; Trivedi et al. 2013). In a recent review (Naylor et al. 2020), soil microbial community responses to global warming and elevated CO_2 concentration were examined and attempted to understand interactions between these components with one another directly or indirectly to influence altering soil microbial community structure and functionality. Modeling studies predict that the Arab region may confront an increase of 2.0 °C to 5.5 °C in the surface temperature by the end of the twenty-first century (AFED 2009), thus soil microbiomes apparently be affected by warming (Jansson and Hofmockel 2020). This situation signifies the inevitable impact of climate change on the soil microbiomes. Soil warming is believed to influence the resident microbial community in a series of events. Soil warming can increase organic carbon decomposition rates and increase microbial population by 40–150% (Sheik et al. 2011). Although the responses between plants and mycorrhizal fungi, nitrogen-fixing bacteria to climate change is fairly documented in relation to functioning and community composition, the response of heterotrophic microbes inhabiting soils to climate change is less understandable (Staddon et al. 2004; Bardgett et al. 2008). A review by Drigo et al. (2008) has described the effect of increased CO_2 on rhizobacterial communities as well as the rhizosphere, endophytic populations may also be impacted (Compant et al. 2009). In another study, it has been reported that microbial respiration declines over time as labile carbon is reduced (DeAngelis et al. 2015; Melillo et al. 2015). Warming has also contributed to enhancing microbiome community diversity and richness including enriching the Acidobacteria and Actinobacteria phyla (Carrel et al. 2019; Rocca et al. 2019) and class Alphaproteobacteria (DeAngelis et al. 2015; Melillo et al. 2015). Similar to warming, elevated CO_2 concentration shows equally direct and indirect influences on soil microbiome (Naylor et al. 2020). Enhanced CO_2 increases respiration rates, carbon cycling processes, and microbial biomass in the short term (Xiong et al. 2015). An increase in aridity as predicted by climate change models decreases the diversity and abundance of bacteria and fungi in dry lands (Maestre et al. 2015). However, a further holistic understanding of how soil microbes and soil-microbe-plant interactions directly and indirectly respond to climate change may be required for the improved climate change model.

Mycorrhizal fungi perform a vital function in plant nutrition, water relations, ecosystem formation, plant diversity, ecosystem functioning, and yield of plants in terrestrial ecosystems (Smith and Read 1997; Quoreshi 2008; Hawkins et al. 2023). Therefore, the mycorrhizal symbiotic system potentially is a crucial link in the chain of response to global climate change (Liu et al. 2023). Nevertheless, little is known about the consequences of climate change that may affect the richness and community of AM fungi connected to different plant communities (Liu et al. 2023) and might respond to land degradation and long-term environmental changes in the desert lands. In the arid environments, mycorrhizal associations are a crucial factor in the survival of plants (Titus et al. 2002; Apple 2010). Arbuscular mycorrhizal (AM) fungi, a ubiquitous type of soil fungi, can form mutual colonization with the roots of over 80% of plant species, and acquire adaptable relationships with their host plants (Madouh and Quoreshi 2023).

Global warming and further human-induced actions have the ability to modify the dynamics and stability of resources in the mutualistic symbiosis between plants and mycorrhizal fungi (Duarte and Maherali 2022). Increased atmospheric CO_2 is widely considered a topic in climate change and influences the growth and performance of mycorrhizal fungi and plants (Dong et al. 2018). Due to inconsistent and mixed findings reported in the published articles, the climate change response model of symbiotic fungi is uncertain. A meta-analysis by Dong et al. (2018) demonstrated that elevated CO_2 enhanced plant biomass, nutrient content, and mycorrhizal fungal growth. AM fungal plants showed a greater increase in biomass and nutrient content compared to ectomycorrhizal (ECM) plants. These data suggest diverse patterns in the growth response of different mycorrhizal fungi under elevated CO_2. A recent study (Liu et al. 2023) investigated the alterations in rhizospheric AM fungal communities and growth performance of maize and wheat under experimental settings with elevated CO_2 and temperature. The results revealed that climate change variables evaluated enhanced rhizosphere AM fungal diversity but conversely, they reduced mycorrhizal colonization of the studied crops. There may be several mechanisms involved by which climate change affects the AM fungal communities either directly or indirectly altering the plant growth (Cotton 2018). This information is indispensable for assessing how natural and anthropogenic climate change drivers affect plant-microbe-soil systems in the Kuwait desert.

Many evidences demonstrate that the microorganisms are shaped by climate change-related interruptions, alter microbial community structure and function, and have consequences on biogeochemical cycling, and carbon pool that may exacerbate or mitigate climate change. Therefore, a more holistic perception of how soil microbiomes respond to climate change is required for further improved climate change models. In a recent review on microorganisms and climate change, Cavicchioli et al. (2019) emphasized the necessity to comprehend how microorganisms affect climate change (including production and consumption of GHGs) but also important to know how microorganisms will be shaped by climate change and other anthropogenic activities in *One Health approach.*

14.5 Improving Ecological Resilience to Climate Change and Practical Management of Desert Ecosystems for Mitigation and Adaptation

Global climate change has become the sternest warning to ecosystem functioning and human development. Adapting to the effects of climate change is becoming an indispensable component of any planning and policy-making procedures at all levels (IPCC 2021). Kuwait is considered most influenced by the negative impacts of global warming, such as extreme temperatures, rainstorms, floods, an erratic rise in the frequency of dust storms, and the SLR (EPA Kuwait 2019). The Kuwait Environment Public Authority (EPA) together with the UN Environment and United Nations Development Program initiated a National Adaptation Plan (EPA Kuwait 2019) in collaboration with other governmental and non-governmental institutions to better organize and develop national adaptive capacity and resilience to decrease vulnerability to the climate change effects.

Proper ecosystem management can perform an effective responsibility in climate change mitigation and adaptation processes. However, the current practices of ecosystem management must be evaluated and amended in the context of the projected climate change scenario to move toward sustainability. Kuwait should undertake a long-term program to assess the extent and drivers of processes leading to ecosystem degradation, desertification, and conversion, as well as undertake an effective plan for the restoration and revegetation of degraded ecosystems and sustainable use of ecosystems. Explore the potential to transfer lessons learned from forest-based mitigation efforts to other ecosystems. Positive mitigation, comprised through ecosystem-based approaches, can also produce a positive feedback loop, as it decreases the hazard of negative impacts of climate change on the ecosystems and their carbon stocks (Epple et al. 2016).

As changes in climate will affect individual plant species in a different manner, an initial screening program should be undertaken to gather information on important attributes of key species present in Kuwait. Grime (1997) provided a list of some of the traits likely to influence plant response to climate change, and these include photosynthetic type, rooting depth, reproductive rate, and life form. Certain combinations of plant traits recur in different species, allowing distinct "plant functional types" (PFTs) to be distinguished. The reduction of complexity due to species diversity into a manageable number of useful PFTs is essential to enable a better prediction of ecosystem and vegetation responses (Grime 1997; Smith et al. 1997). Desert plants are always experienced substantial stress and are subject to huge variations over time due to an unpredictable atmosphere in relation to rainfalls, a relatively short growth period, and extreme aridity.

Plant ecologists and restoration scientists have demonstrated the significance of soil microbial communities for effective vegetation establishment, plant growth, and plant community development. Factors that contribute to the unsuccessful vegetation establishment in arid ecosystems, such as drought, extreme temperature, and nutrient limitations are well known (Vinton and Bruke 1995). Researchers have addressed soil microbial activity and its importance in arid ecosystems, but understanding links to soil biodiversity dynamics and plant-microbes-soil systems in desert ecosystems is poorly known (Herrera et al. 1993; Herman et al. 1995). Research outcomes indicated that establishment of mycorrhizal symbiosis is one of the vital features for the successful development of vegetation on the reclaimed and restored sites; however, the use of mycorrhizal fungal inoculum in the restoration efforts of the degraded desert is not well utilized (Quoreshi 2008; Madouh and Quoreshi 2023). Little is known about how mycorrhizal fungi might respond to land degradation and long-term environmental changes in desert lands. In arid environments, mycorrhizal associations are a crucial factor in the survival of plants (Titus et al.

2002; Apple 2010). Attempts to revegetate arid lands have been made regularly on a small scale without paying any attention to the microorganisms involved, predicting that they will establish and function once the vegetation is established. The selection of suitable plant species and their explicit site-adapted native microbial isolates is likely to extend the success of land reclamation (Quoreshi 2008). Therefore, precise restoration techniques are required to encourage the rapid establishment of initial native plant coverage. The rhizosphere, rhizosheath, a sign of biological/microbial activity in the rhizosphere (root zone), endosphere, and phyllosphere of desert plants display a perfect niche for isolating novel microbes that can be used as bio-fertilizers and bio-control agents for boosting sustainable agriculture and producing high-quality seedling for reforestation and restoration programs in extreme environments in the context of climate change (Alsharif et al. 2020).

14.6 Conclusions and Recommendations

The researchers agreed that global climate change is occurring and that the average temperatures and atmospheric CO_2 concentrations are increasing since 1900. Recent assessment and climate change model reported that the Arab region experienced an irregular escalation in surface air temperature and other climate change events and continue to enhance unprecedentedly. Consequently, climate change and its significance have become an enormous concern and focus of media and political attention, the governments, scientists, and civil societies are now starting to observe the problem more seriously. Arid ecosystems are predicted to be some of the most sensitive systems to global climate change due to their current vulnerability. If temperatures in the region continue to increase, or precipitation gets lower, pressure on biological and physical systems would be intensified. Therefore, Kuwait is not immune to global changes in climatic patterns. Slightly higher mean temperatures and lower rainfall to this already prevailing harsh climate regions of the world are most susceptible to land degradation and desertification, and biodiversity loss. Despite global climate change predictions of greater soil temperature and increased variability in precipitations magnitude and frequency within arid ecosystems, not much efforts are undertaken in Kuwait to investigate the effect of climate change on biodiversity, particularly assessing soil microbiological community responses to current environmental changes projection. There is a necessity to reveal and fully understand terrestrial microbiome feedback responses and the ability to accomplish microbial systems for the adaptation and mitigation of climate change. The IPCC (2021-AR6) report concluded that there are feasible and effective adaptation options identified which can lessen risks of climate change to ecosystems and human population; however, it is also emphasized that there are limits to adaptation and that there is a need for intensified determination in both adaptation and mitigation. New research directions are to be tested to reduce microbial greenhouse gases and mitigate the pathogenic impacts of microbes. This may comprise achieving further controlled experiments on the climate influence on microbial processes, system interdependencies, and responses to human involvements, using microbes and their carbon and nitrogen transformations for useful stable products, improving microbial process data for climate models, and taking the *One Health approach* to study microbes and climate change.

Current research findings confirmed and improved our perception of the significance of climate-resilient development across segments and territories and, as such, calls for the immediate attention of policymakers, scientists, and the public at large. Increase in environmental awareness is required among citizens and expat communities for climate-adapted ecosystem management. Scientists should communicate more effectively with the evidence-based scientific information to policymakers and decision-makers. The Kuwait government has initiated important adaptation measures to mitigate the impact of climate change in different sectors. However, major gaps still exist, and there is a need for coordinated research and continued long-term ecological monitoring programs, and biodiversity conservation strategies to the national adaptation measures. An effective national policy should be considered to reduce fossil fuel carbon emission and adopt renewable energy system for Kuwait to become carbon neutral in oil and gas by 2050 target. The practical challenges confronting the Arab region from climate change are enormous; however, this situation can be prevented if the policymakers of the region act fast and effectively.

Acknowledgements We wish to thank M.T. Sivadasan, Research Assistant at the Desert Agriculture and Ecosystems Program of Environment and Life Sciences Research Center of Kuwait Institute for Scientific Research for her help in formatting, and editing of reference section and the manuscript.

References

Abella SR (2012) Restoration of desert ecosystems. Nat Educ Knowl 4(1):7.6397. T&F Routledge handbook of ecological and environmental restoration, p 160

Abella SR (2017) Persistent establishment of outplanted seedlings in the Mojave Desert. Ecol Restor JSTOR 35(1):16–19

Abella SR, Newton AC (2009) A systematic review of species performance and treatment effectiveness for revegetation in the Mojave Desert, USA. In: Fernandez-Bernal A, De La Rosa MA (eds) Arid environments and wind erosion. Nova Science Publishers, Hauppauge, pp 45–74

AFED (Arab Forum for Environment and Development) (2009) Arab environment 2: impact of climate change on Arab countries. https://wedocs.unep.org/20.500.11822/9631

AGEDI (2016) Final technical: regional desalination and climate change. LNRCCP. CCRG/IO

Al-Blooshi LS, Issa SMG, Ksiksi T (2020) Assessing the environmental impact of climate change on desert ecosystems: a review. Adv Ecol Environ Res 5(2):27–52

Al-Maamary HMS, Kazem HA, Chaichan MT (2017) The impact of oil price fluctuations on common renewable energies in GCC countries. Renew Sust Energ Rev 75:989–1007

Al-Shehabi Y, Murphy K (2017) Flora richness as an indicator of desert habitat quality in Kuwait. J Threatened Taxa 9:9777–9785

Alahamad B, Vicedo-Cabrera AM, Chen K, Garshick E, Bernstein AS, Schwartz J, Koutrakis P (2022) Climate change and health in Kuwait: temperature and mortality projections under different climatic scenarios. Environ Res 17:074001

Alsharif W, Saad MA, Hirt H (2020) Desert microbes for boosting sustainable agriculture in extreme environments. Front Microbiol 11:1666. https://doi.org/10.3389/fmicb.2020.01666

Apple ME (2010) Aspects of mycorrhizae in desert plants. In: Ramawat KG (ed) Desert plants. Springer, Berlin/Heidelberg, pp 121–134

Bardgett RD, Freeman C, Ostle NJ (2008) Microbial contribution to climate change through carbon cycle feedbacks. ISME J 2:2805–2814

Bell CW, Acosta-Martinez V, McIntyre NE, Cox S, Tissue DY, Zak JC (2009) Linking microbial community structure and function to seasonal differences in soil moisture and temperature in a Chihuahuan Desert Grassland. Microb Ecol 58:827–842

Bellard C, Leclerc C, Leroy B, Bakkenes M, Veloz S, Thuiller W, Courchamp F (2014) Vulnerability of biodiversity hotspots to global change. Glob Ecol Biogeogr 23(12):1376–1386. https://doi.org/10.1111/geb.12228

Bernstein L, Peter B, Osvaldo C, Zhenlin C, Renate C, Ogunlade D (2007) An assessment of the Intergovernmental Panel on Climate Change. Report by the section at IPCC Plenary XXVII, Valencia, Spain 52:12–17

Bombi P, Salvi D, Shuuya T, Vignoli L, Wassenaar T (2021) Climate change effects on desert ecosystems: a case study on the keystone species of the Namib Desert *Welwitschia mirabilis*. PLOS ONE 16(11):e0259767

Brown G (2001) Vegetation ecology and biodiversity of degraded desert areas in north-eastern Arabia. Dissertation, Rostock University, Rostock

Brown G (2003) Factors maintaining plant diversity in degraded areas of northern Kuwait. J Arid Environ 54:183–194

Carrel AA, Kolton M, Glass JB, Pelletier DA, Warren MJ, Kostka JE, Iversen CM, Hanson PJ, Weston DJ (2019) Experimental warming alters the community composition, diversity, and N_2 fixation activity of peat moss (*Sphagnum fallax*) microbiomes. Glob Chang Biol 25(9):2993–3004

Cavicchioli R, Ripple WJ, Timmis KN et al (2019) Scientists' warning to humanity: microorganisms and climate change. Nat Rev Microbiol 17:569–586. https://doi.org/10.1038/s41579-019-0222-5

Cherlet M, Hutchinson C, Reynolds J, Hill J, Sommer S, Von Maltitz G (2018) World atlas of desertification: rethinking land degradation and sustainable land management. Publication Office of the European Union, Luxembourg

Coleman DC, Crossley DA (1996) Fundamentals of soil ecology. Academic Press, San Diego

Compant S, Van der Heijden MGA, Sessitsch A (2009) Climate change effects on beneficial plant–microorganism interactions. FEMS Microbiol Ecol 73(2):197–214

Compant S, Clement C, Sessitsch A (2010) Plant growth-promoting bacteria in the rhizo- and endosphere of plants: their role, colonization, mechanisms involved and prospects for utilization. Soil Biol Biochem 42:669–678

Cotton TeA (2018) Arbuscular mycorrhizal fungal communities and global change: an uncertain future. FEMS Microbiol Ecol 94(11). https://doi.org/10.1093/femsec/fiy179

Davies FT, Potter JR, Linderman RG (1992) Mycorrhiza and repeated drought exposure affect drought resistance and extraradical hyphae development of pepper plants independent of plant size and nutrient content. J Plant Physiol 139(3):289–294

DeAngelis KM, Pold G, Topcuoglu BD, Diepen LTAV, Varney RM, Blanchard JL, Melillo J, Frey SD (2015) Long-term forest soil warming alters microbial communities in temperate forest soils. Front Microbiol 6. https://doi.org/10.3389/fmicb.2015.00104

Dong K, Renjin S, Hongdian J, Xiangang Z (2018) CO_2 emissions, economic growth, and the environmental Kuznets curve in China: what roles can nuclear energy and renewable energy play. J Clean Prod 196:51–63

Drigo B, Kowalchuk GA, Yergeau E, Bezemer TM, Boschker HTS, Van Veen JA (2008) Impact of elevated carbon dioxide on the rhizosphere communities of Carex arenaria and Festuca rubra. Glob Chang Biol 13(11):2396–2410. https://doi.org/10.1111/j.1365-2486.2007.01445.x
Duarte AG, Maherali H (2022) A meta-analysis of the effects of climate change on the mutualism between plants and arbuscular mycorrhizal fungi. Ecol Evol 12(1). https://doi.org/10.1002/ece3.8518
Dunnett NP, Willis AJ, Hunt R, Grime JP (1998) A 38-year study of relations between weather and vegetation dynamics in road verges near Bibury, Gloucestershire. J Ecol 86:610–623
EPA Kuwait (2019) Kuwait national adaptation plan 2019–2030. Enhanced climate resilience to improve community livelihood and achieve sustainability. Environmental Protection Authority, Kuwait, p 173
Epple C, García RS, Jenkins M, Guth M (2016) Managing ecosystems in the context of climate change mitigation: a review of current knowledge and recommendations to support ecosystem-based mitigation actions that look beyond terrestrial forests, CBD technical series no. 86. Secretariat of the Convention on Biological Diversity, Montreal, p 55
Fernández-Martínez M, Sardans J, Chevallier F, Ciais P, Obersteiner M, Vicca S, Canadell JG, Bastos A, Friedlingstein P, Sitch S, Piao SL, Janssens IA, Peñuelas J (2019) Global trends in carbon sinks and their relationship with CO_2 and temperature. Nat Clim Chang 9(1):73–79
Fischer EM, Knutti R (2015) Anthropogenic contribution to global occurrence of heavy precipitation and high temperature extremes. Nat Clim Chang 5:560–564. https://doi.org/10.1038/nclimate2617
Goudie AS (2003) The impacts of global warming on the geomorphology of arid lands. In: Alsharhan AS, Wood WW, Fowler A, Abdellatif EM (eds) Desertification in the Third Millennium. Proceedings of an international conference, Dubai, 12–15 February 2000. Lisse AA, Balkema Publishers, Rotterdam, pp 13–19
Grime JP (1997) Climate change and vegetation. In: Crawley MJ (ed) Plant ecology. Blackwell, Oxford, pp 582–594
Halwagy R, Halwagy M (1974) Ecological studies on the desert of Kuwait. II – the vegetation. J Univ Kuwait (Sci) 1:87–95
Hawkins HJ, Cargill RIM, Nuland MEV, Hagen SC, Field KJ, Sheldrake M, Soudzilovskaia NA, Kiers ET (2023) Mycorrhizal mycelium as a global carbon pool. Curr Biol 33:R560–R573
Herman RP, Provencio KR, Herrera-Matos J, Torrez RJ (1995) Resource Island predict the distribution of heterotrophic bacteria in Chihuahuan Desert soils. Appl Environ Microbiol 61:1816–1821
Herrera MA, Salamanca CP, Barea JM (1993) Inoculation of woody legumes with selected arbuscular mycorrhizal fungi and rhizobia to recover desertified Mediterranean ecosystems. Appl Environ Microbiol 59:129–133
Houghton JT, Ding Y, Griggs DJ, Noguer M, van der Linden PJ, Dai X, Maskell K, Johnson CA (2001) Climate change, the scientific basis. Cambridge University Press, Cambridge
Huang J, Ji M, Xie Y, Wang S, He Y, Ran J (2016) Global semi-arid climate change over last 60 years. Clim Dyn 46(3):1131–1150
Ignace D, Huntly N, Chesson P (2018) The role of climate in the dynamics of annual plants in a Chihuahuan desert ecosystem. Biol Sci 19:279–297
IPCC (2019) Summary of policymakers. In: Shukla PR, Skea J, Calvo E, Buendia E, Masson-Delmontte V, Portner HO, Roberts DC, Zhai P, Slade R, Connors S, Diemen RV, Ferrat M, Haughey E, Luz S, Neogi S, Pathak M, Petzold J, Pereira JP, Vyas P, Huntley E, Kissick K, Belkacemi M, Malley J (eds) Climate change and land: an IPCC special report on climate change, desertification, land degradation, sustainable land management, food security and greenhouse gas fluxes in terrestrial ecosystems. IPCC, Geneva
IPCC (2021) In: Masson-Delmonte V, Zhai P, Pirani A, Connors SL, Pean C, Berger S, Caud N, Chen Y, Goldfarb L, Gomis MI, Huang M, Leitzell K, Lonnoy E, Matthews JB, Maycock TK, Waterfield T, Yelecki O, Yu R, Zhou B (eds) Climate change 2021: the physical science basis.

Contribution of Working Group I to the sixth assessment report of the Intergovernmental Panel on Climate Change. Cambridge University Press, Cambridge, pp 745–757
IPCC (2022) Climate change 2022: impacts, adaptation and vulnerability. The Working Group II contribution to the IPCC sixth assessment report. Aassesses the impacts of climate change, looking at ecosystems, biodiversity, and human communities at global and regional levels. Cambridge University Press, Cambridge
IPCC (2023) Summary for policymakers. In: Core Writing Team, Lee H, Romero J (ed) Climate change 2023: synthesis report. A report of the Intergovernmental Panel on Climate Change. Contribution of Working Groups I, II and III to the sixth assessment report of the Intergovernmental Panel on Climate Change. IPCC, Geneva, p 36
IPCC (Intergovernmental Panel on Climate Change) (2007) Climate change 2007: impacts, adaptation and vulnerability. Contribution of Working Group II to the fourth assessment report of the IPCC. Cambridge University Press, Cambridge
Jansson JK, Hofmockel KS (2020) Soil microbiomes and climate change. Nat Rev Microbiol 18(1):35–46
Köppen W (1936) Das geographisca System der Klimate. In: Köppen W, Geiger G (eds) Handbuch. Klimatologie 1. C. Gebr. Borntraeger, Berlin, pp 1–44
Lal R (2004) Carbon sequestration in dryland ecosystems. Environ Manag 33:528–544
Liu Z, Cui D, Liu Y, Wang Y, Yang L, Chen H, Qui G, Xiong Z, Shao P, Luo X (2023) Enhanced ammonia nitrogen removal from actual rare earth element tailings (REEs) wastewater by microalgae-bacteria symbiosis system (MBS): ratio optimization of microalgae to bacteria and mechanism analysis. Bioresour Technol 367:128304. https://doi.org/10.1016/j.biortech.2022.128304
Loarie SR, Duffy PB, Hamilton H, Asner GP, Field CB, Ackerly DD (2009) The velocity of climate change. Nature 462:1052–1055
Ludwig R, Al-Horani FA, de Beer D, Jonkers HM (2005) Photosynthesis-controlled calcification in a hypersaline microbial mat. Limnol Oceanogr 50(6):1836–1843
Madouh TA, Quoreshi AM (2023) The function of arbuscular mycorrhizal fungi associated with drought stress resistance in native plants of arid desert ecosystems: a review. Diversity 15:391
Maestre FT, Delgado-Baquerizo M, Jefferies CT, Sigh BK (2015) Increasing aridity reduces soil microbial diversity and abundance in global drylands. Proc Natl Acad Sci 112(51):15684–15689
Maurer GE, Hallmark AJ, Brown RF, Sala OE, Collins SL (2020) Sensitivity of primary production to precipitation across the United States. Ecol Lett 23(3):527–536. https://doi.org/10.1111/ele.13455
Melillo M, Brunetti MT, Perucacci S, Gariano SL, Guzzetti F (2015) An algorithm for the objective reconstruction of rainfall events responsible for landslide. Landslides 12:311–320
Mellillo JM, McGuire AD, Kicklghter DW, Moore B, Vorosmarty CJ, Schloss AL (1993) Global climate change and terrestrial net primary production. Nature 363:234–240
Merlone A, Al-Dashti H, Faisal N, Cerveny RS, AlSarmi S, Bessemoulin P, Brunet M, Driouech F, Khalatyan Y, Peterson TC, Rahimzadeh F, Trewin B, Abdel Wahab MM, Yagan S, Coppa G, Smorgon D, Musacchio C, Krahenbuhl D (2019) Temperature extreme records: World Meteorological Organization metrological and meteorological evaluation of the 54.0 C observations in Mitribah, Kuwait and Turbat, Pakistan in 2016/2017. Int J Climatol 39(13):5154–5169. https://doi.org/10.1002/joc.6132
Mirzabaev A, Wu J, Evans J, Garcia-Oliva F, Hussein IAG, Iqbal MH, Kimutai J, Knowles T, Meza F, Nedjraoui D, Tena F, Turkeş M, Vazquez RJ, Weltz M (2019) Desertification. In: Shukla PR, Skea J, Calvo Buendia E, Masson-Delmotte V, Portner H-O, Roberts DC, Zhai P, Slade R, Connors S, van Diemen R, Ferrat M, Haughey E, Luz S, Neogi S, Pathak M, Petzold J, Portugal Pereira J, Vyas P, Huntley E, Kissick K, Belkacemi M, Malley J (eds) Climate change and land: an IPCC special report on climate change, desertification, land degradation, sustainable land management, food security, and greenhouse gas fluxes in terrestrial ecosystems. Cambridge University Press, Cambridge

Mirzabaev A, Stringer LC, Benjaminsen TA, Gozalez P, Harris R, Jafari M, Stevens N, Tirado CM, Zakieldeen S (2022) Cross chapter paper 3: deserts, semiarid areas and desertification. In: Portner HO, Roberts DC, Tignor M, Poloczanska ES, Mintenbeck K, Alewgria A, Craig M, Langsdorf S, Loschke S, Moller V, Okem A, Rama B (eds) Climate change 2022: impacts, adaptation and vulnerability. Contribution of Working Group II to the sixth assessment report of the Intergovernmental Panel on Climate Change. Cambridge University Press, Cambridge/New York, pp 2195–2231

Naylor D, Sadler N, Bhattacharjee A, Graham EB, Anderton CR, McClure R, Lipton M, Hofmockel KS, Jansson JK (2020) Soil microbiomes under climate change and implications for carbon cycling. Annu Rev Environ Resour 45(1):29–59

Omar SA (1991) Dynamics of range plants following ten years of protection in arid rangelands of Kuwait. J Arid Environ 21:99–111

Omar SAS, Bhat NR (2008) Alteration of the *Rhanterium epapposum* plant community in Kuwait and restoration measures. Int J Environ Stud 65(1):139–155

Orlowsky B, Seneviratne SI (2012) Global changes in extreme events: regional and seasonal dimension. Clim Chang 110:669–696

Pielke RA, Avissar R, Raupach M, Dolman AJ, Zeng X, Denning AS (1998) Interactions between the atmosphere and terrestrial ecosystems: influence on weather and climate. Glob Chang Biol 4(5):461–475

Quoreshi AM (2008) The use of mycorrhizal biotechnology in restoration of disturbed ecosystem. In: Siddiqui ZA, Akhtar MS, Futai K (eds) Mycorrhizae: sustainable agriculture and forestry. Springer, Dordrecht, pp 303–320. https://doi.org/10.1007/978-1-4020-8770-7_13

Quoreshi AM, Suleiman MK, Kumar V, Anisul Islam M, Ramadan A, Al-Othman ARA, Al-Mulla L, Jacob S, Manuvel AJ, Sivadasan MT, Thomas R, Bhat NR, Zaman S (2018) Investigation of soil microbial communities and vegetation for baseline database development at selected sites in Kuwait desert. Final report, KISR 14865, KFAS project no. P214-42SL-03, pp 217

Quoreshi AM, Suleiman M, Kumar V, Jasmine A, Sivadasan MT, Islam A, Khasa DP (2019) Untangling the bacterial community composition and structure in selected Kuwait desert soils. Appl Soil Ecol 138

Rocca JD, Simonin M, Blaszczak JR, Ernakovich JG, Gibbons SM, Midani FS, Washburne AD, Alex D (2019) The microbiome stress project: toward a global meta-analysis of environmental stressors and their effects on microbial communities. Front Microbiol 9. https://doi.org/10.3389/fmicb.2018.03272

Scanlon BR, Reedy RC, Stonestrom DA, Prudic DE, Dennehy KF (2005) Impact of land use and land cover change on groundwater recharge and quality in the southwestern US. Glob Chang Biol 11(10):1577–1593

Sheik CS, Beasley WH, Elshahed MS, Zhou X, Luo Y, Krumholz LR (2011) Effect of warming and drought on grassland microbial communities. ISME J 5(10):1692–1700

Shree B, Jayakrishnan U, Bhushan S (2022) Impact of key parameters involved with plant microbe interaction in context to global climate change. Front Microbiol 13:1008451. https://doi.org/10.3389/fmicb.2022.1008451

Singh KB, Bardgett RD, Smith P, Reay DS (2010) Microorganisms and climate change: terrestrial feedbacks and mitigation options. Nat Rev 8:779–790

Smith SE, Read DJ (1997) Mycorrhizal symbiosis, 2nd edn. Academic Press, London

Smith T, Shugart H, Woodward F (eds) (1997) Plant functional types: their relevance to ecosystem properties and global change. Cambridge University Press, New York

Smith SD, Huxman TE, Zitzer SF, Charlet TN, Housman DC, Coleman JS (2000) Elevated CO_2 increases productivity and invasive species success in an arid ecosystem. Nature 408:79–82

Staddon PL, Jakobsen I, Blum H (2004) Nitrogen input mediates the effect of free-air CO_2 enrichment on mycorrhizal fungal abundance. Glob Chang Biol 10(10):1678–1688. https://doi.org/10.1111/j.1365-2486.2004.00853.x

Stahlschmitd Z, DeNardo D, Holland J, Kotler B, Kruse-Peeples M (2011) Tolerance mechanism in north American deserts: biological and societal approaches to climate change. J Arid Environ 75:681–687

Sudhersan C, AboEl-Nil M, Hussain J (2003) Tissue culture technology for the conservation and propagation of certain native plants. J Arid Environ 54:133–147

Suleiman MK, Bhat NR, Jacob S, Thomas RR (2011) Germination studies in *Ochradenus baccatus* Delile, *Peganum harmala* L. and *Gynandriris sisyrinchium* Parl. Res J Seed Sci 4:58–63

Suleiman MK, Quoreshi AM, Bhat NR, Manuvel AJ, Sivadasan MT (2019) Divulging diazotrophic bacterial community structure in Kuwait desert ecosystems and their N2-fixation potential. PLoS One 14(12):e0220679. https://doi.org/10.1371/journal.pone.0220679

Sulman B, Phillips R, Oishi A (2014) Microbe-driven turnover offsets mineral-mediated storage of soil carbon under elevated CO_2. Nat Clim Chang 4:1099–1102

Sun Z, Mao Z, Yang L, Liu Z, Han J, Wanag H, He W (2021) Impacts of climate change and afforestation on vegetation dynamics in the Mu Us desert, China. Ecological indicators 129, 108020. Synthesis report. IPCC fourth assessment report. Cambridge University Press, Cambridge

Thuiller W, Lavorel S, Araújo MB (2005) Niche properties and geographical extent as predictors of species sensitivity to climate change. Glob Ecol Biogeogr 14:347–357

Titus JH, Titus PH, Nowark RS, Smith SD (2002) Arbuscular mycorrhizae of Mojave Desert plants. West N Am Nat 62:327–334

Trivedi P, Anderson IC, Singh BK (2013) Microbial modulators of soil carbon storage: integrating genomic and metabolic knowledge for global prediction. Trends Microbiol 21(12):641–651

Vale CG, Brito JC (2015) Desert adapted species are vulnerable to climate change: insights from the warmest region on earth. Global Ecol Conserv 4:369–379

Vinton MA, Bruke IC (1995) Interactions between individual plant species and soil nutrient status in short grass steppe. Ecology 76:1116–1133

Wang S, Zhang Y, Ju W, Chen JG, Ciais P, Cescatti A, Sardans J, Janssens IA, Wu M, Berry J, Campbell E, Fernández-martínez M, Alkama R, Friedlingstein SP, William KS, Yuan W, He W, Lombardozzi D, Kautz M, Zhu D, Lienert S, Kato E, Poulter B, Sanders TGM, Krüger I, Wang R, Zeng N, Tian H, Yuichard Haverd V, Goll D, Peñuelas J (2020) Recent global decline of CO_2 fertilization effects on vegetation photosynthesis. Science 370(6522):1295–1300

Ward D (2009) Biology of deserts. Oxford University Press, Oxford

Warren A, Sud Y, Rozanov B (1996) The future of deserts. J Arid Environ 32:75–89

Xiong J, He Z, Shi S, Kent A, Deng Y, Wu L, Van Nostrand JD, Zhou J (2015) Elevated CO_2 shifts the functional structure and metabolic potentials of soil microbial communities in a C4 agroecosystem. Sci Rep 5:9316. https://doi.org/10.1038/srep09316

Zafar A (2023) Climate change impacts in Kuwait. Climate change, environment, Middle East, pollution. EcoMENA. https://www.ecomena.org/climate-change-kuwait/

Zaman S (1997) Effects of rainfall and grazing on vegetation yield and cover of two arid rangelands in Kuwait. Environ Conserv 24:344–350

Zhou J, Xue K, Xie J, Ye D, Wu L, Cheng X, Fei S, Deng S, He Z, van Nostrand JD, Luo Y (2011) Microbial mediation of carbon-cycle feedbacks to climate warming. Nat Clim Chang 2(2):106–110

Part IV
Agriculture, Food Security and Water Footprint of Crops

Chapter 15
Prospective of Agricultural Farming in Kuwait and Energy-Food-Water-Climate Nexus

Majda Khalil Suleiman and Shabbir Ahmad Shahid

Abstract Kuwait is a desert, arable land, and water-scarce country. Agriculture farming in Kuwait is mainly practiced in Al-Wafra and Abdali, and to a lesser extent in the Sulaibiya area, where both open field and protected agriculture are practiced. The total cultivated area in Kuwait has significantly increased from 9.7 km^2 in 2000 to 146.7 km^2 in 2020. In the open field agriculture, the major crops grown are fodder crops, and to a lesser extent selected vegetable crops during the winter season. The farmers are shifted to protected agriculture mainly growing vegetables (tomato, cucumber, capsicum, lettuce, etc.) throughout the year. The groundwater of various salinity levels is used in the open field agriculture, whereas in the protected agriculture low salinity water obtained from reverse osmosis units is used. Due to the harsh climate, protected agriculture is energy-intensive and reflects integrated linkages "water-energy-food-climate nexus", which is described in this publication. Small, medium, and large farms typology and crops grown are presented. Major soil types and their characteristics in the farming area are described. Threats to the soils in the farming area are highlighted and best soil management practices, including potential climate smart agriculture, regenerated practices, and adaptation to climate change impacts in the context of local conditions, are discoursed.

Keywords Protected agriculture · Open field agriculture · Fodders · Farms typology · Vegetables · Reverse osmosis · Al-Wafra · Abdali · Nexus

M. K. Suleiman · S. A. Shahid (✉)
Desert Agriculture and Ecosystems Program, Environment and Life Sciences Research Center, Kuwait Institute for Scientific Research, Safat, Kuwait
e-mail: mkhalil@kisr.edu.kw; sshahid@kisr.edu.kw

M. K. Suleiman, S. A. Shahid (eds.), *Terrestrial Environment and Ecosystems of Kuwait*, https://doi.org/10.1007/978-3-031-46262-7_15

15.1 Introduction

Geographically, Kuwait is situated at latitude 28° 30′N and 30° 05′N and longitude 46° 33′E and 48° 35′E. The total area of the State of Kuwait is 17,818 km^2 of land and about 1000 km^2 of off-shore islands (Amr 2021). Kuwait is located in the most northwestern corner of the Arabian Gulf. It borders with Saudi Arabia in the west and south and Iraq in the north. Kuwait is hyper-arid, water, and arable land-scarce country. Summer temperature is high and winter is mild. Annual average rainfall is less than 110 mm and there are no surface water resources. Dust and dust storms are common throughout the year. Due to the arid climate and loose sandy nature of Kuwait deserts, it is not possible to form soil, where strong wind erosion also limits soil formation. The soil survey of Kuwait (KISR 1999; Omar and Shahid 2013) depicted eight soil great groups of soil taxonomy (Soil Survey Staff 2022; Shahid and Omar 2022). Among the eight soil great groups, five (*aquisalids, torriorthents, petrogypsids, petrocalcids,* and *haplogypsids*) are permanently unsuitable for agriculture production due mainly to high salt contents (*aquisalids, torriorhents*), hard pan within 1 meter from soil surface (*petrogypsids, petrocalcids*), and high gypsum (*haplogypsids*). Other three soil subgroups (*haplocalcids, torripsamments,* and *calcigypsids*) are moderately suitable for agriculture and are currently used or considered for agriculture extension (Omar and Shahid 2013; Shahid and Omar 2022). Due to harsh climatic conditions (high temperature, low rainfall) and water and arable land scarcity, agriculture is practiced both in open field (OF) and protected agriculture (PA). The former includes production of fodders and few vegetable crops in winter, the latter focuses on vegetable crops in controlled greenhouse conditions (protected agriculture).

In general, the major gaps in the water sector include water scarcity, water management, and the lack of modern irrigation technologies in agricultural farms. In the Gulf Cooperation Council (GCC) countries the agricultural land is very scarce, and cultivated land amounts to 3.9% of the GCC's total area, compared with the global average of 10.6% (Collins 2017). Due to the prevailed hot environmental conditions, poor soil capacity, agriculture in Kuwait is limited and mainly practiced in Al-Wafra in the south, Abdali in the north and Sulaibiya in the center. As reported by Al-Nasser and Bhat (1998) about 85% of the PA is carried out in uncooled (57%) and cooled (28%) plastic tunnels, with the remaining 15% in cooled greenhouses covered with fiberglass, glass, or acrylic material. They further emphasized that cucumber and tomato are the two main crops grown in PA, accounting for approximately 90% of the total area.

The government is making significant efforts to invest in agriculture sector to improve local crop production for food security. The most recent one is funded by the Supreme Council of Kuwait-Government Initiative P-KISR-17 food security project (Suleiman 2020). Through this project, multiple tasks are implemented including soil, plants, energy, water, fisheries, and renewable energy.

This chapter describes agriculture farming in Kuwait, constraints to crop production and mitigation measure to improve crop and water productivities, and the nexus between food-water-energy-climate.

15.2 Weather and Climate

Kuwait is hot and has arid desert climate. The June–August are the hottest months (average 42–48 °C) and October–February relatively mild temperature. There is significant difference in temperature between summer and winter months. The rainfall is not uniform, it occurs mainly in winter months Jan-April, occasionally in summer due to climate change, otherwise summers are dry. Mean annual rainfall is less than 110 mm. However, in Jan. 2023 very high rainfall (70 mm) was recorded in 1 day which caused flash floods in low-lying areas in Kuwait.

15.3 Temporal Progression of Farming Areas in Kuwait (2–5 Decades)

There are three distinct farming areas in Kuwait, (i) Abdali in the north, (ii) Al-Wafra in the south, and (iii) Sulaibiya in the center of Kuwait. There are 3913 small, medium, and large vegetable and crop farms in Kuwait (Central Statistics Bureau State of Kuwait 2019). These farms are located in Al-Wafra (1889), Abdali (1945), and Sulaibiya (44) including 35 in other areas. The largest farming area is in the south of Kuwait bordering with Kingdom of Saudi Arabia and to a similar extent is the agriculture farming area in the north (Abdali). Agriculture and fishing contribute 0.6% to national Goss Domestic Product (GDP) of 2017 with a growth rate of 5.5% at current prices (Central Statistics Bureau 2019).

According to *emisk* (https://gisportal.emisk.org/arcgis/apps/MapJournal) the Al-Wafra agriculture area has grown significantly over the past four decades (1973–2016). The farming area in Al-Wafra spreads over 8 km^2 in 1973, which increased to 94, 171, and 205 km^2 in the years 1991, 2002, and 2016 respectively. That is many folds increase over the 1973 area. The drawback in this prediction is that these areas include both the actual farm area (cultivated) and the empty area between the farms (non-cultivated area). To refine the predictions on the actual farming area (cultivated) and to have recent information about the temporal growth of Al-Wafra and Abdali farming areas, we downloaded the Landsat images in the month of August for 3 years (2000, 2013, and 2022). The farming area boundaries were established using the GIS tool based on the 2022 Landsat images (covered both by farms and empty areas between the farms) which come up as 298.2 and 276.8 km^2 in Abdali and Al-Wafra areas respectively (Table 15.1). After this preliminary assessment, the cultivated areas both in the Abdali and Al-Wafra area were assessed using Natural Deviation Vegetation Index (NDVI). The results (Table 15.1) clearly show that since 2000, the actual cultivated area is significantly increased, e.g., from 5.2 km^2 to 72.2 km^2 in Abdali and 4.5 km^2 to 74.5 km^2 in Al-Wafra, almost in similar extent increase in both the farming areas. The latest satellite views of both the agricultural areas which we used to predict total farming and cultivated areas are shown in Fig.15.1.

Table 15.1 Total farming and cultivated area in Abdali and Al-Al-Wafra

Year	Abdaly			Al-Wafra		
	Total farming area in km^2 (2022)	NDVI km^2	Non NDVI km^2	Total farming area in km^2 (2022)	NDVI km^2	Non NDVI km^2
2022	298.2	72.2	226	276.8	74.5	202.3
2013		43.8	254.4		40.7	236.1
2000		5.2	293		4.5	272.3

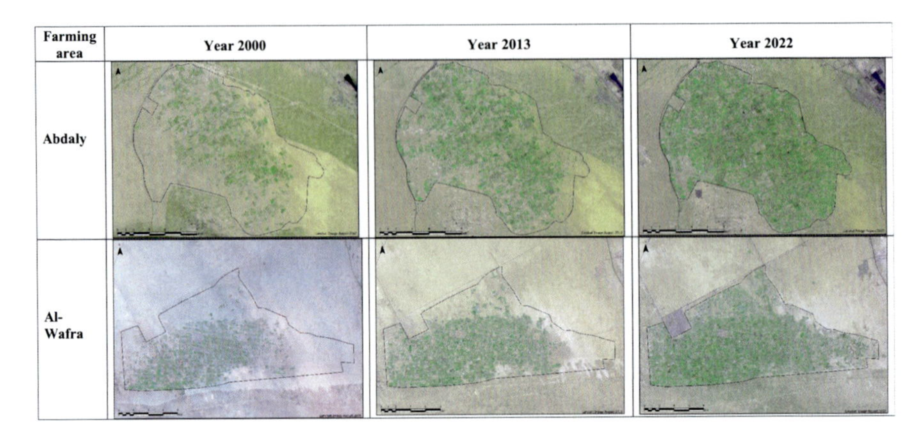

Fig. 15.1 Landsat images of Abdali and Al-Al-Wafra agricultural areas showing the actual cultivated areas (NDVI) with respect to non-farming (non-NDVI)

The results presented in Table 15.1 about the cultivated area were validated with those reported in the Annual Agricultural Statistics (Central Statistics Bureau State of Kuwait 2019). It describes results by Governorate wise, i.e., Al-Ahmadi Governorate (Al-Al-Wafra) and Al-Jahra Governorate including Al-Khuwaisat, Al-Sabiya, Al-Sulaibiya, Al-Abdali, and Um Al-Shagaya. Total cultivated area in Al-Al-Wafra Governorate is 41,972 donums (4197 ha ~ 42 km^2), whereas in Abdali cultivated area is 55,549 donums (5555 ha ~ 55.6 km^2). These cultivated areas are slightly lower than the one predicted through NDVI (Table 15.1) in the year 2022, however, it closely matches to 2013 predictions.

15.4 Agricultural Farms Typology and Crops Production

There are three types of farms based on farm size, (i) small size (2–5 ha), medium size (5–10 ha), and large size (10–200 ha). The major crops grown in small size farms are dates, vegetables, and forages. The medium size farms are used for OF agriculture and PA. The PA is a modification of the natural environment to suit condition for optimal plants growth and protection from biotic and abiotic stress factors. The construction of greenhouses in agriculture farms using the diversified materials based on the country-specific climatic conditions is the most practicable approach to attain the goals of PA.

The PA is gaining international recognition and globally (115 countries practice greenhouse cultivation) the area under greenhouses has exponentially been increased for the past three decades (Singh et al. 2022). China is on the top with 3.5 million hectares of greenhouse area (96% used for fresh vegetables and hybrid seed production) followed by the Republic of Korea (Kacira 2011; Nair and Barche 2014). The Public Authority for Agriculture Affairs and Fish Resources (PAAFR) encourages the expansion of greenhouse production. The PAAFR also issued technical greenhouse specifications for farmers' compliance and an approval from PAAFR before construction (FAO 2021). Under PA cultivation of fruit vegetables can be grown successfully, provided the PA is managed scientifically, where good quality seeds, productive soil media are used and crop-based nutritional balance is maintained during the growth season, the farmers can get substantial income, good examples is from India (Singh et al. 2022) where farmers obtained double income by practicing PA.

The PA cultivation technology (Singh and Solanki 2014) is highly dependent upon intelligent implementation of PA for vegetable cultivation. Vegetables production under PA provides high water and nutrient use efficiency. This technology has very good potential especially in peri-urban areas and can be profitably used for growing high-value vegetable crops like tomato, cherry tomato, colored peppers, cucumber, healthy, and virus-free seedlings in agri-entrepreneurial models. The range of technologies is categorized under PA, starting from conventional plastic tunnels (low cost) to high-cost hydroponic and high-tech Greenhouses. In Kuwait, the large size farms are mainly specialized in date palms production and PA (Zekri and Zaibet 2019). Farms typology in Kuwait is described in Table 15.2.

15.5 Soil Types and Classification in the Farming Areas

The review of soil classification work in the farming areas indicates that no detailed soil survey has been completed covering the entire farming area in Kuwait. However, at demonstration farm level (50 ha) soil survey has been completed in both farming areas, which at the time of survey was not used for agriculture, but were planned to develop these areas as demonstration farms to display high technologies for farmer's adoption (Shahid and Omar 1999; Shahid et al. 2004). These areas are representative of general soil types in the Abdali and Al-Al-Wafra farming areas.

15.5.1 Soil Subgroups in Abdali Farming Area

Three soil subgroups were mapped in the 50 ha area. These three soil subgroups are, (i) *Typic Calcigypsids*, (ii) *Typic Haplocalcids*, and (iii) *Leptic Haplogypsids* (Shahid and Omar 1999). These are briefly described in the following section.

Table 15.2 Farms typology and crops grown in open field and protected agriculture in the farming areas of Kuwait

Large-sized farms	Medium-sized farms	Small-sized farms
Open field agriculture		
An average size of large farms is of 49 ha. Only 6% area of large farms is cropped. The main constraint in increasing cropped area is freshwater scarcity. Potato production is by far the most important crop covering 23% of the cropped area followed by pumpkin and okra with 13% and 8.5% respectively.	Open field agriculture in medium-sized farms uses land more intensively (18.2%) relative to large farms (6%). A large variety of crops is observed in the medium-sized farms including onion occupying 32%, pumpkin and cabbage (10% each), and strawberry occupying 7% of the cropped area.	Open field agriculture produces 9 crops and uses about 23.3% of the available arable land, which is higher than the large-sized farms (6%). Forage are the priority crops, where alfalfa covering 26%, blue panic 23%, and barley 14% of the farm area, cumulatively making 63% of the cropped area. The remaining area is reserved for vegetables with pepper as a dominant crop representing 22% of the cropped area.
Protected agriculture		
On average there are 93 greenhouses (average area per greenhouse is 329 m^2) per farm. The cropped area represents 6.23% of the farms' area. Tomato is grown on 50% and cucumber 15%, and ornamental plants on 13% of the area. Pepper and beans represent 6% of the area each.	In medium-sized farms, there are on average 50 greenhouses per farm covering an area of 375 m^2 per greenhouse. The PA uses 23.7% of the land which is slightly higher than the OF medium farms (18.2%). Tomato is prime crop followed by cucumber occupying almost 50% of the cropped protected area.	The average number of greenhouses is 24 per farm covering an area of 284 m^2 per greenhouse. All greenhouses cover 14% area of the farms. Shortage of labor and affordability being the main reason to intensify crops in small-sized farms.

Source: Zekri and Zaibet (2019)

15.5.1.1 Typic Calcigypsids

The *calcigypsids* present both the calcic and gypsic horizons within the upper 100 depths (Fig. 15.2a). Calcigypsids are unsuitable for agriculture due to high gypsum contents at various depths within the upper 100 cm. However, these soils can be used for shallow-rooted crops provided the roots do not reach to the gypsum layer, where up to 60% gypsum was found. The soils are very deep (no water table), moderately slow permeable, and are well drained without any hardpan within the upper 2 m depth. At the time of the survey annual grass "*Stipa grostis*" was the main vegetation type grown in this soil types. Typical soil profile is shown in Fig. 15.2a. General characteristics and hydrological properties of the typical soil profiles are shown in Tables 15.3 and 15.4 respectively. The soil texture ranges from sand, loamy sand to sandy loam. Of 50 ha at Abdali farm, typic calcigypsids were mapped in 25 ha.

Fig. 15.2 Typical soil profiles from agriculture farming areas in Kuwait, Abdali farming area (**a**, **b**, **c**) and Al-Wafra farming area (**b**, **d**)

15.5.1.2 Typic Haplocalcids

The *haplocalcids* present the calcic subsurface diagnostic horizon within the upper 100 cm depth. The $CaCO_3$ equivalents range between 5% and 20% with occasional concentrations of 1–2 cm diameter (Fig. 15.2b). The soils are very deep and moderately slow permeable. The *Typic Haplocalcids* are the best among all other soils mapped in Kuwait for agricultural use. However, according to land use evaluation (FAO 1976) *haplocalcids* are classified as moderately suitable for agriculture, the main constraint is sandy texture and low clay content. The soil texture ranges from loamy sand to sandy loam. Overall the profile is very slightly saline. Of 50 ha at Abdali farm, *typic haplocalcids* were mapped in 12 ha.

15.5.1.3 Leptic Haplogypsids

These soils have a gypsic subsurface diagnostic horizon within the upper 18 cm soil surface (Fig. 15.2c). These soils have very well-developed deeper gypsum horizon with a considerable amount of gypsum (60%), that will be dissolved with irrigation water and can cause soil subsidence. Although these soils are very deep (no hardpan and water table within 2 m), moderately permeable, and are well drained, the presence of shallow gypsum layer (within upper 18 cm) is the main cause for not recommending for general agriculture purpose. The soil texture ranges from sand to loamy. Of 50 hectares at Abdali farm, *Leptic haplogypsids* were mapped in 13 ha.

15.5.2 Soil Subgroups in Al-Wafra Farming Area

Based on the soil classification criteria (Soil Survey Staff 2022; Shahid and Omar 2022) two soil subgroups were mapped in the 50 ha area at the Al-Wafra demonstration farm. These are (i) *Typic Torripsamments*, and (ii) *Typic Haplocalcides*. Of 50 ha *Typic torripsamments* were mapped in 48 ha, and *Typic haplocalcids* in 2 ha.

15.5.3 Chemical Characteristics of Soils Mapped at Abdali and Al-Wafra Demonstration Farms

Weighted average (100 cm) of important soil properties in the above soil types is shown in Tables 15.3 and 15.4. Table 15.3 clearly shows that the native soils in the farming areas of Abdali and Al-Wafra are non-saline (*Typic torripsamments*), very slightly saline (*Typic Calcigypsids, Typic Haplocalcids*), and slightly saline (*Leptic Haplogypsids*) and non-sodic (Soil Science Division Staff 2017). The pHs is between slightly alkaline to moderately alkaline range (7.6–8.2).

Table 15.3 General soil characteristics of soil types in Abdali and Al-Wafra farming area

Soil types	ECe	pHs	SAR	Gypsum	$CaCO_3$	Texture	Farming area
	mS/cm		$(mmoles/L)^{0.5}$	%			
Typic Calcigypsids	2.86	7.66	0.48	15.3	2.73	Sandy loam	Abdali
Typic Haplocalcids	2.52	7.93	1.46	1.07	11.7	Sandy loam	Abdali/ Al-Wafra
Leptic Haplogypsids	5.50	7.81	4.2	36.2	2.51	Loamy sand	Abdali
Typic Torripsamments	0.48	8.21	0.81	0.00	3.50	Sand	Al-Wafra

ECe electrical conductivity of soil saturation extract, *pHs* pH of saturated soil paste, SAR sodium adsorption ratio

Table 15.4 Hydraulic properties of soil types in Abdali and Al-Al-Wafra farming area

Soil types↓	Sand	Silt	Clay	FC[a]	PWP[b]	AW[c]	Drainage rate[d]	Farming area
	%			cm^3 water/cm^3 soil			mm/h	
Typic Calcigypsids	78	9	13	0.185	0.100	0.085	17.7	Abdali
Typic Haplocalcids	82	6	12	0.176	0.096	0.080	21.1	Abdali/ Al-Wafra
Leptic Haplogypsids	82	11	7	0.156	0.071	0.084	48.4	Abdali
Typic Torripsamments	95	2	3	0.109	0.043	0.066	132.6	Al-Wafra

[a]Field capacity (FC) – the water content when the soil is saturated but not yet overflowing
[b]Permanent wilting point (PWP) – the water level at which water remaining in the soil is held so tightly by capillary action that it is unavailable to plants
[c]Available water (AW)
[d]Drainage rate – the speed water travels through saturated soil

15.5.4 Hydraulic Properties of Soils Mapped at Abdali and Al-Wafra Demonstration Farms

The hydraulic properties are presented in Table 15.4. The available water is low and drainage capacity ranged from low to medium in *Typic Calcigypsids, Typic Haplocalcids, Leptic Haplogypsids,* and high (*Typic torripsamments*) (Table 15.4). Particle size distribution is dominant in sand fraction in all soil profiles. Overall the soil texture is coarse ranging from sand to sandy loam.

15.5.5 Water Use in Agriculture Farms

Based on the criteria of water availability and consumption Kuwait may face high water scarcity risk in future. This is due to no permanent rivers or lakes, and groundwater is the only natural water resource, with low aquifer recharge due to scanty rainfall. The immediate risks to agricultural production in Kuwait will be due to lack of surface water and high salinity in the groundwater.

The trend of agriculture water consumption shows a significant increase from 80 million cubic meters (MCM)/yr. in 1990, to 221 MCM (2000) and 513 MCM (2010). Whereas the groundwater withdrawal rate in 2015 was 255 MCM/yr, of which 54% was used for agriculture, 44% for municipal purposes, and 2% for industrial purposes (Ismail 2015). Kuwait relies on desalination, and recently on treated wastewater (TWW), to provide water for all sectors including agriculture.

Flood irrigation is the most common irrigation method used on 63% of the agricultural land resulting into high water losses and low irrigation efficiencies (Al-Zubari 2003; Al-Zubari et al. 2017). The higher evaporative demand due to the increase in temperature as a consequence of climate change is expected to increase irrigation demand and water stress. FAO's latest projections indicate that global agricultural production must grow by 70% by 2050 in order to feed an additional 2.3 billion people. The projections indicate that most gains in production will be achieved by increasing yield growth and cropping intensity on existing farmlands rather than by increasing the amount of land brought under agricultural production. Hunter et al. (2017) presented an analysis and reported an increase of about 25–70% above current production levels may be sufficient to meet 2050 crop demand. This will significantly affect existing groundwater flow and storage, and diminish the chances of recharge, particularly in arid landscapes.

In Kuwait, there are three types of water used in agriculture farms, (i) groundwater-brackish, (ii) treated waste water-tertiary treated and ultra-filtered water (RO water), (iii) brackish water based desalinated water. In addition, large-scale desalination plants produce 2.432 million m^3 water on a daily basis. Partially, this water is used for agriculture to grow cash-high values crops (vegetables) and domestic purposes. Darwish et al. (2009) warned that more than one-tenth of oil production in Kuwait was used in the desalination plants in 2003 and that this number is almost doubling every 10 years.

In addition to groundwater, the Treated Sewage Effluent (TSE) is a growing water resource in Kuwait. The safe use of TSE can compensate the depleted groundwater resources by direct use in agriculture or through aquifer recharge, if considered safe from soil quality protection and human health perspectives. In general, in the GCC countries these two practices are generally used. Furthermore, using TWW would alleviate the burden of over-drafting limited groundwater resources and avoid the degradation of groundwater quality from seawater intrusion (Abulibdeh et al. 2019).

A review of wastewater produced and used in the GCC countries has revealed that, (i) the average wastewater treatment rate is no more than 56%, while the mean reuse rate does not exceed 30%, and most of that is used for landscape irrigation, (ii) if 50% of the wastewater in GCC countries were treated and used for the agricultural sector, it would supply 11% of the total GCC water demand and 14% of the agricultural water supply demand, and reduce non-renewable (fossil) water withdrawals by 15% by 2020 (Aleisa and Al-Zubari 2017).

The expansion in the utilization of the TWW as a strategically alternative source to meet the GCC countries' future demands is one of the main strategic objectives and policies in the 2016–2035 GCC Unified Water Strategy (UWS) (Al-Zubari et al. 2017). A recent study (Tashtush et al. 2023) indicated that the potential of the generated wastewater, if properly treated and fully reused, would completely fulfill all the agricultural water needs in Bahrain, Kuwait, and Qatar (averaged 50% of the total water demands in the GCC countries).

The use of brackish water has converted many farms to saline and their productivity has declined. At extreme salinity levels of both water and soil, the farms may be abandoned for further OF agriculture. Brackish water based on the water salinity level can be grouped into three categories, i) fresh (<1000 mg/L) water mostly used for drinking, which is abstracted from Rawdatian and Umm Al Eish fields; (ii) brackish (1000–7000 mg/L), (iii) and saline (7000–20,000 mg/L). Brackish water can be used to irrigate crops based on their salinity tolerance levels (Shahid et al. 2018). The sustainable use of brackish water requires root zone salinity monitoring and leaching of salts to maintain root zone salinity below the salinity threshold level of crop grown. Periodic groundwater salinity monitoring in Al-Wafra has revealed that in 1989, the water salinity of 50% wells presented salinity >7500 mg/L, which increased by 75% & 85% wells in 1997 and 2002 respectively (Zekri and Zaibet 2019).

15.5.6 Threats to Agriculture in Farming Area: Global Context and Kuwait Conditions

Out of the 10 main threats (erosion, soil organic carbon loss, nutrient imbalance, acidification, contamination, waterlogging, compaction, sealing, salinization, and loss of soil biodiversity) to soil functions (FAO-ITPS 2015), the soils of Kuwait are vulnerable to nine except acidification. Among these threats, soil erosion is one of the most serious threats facing world food production. In Kuwait, 63.5% of the total land area is degraded (Kuwait Voluntary National Review 2019), specifically, this review states "loss of agriculture soil due to the nature of the arid country's soil salinity and increased levels of groundwater salinity used in irrigation".

The loss of soil due to erosion is widespread and reduces the productivity of all natural ecosystems as well as agricultural, forest, and pasture ecosystems (Pimentel et al. 1995; Troeh et al. 2004). About 80% of the world's agricultural land suffers moderate to severe erosion, while 10% experiences slight erosion (Lal 1994; Speth 1994). Each year about 10 million ha of cropland are lost due to soil erosion, thus reducing the cropland available for world food production (Pimentel and Burgess 2013). Aside from farm abandonment, erosion also reduces farm productivity of leftover farmland which is compensated by the addition of costly N and P fertilizers (Pimentel et al. 1995; Young 1998; Lal 2006; Pimentel 2006). Pimentel and Burgess (2013) further emphasized that overall, soil is being lost from agricultural areas 10–40 times faster than the rate of soil formation, thus jeopardizing humanity's food security. Annual cereal production losses due to topsoil erosion are estimated to be of the order of 7.6 million tons (FAO 2022a). According to OECD the annual soil loss rate of 11 tons/ha is considered critical for crop losses (Borrelli et al. 2020). The FAO (2019) considers erosion the greater challenge to sustainable soil management, whereby up to 60% of losses to crop production are due to erosion only.

Similar to the soil losses globally, Kuwait agriculture farm soils are also vulnerable to various threats including wind erosion, nutrient imbalance, waterlogging, and salinization, thus improvising the farm's capacity to produce food to their full potential. The specific constraints to farming areas in Kuwait are given below:

- Sandy texture, low clay, and organic matter contents leading to low nutrient and moisture retention and high losses through leaching and polluting groundwater.
- High soil pH, the optimum soil pH range is between 6.7 and 7.3 where most of the nutrients are available to plants, at high pH they are fixed in soil and become unavailable.
- Soil salinity, the accumulation of salts in soil causes plant stress. High soil and water salinity limits crop choices. Placing seeds in less salt zone is good practice for salinity management.
- Shallow soil depth (*Leptic Haplogypsids*) where gypsum accumulates at less than 18 cm from soil surface (Abdali area).
- High gypsum content (*Leptic Haplogypsids*).
- Poor soil structure, sand-loamy and sand textured soils are loose and fragile, thus are highly vulnerable to wind erosion and losing productive surface soil.

15.5.7 Soil Management Options for Sustainable Agriculture Production

Sustainable soil management (SSM) requires integrated best soil management practices tested and proven successful under similar soil and environmental conditions for adoption in Kuwait, provided if these practices are not already established or adopted in Kuwait. The soils of Kuwait require adoption of proper tillage practices (low to no till) to conserve soil moisture and organic carbon, water-saving techniques, crop residue management to transform to organic fertilizers, irrigation and weed management as well as crop rotation by including crops having the character of biological nitrogen fixation (BNF) to maintain soil nitrogen. Under Kuwait conditions, use of soil amendments, balanced fertilizers, tillage practices, water management, and crop rotation are the key to SSM. Avoiding these practices may lead to reduced soil quality to provide functions for optimum crop production. Therefore, the objectives of all these practices should be to increase soil and water productivities in terms of yields and value per unit of land and water. Following are the key soil management options to improve the productivity of farming areas in Abdali and Al-Wafra.

- The soils which are not suitable for crop production, such as *Leptic Haplogypsids* (Abdali farming area) can be used for other purposes, such as farm machinery sheds, poultry and animal sheds, storage for fertilizers, pesticides, and other materials.
- Use organic-based fertilizers [plants based compost, animal manure (sheep/poultry/cow)] to improve soil organic matter content, plant nutrients, and soil carbon sequestration.
- Due to sandy soil texture, the infiltration rate will be very high, which needs to be controlled by small but frequent irrigation to offset plants water requirements, and to reduce leaching of nutrients, etc.
- Use crops-based modern irrigation systems (drip, sprinkler, subsurface drip, etc.) to optimize water usage and to improve water productivity (crop per drop).
- Adopt 4R Nutrient stewardship for sustainable use of fertilizers. 4 R (Right – time, source, place, and rate).
- Make crop selection based on their salt-tolerance levels.

15.6 Prospective Agriculture Globally and in Kuwait in the Light of Climate Change and Soil Salinization

Salinity is the presence of soluble salts more soluble than gypsum ($CaSO_4.2H_2O$) in soil. Electrical conductivity measured in the extract from saturated soil paste (ECe) is globally accepted a standard method (US Salinity Laboratory Staff 1954). There are two types of soil salinities, (i) *Primary* – naturally saline soil, (ii) *Secondary* – saline soil formed by the activities of man such as improper use of resources. Soil salinity is a significant problem globally and is among the most important problems agriculture is facing. In Kuwait primary salinity occurs in the coastal land due to seawater intrusion and subsequent evaporation. The secondary salinity is in the agricultural farms irrigated with brackish/saline waters of various salinity levels (Shahid et al. 1998, 2002; Hamdallah 1997; Hachicha and Abdelgawad 2003; Al-Rashed and Al-Senafy 2004; Al-Menaie et al. 2018; Shahid 2022). Actual extent (aerial coverage) of both types of salinity in Kuwait is not yet known, but needs to be determined for their optimum uses.

The Global Map of Salt-affected Soils v1.0 indicates that more than 424 million ha of topsoil (0–30 cm) and 833 million ha of subsoil (30–100 cm) are salt-affected (FAO 2022b). Soil salinity is estimated to take up to 1.5 million ha of cropland out of production each year (FAO-ITPS 2015). At the global level, it is estimated to take 0.3–1.5 million ha of farmland out of production each year and reduce productivity for a further 20–46 million ha (FAO-ITPS 2015). However, about 380 million ha of salt-affected soils could be restored for agriculture (Lambers 2003). Globally salinity effects 62 million ha (20% of total irrigated lands) in over 75 countries with a value of annual global economic losses at USD $27.3 billion (Qadir et al. 2014) that is 442 US$/ha. According to a study by United Nations University's Canadian-based Institute for Water, Environment and Health (UNU-IWEH 2014), globally, on a daily basis for more than 20 years, an average of 2000 hectares of irrigated land in arid and semi-arid areas worldwide have been degraded by salt, a problem occurring in 75 countries. At this pace it will take 100 years to lose 1/4 of total irrigated agriculture land which is currently practiced on 310 million ha, by then population may increase many folds.

The salinity problem will be exacerbated with the climate change impact due to the increase of temperature and evapotranspiration, thus, leading to bring salts from lower soil zone to surface. In parallel to soil salinization low aquifer recharge due to decreased rainfall also results into saline groundwater which is non-conducive to plant growth. To address salinity hazard limiting crops/agriculture it is essential to understand current salinity status at national level and monitoring over a period to understand soil salinity dynamics leading to develop national salinity management strategy. We aim to accomplish this strategy on a priority basis in Kuwait (Shahid et al. 2022).

15.6.1 *Causes of Soil Salinity Development in Irrigated Agriculture Fields: Global and Kuwait Contexts*

Climate change-induced increased temperature is likely to increase evapotranspiration and soil erosion, and reduce groundwater recharge and soil moisture for plant growth, leading to a higher incidence of soil salinization especially in hot-dry arid regions like Kuwait. The plants need water to grow and to offset the crop water requirements. The scarcity of freshwater necessitates the use of marginal quality water, generally high in salts, which creates soil salinity. Increased salinity in the root zone increases water stress and physiological drought in crops due to the inability of plants to extract water from soil and decreases crop yield, because salt tolerance of a specific crop depends on its ability to extract water from salinized soils (Choukr-Allah et al. 2023).

15.6.2 Best Soil Salinity Management Options Under Kuwait Environmental Conditions

Prolonged soil salinity problem in irrigated agriculture farms can cause negative effects on food security and the environment. In Kuwait, prevention and management of soil salinity requires integrated approach including but not necessarily limited to:

- Adequate drainage system if Gatch is identified in the farming areas to avoid salinity build up above the gatch.
- Salts leaching prior to seeding "*reclamation leaching*".
- Salt tolerant and market-oriented crop selection (Choukr-Allah et al. 2023). For guidance at what root zone soil salinity (ECe dS/m) various salt tolerant crops loss % yield to various extents, use Table 15.5a–d.
- Seeding at minimum salt zone using appropriate tillage practices (Shahid 2013).
- Crop-specific suitable irrigation system (drip, sprinkler, pivot, etc.) to be carefully used to avoid salts induced foliar damage.
- Apply extra water above crop water requirement to maintain root zone soil salinity below threshold salinity.
- Regular monitoring of root zone salinity and subsequent leaching if root zone salinity is higher than the crop threshold salinity level.
- Reuse drainage water for crops matching salt tolerance levels.
- Where irrigation water is brackish/saline and not suitable for specific crop, blend it with desalinated or treated sewage effluent (TSE)-restricted use, to desired water salinity.
- Use desalinated water at the initial stage of plant growth, later switch to brackish water of reasonable salinity.

15.7 Investing in Soils Is Key to Sustainable Soil Management

The investment in soils to achieve sustainable soil management (SSM) should be based on diagnostics of the problems limiting crop production in a specific area or at national level. This will lead to context-specific adaptation measures for soil investment planning. The key objective of SSM is to improve soil productivity for crops intensification for food and nutrition security. Feed the soil to feed the plant through balanced nutrients input is the basic principle of SSM from the crops intensification point of view to improve domestic food production and to reduce foreign dependence on food import. This is how we see SSM works for UN SDG 2 to *End Hunger* by 2030.

Table 15.5 Relative yield reduction of important field, vegetables, forage, and fruit crops as affected by soil salinity (ECe)

	Percent yield reduction at different soil salinity levels			
Crops ↓	0%	10%	25%	50%
Root zone salinity →	ECe dS/m			
(a) Important field crops: relative yield decrease				
Barley	8.0	10.0	13.0	18.0
Sugarbeet	7.0	8.7	11.0	15.0
Safflower	5.3	6.2	7.6	9.9
Soybean	5.0	5.5	6.3	7.5
Maize	1.7	2.5	3.8	5.9
Cowpea	1.3	2.0	3.1	4.9
(b) Important vegetable crops: relative yield decrease				
Beets, red	4.0	5.1	6.8	9.6
Broccoli	2.8	3.9	5.5	8.2
Tomato	2.5	3.5	5.0	7.6
Cucumber	2.5	3.3	4.4	6.3
Spinach	2.0	3.3	5.3	8.6
Celery	1.8	3.4	5.8	9.9
Cabbage	1.8	2.8	4.4	7.0
Potato	1.7	2.5	3.8	5.9
Pepper	1.5	2.2	3.3	5.1
Lettuce	1.3	2.1	3.2	5.1
Radish	1.2	2.0	3.1	5.0
Onion	1.2	1.8	2.8	4.3
Carrot	1.0	1.7	2.8	4.6
Beans	1.0	1.5	2.3	3.6
(c) Important forage crops: relative yield decrease				
Wheat grass, tall	7.5	9.9	13.0	19.0
Wheat grass, crested	7.5	9.0	11.0	15.0
Bermuda grass	6.9	8.5	11.0	15.0
Barley, hay	6.0	7.4	9.5	13.0
Rye grass, perennial	5.6	6.9	8.9	12.0
Vetch, common	3.0	3.9	5.3	7.6
Sudan grass	2.8	5.1	8.6	14.0
Cowpea	2.5	3.4	4.8	7.1
Sesbania	2.3	3.7	5.9	9.4
Alfalfa	2.0	3.4	5.4	8.8
Love grass	2.0	3.2	5.0	8.0
Corn fodder	1.8	3.2	5.2	8.6
Berseem, clover	1.5	3.2	5.9	10.0
Clover	1.5	2.3	3.6	5.7

(continued)

Table 15.5 (continued)

Crops ↓	Percent yield reduction at different soil salinity levels			
	0%	10%	25%	50%
Root zone salinity →	ECe dS/m			
(d) Important fruit crops: relative yield decrease				
Date palm	4.0	6.8	11.0	18.0
Fig, olive	2.7	3.8	5.5	8.4
Grapefruit	1.8	2.4	3.4	4.9
Orange	1.7	2.3	3.3	4.8
Lemon, apple	1.7	2.3	3.3	4.8
Peach	1.7	2.2	2.9	4.1
Apricot	1.6	2.0	2.6	3.7
Grape	1.5	2.5	4.1	6.7
Almond	1.5	2.0	2.8	4.1
Plum, prune	1.5	2.1	2.9	4.3
Strawberry	1.0	1.3	1.8	2.5

Source: Ayers and Westcot (1985), Zaman et al. (2018), and Choukr-Allah et al. (2023)

Sustainable Soil Management Practices Through Investing in Soils A Big Step to Achieving the Sustainable Development Goal 2 in Kuwait

In addition to investment in soil for sustainable services, this is the time to introduce the farmers with innovative approach of regenerative agriculture (RA) which is gaining international recognition and popularity globally. Since RA is new approach therefore is no general consensus on the definition of RA, but definition is evolving. However, in the broader context RA is an approach that seeks to work with natural systems to restore and enhance the biodiversity, and soil fertility of farmed land (Savills Research 2021). The use of RA concept increases the resilience of ecological systems, rather than extracting from these systems solely to achieve market returns. The ultimate goal of RA is to protect and restore the soils rather degrading through overuse without repairing the damage due to overexploitation. Five core principles of RA are: (i) minimize soil disturbance (to conserve soil moisture and minimize organic carbon decomposition), (ii) maximize species diversity (monocropping depletes soil fertility) by crop rotation with crops who can fix soil e.g., biological nitrogen fixation (BNF) by cropping leguminous crops, (iii) keep the soil covered and build its organic matter (exposure of organic matter will lead to decomposition of organic matter and carbon dioxide emission leading to global warming, covering the soil will conserve organic carbon and improve soil health through buildup of soil structure leading to increase soil moisture retention), (iv) maintain living roots all year round (living roots provide conducive environment in the rhizosphere to increase soil biology population which take part in many bio-chemical

Table 15.6 Comparison of industrial and regenerative approach

Industrial or conventional approach	Regenerative approach
Disconnected	Interconnected
Extractive	Value added
Control nature	Works with nature
Higher chemical inputs	Lower chemical inputs
Monoculture	Diversity
Carbon emission	Carbon fixation
Moisture depletion	Moisture conservation
Soil degradation	Soil build up
Lower soil biology population	Higher soil biology population
High cost of production	Low cost of production

reactions and nutrient release and availability to plants, (v) integrate livestock (integration of livestock will help improve soil through using the animal manure as organic soil amendment and to improve soil fertility). Future agricultural systems should combine traditional and nature-based solutions, novel technologies including artificial intelligence, and microbiome-based precision farming (WWF 2020). Savills Research (2021) made a brief comparison between conventional called industrial and regenerative approach (Table 15.6) which has been modified in this chapter to present a wider picture of the RA approach.

It is to be noted that all activities in RA are sustainable in contrast to conventional practices. A regenerative system fixes the root cause of the problem and then renews its growth potential, whereas sustainability focuses on not letting the problem get any greater (Savills Research 2021).

15.8 Adoption of Climate Smart Agricultural Practices in the Light of FAO Recommendations

To implement the 2030 SDG agenda, the FAO (2017) recommended five principles to balance social, economic, and environmental considerations.

- Improving efficiency in the use of resources is crucial to sustainable agriculture
- Sustainability requires direct action to conserve, protect, and enhance natural resources
- Agriculture that fails to protect and improve rural livelihoods, equity, and social well-being is unsustainable
- Enhanced resilience of people, communities, and ecosystems is key to sustainable agriculture
- Sustainable food and agriculture requires responsible and effective governance mechanisms

These are the general principles likewise applicable for Kuwait, in addition to the rest of the world agricultural communities, and therefore strongly recommended to lead towards achieving climate resilient sustainable agriculture in Kuwait. The non-compliance of five principles globally may lead to not produce 50% extra food and other agricultural products by 2030. Adoption of integrated agriculture production system (IAPS) in Kuwait is another viable option to sustain agriculture, conserve soils, and be climate-friendly and cost-effective.

In IAPS the products and by-products in the farm are recycled-reused in a cycle. Where, the product-byproduct of one production system may be used as input in the cycle. In such linked cycle each production component is supportive and mutually dependent, for example in the case of Kuwait the IAPS can be framed as farm land-forages-livestock-animal manure/compost–soil conservation-low fertilizer requirements–crops intensification–soil carbon sequestration-climate mitigation–environment protection, etc.

15.9 Constraints to Cop Production in Kuwait and Suggested Mitigation Measures

There are multiple constraints to crop production in Kuwait, these are summarized in Table 15.7 along with Kuwait-specific mitigation measures for crop improvement.

Table 15.7 Threats to crops' production and mitigation measures

Threats to crop production	Mitigation measure
Erosion	Erect Kuwait environment adopted trees-based shelterbelts in the farm to protect from wind erosion Soilization – use plants-based material "Carboxy methyl cellulose" to stabilize sandy soil surface and develop soil structure
Fertility	Increase the use of organic fertilizers (compost, manures) to improve soil organic matter, nutrient status, and soil structure development to improve moisture and nutrients retention capacities of farm soils Adopt 4R nutrients stewardship to manage soil fertility
Salinization	Grow crops tolerant to salts Use modern irrigation systems (drip, sprinkler) based on crops type Drip irrigation is the most efficient system in terms of saving water, managing root zone salinity and no direct leaf salt burn Use the concept of leaching fraction to leach salts from the root zone by using extra water above crop water requirement
Water scarcity	Grow crops which are drought-tolerant Bring the marginal quality water (treated sewage effluent-reclaimed water) for use in crop production systems to release pressure on freshwater resources

15.10 Soils and the UN-Sustainable Development Goal 2

Among the UN-17SDGs, SDG 2 directly relates to agriculture, food production, and soils, i.e., "*To end hunger, achieve food security and improved nutrition and promote sustainable agriculture*". Target 2.3 expects double agricultural productivity and the incomes of small-scale food producers by 2030, whereas target 2.4 ensures sustainable food production systems and implements resilient agricultural practices that progressively improve land and soil quality.

The Organization for Economic Co-operation and Development OECD/FAO outlook for 2021 (OECD and FAO 2021) expects crop yield growth to account for 88% of crop production increases to 2030. This ambitious increase in crops yield is likely to come from improvement of genetic material, investment in soil health, and production technologies. In contrast to the ambitious thinking of increasing crop productivity, however, a reverse trend of growth rate has been observed, where during 2011–2019 growth rate slowed to an average of about 2% and is attributed largely to slowing agricultural productivity in developing countries including Brazil, China and India (Fuglie et al. 2021). This growth slowing down trend in agricultural productivity highlights the impact of human-induced climate change (Ortiz-Bobea et al. 2021). Among other immediate actions to be taken to achieve the SDGs, it is believed that the water, energy, and food security nexus (WEF nexus) can help monitor and achieve the SDGs (Stephan et al. 2018). In view of this, the WEF nexus and the SDGs in the GCC region need to be addressed in an integrated way (Abulibdeh et al. 2019). Considering the food security issue the government of Kuwait has already realized and developed Kuwait food security and investment strategy (ICBA-KIA 2014a, b) and highlighted the need to make local and international investment in crop production and to develop international trade with potential countries.

15.11 Cost of Water Production Through Desalination of Salty Water

There is no fixed cost to desalinate the salty water. The cost depends on many factors, such as, but not necessarily limited to the following (Soliman et al. 2021).

- Source of energy (hydraulic, solar, oil, biogas, etc.)
- Scale of desalination unit (small or large) and maintenance cost
- Quality of the feed water (water salinity)
- Cost of brine disposal
- Distance of the sea to the desalination facility (water transport cost)
- Quality of desalinated water (water salinity target)
- The use of desalinated water (drinking, industrial, agriculture)
- Cost to abate environmental impact

15.12 Production of Desalinated Water and Energy Consumption to Produce Desalinated Water (kWh/m^3 of Water)

The recently published report (EBRC-KISR 2019) considered desalination Kuwait's only reliable option to meet the future water consumption needs of its population and economy, and in 2015 Kuwait produced the third greatest amount of desalinated water in the GCC, after Saudi Arabia and UAE. Seventy-seven percent of all desalinated water used in the region is used by Kuwait, Saudi Arabia, and United Arab Emirates (Bazza 2005; FAO 2019). The EBRC-KISR (2019) report further states that in 2015, multi-stage flash (MSF), which requires extensive amounts of process heat for the desalination process, accounted for 84% of desalination in Kuwait following by 6% (MED) and 10% (RO), whereas according to Kuwait Energy Outlook, the share of MSF is expected to fall to 39% by 2035. The share of desalinated water produced using the reverse osmosis process increases but still accounts for just 13%, and MED 39% of total production capacity in 2035 (EBRC-KISR 2019). Worldwide, reverse osmosis accounts for about 65% of desalinated water production.

Darwish et al. (2003) compared the energy consumption for different desalination systems used in Kuwait and GCC countries. The MSF system uses 25 kWh/m^3 the only system used in Kuwait in 2003. In other GCC countries, the average energy consumed by the reverse osmosis (RO) system is 5 kWh/m^3, and by the Multi-Effect Boiling (MEB) is in the range of 12 kWh/m^3 when steam is extracted from steam turbines at low availability. These are the old estimates, with the improvement in technology the energy consumption has reduced significantly.

The EBRC-KISR (2019) gave an overview of energy consumption by three currently used desalination methods (MFS, MED, RO), on average the current design of the MSF process requires 25 kWh of heat input and 3.5 kWh of electricity input (total 28.5 kWh) per cubic meter of desalinated water produced. The multi-effect desalination process requires about 12 kWh of heat and 1.5 kWh of electricity per cubic meter (total 13.5 kWh). Energy consumption needs in the RO process are even lower, some 6.5 kWh of electricity input per cubic meter. What makes the RO process less appealing is the high cost of membrane replacement, the extensive feed treatment and lower plant factors (EBRC-KISR 2019). Amir Rubinstein (https://www.quora.com/profile/Amir-Rubinstein-2) a Former Reverse Osmosis Seawater Desalination Process Engineer gave an estimated gross power consumption of 3.3–3.7 kWh/m^3 that is 1000 L of water for large scale production in the Eastern Mediterranean. This energy includes entire plant's consumption, as well as pumping and distribution to the reservoir. Another energy consumption estimate is 1.2 kWh/m^3 for 50% recovery, in addition, specific power consumption in the MED system is below 2 kWh/m^3 of distillate, whereas, power consumption for the MSF system is typically 4 kWh/M^3 (Ghaffour et al. 2013).

15.13 Energy-Water-Food-Climate Nexus (Kuwait Scenario)

Kuwait is located in one of the most arid regions (hot arid climate) and is scarce in water and arable land like other GCC countries. Kuwait is rich in fossil-fuel-based energy resources like other GCC countries (World Resource Institute 2015; World Bank 2013), therefore, if sufficient energy sources are allocated to produce water through desalination, this can compensate water scarcity to a large extent (Al-Saidi et al. 2016); however, this practice may cause environmental issues like GHG emission which need to be abated sustainably without compromising the integrity of the environment.

The local food production requires high inputs of fertilizers and water. Most of the food is imported from other countries, partially, the food demand is met through local production. There are no surface water sources (rivers, dams, freshwater lakes, etc.) in Kuwait. The crops are grown using the groundwater through installed wells, where the water salinity varies from medium to very high. At these levels of water salinity, the salt-sensitive crops (vegetables, fruits) cannot give economic production; therefore, farmers have installed small-scale desalination units to produce fresh water quality to grow salt-sensitive crops in OF or PA. In any case the energy is used to abstract groundwater or to produce desalinated water. In this case, the energy-water-food-climate forms a nexus at the heart of sustainable agriculture development, because, agriculture is the largest consumer of the world's freshwater resources, and water is used to produce most forms of energy. Demand for all the components of nexus is likely to be increased due to ever increasing population, impact of climate change on water resources, and global warming (high demand of water due to drought). The treated wastewater "*treated sewage effluent*-TSE" is now the potential source of water in Kuwait and other countries for restricted crop production and to reduce water scarcity, expand irrigated agriculture for food security as well as the use for landscaping in urban areas to improve landscapes aesthetic value. However, it should be noted that unless the wastewater is appropriately treated, quality assured, its use in agriculture poses serious risks to public health and the environment (FAO 2019). The water-energy-food nexus is considered a new approach in support of food security and sustainable agriculture (FAO 2014).

The nexus of the water, energy, and food sectors is closely linked where an impact on one sector does have impact on the other two sectors. Hence, a cross-sectoral and dynamic perspective between the WEF sectors is essential (Aliewi and Alomirah 2020), an interconnectivity between the WEF nexus has been shown by Al-Zubari et al. (2018). The FAO (2014) considers the WEF nexus: a new approach in support of food security and sustainable agriculture, and Endo et al. (2015) described the methods of the WEF nexus. However, Majgl et al. (2015) have shown interactions among nexus sectors are ought to be dynamic.

Very few studies are reported from the Arab region on the nexus and are cited by Al-Zubari et al. 2018). Al-Zubari et al. (2018) indicated WEF nexus has become the center of global policy, development and research in order to meet the ever-increasing demand on water, energy, and food against strong resource limits. Recently, Aliewi and Alomirah (2020) have recommended an integration of good agricultural practices that combine technology such as smart irrigation, green energy, optimization through modeling, crop selection for better water productivity, improved efficiency, alternative water resources, land management, and food trade is needed to address the major challenges to sustainable agricultural development in Kuwait. This integration and the modeling optimization require understanding of the inter-linkages between the WEF resources which allows the quantification of the trade-offs between WE & F (Mohtar et al. 2014; Daher and Mohtar 2015; Mohtar and Daher 2016; Amy et al. 2017; Mohtar 2017; Mortada et al. 2018; Degirmencioglu et al. 2019; Lee et al. 2019).

Another potential example of *energy-water-food-climate nexus* can be the recycling of wastewater and subsequent production of energy from biogas generated in wastewater treatment plants. The energy generated from wastewater treatment plants can be used for desalination. Kuwait is the most efficient country in terms of wastewater treatment, treating 28% of the water withdrawn. The energy consumption from any source emits GHGs and increases the concentration in the atmosphere and the temperature that may cost significantly to abate the environmental impact (Moossa et al. 2022). In order to offset the impact of GHGs emitted from the energy used in water production, a parcel of green land (trees/grasses/crops) is required to sequester carbon from the atmosphere. This creates link between energy-water-food-climate, and this nexus needs to be fully analyzed and understood before upscaling crop production locally or internationally.

The water footprint is the total volume of freshwater used to produce the product (Hoekstra et al. 2009). The "water footprint" introduced by Hoekstra (2003) and subsequently elaborated by Hoekstra and Chapagain (2008) provides a framework to analyze the link between human consumption and the appropriation of the globe's freshwater. Three types of water footprints are used (Mekonnen and Hoekstra 2010) in terms of predicting the water required to produce the product (crops), that is, (i) *blue water footprint* – volume of surface and groundwater consumed (evaporated) as a result of the production of a good, (ii) the *green water footprint* refers to the rainwater consumed, and (iii) *grey water footprint* of a product refers to the volume of freshwater that is required to assimilate the load of pollutants based on existing ambient water quality standards. 1 kg of wheat requires about 1000 L of water, and 15,500 L of water is required to produce 1 kg of edible beef (D'Silva 2011). The energy-food-water-climate nexus (GFN 2011) is shown below (cf. Shahid and Ahmed 2014).

At present, only a narrow range of crops is grown in PA, primarily cucumber and tomatoes; however, tomato under PA is less profitable than cucumber, and neither of these crops can be grown in summer months without cooling (FAO 2021). The FAO (2021) reported 1 m^3 of water produces 22 kg of cucumber.

Energy-Food-Water-Climate Nexus

The GFN (2011) clearly shows the nexus (complex) between the resources used and generated. It shows 20,000 L of water is required to produce 1 kg of beef, 60 kWh is required to produce 20,000 L of desalinated water, and 140 global square meters of land is required for a year to absorb the CO_2 of diesel used to generate 60 kWh of electricity. This can be illustrated in a simple equation:

20,000 L of water = 1 kg of beef = 60 kWh of energy consumed = 140 m^2 land for 1 year is required to absorb GHGs emitted during the production of 20,000 L of desalinate water.

cf. Shahid and Ahmed (2014)

Energy-Food-Water-Climate Nexus
Kuwait Scenario – Fresh Tomato and Cucumber

The average energy (kWh) required for three desalination systems (MFS-MED-RO) currently practiced in Kuwait (MFS 84%; MED 6%; RO 10%) to produce 1 cubic meter of water is 28.5, 13.5 and 6.5 kWh respectively. Currently, 608.5 Million Cubic Meter water is produced per year (i.e., MSF 511 MCM, MED 36.5 MCM & RO 60.85 MCM) as reported in EBRC-KISR (2019).

Therefore, based on these values average kWh to produce 1 cubic meter of water in Kuwait is 25.40 kWh/m^3 (Table 15.7). As per FAOSTAT data on the tomato crop, to produce 1 ton of tomato (fresh or chilled) 78 m^3 blue water is used. This can be put into equation.

Example 1: 78 m^3 of water = 1000 kg of tomato = (78 × 25.4 = 1981.2 kWh of energy) = 4622.8 m^2 land is required for 1 year to absorb GHGs emitted during the production of 78 m^3 of desalinate water.

As per FAOSTAT data on the cucumber crop, to produce 1 ton of cucumber 45.45 m^3 blue water is used. This can be put into the equation.

Example 2: 45.45 m^3 of water = 1000 kg cucumber = (45.45 × 25.4 = 1154.43 kWh energy) = 2694 m^2 land is required for 1 year to absorb GHGs emitted during the production of 45.45 m^3 of desalinate water (Table 15.8).

Table 15.8 Comparison of energy consumption between three desalination systems used in Kuwait

Desalination system	Energy kWh/m^3 by the system	Water produced by each system/year (million cubic meters)	Total energy used by each system (million kWh)
MSF	28.5	511.1	14,566
MED	13.5	36.5	492.75
RO	6.5	60.9	395.85
Total		608.5	15,454.6
Average per m^3			25.40 kWh

EBRC-KISR (2019)

15.14 Conclusions and Recommendations

Hot and dry climatic conditions, infertile sandy soils, water scarcity, soil erosion, and salinization are observed as the major constraints to Kuwait agriculture and food security. Therefore, it is essential to complete soil/salinity mapping in the farming areas for informed decisions on crop land use based on the capacity of soils to produce food. In Kuwait this effort was made, where the farm level survey of 50 ha each in Al-Wafra and Abdali farming areas has revealed diversity of soil types with different suitability for crop production. The area in the farm which is unsuitable for crop production (shallow depth soil) is recommended to be used for infrastructure (poultry, animal sheds, machinery sheds, and to construct greenhouse), and the area in the farm where deep soils exist can be used for OF agriculture. Integrated farming system (crops-animal-poultry-aquaculture) can help animal waste (manure) for use in farm soils to improve soil tilth and fertility status of soils. In return, the animals in the farms can be fed with forages grown in the farms. In addition, Kuwait farmers should continue intensification by using a combination of best and improved management practices, tested and approved technologies in integrated manner including soil, plants (improved crop varieties), expanded modern irrigation systems, and soil fertility (integrated soil fertility management). Adoption of conservation agriculture practices, such as mulching, low-no tillage can help conserve soil moisture. Crops to be rotated with varieties having biological nitrogen fixation character. Recycling of plant-based farm residue to compost should be included in the integrated farm management policy. It is recommended to use reclaimed water (treated sewage water) as alternative water and nutrient resources to reduce the energy of desalinated water and the cost of fertilizers. This will increase the circular economy and reinforce the approach of WEFC Nexus. It is also recommended to keep the balance between the food-water-energy-climate nexus to avoid provoking the integrity of each nexus component, e.g., at the most the emission of GHG from the nexus affecting the environment.

Acknowledgements We would like to thank Mr. Ahmed Abdulhadi, Systems and Software Development Department (SSDD), Science and Technology Sector (STS), Kuwait Institute for Scientific Research Kuwait for preparing the map to predict the temporal increase in total and the cultivated areas in Abdali and Al-Wafra.

References

Abulibdeh A, Zaidan E, Al-Saidi M (2019) Development drivers of the water-energy-food nexus in the Gulf Cooperation Council region. Dev Pract. https://doi.org/10.1080/09614524.2019.1602109

Aleisa E, Al-Zubari W (2017) Wastewater reuse in the countries of the Gulf Cooperation Council (GCC): the lost opportunity. Environ Monit Assess 189:553

Aliewi A, Alomirah H (2020) Assessment of the significance of water-energy-food nexus for Kuwait. In: Kumar M et al (eds) Resilience, response, and risk in water systems, Springer transactions in civil and environmental engineering, pp 357–367. https://doi.org/10.1007/978-981-15-4668-6_19

Al-Menaie H, Al-Ragom A, Al-Shatti A, Babu MA, Wahbi A (2018) Kuwait. In: IAEA-TECDOC-1841 (challenges and opportunities for crop production in dry and saline environments in ARASIA member states). International Atomic Energy Agency, Vienna, pp 42–46

Al-Nasser AY, Bhat NR (1998) Protected agriculture in the State of Kuwait. Kuwait Institute for Scientific Research, Safat. https://www.researchgate.net/publication/237258098_Protected_Agriculture_in_the_State_of_Kuwait

Al-Rashed M, Al-Senafy M (2004) Assessment of groundwater salinization and soil degradation in Abdally Farms, Kuwait. Agric Mar Sci 9(1):17–19

Al-Saidi M, Birnbaum D, Buriti R, Diek E, Hasselbring C, Jimenez A, Desiree W (2016) Water vulnerability assessment for MENA countries considering energy and virtual water interactions. Procedia Eng 145:900–907. https://doi.org/10.1016/j.proeng.2016.04.117

Al-Zubari WK (2003) Alternative water policies for the Gulf Cooperation Council countries. Dev Water Sci 50:155–167

Al-Zubari W, Al-Turbakb A, Zahid W, Al-Ruwis K, Al-Tkhais A, Al-Muatazb I, Abdelwahabd A, Murad A, Al-Harbi M, Zaher Al-Sulaymanig Z (2017) An overview of the GCC Unified Water Strategy (2016–2035). Desalin Water Treat 81:1–18

Al-Zubari W, ElSadek A, Mohamed A (2018) The water-energy-food nexus in the Arab region, Arabian Gulf University. A policy paper published by the League of Arab States (LAS), with technical and financial support from the Deutsche Gesellschaft fur Internationale Zusammenarbeit (GIZ)

Amr ZS (2021) The state of biodiversity in Kuwait. IUCN/Environmental Public Authority, Gland/Kuwait

Amy G, Ghaffour N, Li Z, Francis L, Linares RV, Missimer T, Lattemann S (2017) Membrane-based seawater desalination: present and future prospects. Desalination 401:16–21. https://doi.org/10.1016/j.desal.2016.10.002

Ayers RS, Westcot DW (1985) Water quality for agriculture: FAO irrigation and drainage. FAO, Rome, p 174

Bazza M (2005) Policies for water management and food security under water-scarcity conditions: The Case of GCC Countries. Paper presented at 7th Gulf Water Conference. Kuwait, Water Science and Technology Association, pp 19–23

Borrelli P, Robinson DA, Panagos P, Lugato E, Yang JE, Alewell C, Wuepper D, Montanarella L, Ballabio C (2020) Land use and climate change impacts on global soil erosion by water (2015–2070). Proc Natl Acad Sci U S A 117(36):21994–22001

Central Statistics Bureau State of Kuwait (2019) Annual agricultural statistics. Central Statistics Bureau State of Kuwait, Kuwait, p 54

Choukr-Allah R, El Mouridi Z, Benbessis Y, Shahid SA (2023) Salt-affected soils and their management in the Middle East and North Africa (MENA) region: a holistic approach. In: Choukr-Allah R, Ragab R (eds) Biosaline agriculture as a climate change adaptation for food security. Springer, pp 13–45. https://doi.org/10.1007/978-3-031-24279-3_2

Collins G (2017) Carbohydrates, H_2O, and hydrocarbons: grain supply security and the food-water-energy nexus in the Arabian Gulf region. James A. Baker III Institute for Public Policy of Rice University, Houston

D'Silva J (2011) Food price rises and the meat connection. Newsl World Forum Clim Change Agric Food Secur (WFCCAFS) 1(2):2

Daher B, Mohtar R (2015) Water-energy-food (WEF) Nexus Tool 2.0: guiding integrative resource planning and decision-making. Water Int 40(5):1–24. https://doi.org/10.1080/02508060.2015.1074148

Darwish M, Al Asfour F, Al Najem M (2003) Energy consumption in equivalent work by different desalting methods: case study for Kuwait. Desalination 152(1):83–92. https://doi.org/10.1016/S0011-9164(02)01051-2

Darwish MA, Al-Najem NM, Lior N (2009) Towards sustainable seawater desalting in the Gulf area. Desalination 235:58–87

Degirmencioglu A, Mohtar RH, Daher B, Ozgunaltay-Ertugrul G, Ertugrul O (2019) Assessing the sustainability of crop production in the Gediz Basin, Turkey: a water, energy, and food nexus approach. Fresenius Environ Bull 28(4):2511–2522

EBRC-KISR (2019) Sustaining prosperity through strategic energy management. Joint report published by Kuwait Institute for Scientific Research, New Kuwait, UNDP and General Secretariat of the Supreme Council for Planning and Development, p 84

Endo A, Burnett K, Orencio P, Kumazawa T, Wada C, Ishii A, Tsurita I, Taniguchi M (2015) Methods of the water-energy-food nexus. Water 7:5806–5830. https://doi.org/10.3390/w7105806

FAO (1976) A framework for land evaluation, FAO soil bulletin 32. Food and Agriculture Organization, Rome

FAO (2014) The water–energy–food nexus: a new approach in support of food security and sustainable agriculture. Food and Agriculture Organization of the United Nations, Rome

FAO (2017) Climate smart agriculture sourcebook. Overview-significant development, summary, 2nd edn. isbn:978-92-5-109988-9

FAO (2019) In: Pennock D, Lefèvre C, Vargas R, Pennock L, Sala M (eds) Soil erosion: the greatest challenge for sustainable soil management, Rome. www.fao.org/3/ca4395en/ca4395en.pdf

FAO (2021) Protected agriculture in Kuwait. In: Unlocking the potential of protected agriculture in the GCC countries: saving water and improving nutrition, Cairo, p FAO, 216. Licence: CC BY-NC-SA 3.0 IGO, pp 68–72

FAO (2022a) The state of the world's land and water resources for food and agriculture – systems at breaking point. Main report. Rome. https://doi.org/10.4060/cb9910en

FAO (2022b) Global map of salt-affected soils (GSASmap). In: Global Soil Partnership. Rome. Cited 9 February 2022. www.fao.org/global-soil-partnership/gsasmap/en. Modified to comply with UN. 2020. Map of the World. https://www.un.org/geospatial/file/3420

FAO and ITPS (2015) Status of the world's soil resources: Main report. Rome. www.fao.org/3/i5199e/i5199e.pdf

Fuglie K, Jellife J, Morgan S (2021) Slowing productivity reduces growth in global agricultural output. In: Amber Waves. 28 December 2021. United States Department of Agriculture Economic Research Service. www.ers.usda.gov/amber-waves/2021/december/slowing-productivity-reduces-growth-in-global-agricultural-output

Ghaffour N, Missimer TM, Amy GL (2013) Technical review and evaluation of the economics of water desalination: current and future challenges for better water supply sustainability. Desalination 309:197–207. https://doi.org/10.1016/j.desal.2012.10.015

Global Footprint Network (GFN) (2011) What happens when infinite-growth economy runs into a finite planet. Global Footprint Network 2011 annual report. Oakland, California 94607-3510 USA

Hachicha MM, Abdelgawad G (2003) Aspects of salt-affected soils in the Arab World. Saltmed Workshop, Cairo, December 8–10, 2003. Available at http://www.nwl.ac.uk/research/cairo-workshop/papers/hachicha2.pdf

Hamdallah G (1997) An overview of the salinity status of the near east region. Proceedings of the Workshop on Management of Salt-Affected Soils in the Arab Gulf States, Abu Dhabi, United Arab Emirates, October 29–November 2, 1995. Food and Agriculture Organization of the United Nations, Regional Office for the Near East, Cairo, pp 1–5

Hoekstra AY (2003) Virtual water trade: proceedings of the international expert meeting on virtual water trade, Delft, The Netherlands, 12–13 December 2002, Value of water research report series no. 12. UNESCO-IHE, Delft. www.waterfootprint.org/Reports/Report12.pdf

Hoekstra AY, Chapagain AK (2008) Globalization of water: sharing the planet's freshwater resources. Blackwell Publishing, Oxford

Hoekstra AY, Chapagain AK, Aldaya MM, Mekonnen MM (2009) Water footprint manual: state of the art 2009. Water Footprint Network, Enschede

Hunter MC, Smith RG, Schipanski ME, Atwood LW, Mortensen DA (2017) Agriculture in 2050: recalibrating targets for sustainable intensification. Bioscience 67(4):386–391

ICBA-KIA (2014a) Kuwait food security and investment strategy assessment report. Annex 12: task 12L state of art agriculture technologies and methods to enhance vegetation and agriculture production. Sub-task: crop production. Kuwait, p 30

ICBA-KIA (2014b) Kuwait food security and investment strategy – assessment report – task 12: state of art agricultural technologies and methods to enhance vegetation and agriculture production. Kuwait Investment Authority, Kuwait

Ismail H (2015) Kuwait: food and water security. Strategic Anal Paper Future Directions International. https://water.fanack.com/kuwait/water-use-in-kuwait/. Accessed on 13 June 2023

Kacira M (2011) Greenhouse production in US, status, challenges, and opportunities. Proceedings of GIGR international symposium 2011 sustainable bioproduction-water, energy, and food, Tokyo, 19-23 September 2011

KISR (1999) Soil survey for the State of Kuwait: reconnaissance survey, vols II & III. Kuwait Institute for Scientific Research, Kuwait

Kuwait Voluntary National Review (2019) New Kuwait – report on the implementation of the 2030 agenda to the UN high-level political forum on sustainable development. Kuwait

Lal R (1994) Water management in various crop production systems related to soil tillage. Soil Tillage Res 30:169–185

Lal R (2006) Enhancing crop yield in the developing countries through restoration of soil organic carbon pool in agricultural lands. Land Degrad Dev 17:197–209

Lambers H (2003) Introduction: dryland salinity: a key environmental issue in southern Australia. Plant Soil 257:5–7

Lee SH, Mohtar RH, Yoo S (2019) Assessment of food trade impacts on water, food, and land security in the MENA region. Hydrol Earth Syst Sci 23:557–572. Copernicus Publications EGU. https://doi.org/10.5194/hess-23-557-2019

Majgl A, Eard J, Pluschke L (2015) The water-food-energy nexus-realising a new paradigm. J Hydrol:530–540. https://doi.org/10.1016/j.jhydrol.2015.12.033

Mekonnen MM, Hoekstra AY (2010) The green, blue and grey water footprint of crops and derived crop products, Value of water research report series no. 47. UNESCO-IHE, Delft. http://www.waterfootprint.org/Reports/Report47-WaterFootprintCrops-Vol1.pdf

Mohtar RH (Moderator) (2017) Opportunities in the food-energy-water nexus. In: White paper for the ASABE global initiative—special session. ASABE Conference in Spokane, WA, 17 July 2017

Mohtar RH, Daher B (2016) Water-energy-food nexus framework for facilitating multi-stakeholder dialogue. Water Int 41(5):1–7. https://doi.org/10.1080/02508060.2016.1149759

Mohtar RH, Daher B, Mekki I, Chaibi T, Chebbi RZ, Salaymeh A (2014) The water, energy and food (WEF) nexus project: a basis for strategic planning for natural resources sustainability-challenges for application in the MENA region. Geophys Res Abstr 16:EGU2014-15330

Moossa B, Trivedi P, Saleem H, Zaidi SJ (2022) Desalination in the GCC countries – a review. J Clean Prod 357(1–3):131717. https://doi.org/10.1016/j.jclepro.2022.131717

Mortada S, Abou NM, Yassine A, El Fadel M, Alamiddine I (2018) Towards sustainable water-food nexus: an optimization approach. J Clean Prod 178:408–418

Nair R, Barche S (2014) Protected cultivation of vegetables–present status and future prospects in India. Indian J Appl Res 4(6):245–247. https://doi.org/10.1016/j.jclepro.2018.01.020

OECD (Organization for Economic Co-operation and Development) & FAO (2021) OECD-FAO agricultural outlook 2021–2030. OECD Publishing, Paris. https://doi.org/10.1787/19428846-en

Omar SAS, Shahid SA (2013) Reconnaissance soil survey for the State of Kuwait. In: Shahid SA, Taha FK, Abdelfattah MA (eds) Developments in soil classification, land use planning and policy implications: innovative thinking of soil inventory for land use planning and management of land resources. Springer, Dordrecht, pp 85–110

Ortiz-Bobea A, Ault TR, Carrillo CM, Chambers RG, Lobell DB (2021) Anthropogenic climate change has slowed global agricultural productivity growth. Nat Clim Chang 11(4):306–312

Pimentel D (2006) Soil erosion: a food and environmental threat. Environ Dev Sustain 8:119–137

Pimentel D, Burgess M (2013) Soil erosion threatens food production. Agriculture 3(3):443–463. https://doi.org/10.3390/agriculture3030443

Pimentel D, Harvey C, Resosudarmo P, Sinclair K, Kurz D, McNair M, Crist S, Sphpritz L, Fitton L, Saffouri R, Blair R (1995) Environmental and economic costs of soil erosion and conservation benefits. Science 267:1117–1123. https://doi.org/10.1126/science.267.5201.1117

Qadir M, Quillérou E, Nangia V, Murtaza G, Singh M, Thomas RJ, Drechsel P, Noble AD (2014) Economics of salt-induced land degradation and restoration. Nat Res Forum 38(4):282–295

Savills Research (2021) Regenerative agriculture. UK-Rural February 2021. www.savills.co.uk/insight-and-opinion/

Shahid SA (2013) Irrigation-induced soil salinity under different irrigation systems – assessment and management, short technical note. Clim Change Outlook Adapt Int J 1(1):19–24

Shahid SA (2022) Innovative thinking and use of salt-affected soils in irrigated agriculture. In: FAO. 2022. Halt soil salinization, boost soil productivity – Proceedings of the Global Symposium on Salt-affected Soils. 20–22 October 2021. Rome, pp 292–293. https://doi.org/10.4060/cb9565en

Shahid SA, Ahmed M (2014) Changing face of agriculture in the Gulf Cooperation Council countries. In: Shahid SA, Ahmed M (eds) Environmental cost and face of agriculture in the Gulf Cooperation Council countries: fostering agriculture in the context of climate change, pp 1–26. https://doi.org/10.1007/978-3-319-05768-2_1

Shahid SA, Omar SAS (1999) Order 1 soil survey of the demonstration farm sites with proposed management. Kuwait Institute for Scientific Research, Kuwait, pp viii + 144. KISR No 5463. isbn:0 957700369

Shahid SA, Omar SAS (2022) Kuwait soil taxonomy. Springer, Cham, p 149

Shahid SA, Omar SAS, Grealish G, King P, El-Gawad MA, Al-Mesabahi K (1998) Salinization as an early warning of land degradation in Kuwait. Probl Desert Dev 5:8–12

Shahid SA, Omar SAS, Jamal ME, Shihab A, Abo-Rezq H (2004) Soil survey for farm planning in northern Kuwait. Kuwait. J Eng Sci 31(1):43–57

Shahid SA, Abo-Rezq H, Omar SAS (2002) Mapping soil salinity through a reconnaissance soil survey of Kuwait and geographic information system, Annual research report, report no. KISR6682. Kuwait Institute for Scientific Research, Kuwait, pp 56–59

Shahid SA, Burezq HA, Baron HJ (2022) Farm land salinity risk assessment to develop a management strategy for food security. Proposal. Kuwait Institute for Scientific Research, Kuwait. KISR # 17167. Kuwait Foundation for the Advancement of Sciences-KFAS code:PN22-41SE-1668, p 51

Shahid SA, Zaman M, Heng L (2018) Salinity and sodicity adaptation and mitigation options. In Zaman M, Shahid SA, Heng L (eds) Guidelines for salinity assessment, mitigation and adaptation using nuclear and related techniques. Springer, Cham, pp 55-89

Singh B, Solanki R (2014) Protected cultivation technologies for vegetable cultivation under changing climatic conditions. In: Chaudhary ML, Patel VP, Siddiqui MW, Mahdi SS (eds) Climate change: the principles and applications in horticultural science. CRC Press, Taylor & Francis Group, Boca Raton, pp 106–114

Singh RK, Singh AK, Singh AK, Singh N, Singh A, Kalia A, Gautam US (2022) Doubling the farmers income through protected cultivation in Bundelkhand region of Uttar Pradesh. In: Horticulture science for doubling farmers income. Brillion Publishing, New Delhi, pp 188–210

Soil Science Division Staff (2017) Soil survey manual. United States Department of Agriculture Handbook No. 18

Soil Survey Staff (2022) Keys to soil taxonomy, 13th edn. US Department of Agriculture, Natural Resources Conservation Service, US Government Printing Office, Washington, DC

Soliman MN, Guen FZ, Ahmed SA, Saleem H, Khalil MJ, Syed Javaid Zaidi SJ (2021) Energy consumption and environmental impact assessment of desalination plants and brine disposal strategies. Process Saf Environ Prot 147:589–608. https://doi.org/10.1016/j.psep.2020.12.038

Speth JG (1994) Towards an effective and operational international convention on desertification; international convention on desertification. New York, International Negotiating Committee, United Nations

Stephan RM, Mohtar RH, Daher B, Embid Irujo A, Hillers A, Ganter JC, Karlberg L, Martin L, Nairizi S, Rodriguez DJ, Sarni W (2018) Water-energy-food nexus: a platform for implementing the sustainable development goals. Water Int 43(3):472–479

Suleiman MK (2020) Pilot study for utilization of modern technologies in sustainable local food production in support of Kuwait's food security. Executive summary. GI P-KISR-17 project proposal. Kuwait Institute for Scientific Research, Kuwait

Tashtush FM, Al-Zubari WK, Al-Haddad AS (2023) The use of non-conventional water resources in agriculture in the Gulf Cooperation Council countries: key challenges and opportunities for the use of treated wastewater. In: Choukr-Allah R, Ragab R (eds). Springer, Biosaline agriculture as a climate change adaptation for food security, pp 285–322. https://doi.org/10.1007/978-3-031-24279-3

Troeh FR, Hobbs AH, Donahue RL (2004) Soil and water conservation: for productivity and environmental protection. Upper Saddle River, Prentice Hall

UNU-IWEH (2014). https://www.iwmi.cgiar.org/News_Room/Press_Releases/2014/pdf/press-release_economics_of_salt-induced_land_degradation_and_restoration.pdf. Accessed 13 June 2023

US Salinity Lab Staff (1954) Diagnosis and improvement of saline and alkali soils. USDA Handbook 60

World Bank (2013) Agricultural land (% of land area). http://data.worldbank.org/indicator/AG.LND.AGRI.ZS/countries/1W-A?display=default.Last. Accessed 13 June 2023

World Resource Institute (2015) Ranking the world's most water-stressed countries in 2040. www.wri.org/blog/2015/08/ranking-world's-most-water-stressed-countries-2040. Accessed 13 June 2023

WWF (2020) Soil biodiversity: saving the word beneath our feet. In: Almond REA, Grooten M, Petersen T (eds) Living planet report – bending the curve of biodiversity loss. WWF, Gland

Young A (1998) Land resources: now and for the future. Cambridge University Press, Cambridge

Zaman M, Shahid SA, Heng L (2018) Irrigation systems and zones of salinity development. In: Zaman M, Shahid SA, Heng L (eds) Guideline for salinity assessment, mitigation and adaptation using nuclear and related techniques. Springer, pp 91–111

Zekri S, Zaibet L (2019) Pre-feasibility study of establishing an integrated model farm utilizing modern technologies for local agricultural commodities in Kuwait. Unpublished feasibility report, Kuwait Institute for Scientific Research Kuwait, p 78

Chapter 16
Agricultural Water Footprint of Major Crops Grown in Kuwait Compared to the World Average: A Review

Majda Khalil Suleiman and Shabbir Ahmad Shahid

Abstract Analysis of water footprint (green, blue, gray) of major crops grown in Kuwait and in the world is completed. The data will be eye opening for the potential consumers to realize the volume of water used to produce a specific quantity of crop or product. Authentic sources, e.g., FAOSTAT, water footprint organization, and recently published literature are reviewed to synthesize the information. In Kuwait, diversified crops are grown in open field and protected agriculture. The crops grown in open field in Kuwait (barley, olives, watermelons, grapes, oranges, limes, etc.) present higher water footprint of production relative to global average. Of 24 key crops grown in Kuwait, 14 crops (maize, potatoes, cabbages, lettuce, spinach, tomatoes, cauliflowers, pumpkins, aubergines, peppers, onion, beans, strawberries, and dates) have less water footprint of production than the global average of the same crops. The concept of virtual water footprint is defined and imported and exported water footprint of a nation are presented. Globally green water footprint consumption in agriculture is the highest among gray and blue water footprints, whereas in Kuwait blue water footprint consumption in agriculture is the highest. Water footprint of animal production is given briefly. These data will be useful for future considerations to improve water footprints of various crop products in Kuwait. Water footprint data can provide valuable information to governments and can help to achieve sustainable development goals.

Keywords Water footprint · Green · Blue · Gray · Virtual water · Consumption · Production · Animals

---

M. K. Suleiman · S. A. Shahid (✉)
Desert Agriculture and Ecosystems Program, Environment and Life Sciences Research Center, Kuwait Institute for Scientific Research, Safat, Kuwait
e-mail: mkhalil@kisr.edu.kw; sshahid@kisr.edu.kw

M. K. Suleiman, S. A. Shahid (eds.), *Terrestrial Environment and Ecosystems of Kuwait*, https://doi.org/10.1007/978-3-031-46262-7_16

16.1 Introduction

Water is an essential part of human life, a source to grow crops for fuel, fiber, and food and to keep the ecosystems healthy to function and provide the ecosystem services on a sustainable basis. Agriculture is by far the largest user of water (Mekonnen and Gerbens-Leense 2020), and according to OECD (2010), approximately 70 percent of global freshwater withdrawal is used in agriculture, and a large share goes on feed and livestock production (Opio et al. 2011). In addition to water use in agriculture for food production, each component of life requires specific volume of water to survive or to produce goods, e.g., different crops require different volumes of water to produce a specific quantity of food (water footprint) under different geographical and environmental conditions. Water footprint (WF) can be classified based on the sources of water and uses. The WF can be divided into three types of water uses: blue (B), green (G), and gray (Gr) WF. The GWF relates to the rainwater stored in soil and evaporates/transpires through crop growth. The BWF is associated with the water sourced from lakes, rivers, and reservoirs and withdrawn from the ground for use in crop fields. The GrW is referred to the volume of water polluted during the production process in the agriculture sector or other industries as well as wastewater from household use. The GrWF is the volume of water required to dilute pollutants to such an extent that the water quality reaches an acceptable level. In the context of the above explanation, it can be deduced that there is no GWF for household and domestic use, but both the BWF and GrWF are part of household and domestic uses.

Kuwait is a hyper-arid country, where there is low rainfall and high evaporation rates, thus, crop production requires daily application of water to offset the water requirements of a crop. The water requirement is met by using different quality waters for specific crops, these are groundwater, desalinated water (DW), treated wastewater (TWW), and high-quality ultra-pure reverse osmosis (RO) water. The cost of production of each type of water varies depending upon the technology used, energy consumption, and efficiency of the technology. In Kuwait due to harsh climatic conditions, open field agriculture is practiced for selected forage crops, and vegetables are grown in open field during winter months. The high-value cash crops (vegetables like cucumber, tomato, capsicum, lettuce, greens) are produced under temperature-controlled greenhouses (protected agriculture). Therefore, the WF of various crops grown in open field or protected agriculture vary significantly.

The country's BWF for food items can be divided into 2, internal and external. For the countries that have water and arable land scarcity within their national boundaries, their main component of BWF depends on other countries. These countries import food and also at the same time import virtual water, and the export countries make trade with other countries (Hoekstra 2003) by selling the food products and the water embedded in these products (physical water and water used to produce these products). Thus, relieving the pressure on the domestic water resources of the countries importing the food. Therefore, international trade in food implies international flows of virtual water (Chapagain and Hoekstra 2003). In this chapter, efforts have been made to synthesize the information on different WF of various crops in Kuwait and the world contributing to food security.

16.2 The State of World Water

The planet Earth is covered by 75% water (1.386 billion km^3) and 25% land by area (Eakins and Sharman 2010; Gleick 1993). About 97.5% of water on the planet is salty. The remaining water is fresh (2.5%) which is locked up in glaciers and ice caps, deep aquifers underground. Considering the 2.5% freshwater as 100%, this can be divided into glaciers and ice caps (68.7%), groundwater (30.1%) and surface water (1.2%). Similarly, if we consider 1.2% surface/other freshwater as 100%, this can be divided into atmosphere (3%), living things (0.26%), rivers (0.48%), swamps and marshes (2.6%), soil moisture (3.8%), lakes (20.9%), and ground ice and permafrost (69.0%). Global freshwater withdrawal has increased nearly sevenfold in the past century (Gleick 2000). Figure 16.1 presents the global distribution of water resources. About 1% water is renewed hydrologically every year, and the water renewal capacity varies in different countries and the regions based on the amount of rainfall and aquifer recharge capacity. Therefore, the freshwater resources vary in different countries. As an example, Kuwait has low rainfall (< 110 mm/year) and recharge capacity is low compared with countries in the humid zone like the United Kingdom, where green water resources are significant, although there is prolonged drought in the UK due to climate change impact which affected the agriculture sector. Thus, the aquifer recharge is reduced in recent times due to decrease in rainfall.

Considering the uneven distribution of water resources in different regions and the impact of climate change a high population will be in water stress (when annual water supplies drop below 1700 m^3 per person), water scarcity (when annual water supplies drop below 1000 m^3 per person), and absolute water scarcity (when water supply is below 500 m^3 per person) (WWAP 2012). The forecasted increase in the

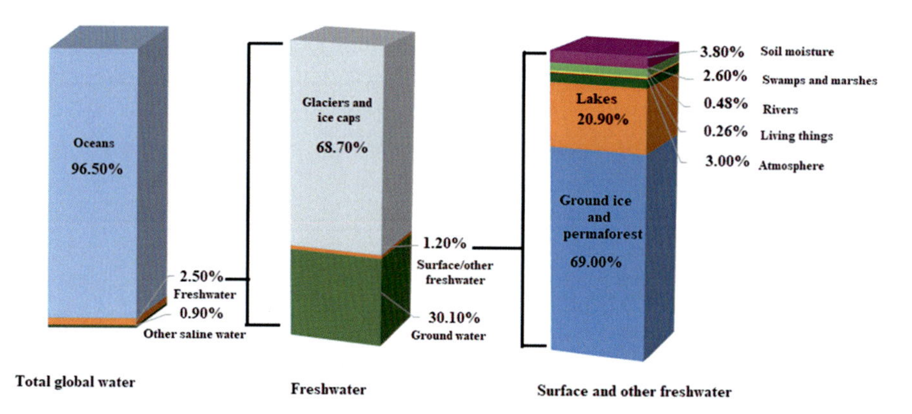

Fig. 16.1 The state of world water resources

global population (10 billion in 2050) will result in available freshwater resources being reduced by half to 6300 m^3 per capita by the mid twenty-first century (Lutz et al. 1997; Ringler et al. 2010).

16.3 Water Resources of Kuwait

Water, like energy, is a key input into national and global economies. Water can be a local or global collective resource, due mainly to national consumption and international trade of goods to meet the needs of the world's populations. The United Nations warns that water use is growing at twice the rate of population growth. Unless this trend is reversed, two-thirds of the global population will face water "stress" by 2025. In addition, the World Health Organization has stated that by 2025, half of the world's population will be living in water-stressed areas, and issues related to water are intrinsically linked to climate change.

In Kuwait, there are three main sources of water (Table 16.1), desalinated water, groundwater, and treated wastewater (TWW) in addition to one minor (ultra-filters reverse osmosis) source. The green water is scarce based on the low rainfall. The groundwater is coming from two aquifers, the upper (Kuwait Group) and the lower (Dammam Formation) (Al-Senaf and Abraham 2004). In general, water requirements of Kuwait are met through three main resources covering 61% desalinated water, 20% groundwater, and 19% TWW (https://water.fanack.com/kuwait/water-resources-in-kuwait/).

16.3.1 Desalinated Water (DW)

The seawater recycling through desalination is the main source of freshwater, in addition to groundwater (GW) and treated wastewater (TWW). The DW is the primary source of water providing 61% of freshwater for drinking, agriculture, and other water-dependent industries. There are eight desalination plants, with total installed and consumption capacities of 3.11 and 2.0 million m^3 day^{-1} (year 2019) respectively (MEW 2020). However, the annual production of DW is about 717.90 Mm^3 in 2019 (MEW 2020).

Table 16.1 Water produced from different sources in Kuwait

Source of water	Water generated from different sources in Kuwait
Groundwater	240.6 million m^3 year^{-1}, 2019 (MEW 2020)
Desalinated water	717.90 million m^3 year^{-1}, 2019 (MEW 2020)
Treated wastewater	1 million m^3 day^{-1} ~ 365 million m^3 year^{-1} (Al-Shammari and Shahalam 2006)
Total	1323.5

16.3.2 Groundwater (GW)

This water is drawn from a number of groundwater wells. The salts level ranges between brackish and saline (Al Ali 2008; EPA 2019; MEW 2020). Although the groundwater is salty to various levels based on the location of abstraction, it is mainly used for agricultural purposes. This may cause soil salinization and reduce crops yield based on the water salinity and salt-tolerance level of crop grown (Shahid et al. 2018; Zaman et al. 2018). In addition, the farmers have also installed small-scale RO units (feeding groundwater) to have high-quality water to use for high valued-cash crops (vegetables and fruits) in protected agriculture. Variable quantities of groundwater were produced in different years [(e.g., 143 Mm^3 $year^{-1}$ 1990 Vs 240.6 Mm^3 $year^{-1}$ in 2019 (0.27 Mm^3 daily consumption in 2019)] (MEW 2020). The brackish groundwater salinity ranges between minimum 2800 mg/l in the Shagaya field, and maximum in the Wafra field (5500 mg/l) (MEW 2020).

16.3.3 Treated Waste Water (TWW)

This water is produced through recycling domestic and industrial wastewater. The water produced in Kuwait is treated at tertiary level (25%) and RO quality (75%). About 58% of RO quality water is reused and 31% not used (Aleisa and Alshayji 2019). The use of TWW (restricted/unrestricted) overtime will reduce pressure on the freshwater resources. The TWW is considered a viable future source of water for use in Agriculture, urban landscaping, and perhaps expansion of forestation in Kuwait to mitigate climate change to increase carbon sequestration and to reduce per capita carbon emission in Kuwait. Aleisa and Alshayji (2019) presented the latest information about the use of TWW (agriculture 19% of all water use, 270,000 m^3 for landscaping and fodder production) whereas RO level water quality TWW (318,000 m^3 day^{-1}) is used for crop production and in natural reserves. The combined (desalination-ultrafiltration) facility is unique in Kuwait established at Sulaibiya to treat wastewater to high quality (Hamoda 2013).

16.3.4 Recent Information of Water Consumption in Kuwait

An effort has been made to get access to recent water consumption data (m^3 $year^{-1}$ $capita^{-1}$) for Kuwait (UWSS 2016). This provides a trend of water consumption between 2010 and 2016. The data from 2017 to 2020 was obtained from official records of Gulf Cooperation Council (GCC) countries, finally a projection was made for the duration of 2021 to 2023 based on a 3.6% annual growth rate (Fig. 16.2). It is to be noted that the water supplies include groundwater well abstraction, desalinated water, TWW and RO-ultrafiltered water (personal communication with Dr. Amjad Aliewi Senior Research Scientist Water Research Center Kuwait, KISR).

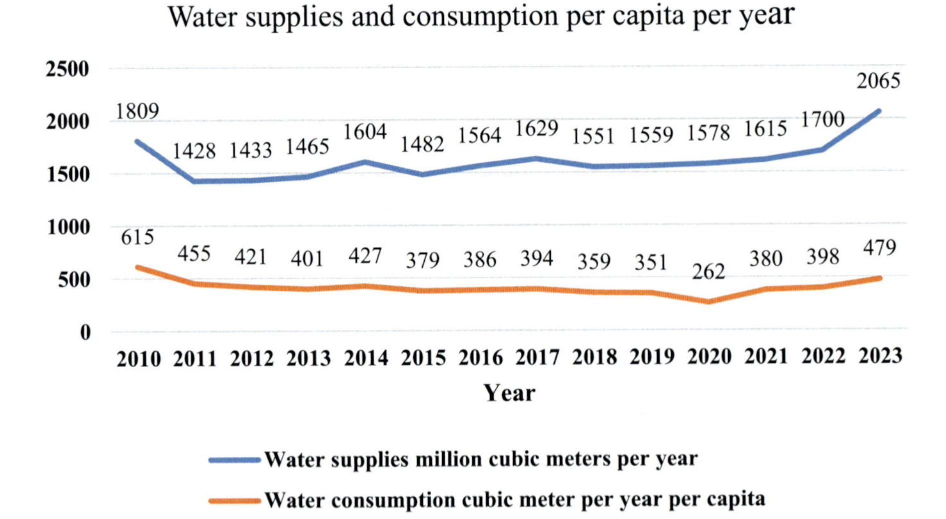

Fig. 16.2 Trend showing annual total water supplies and consumption per year per capita

16.4 Green, Blue, and Gray Water Footprints (WF)

To have a clear understanding of three water footprints (WF), a brief description is given below.

Water Footprint (WF) – defined as the sum of the volume of freshwater (blue, green, and gray) used to produce the product along its supply chain (Hoekstra et al. 2009). The WF concept was developed by Hoekstra in 2002 (Hoekstra 2003).

Green Water Footprint (GWF) – the rainwater and soil moisture consumed by plants and crops in their cultivation.

Blue Water Footprint (BWF) – surface and groundwater used for irrigation (i.e., evaporated or incorporated into a crop) during cultivation.

Gray Water Footprint (GrWF) – freshwater required to dilute pollution and bring the water resource up to safe water quality standards.

16.4.1 Global Trend of Green, Blue, and Gray Water Footprints of Various Crops

Technological innovations and upgradation of software packages have made it possible to enter and analyze larger data in excel sheets with in-built complex formulas to allow making final predictions. The WF is a relatively new concept and WF assessment is a new tool. Mekonnen and Hoekstra (2010, 2011a) are the pioneers to introduce these concepts and have analyzed larger data (1996–2005) about three water uses (green, blue, gray) in global crop production. They reported, during 1996–2005 crop production has consumed 7404 Giga cubic meters per year (Gm^3 year^{-1}) constituting 78%, 12%, and 10% green, blue, and gray water respectively. The water used by different crops is shown in Fig. 16.3, illustrating wheat crop is consuming 15% of total crop production WF (1087 Gm^3 year^{-1}, 70% green, 19% blue, and 11% gray), followed by rice (992 Gm^3 year^{-1}), maize (770 Gm^3 year^{-1}), and fodder crops (693 Gm^3 year^{-1}) as reported by Mekonnen and Hoekstra (2011b). The share of green water use in crop production is significantly different between rainfed (4701 Gm^3 year^{-1}) and irrigated agriculture (1070 Gm^3 year^{-1}). Globally, green water contributes significantly (86.5%) both in rainfed and irrigated agriculture (Mekonnen and Hoekstra 2011b). Since there is limited rainfall in arid and semi-arid regions, like Kuwait, in such region countries BWF is higher than the GWF and GrWF. Globally rain-fed agriculture has a WF of 5173 Gm^3 year^{-1} (91% green, 9% gray) and irrigated agriculture has a WF of 2230 Gm^3 year^{-1} (48% green, 40% blue, 12% gray) (Mekonnen and Hoekstra 2011b).

Although the WF subject is fairly new, recently a number of studies have been completed specifically on GW and BW. With regard to studies on GrWF in agriculture, globally no studies have been reported (Mekonnen and Hoekstra 2011a); however, a number of studies have been conducted on GW and BW on agriculture [(blue water, Seckler et al. 1998), green water consumption (Rockström et al. 1999; Rockstrom and Gordon 2001), and consumption of blue water (Shiklomanov and Rodda 2003; Hoekstra and Hung 2002)].

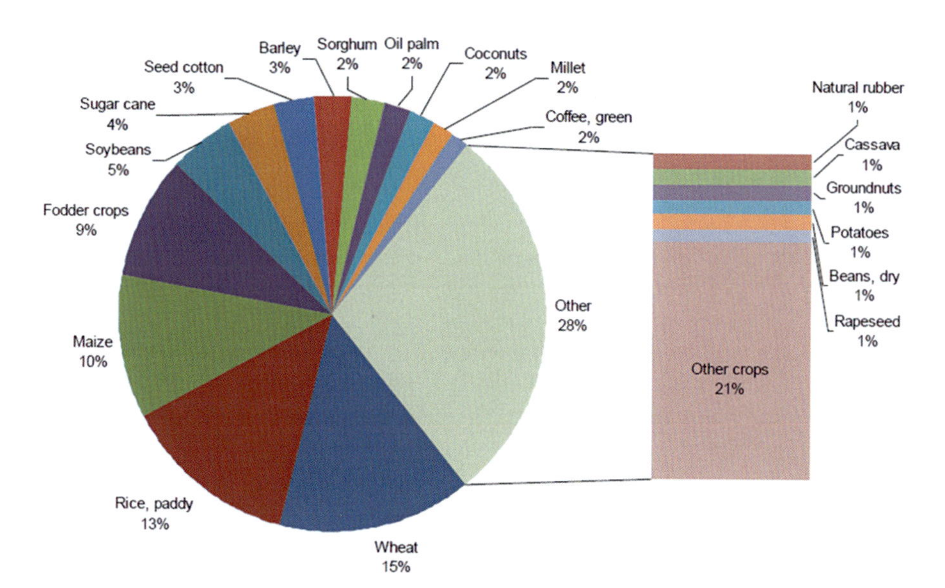

Fig. 16.3 Contribution of different crops to the total water footprint of crop production (1996–2005). (Source Mekonnen and Hoekstra (2011b), Fig. 1

16.4.2 Water Footprint Assessment Methodologies

As highlighted earlier the WF is a relatively new concept and WF assessment is a new tool. Over time the assessment procedures may be refined and better predictions made for WF data assessment and interpretation. Hoekstra et al. (2011) provided the first WF assessment guidelines "*The Water Footprint Assessment Manual*". We used these guidelines in assessing the various WFs for different crops grown in Kuwait. The data used is from an authentic source (FAOSTAT) as reported by Hoekstra et al. (2011). In this section, these methodologies are briefly described without going into further detail. Here few terms are used with reference to Kuwait WFs and crop grown, nationally, and imported.

16.4.3 Total National Crop Water Footprint Accounting

The focus of this chapter is on crops grown internally or crops products imported externally. Therefore, we are not going to address other commodities and associated WFs. The total water consumed to grow crops nationally (internal WF) and the water imported (virtual) by importing food from other countries (external WF) can be expressed as (Eq. 16.1).

$$\sum WF_{\text{consumption national}} = \sum WF_{\text{consumption internal}} + \sum WP_{\text{consumption external}} \tag{16.1}$$

These WFs include all three types (green, blue, and gray) of waters used to grow crops nationally, or crop products imported, and can be presented on the basis of volume per unit time or crop. The external WF consumption is the virtual water – *water used in food exporting country to grow crops* and consumed by the country of import. However, if the country under consideration is also exporting food to other countries, that water (virtual water export) should be minus from the total water consumption (Eq. 16.2).

$$\begin{aligned} \sum WF_{\text{national consumption}} = & \sum WF_{\text{consumption internal}} + \sum WP_{\text{consumption external}} \\ & - \sum WP_{\text{virtual water export}} \end{aligned} \tag{16.2}$$

However, if a crop is grown internally by using all three sources of waters (green, blue, gray) then Eq. 16.3 can be used. Briefly, the national WF of consumption (internal and external) is shown in Fig. 16.4.

$$\begin{aligned} \sum WF_{\text{crop processing}} = & \sum WF_{\text{crop processing green}} + \sum WF_{\text{crop processing blue}} \\ & + \sum WF_{\text{crop processing gray}} \end{aligned} \tag{16.3}$$

Similarly, WF of any product, crop, area, or region can be calculated provided the data is available to compute into the specific equation (Hoekstra et al. 2011).

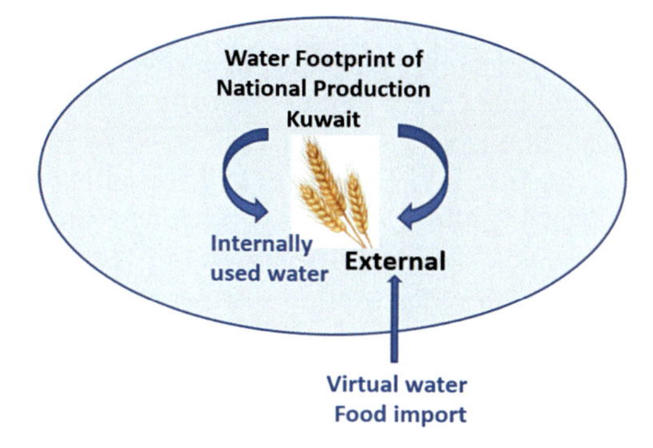

Fig. 16.4 Water footprint of national production, Kuwait

More details can be taken from Mekonnen and Hoekstra (2011a) who have presented a clear picture of three WFs of crops and derived crop products.

16.5 Why National Level Crop Water Footprint Consumption Assessment?

The national WF provides information how much water (green, blue, gray) has been consumed to grow crops against crop production. The local crop production is then compared with the national demand and guide the government to develop trade partnership with other countries where the needed food products can be imported to meet the food demand. This is the case with most of the water-scarce countries in the Middle East and North African (MENA) region including GCC countries (Bahrain, Kuwait, Oman, Qatar, Saudi Arabia, and United Arab Emirates). Thus, it is crucial for these countries to keep eyes both on food and virtual water and to maintain ties with potential food exporting countries to assure intake of water-intensive and needed food commodities, which cannot be produced locally, due to many reasons including water and arable land scarcity and harsh climatic conditions. The government of Kuwait has already realized and developed Kuwait food security and investment strategy (KIA 2014) and highlighted the need to make local and international investment in crop production and to develop international trade with potential countries.

16.5.1 Total Water Footprint Related to Crop Production in Kuwait

Search of literature revealed a comprehensive information about the total WF related to crop production at national levels (Mekonnen and Hoekstra 2010) presented on average per year over the period of 1996–2005. This was the latest information available at the time of chapter preparation. Mekonnen and Hoekstra (2010) have used the data from an authentic source (FAOSTAT). They reported total water used (green, blue, gray) in crop production in Kuwait was 69 million cubic meters per year (Mm^3 $year^{-1}$) during a period of 1996–2005 (Table 16.2).

Overall 69.57% of the total WF is related to crop production through irrigated systems, and only 15.94% is coming from the rain source (Table 16.2). Similar percentage of water is reported by World Bank (2011) that globally agriculture is accounting 70% of freshwater withdrawal (AQUASTAT 2023). It has also been reported that agriculture and especially irrigated agriculture is the sector with by far the largest consumptive water use and water withdrawal (AQUASTAT 2023).

Gray water contribution to crop production is minimum. It should be noted that the above results are based on the data availability during 1996–2005, after this period significant developments have been made in Kuwait in the water sector, therefore, the above results should be taken as indicative of water resources used in the past in Kuwait in the crop production sector. However, it is clear from Table 16.2 that the contribution of green water in agriculture is the least in Kuwait, mainly due to low rainfall in Kuwait and its geographic location in the arid region. At the time of chapter preparation, the data reported by Mekonnen and Hoekstra (2010) was the latest in the subject matter. We intend to synthesize similar information in Kuwait in the coming years when sufficient data become available.

16.5.2 *The Water Footprint of National Production in Kuwait*

Total national WF of production includes water used internally (agriculture, industry, domestic) to produce goods and external WF (water imported through goods-virtual water). The WF of production measures the pressure on local water resources and can be used to determine if they are being used in a sustainable way. Understanding how much of that WF is within national borders (internal) and the amount of the WF that is elsewhere (food exporting country) is a first step toward

Table 16.2 Average annual water footprint (green, blue, gray) related to crop production in Kuwait (1996–2005) Mekonnen and Hoekstra (2010)

Water footprint type	Water footprint (Mm^3 $year^{-1}$)	% contribution
Green (rainfed)	1.80	2.61
Green (irrigated)	9.20	13.33
Total green	**11.00**	**15.94**
Blue (irrigated)	48.0	69.57
Gray	10.0	14.49
Grand total	**69.0**	100.00

assessing the country's external water dependence and its influence on food and other forms of security (www.waterfootprint.org).

Mekonnen and Hoekstra (2011b) has reported comprehensive information on national WF of many countries including Kuwait (1996–2005). They divided the WF of national production into four main categories:

- WF of agriculture production.
- WF of grazing.
- WF of industrial production.
- WF of domestic water supply.

At the time of chapter preparation, the latest information available is reported. It is believed that in future such information will be updated globally and nationally. However, the currently reported information should be taken as an indicative of water uses at national level (Table 16.2) and will be based for future comparisons. The current information gives a guide for rationale uses of water resources to keep the balance and sustainability in water uses in Kuwait. The total WF of production in Kuwait is 564 Mm^3 year^{-1} (green 80, blue 94, and gray 390 Mm^{3}year^{-1}). The relative percent distribution of three WFs in Kuwait is shown in Table 16.3 showing the dominance of GrWF contribution in national production (Mekonnen and Hoekstra 2011a, b).

16.5.3 Average Water Footprint Per Ton of Crop or Derived Crop Products in Kuwait

The water consumed per unit production of crop (liters/kg) can be determined by recording the volume of water used and quantity produced. This will show how efficiently the product has been produced. The information of WF of key crops grown in Kuwait is taken from Mekonnen and Hoekstra (2011a), who have compiled the information from authentic source-FAOSTAT sources. The FAOSTAT is

Table 16.3 Breakdown of different water footprints of national production in Kuwait

Water footprint →	Green WF	Blue WF	Gray WF
Type of national production ↓	.... million m^3 year^{-1}		
Water footprint of crop production	11.0	48.0	10.0
Water footprint of grazing	69.0	–	–
Water footprint animal water supply	–	5.0	–
Water footprint of industrial production	–	1.0	19.8
Water footprint of domestic water supply	–	40.1	360.5
Total	**80.0**	**94.1**	**390.3**
Grand total water FP of national production	**564.3**		

Mekonnen and Hoekstra (2011a)

maintained by the statistics division of the UN-FAO and provides free access to food and agriculture statistics (including crop, livestock, and forestry sub-sectors) of various countries and territories. The FAOSTAT directly works with the countries to obtain the data. The FAOSTAT provides high-quality statistics at global and national levels. Table 16.4 and Fig. 16.5 clearly show different levels of water used to grow different crops, some consume higher volumes and others less. The heterogeneity depends on the crop requirement and geographical location where the crops are grown. Although grapes and olives are grown in small areas in Kuwait, their water consumption is the highest among all crops grown in Kuwait.

Table 16.4 Average water footprint (liters/kg) of crops or derived crop products in Kuwait

Crop-product description (FAOSTAT code) ↓	Green WF liters/kg	Blue WF liters/kg	Gray WF liters/kg	Total WF liters/kg
Barley (44)	929	2258	858	4045
Maize (corn) (56)	41	208	–	249
Potatoes (116)	12	184	91	287
Olives (260)	1955	10,592	–	12,547
Cabbages (358)	19	183	32	234
Lettuce and chicory (372)	5	146	31	182
Spinach (373)	7	180	35	222
Tomatoes (388)	14	78	20	112
Cauliflowers and broccoli	7	157	33	197
Pumpkins, squash, and gourds (394)	39	234	58	331
Cucumbers (397)	22	119	27	168
Aubergines (eggplant) (399)	14	159	29	202
Peppers-capsicum (401)	24	36	23	83
Onions (402)	56	224	49	329
Garlic (406)	117	573	126	816
Beans (greens) (414)	23	286	78	387
Carrots and turnips (426, 428)	26	235	31	292
Okra (430)	57	573	131	761
Vegetable fresh/chilled (463)	23	49	27	99
Oranges (490)	266	1333	–	1599
Lemons and limes (497)	180	899	–	1079
Strawberries (544)	42	189	–	231
Grapes (560)	1090	16,309	–	17,399
Water melons (567)	67	297	–	364
Dates (577)	337	1688	–	2025

Mekonnen and Hoekstra (2010)

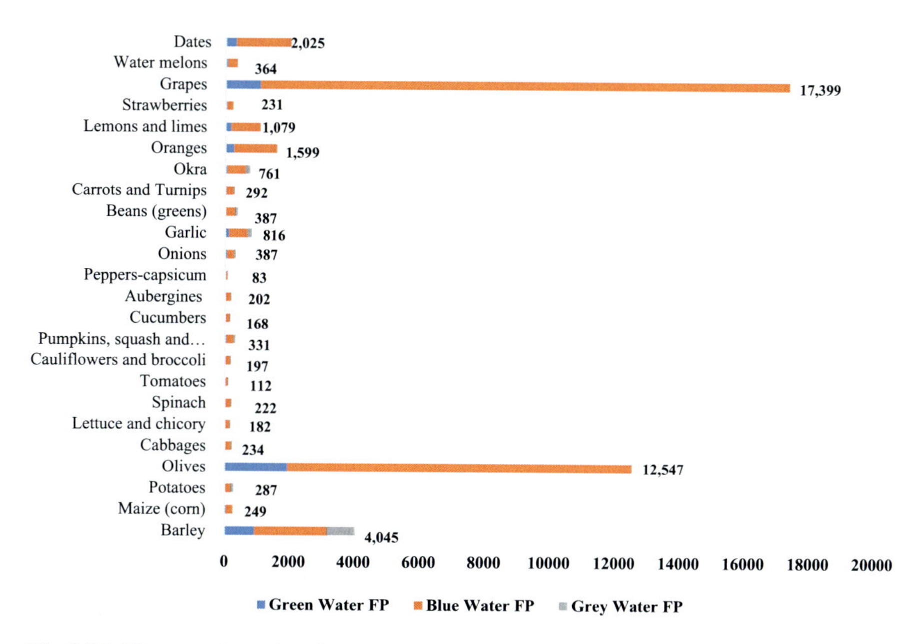

Fig. 16.5 The water footprint of selected crops grown in Kuwait (liters of water per kg of product). (Data source from Mekonnen and Hoekstra (2010))

16.5.4 Average Global Water Footprint (Liter Per Kg) of Crop or Derived Crop Products

In this section, we retrieved the information about green, blue, and gray WF from Mekonnen and Hoekstra (2010) for the common crops grown globally (Table 16.5, Fig. 16.6). Generally, globally the GWF is higher than the BWF and GrWF. In the case of Kuwait, the BWF is higher than the GWF and GrWF. The global trend is GWP > GrWF > BWF, whereas the trend for Kuwait is different (BWF > GrWF > GWF).

16.5.5 Comparison of Water Footprint of Crops in Kuwait and Global Average

The comparison of total WF of common crops grown globally (Mekonnen and Hoekstra 2011b) and in Kuwait (Mekonnen and Hoekstra 2010) revealed few crops grown in Kuwait present higher total WF (barley, olives, onions, garlic, carrots/

Table 16.5 Average global water footprint (liter/kg) of various products

Crop-product description (FAOSTAT code) ↓	Green WF	Blue WF	Gray WF	Global WF
	..........liters per kg			
Barley (44)	1213	79	131	1423
Maize (corn) (56)	947	81	194	1222
Potatoes (116)	191	33	63	287
Olives (260)	2470	499	45	3015
Cabbages (358)	181	26	73	280
Lettuce and chicory (372)	133	28	77	237
Spinach (373)	118	14	160	292
Tomatoes (388)	108	63	43	214
Cauliflowers and broccoli	189	21	75	285
Pumpkins, squash, and gourds (394)	228	24	84	336
Cucumbers (397)	206	42	105	353
Aubergines (eggplant) (399)	234	33	95	362
Peppers-capsicum (401)	240	42	97	379
Onions (402)	176	44	51	272
Garlic (406)	337	81	170	589
Beans (greens) (414)	320	54	188	561
Carrots and Turnips (426)	106	28	61	195
Okra (430)	474	36	65	576
Oranges (490)	401	110	49	560
Lemons and limes (497)	432	152	58	642
Strawberries (544)	201	109	37	347
Grapes (560)	425	97	87	608
Watermelons (567)	147	25	63	235
Dates (577)	930	1250	98	2277

Source: Mekonnen and Hoekstra (2011b)

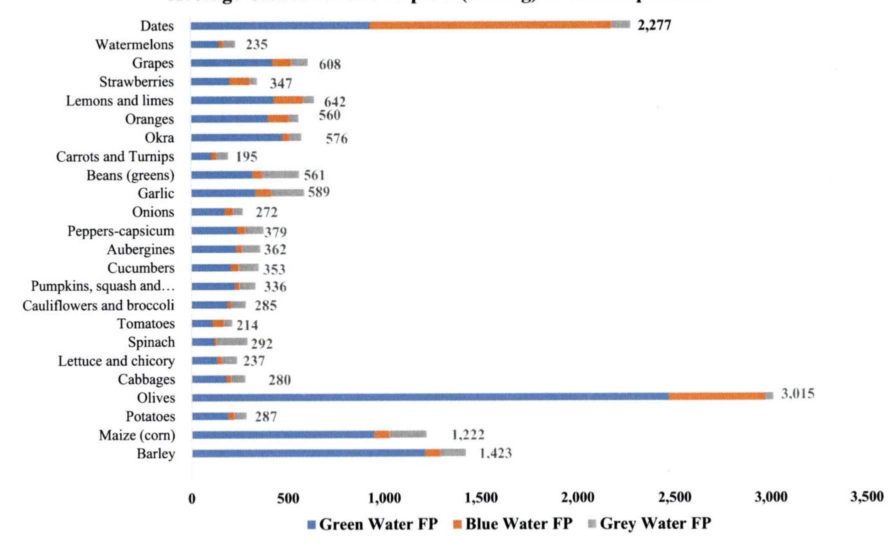

Fig. 16.6 The global average water footprint of selected crops (liters of water per kg of product). (Data source from Mekonnen and Hoekstra (2011b))

turnips, okra, oranges, lemons and limes, grapes, and watermelons) compared to global average of same crops (Table 16.6). It appears that these crops are grown in open field and require high water for their growth to compensate hot weather. Other crops showing lower WF in Kuwait than global average are usually grown in protected agriculture under controlled environmental condition. In both ways there are opportunities to improve the water use efficiencies leading to improved water productivities (*crops per drop*). The analyses show that, of 24 key crops grown in Kuwait, 14 crops (maize, potatoes, cabbages, lettuce, spinach, tomatoes, cauliflowers, pumpkins, aubergines, peppers, onion, beans, strawberries, and dates) have less WF of production than the global average of the same crops (Table 16.6, Fig. 16.7). This can be related to lower water consumption due to some vegetables crops and leafy greens are grown under controlled conditions (protected agriculture) thus consuming less water compared to open field crops stated globally. In Kuwait, due to harsh climate conditions, a number of farmers have shifted to protected agriculture to avoid the harmful effects on crops of the high temperatures, winds, and the lack of moisture (FAO 2021).

Table 16.6 Kuwait and global average of water footprint of crops and derived crops products

Crop-product description	[1]Kuwait average WF (liters /kg)	[2]Global average WF (liters/kg)
Barley	4045	1423
Maize (corn)	249	1222
Potatoes	287	287
Olives	12,547	3015
Cabbages	234	280
Lettuce and chicory	184	237
Spinach	222	292
Tomatoes	112	214
Cauliflowers and broccoli	197	285
Pumpkins, squash, and gourds	331	336
Cucumbers	167	353
Aubergines (eggplant)	202	362
Peppers (capsicum)	83	379
Onions	329	272
Garlic	816	589
Beans (greens)	387	561
Carrots and turnips	292	195
Okra	761	576
Oranges	1599	560
Lemons and limes	1079	642
Strawberries	231	347
Grapes	17,399	608
Watermelons	364	235
Dates	2025	2277

[1]Mekonnen and Hoekstra (2010); [2]Mekonnen and Hoekstra (2011b)

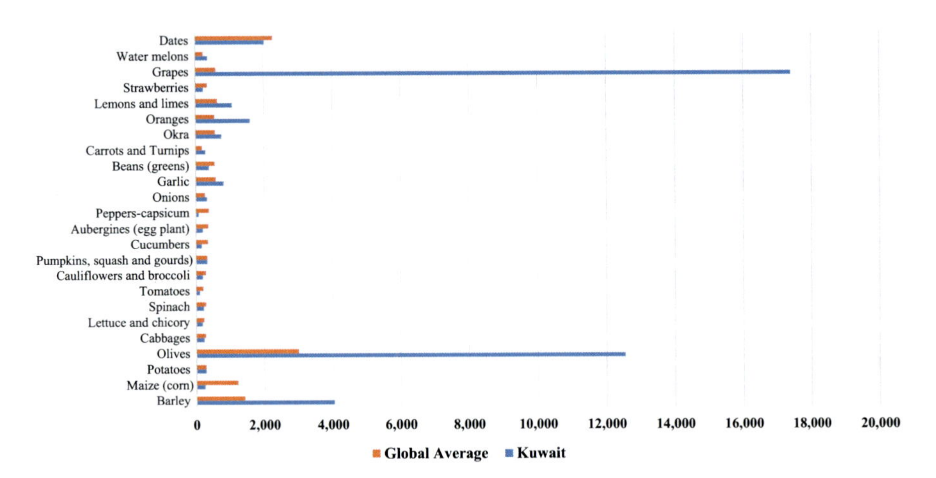

Fig. 16.7 Comparison of average global and Kuwait water footprint of key crops (liters per kg). (Data source: Mekonnen and Hoekstra (2010, 2011b))

In Kuwait, the cultivation of fruit vegetables is the dominant category followed by leafy vegetables grown in protected agriculture. This shift from open field agriculture to protected agriculture is likely to be the cause of efficient water use in Kuwait. In contrast, the crops grown in open field in Kuwait (barley, olives, water melons, grapes, oranges, limes, etc.) present higher WF of production relative to global average. This is due to harsh climatic condition and frequent irrigation. From the above results, it is apparent that water saving in Kuwait can be made through the adoption of protected agriculture. However, the protected agriculture requires significant energy to maintain the temperature in the greenhouses based on the crops grown.

16.5.6 Water Footprint of National Consumption Per Year Per Capita

A nation's WF can be viewed from production and consumption perspectives. The WF of consumption is the volume of water used to produce goods and services consumed by the national population. This water has two directions (internal and external), the internal represents the water consumed locally within national boundaries to produce product and the external represents the imported water (water used to produce imported products in the country of export), it is called *virtual water*.

The imbalances of food production in many countries due mainly to water and arable land scarcity necessitates the countries to go for global food trade in food products to meet the demand of the population. The countries exporting the food used the water to produce the product, thus, in addition to the product, these countries are also exporting the water (hidden resource). Because to produce the same product internally same or more volume of water will be required to produce the same product locally, this hidden water is called the virtual water. Therefore, the WF of national consumption is partly inside the country and partly outside of it, depending on whether the products are locally produced or imported and to what extent. Both the WF of production and consumption from internal and external sources provide the information to policymakers to set up their water management policies related to water management, food security and to develop international food trade with countries having high potential of food production for long-term food supply. Mekonnen and Hoekstra (2011a) synthesized the information about WF of national consumption (1996–2005) for many countries including Kuwait (Table 16.7). They reported green, blue, and gray WF of consumption for three consumption categories, i) WF of consumption of agricultural crops, ii) WF of consumption of industrial products, and iii) WF of domestic water consumption (Table 16.7). At the time of reporting, latest information available is presented in this chapter. Tables 16.7 and 16.8 show there is very high total WF of consumption year^{-1} capita^{-1} (2072.3 m^3) in Kuwait, consisting of i) green (1419.5 m^3, ii) blue 291.1 m^3, and iii) gray 361.0 m^3. Although this information is older, but an indicative to develop water and food policies for national development. The percent relative distribution of internal and external WF with respect to WF type (green, blue, gray) is shown in Fig. 16.8. Figure 16.8 shows very high external WF of consumption compared to internal WF consumption. This makes sense as most of the food and other products are imported in Kuwait.

Table 16.7 Breakdown of different water footprints of national consumption year^{-1} capita^{-1} of Kuwait

Water footprint Type→	Green WF	Blue WF	Gray WF
	.... m^3 year^{-1} capita^{-1}.......		
Type of national production ↓			
Water footprint of consumption of agricultural products (internal)	6.6	22.9	6.4
Water footprint of consumption of agricultural products (external)	1412.9	246.0	138.2
Water footprint of consumption of industrial products (internal)	0	0.2	4.3
Water footprint of consumption of industrial products (external)	0	4.9	51.9
Water footprint of domestic water consumption	0	17.8	160.1
Total	**1419.5**	**291.9**	**360.9**
Grand total	**2072.3**		

Mekonnen and Hoekstra (2011a)

Table 16.8 Total internal and external water footprint of national consumption (m^3 $year^{-1}$ $capita^{-1}$) in Kuwait

Water footprint type↓	m^3 $year^{-1}$ $capita^{-1}$	m^3 $year^{-1}$ $capita^{-1}$	m^3 $year^{-1}$ $capita^{-1}$
	Internal	*External*	*Total*
Green	6.6	1412.9	1419.5
Blue	40.9	251.0	291.9
Gray	170.8	190.2	361.0
Total	**218.3**	**1854.1**	**2072.4**

Data source: Mekonnen and Hoekstra (2011a)

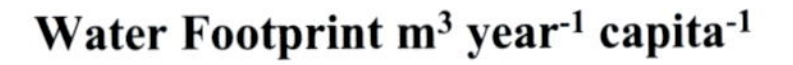

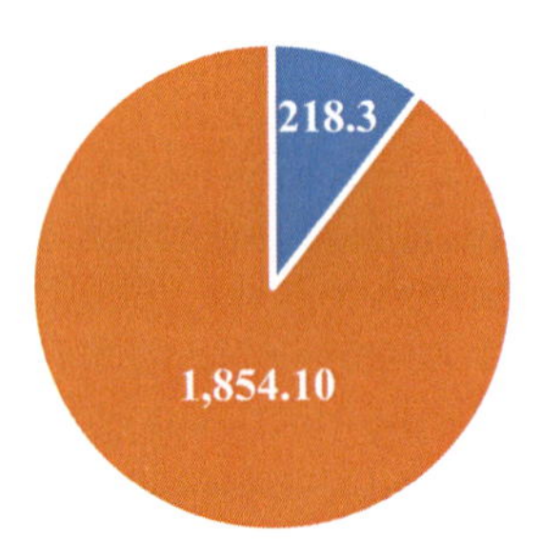

Fig. 16.8 Internal and external water footprint of consumption (m^3 $year^{-1}$ $capita^{-1}$ (Kuwait)

16.5.7 Water Footprint of Animal Products

The livestock sector is currently utilizing nearly 35 percent of total cropland and about 20 percent of blue water for feed production (Opio et al. 2011; FAO 2019). In addition, livestock sector uses an equivalent of 11,900 km^3 of freshwater annually, that is approximately 10 percent of the annual global water flows (estimated at 111,000 km^3) (Deutsch et al. 2010). Weindl et al. (2017) estimated 2290 km^3 of green water and 370 km^3 of blue water attributed in the year 2010 to feed production on cropland. Other estimates present over 90 percent of the water consumption in livestock and poultry production associated with feed production (Mekonnen and Hoekstra 2012; Legesse et al. 2017).

Meat is an essential part of human diet and source of proteins. Its global production and consumption is increasing with the increase of global population. The FAO (2005) has warned that meat production has doubled in the 1980–2004 period, and further doubling in the period 2000–2050 is projected (Steinfeld et al. 2006). Recent data showed global demand for meat is growing: over the past 50 years, meat production has more than tripled. The world now produces more than 340 million tons of meat each year. With business as usual, it is projected that by 2050 global meat consumption will reach 460 million tons. The increase in meat consumption means increase in water consumption as the animal products are water-intensive (Chapagain and Hoekstra 2003; Pimentel et al. 2004). Currently, globally, dairy sector consumes 19% of the water in the livestock sector (Velarde-Guillén et al. 2022). The animals require water over the entire growing period including all types of water (green, blue, grey) used to raise the animals producing pastures and fodders to feed animals, etc. Simultaneous to increase in meat production, scientists also argue on high water consumption and expected environmental consequences (Sutton et al. 2011) like the contribution of livestock to the emission of greenhouse gasses (Bouwman et al. 2011). Therefore, it is pertinent to evaluate the water consumption by animals to produce a quantity of meat. The animal WF could be indirect (water used to produce fodder/feed) and direct (drinking and service water consumed) water consumption (Chapagain and Hoekstra 2003, 2004). Mekonnen and Hoekstra (2012) reported global average of WF of meat production, e.g., chicken meat (4300 liters/kg), goat meat (5500 liter/kg), sheep meet (10,400 liters/kg), and beef (15,400 liters/kg). The major component of total water footprint in animal sector is from feeds consumption, accounting to 98%. The share of total WF for feed production through growing crops (1463 Giga m^3 $year^{-1}$) is 20% of the total global water footprint (7404 Giga m^3 $year^{-1}$) of total crop production (Mekonnen and Hoekstra 2011a). It is interesting to note that the total global share of WF for animal production is 2422 Giga m^3 $year^{-1,}$ which is 29% of the total WF of total agriculture production (8363 Giga m^3 $year^{-1}$). It is also found (Mekonnen and Hoekstra 2012) that different animals consume different volumes of water during the whole life, e.g., beef cattle (1889 m^3/animal, average life 3 years), dairy cattle (20,558 m^3/animal, average life 10 years), sheep (141 m^3/animal, average life 2.1 years), and goats (76 m^3/animal, average life 2.3 years).

16.6 Conclusions and Recommendations

The review of literature with regard to water footprint of Kuwait and the global average has shown interesting data on three water footprints (green, blue, and gray). Globally green water footprint is the highest for crop production, whereas in Kuwait blue water footprint is the highest in agriculture use. The comparison of water footprints for various crops and their products has revealed that number of vegetable crops in Kuwait are more water efficient than the same products globally, this is due to shift of farmers from open field to protected agriculture (temperature-controlled greenhouses). Kuwait is situated in hyper-arid climate where annual water losses (evapotranspiration) are higher than the renewal; therefore, Kuwait depends mainly on desalinated water for domestic and other use, as well as treated waste water and reverse osmosis water for agriculture production. Kuwait is an arable land and water-scarce country and its most of the food demand is met through food import from food-surplus countries. This has dual benefits, food import and the import of virtual water (the water used in the exporting country to grow the crops being exported to other countries). There is a need to initiate national-level projects to update the water footprint information on crops and animal products to use as a guide to develop water and food security policies.

Acknowledgements We would like to extend our thanks to Professor Mesfin M Mekonnen for granting permission to use the data on water footprints and Fig. 16.3 from his publication (Mekonnen and Hoekstra 2011b) cited in the references. The data and figure are cited in the text and added in the reference section. His approval gave us confidence to prepare the chapter on various aspects of water footprints.

References

Al Ali EH (2008) Groundwater history and trends in Kuwait. Sustainable Irrigation Management, Technologies and Policies II

Aleisa E, Alshayji K (2019) Analysis on reclamation and reuse of wastewater in Kuwait. J Engineering Res 7(1):1–3

Al-Senaf M, Abraham J (2004) Vulnerability of groundwater resources from agricultural activities in southern Kuwait. Agric Water Manag 64:1–15

Al-Shammari S, Shahalam A (2006) Effluent from an advanced wastewater treatment plant – an alternate source of non-potable water for Kuwait. Desalination 196(1):215–220

AQUASTAT – FAO's global Information System on Water and Agriculture. (https://www.fao.org/aquastat/en/data-analysis/irrig-water-use) Last accessed on 4 May 2023

Bouwman L, Goldewijk KK, Van Der Hoek KW, Beusen AHW, Van Vuuren DP, Willems J, Rufino MC, Stehfest E (2011) Exploring global changes in nitrogen and phosphorus cycles in agriculture induced by livestock production over the 1900-2050 period. Proceed National Academy Sci USA. https://doi.org/10.1073/pnas.1012878108

Chapagain AK, Hoekstra AY (2003) Virtual water flows between nations in relation to trade in livestock and livestock products. Value of water research report series no. 13. Available online: [www.waterfootprint.org/Reports/Report13.pdf]. Delft, The Netherlands: UNESCO-IHE

Chapagain AK, Hoekstra AY (2004) Water footprint of nations, 16th edn. UNESCO-IHE: Delft, The Netherlands

Deutsch L, Falkenmark M, Gordon L, Rockstrom J, Folke C (2010) Water-mediated ecological consequences of intensification and expansion of livestock production. In: Steinfeld H, Mooney H, Schneider F, Neville LE (eds) Livestock in a changing landscape. Drivers, consequences and responses,. Island Press, London, pp 97–110

Eakins BW, Sharman GF (2010) Volumes of the World's Oceans from TOPO1. NOAA National Geophysical Data Center, Boulder, CO

Environment Public Authority (2019). First biennial update report of the State of Kuwait

FAO (2005) Livestock policy brief 02. Food and Agriculture Organization, Rome

FAO (2019) Water use in livestock production systems and supply chains – guidelines for assessment (version 1). Livestock Environmental Assessment and Performance (LEAP) Partnership, Rome, p 96

FAO (2021) Unlocking the potential of protected agriculture in the GCC countries: saving water and improving nutrition. FAO, Cairo, p 216

Gleick PH (1993) Water in crisis: Chapter 2. Oxford University Press

Gleick PH (2000) The changing water paradigm: a look at twenty-first century water resources development. Water Int 25(1):127–138

Hamoda M (2013) Advances in wastewater treatment technology for water reuse. J Engineering Res 1(1):1–27

Hoekstra AY (2003) Virtual water trade. Proceedings of the international expert meeting on virtual water trade, Delft, The Netherlands, 12–13 December 2002, value of water research report series no.12, UNESCO-IHE, Delft, The Netherlands (2003)
Hoekstra AY, Chapagain AK, Maite M, Aldaya MM, Mekonnen MM (2011) The water footprint assessment manual - setting the global standard. First published in 2011 by Earthscan. Earthscan ltd, Dunstan house, 14a St cross street, London EC1N 8XA, UK
Hoekstra AY, Hung PQ (2002) Virtual water trade: a quantification of virtual water flows between nations in relation to international crop trade. Value of water research report series no. 11, UNESCO-IHE, Delft, The Netherlands
Kuwait Investment Authority-KIA (2014) Kuwait food security and investment strategy – assessment report – task 12: state of art agricultural technologies and methods to enhance vegetation and agriculture production. Kuwait Investment Authority, Kuwait
Legesse G, Ominski K, Beauchemin K, Pfister S, Martel M, McGeough E, Hoekstra A, Kroebel R, Cordeiro M, McAllister TA (2017) Quantifying water use in ruminant production: a review. J Anim Sci 95:2001–2018
Lutz W, Sanderson W, Scherbov S (1997) Doubling of world population unlikely. Nature 387:803–805
Mekonnen MM, Gerbens-Leense W (2020) Review-the water footprint of global food production. Water 12(269). https://doi.org/10.3390/w12102696
Mekonnen MM, Hoekstra AY (2010) The green, blue and grey water footprint of crops and derived crop products. Value of water research report series no. 47, UNESCO-IHE, Delft, The Netherlands. http://www.waterfootprint.org/Reports/Report47-WaterFootprintCrops-Vol1.pdf
Mekonnen MM, Hoekstra AY (2011a) National water footprint accounts: the green, blue and grey water footprint of production and consumption, Value of water research report series no.50, UNESCO-IHE, Delft, the Netherlands. http://www.waterfootprint.org/Reports/Report50-NationalWaterFootprints-Vol1.pdf
Mekonnen MM, Hoekstra AY (2011b) The green, blue and grey water footprint of crops and derived crop products. Hydrol Earth Syst Sci 15(5):1577–1600
Mekonnen MM, Hoekstra AY (2012) A global assessment of the water footprint of farm animal products. Ecosystems 15:401–415. https://doi.org/10.1007/s10021-011-9517-8
Ministry of Electricity and Water (MEW) (2020) Statistical yearbook water edition 2019
Opio C, Gerber P, Steinfeld H (2011) Livestock and the environment: addressing the consequences of livestock sector growth. Adv Anim Biosci 2:601–607
Organisation for Economic Co-operation and Development (OECD) (2010) Sustainable management of water resources in agriculture
Pimentel D, Berger B, Filiberto D, Newton M, Wolfe B, Karabinakis E, Clark S, Poon E, Abbett E, Nandagopal S (2004) Water resources: agricultural and environmental issues. Bioscience 54(10):909–918
Ringler C, Bryan E, Biswas A, Cline SA (2010) Water and food security under global change, global change: impacts on water and food security. Springer, pp 3–15
Rockstrom J, Gordon L (2001) Assessment of green water flows to sustain major biomes of the world: implications for future ecohydrological landscape management. Phys Chem Earth 26(11–12):843–851. https://doi.org/10.1016/S1464-1909(01)00096-X
Rockström J, Gordon L, Folke C, Falkenmark M, Engwall M (1999) Linkages among water vapor flows, food production, and terrestrial ecosystem services. Conservation Ecology 3(2):5. [online] URL: http://www.consecol.org/vol3/iss2/art5/. Last accessed 19 July 2923
Seckler D, Amarasinghe U, Molden DJ, de Silva R (1998) World water demand and supply, 1990–2025: scenarios and issues, IWMI research report 19. IWMI, Colombo, Sri Lanka
Shahid SA, Zaman M, Heng L (2018) Introduction to soil salinity, sodicity, and diagnostics techniques. In: Zaman M, Shahid SA, Heng L (eds) Guidelines for salinity assessment, mitigation and adaptation using nuclear and related techniques. Springer, Cham, pp 1–42
Shiklomanov IA, Rodda JC (eds) (2003) World water resources at the beginning of the twenty-first century. Cambridge University Press, Cambridge, UK

Steinfeld H, Gerber P, Wassenaar T, Castel V, Rosales M, de Haan C (2006) Livestock's long shadow: environmental issues and options. Food and Agriculture Organization, Rome, p 390

Sutton MA, Oenema O, Erisman JW, Leip A, Van Grinsven H, Winiwarter W (2011) Too much of a good thing. Nature 472(7342):159–161

UWSS (2016) GCC unified water strategy (2016–2035) King Abdullah Institute for research and consulting studies: executive summary report. King Saud University, Kingdom of Saudi Arabia

Velarde-Guillén J, Viera M, Gómez C (2022) Water footprint of small-scale dairy farms in the central coast of Peru. Trop Anim Health Prod 55(1). https://doi.org/10.1007/s11250-022-03437-8

Weindl I, Bodirsky BL, Rolinski S, Biewald A, Lotze-Campen H, Müller C, Dietrich JP, Humpenöder F, Stevanović M, Schaphoff S (2017) Livestock production and the water challenge of future food supply: implications of agricultural management and dietary choices. Glob Environ Chang 47:121–132

WWAP (World Water Assessment Programme) (2012) World Water Development Report 4

Zaman M, Shahid SA, Heng L (2018) Irrigation systems and zones of salinity development. Chapter 4. In: Zaman M, Shahid SA, Heng L (eds) Guidelines for salinity assessment, mitigation and adaptation using nuclear and related techniques. Springer, Cham, pp 91–111

Index

A
Abdali, 29, 43, 44, 58, 149, 364–371, 373, 374, 386
Agriculture
  farming, 364, 365, 369, 372–373, 379, 386
Agroforestry, 83, 85, 112, 114
Alluvial plain
  ecosystem, 248, 312
Animal diversity, 41
Anthropogenic
  disturbances, 270, 330, 336, 337, 346, 348
Aquifers
  Dammam, 143, 144, 146–149, 174, 175, 198
  Kuwait, 43, 188, 200, 209
Arabian
  Gulf, 22, 25, 27, 33, 34, 37, 38, 42, 45, 49, 51, 58, 151, 154, 162, 175, 312, 325, 342, 344, 345, 364
  oryx, 313, 314, 323, 325
  Peninsula, 22, 25, 30, 48, 69, 120, 142, 248, 266, 314, 323, 325, 342, 345
  wolf, 313
Arbuscular mycorrhiza (AM), 14, 349–351
Arid
  desert ecosystem, 50, 248–249, 261, 332, 345–347, 352
Aridisols, 7, 18, 44, 123, 249, 266
Armor layer, 126, 127, 129
Artificial recharge, 156, 159–160, 162

B
Bahrain, 22, 106, 109, 110, 113, 314, 372, 401
Barchan dunes, 37, 255–257, 261
Biocapacity, 94–97, 99–109, 112, 116
Biodiversity, 14, 22, 25, 41–42, 45, 47, 52, 70, 76–81, 83, 85, 88, 89, 112, 120–123, 191, 248, 259, 260, 266, 267, 270, 271, 306–308, 312–314, 319, 323–325, 334–339, 344–348, 350, 352–354, 372, 378
Bioremediation, 13, 17
Brackish water, 4, 9, 13, 80, 121, 131, 146, 148, 149, 154, 155, 177, 186, 188, 198, 200, 201, 218, 224, 226, 227, 231–234, 236–241, 371, 372, 376

C
Calcigypsids, 9, 11, 266, 364, 367–371
Captive-breeding, 322, 325
Carbon
  demand, 94, 97
  footprint (CF), 61, 71–81, 86, 97, 101, 103, 104, 106, 113, 116
Carnivores, 321, 322, 325
Circular economy (CE), 80, 160–162, 386
CITES, 325
Climate
  change, 45, 58, 81, 101, 120, 144, 266, 342, 365
  change adaptation, 81–84, 88
  change mitigation, 86, 89, 112, 352
  resilience, 70, 71, 88, 351–353
  smart agriculture, 112, 116
  warming, 345
Clouds, 59, 63–64, 87
Combat desertification, 50, 110, 114–115, 121, 122, 325

M. K. Suleiman, S. A. Shahid (eds.), *Terrestrial Environment and Ecosystems of Kuwait*, https://doi.org/10.1007/978-3-031-46262-7

Confined aquifer, 200, 209–210, 212
Conservation
agriculture (CA), 83, 112, 114, 386
Contamination, 13, 46, 126, 133, 320, 321, 331
Conventions, 50, 58–59, 121, 266, 324, 325, 339
Coral reefs, 38, 74, 77, 78, 89
Creep, 123, 124, 135
Crop
diversification, 83
intensification, 376, 386
lands, 94, 368, 371, 373, 374, 378, 384, 386, 401, 411
production, 4, 11, 12, 15, 17, 83, 112, 121, 364, 368, 371, 373, 374, 376, 380, 381, 383, 384, 386, 394, 397, 399, 401–403, 410, 411

D
Decades, 52, 60–62, 71, 72, 77, 81, 94, 95, 99–102, 105, 106, 120, 154, 163, 314, 323, 325, 326, 333, 334, 344, 345, 365–367
Desalination, 37, 42, 45, 47, 50, 51, 58, 71, 79, 142–145, 150–151, 155, 158, 159, 162, 163, 371, 381–386, 396
Desert
animals, 41, 313, 314, 317
camping, 132
plain ecosystem, 250, 251
vegetation, 314, 346, 347
Desertification, 13, 50–51, 101, 113, 114, 120–135, 332, 335, 343–345, 348, 352, 353
Dibdibba, 27, 28, 32, 34, 35, 39, 174, 175, 177, 183–186, 191
Digital elevation model (DEM), 24
Dispersivity, 203, 204, 213, 217, 218, 227, 229–231, 239, 241
Drought, 8, 23, 25, 26, 41, 42, 45, 47, 50–51, 59, 60, 70, 74–77, 79–81, 84, 85, 89, 101, 121, 122, 173, 248, 260, 266, 267, 274, 314, 315, 330, 343, 345–347, 349, 352, 375, 383, 395
Dunes, 35–37, 39, 225, 255, 257, 270, 271, 330
Dust, 10, 48, 51, 67, 69–70, 124, 134, 332, 352, 364
Dust storms, 26, 35, 47–50, 61, 69–71, 85, 121, 124, 126, 135, 248, 332, 352, 364
Dynamics, 4, 6, 9, 61, 62, 64–68, 71, 83, 100, 101, 103–105, 173, 253, 257, 261, 347, 348, 351, 352, 375, 383

E
ECe, 12, 172, 188, 370, 374, 376–378
Ecological
degradation, 98
footprint (EF), 94, 156
fragmentation, 316
imbalances, 45, 52
interactions, 78, 348
resilience, 351–353
restoration, 89
Economic benefits, 81, 86–87, 270
Ecosystem
functions, 122, 261, 330–336, 338, 348–351
resilience, 98–99
services, 13, 18, 70, 95, 97, 110, 122, 141, 172, 191, 226, 270, 274, 330, 331, 333, 394
Electrical conductivity (EC), 147, 172, 176, 370, 374
Entisols, 7, 8, 11, 18, 44, 123, 249, 266
Environmental conservation, 85, 88
Erosion, 8, 10, 12–14, 16, 17, 27, 31, 46, 49, 70, 74, 83, 114, 120, 123, 125–129, 131, 132, 135, 253, 267, 270, 332, 346, 348, 372, 373, 380
Escarpment, 23, 27, 28, 30–33, 38, 249, 253–256, 261, 266, 271, 312, 330

F
Farms typology
large, 366–368
medium, 366, 368
small, 366, 368
Feeding habits, 312, 317
Flash floods, 10, 47, 49, 70, 123, 125, 131, 365
Fog, 64–65
Food
resources, 86, 314, 315, 317, 320, 330, 383, 384, 386, 394, 402, 409
security, 12, 18, 50, 52, 70, 79–82, 86, 89, 94, 110, 111, 121, 122, 135, 364, 373, 376, 381, 383, 386, 395, 401, 403, 409, 412
Footprint, 71, 94–116, 153, 155–156, 384, 394–412
Forages, 51, 94, 317, 331, 333, 366, 368, 377–378, 386, 394
Forests, 22, 50, 94, 97–100, 102, 103, 110, 121, 331, 332, 373
Fruit, 67, 367, 377–378, 383, 397, 408

G
Genetic
  adaptation, 84
  disorders, 319
  diversity, 270, 314
Geographical information system (GIS), 22, 30, 186, 321, 365
Geography, 8, 22–52
Geology, 24, 30, 31, 147
Geomorphology, 25, 27–29, 40, 46, 249
Glaciers, 74, 395
Global
  climate change, 47–51, 59, 330, 332, 336, 338, 342–345, 348–351, 353, 371, 383, 395, 396
  hectares, 95, 97, 99, 100, 103, 104, 106, 107, 375
  warming, 47–51, 58, 59, 63, 66, 71, 74, 319, 320, 342–344, 351, 383
Governance, 80, 110, 122, 161–163, 379
Granivorous rodents, 315
Grasses, 172, 248, 272, 274, 275, 277, 279–282, 284–287, 289, 291–293, 296–304, 342, 368, 377, 384
Grazing
  animals, 318
  controlled, 335
  lands, 97, 103, 110, 114, 318, 319
  uncontrolled, 248, 334, 338
Greenhouse gas (GHG), 46, 52, 58–60, 70–81, 84, 86, 87, 89, 97, 98, 112, 342–344, 346, 349, 351, 383–386, 411
Gross domestic product (GDP), 45, 121, 158, 365
Groundwater
  aquifers, 50, 142, 146, 148, 156, 179, 183, 186, 198, 204, 216, 371, 372, 375, 395, 396
  recharge, 80, 177, 179, 180, 183, 371, 372, 375, 395
  resources, 144–149, 198, 225, 371, 372, 395, 396
  salinization, 375, 397
  wells, 79, 183, 184, 331, 372, 383, 397
Gulf Cooperation Council (GCC), 22, 42, 94, 95, 106, 109, 110, 112, 113, 116, 121, 144, 153, 155, 157, 163, 323, 325, 339, 364, 372, 381–383, 397, 401
Gypsum, 9, 11, 17, 27, 32, 34, 35, 39, 43, 44, 132, 312, 364, 368–370, 373, 374

H
Habitat
  fragmentation, 77, 266, 313, 316, 319, 320
  losses, 77
  utilization, 315
Haloxylon salicornicum, 41, 133, 250, 251, 253, 257, 269–271, 290, 332, 333
Haplocalcids, 12, 266, 364, 367, 369–371
Haplogypsids, 9, 11, 266, 364
Harsh environments, 314, 345
Homogenization, 10–11
Honey badger, 313
Humidity, 22, 29, 47, 60, 64, 66, 67, 75, 81, 83, 342, 347
Hyper-arid, 8, 22, 26, 45, 47, 58, 89, 94, 122, 248, 330, 332, 342, 345, 364, 394, 411

I
Indigenous
  ecosystems, 172–191, 320
  plant communities, 346
Industrial waste, 267–270
Intergovernmental Panel on Climate Change (IPCC), 58, 89, 113, 121, 342–346, 348, 349, 351, 353
International Union of Conservation Nature (IUCN), 266
Invasive species, 78, 314, 320
Isochlors, 214–216, 227, 229–231, 237–241

J
Jaber Al-Ahmad, 180–183, 224, 227, 231, 232, 236–237, 240–241
Jal Az Zour, 23, 27, 28, 30–36, 38, 125, 131
Jerboas, 317

K
Kuwait
  adaptation plan, 71, 79, 84, 352
  vision, 111, 115–116

L
Land
  degradation, 49, 120, 125, 129–133, 312, 317, 332, 336
  degradation neutrality, 121–122, 135
  filling, 15, 122, 334–335, 338
Landforms, 22, 27–33, 42, 51, 52, 249, 252, 270

Leptic haplogypsids, 9, 11, 367, 369–371, 373, 374
Losses, 9, 10, 14, 16, 47, 49, 51, 64, 74, 77, 78, 83, 89, 98, 114, 120–127, 129, 130, 132, 134, 156, 157, 163, 216, 225, 248, 260, 266, 267, 270, 274, 307, 313, 315, 317–320, 322, 324, 330–334, 337, 344, 346, 353, 371–373, 375, 376, 411

M
Mammals, 41, 77, 249, 312–314, 319, 333
MFS, 382, 385
Microbial diversity, 342, 349
Microbiomes, 343, 348–351, 353
Model, 157–159, 161, 198–218, 224–241, 343, 349–351, 353, 354, 367
Model verification, 209–218
Multi effect distillation (MED), 150, 382, 385, 386
Multi-stage flash (MSF), 150, 382, 385, 386

N
National
  adaptation plan, 71, 79, 84, 352
Native
  plants, 52, 252, 261, 266, 267, 270–271, 274, 306, 308, 314, 330, 332–335, 337, 338, 346–348, 352
Natural habitats, 88, 312, 313, 317, 319, 320, 323, 325, 339
Nature reserve, 42, 134, 336, 337
Nebkhas, 5, 23, 32, 33, 35–38, 126, 129, 251, 252, 255–257, 261
Nexus, 155, 364–386
Normalized difference vegetation index (NDVI), 347, 365, 366
Numerical modelling, 226

O
Oil drilling, 319
Oman, 22, 106, 109, 110, 113, 314, 401
Open field
  agriculture, 17, 364, 394, 411

P
Pakistan, 199–201, 204, 209, 216, 218, 224–241
Perennial
  herbaceous, 248, 249, 267, 346
  plants, 251, 272, 273, 333, 342
  plants cover, 335
Petrocalcids, 12, 266, 364
Petrogypsids, 9, 11, 266, 364
Phytoremediation, 13, 17, 270, 271, 277, 279, 280
Plants, 4, 22, 59, 114, 129, 142, 172, 173, 188, 266, 312, 330, 342, 364, 396
Populations, 8, 45, 46, 48–50, 58, 64, 70, 71, 74, 77–80, 85, 89, 94, 101, 105, 109, 113, 116, 121, 122, 142, 150, 152, 153, 156, 250, 251, 266, 267, 270, 306, 307, 313, 317, 319, 320, 322, 323, 325, 331, 332, 334, 335, 343, 346, 348–350, 353, 375, 378, 379, 382, 383, 395, 396, 408–410
Predation, 315–317, 345
Protected agriculture (PA), 17, 364, 366–368, 383, 384, 394, 397, 407, 408, 411

Q
Qatar, 22, 106, 109, 110, 113, 155, 314, 372, 401
Quality
  dynamic, 4
  inherent, 4

R
Rainfall, 7, 8, 49, 58–60, 64–66, 70, 77, 79–81, 83, 94, 101, 120, 125, 129, 131, 134, 142, 143, 173, 177, 187–191, 248, 249, 330, 332, 342–344, 347, 352, 353, 364, 365, 371, 375, 394–396, 399, 402
RASIM, 200, 201, 204, 209, 210, 212, 214–218, 224, 226, 228, 229, 232
Regenerative agriculture (RA), 378, 379
Rehabilitation, 89, 101, 113, 125, 134, 323, 337–338
Reintroduction programs, 314, 322–323, 325
Remediation, 133, 335
Renewable energy, 158, 159, 162, 364
Renewable water resources, 79–81
Restoration, 71, 88, 89, 134, 260, 314, 315, 320–324, 345, 347, 352, 353
Revegetation, 13, 101, 114, 123, 133, 134, 320, 335–337, 352
Reverse osmosis (RO), 43, 112, 150–152, 159, 371, 382, 385, 386, 394, 396, 397, 411
Rhanterium epapposum, 38, 41, 133, 250, 253, 257, 269, 299, 332, 333
Rhizosphere, 350, 351, 353, 378

S

Sabkha
coastal, 32–34, 258
inland, 33, 131, 258
Saline water
upconing, 198, 209, 218, 225, 226, 228, 229, 232, 233, 235, 238–239, 241
water intrusion, 212–216
Salinity
threshold, 17, 372, 376
tolerance, 372, 375
tolerant plants, 12, 13, 52, 115, 259, 376
Saltation, 123–125, 135
Sand
cat, 313
dunes, 5, 8, 23, 27, 30, 32, 35–39, 49, 123, 249, 255–258, 261, 270, 271, 312, 317, 330
Sandy
ecosystem, 74, 114, 134, 249
texture, 18, 58, 249, 266, 368, 369, 371, 374
Saudi Arabia, 22, 25, 35, 36, 48, 69, 79, 106, 109, 110, 113, 143, 146, 175, 176, 342, 364, 365, 382, 401
Scavenger wells, 198–219, 224–241
Scopus database, 142–144
Sea level rise (SLR), 46, 47, 49–50, 59, 70, 72–75, 121, 173, 344, 345, 352
Shoreline, 35, 46, 49, 72–74, 255
Sindh Province, 224–226, 228, 229, 231–236, 238–241
Snakes, 41, 312
Soil
classification, 4, 5, 11, 123, 367, 370
composition, 6, 18, 267, 270, 348–350
contamination, 6, 14, 122, 331, 349, 372
crusts, 126, 128
degradation, 5, 50, 85, 114, 126, 172, 191, 270, 331, 332, 337, 346, 347, 349, 350, 352, 353, 372
dynamic quality, 4
erosion, 7, 13, 16, 59, 64, 67, 70, 71, 76, 82, 120, 121, 127–128, 332, 346, 372, 375, 386
fertility, 16, 70, 122, 248, 267, 271, 272, 306, 346, 349, 378, 380, 386
inherent quality, 4
particles, 13, 14, 120, 127, 128, 131, 135, 306, 371
quality, 4, 18, 52, 111, 172, 270, 367, 372, 374, 375, 381, 394
reclamation, 13, 17, 353
remediation, 133
resources, 5, 15, 110, 121, 271, 345, 349, 367, 374
salinization, 9–10, 51, 172, 173, 372–376, 386
survey, 4, 5, 7, 18, 44, 110, 123, 172, 249, 364, 367, 368, 370, 386
taxonomy, 5, 18, 364
warming, 337, 344, 345, 350, 378
Solar, 26, 58, 59, 62–63, 71, 85, 87, 159, 248, 333, 381
Species
endangered, 325, 337, 338
extinct, 313, 314, 320, 322, 337
invasive, 78, 314, 320
rare, 313, 322, 337, 338
Sulaibiya, 43, 51, 58, 112, 146, 148, 149, 151, 152, 364, 365, 397
Sunshine, 63, 81
Suspension, 123, 124, 135
Sustainable
soil management, 18, 373–374, 376–379
Sustainable Development Goal (SDG), 111, 115, 116, 142, 161, 376, 379, 381

T

Temperature, 8, 22, 26, 27, 30, 37, 41, 45–47, 50–52, 58–81, 83, 85, 89, 94, 103, 112, 120, 121, 129, 134, 153, 188, 248, 260, 266, 312, 314, 330, 332, 333, 336–338, 342–345, 347–353, 364, 365, 371, 375, 384, 407, 408
Terrestrial
ecology, 345
ecosystems, 7, 47, 123, 133, 249–250, 260, 312–326, 330–339, 342, 344, 345, 348, 350, 353
environments, 22, 24, 41, 42, 50, 52, 123, 129–133, 323, 332, 336, 337, 345, 349, 350
habitats, 249–250, 260–261, 313, 323, 325, 336
Tertiary, 30, 152, 397
Torripsamments, 266, 364, 370, 371
Total dissolved solid (TDS), 43, 146, 148, 149, 151, 155, 177–184, 186, 188, 226
Transitional zone, 177–180, 183, 184, 224–241
Treated sewage effluent (TSE), 112, 142, 372, 376
Treated waste water (TWW), 43, 50, 58, 79–81, 155, 162, 173, 371, 372, 394, 396, 397, 411

U
Ultrafiltration (UF), 43, 151
Unexploded ordnance (UXO), 331
United Arab Emirates, 22, 109, 110, 113, 124, 382, 401
Upper Dammam aquifer, 176–181
Urbanization, 15, 45, 46, 105, 154, 266, 313, 319

V
Vegetation
  conservation, 274, 337, 338
  degradation, 260, 315, 320, 330, 333
  diversity, 260, 261, 267–270, 273, 315, 320, 345, 348, 352
  dynamics, 257, 347, 348
  losses, 101
  monitoring, 348
Vertical salinity profile, 173–185, 187–189, 191
Virtual water, 106, 153, 155–156, 163, 394, 400, 401, 408, 409, 411

W
Wafra
  farms, 149
Wastewater, 163, 394, 397
Water
  logging, 13–14, 131, 132, 188, 216, 225, 372, 373
  productivity, 142, 149, 172, 364, 372, 374, 384, 407
  resources, 23, 25, 83, 88, 141–163, 191, 270, 314, 347, 364, 380, 383, 385, 395–398, 402, 403
  scarcity, 23, 44, 51, 70, 74, 80, 101, 156, 157, 248, 344, 364, 371, 375, 383, 386, 394, 395, 401, 408
  security, 157
Water footprint
  blue, 384, 394, 398, 402, 411
  green, 384, 394, 398, 402, 411
  grey, 384, 411
Water footprint consumption
  external, 156
  internal, 156
Weather, 22, 26, 35, 59–72, 74–77, 80–83, 86, 89, 112, 154, 248, 270, 344–346, 348, 365, 407
Wind
  erosion, 5, 7, 8, 10, 14, 48, 51, 58, 114, 120, 123, 126, 127, 132–135, 348, 364, 373, 380
  speed, 36, 48, 59, 64, 66–68, 85, 124, 255, 332

Y
Yield, 10, 14, 16, 76, 81, 121, 131, 157, 172–175, 185–188, 191, 201, 217, 218, 226, 227, 343, 350, 371, 374–378, 381, 397

Z
Zagros mountains, 27
Zero liquid discharge (ZLD), 159
Ziziphous, 133
Zone
  arid, 198, 375
  humid, 395
  semi-arid, 198
Zygophyletum, 267–269
Zygophyllum, 35, 37, 38, 41, 257, 269, 271